Advances in Intelligent Systems and Computing

Volume 262

Series editor

Janusz Kacprzyk, Polish Academy of Sciences, Warsaw, Poland
e-mail: kacprzyk@ibspan.waw.pl

For further volumes:
http://www.springer.com/series/11156

About this Series

The series "Advances in Intelligent Systems and Computing" contains publications on theory, applications, and design methods of Intelligent Systems and Intelligent Computing. Virtually all disciplines such as engineering, natural sciences, computer and information science, ICT, economics, business, e-commerce, environment, healthcare, life science are covered. The list of topics spans all the areas of modern intelligent systems and computing.

The publications within "Advances in Intelligent Systems and Computing" are primarily textbooks and proceedings of important conferences, symposia and congresses. They cover significant recent developments in the field, both of a foundational and applicable character. An important characteristic feature of the series is the short publication time and world-wide distribution. This permits a rapid and broad dissemination of research results.

Jorge Freire de Sousa · Riccardo Rossi
Editors

Computer-based Modelling and Optimization in Transportation

Springer

Editors
Jorge Freire de Sousa
UGEI-INESC TEC
University of Porto
Porto
Portugal

Riccardo Rossi
Department of Civil, Environmental and Architectural Engineering
University of Padova
Padova
Italy

ISSN 2194-5357 ISSN 2194-5365 (electronic)
ISBN 978-3-319-04629-7 ISBN 978-3-319-04630-3 (eBook)
DOI 10.1007/978-3-319-04630-3
Springer Cham Heidelberg New York Dordrecht London

Library of Congress Control Number: 2014932286

Printed on acid-free paper

Springer is part of Springer Science+Business Media (www.springer.com)

Preface

This volume brings together works resulting from research carried out by members of the EURO Working Group on Transportation (EWGT) and presented during meetings and workshops organized by the Group under the patronage of the Association of European Operational Research Societies in 2012 and 2013.

The main targets of the EWGT include providing a forum to share research information and experience, encouraging joint research and the development of both theoretical methods and applications, and promoting cooperation among the many institutions and organizations, which are leaders at national level in the field of transportation and logistics.

The primary fields of interest concern operational research methods, mathematical models, and computation algorithms, to solve and sustain solutions to problems mainly faced by public administrations, city authorities, public transport companies, service providers, and logistic operators. Related areas of interest are: land use and transportation planning, traffic control and simulation models, traffic network equilibrium models, public transport planning and management, applications of combinatorial optimization, vehicle routing and scheduling, intelligent transport systems, logistics and freight transport, environment problems, transport safety, and impact evaluation methods.

In this volume, attention focuses on the following topics of interest:

- Decision-making and decision support
- Energy and environmental impacts
- Urban network design
- Optimization and simulation
- Traffic modeling, control and network traffic management
- Transportation planning
- Mobility, accessibility, and travel behavior
- Vehicle routing

The complexity of the problems analyzed made it difficult to give a complete picture of the above aspects but, in the opinion of the editors, the works presented here will help readers to go more deeply into some significant subjects with the aid of experts' viewpoints of the problems. The high standard of the papers submitted was ensured by a large and competent scientific committee and by a selective reviewing process.

To end this preface, special thanks go to all those who contributed to this book, including authors, reviewers and Springer, and in particular to Dr. Thomas Ditzinger (Springer, Applied Sciences and Engineering).

Jorge Freire de Sousa
Riccardo Rossi

Contents

Part VI Transportation Planning

Part VII Mobility, Accessibility and Travel Behavior

Part VIII Vehicle Routing

Part I
Decision Making and Decision Support (in Transportation)

In the transportation field—as in many others—it is important to be able to guarantee that the *decision-making* has been systematically followed, to ensure that the right decisions are taken. Identifying needs, checking the availability of resources, formulating frameworks of analysis, and evaluating methods are just some of the aspects involved in this process, often reinforced by *decision support* techniques. *Hankach and Lepert* introduce a decision support tool for evaluating maintenance strategies on road networks with incomplete data. *Ambrosino and Siri* provide a general formulation for proper train load planning for maritime container terminals, as well as two other formulations for specific cases in which some unproductive operations or movements are not permitted. *Haasis et al.* develop a rule-based decision support system for improving the energy efficiency of passive temperature-controlled means of transport, and apply it to the case of liquid aluminium transport in Germany. *Bandeira et al.* present an integrated numerical computing platform, using microscopic traffic and emission models, as a tool for traffic assignment taking into account eco-routing purposes: the main aim of this work was to identify the best traffic volume distribution which allows the environmental costs for a given corridor with predetermined different alternative routes to be minimized. *Mendes-Moreira and Freire de Sousa* discuss evaluation of the impact of changes, before they are made, in global operational planning in real-life conditions, and present a framework for their evaluation.

A Decision Support Tool for Evaluating Maintenance Strategies on Roads Networks with Incomplete Data

Pierre Hankach and Philippe Lepert

Abstract Choosing a maintenance strategy is a major challenge for road operators because of its implications on the required budget and the quality of the road network. However, a major problem has to be solved before we can compare the performance of different strategies: on many real networks, essential data for simulating the evolution of the distress of the pavement for the coming years and hence simulating the strategies in order to evaluate them, are missing or highly uncertain. In this chapter, we introduce a decision support tool for evaluating maintenance strategies on roads networks with incomplete data. First, this tool is used to build a Virtual Road Network (VRN) that represents the real network, and on which plausible values are allocated to missing or uncertain data. Then, it is used to simulate alternative strategies. To this end, the distress state of the pavement is evolved using an appropriate evolution model and each strategy is applied in order to determine the associated maintenance interventions. The different strategies are thus compared (on the network scale) on the criteria of cost and resulting distress state of the pavement in order to choose the best solution.

Keywords Pavement · Damage · Distress · Maintenance · Strategies · Simulation · Evolution models · Network

1 Introduction

Defining a maintenance strategy is a very important task for road mangers. In fact, both the condition of the road network and the required maintenance budget depend on the chosen strategy. Therefore, evaluating and comparing the cost and

P. Hankach (✉) · P. Lepert
IFSTTAR, Route de Bouaye, Bouguenais 44344, France
e-mail: pierre.hankach@ifsttar.fr

P. Lepert
e-mail: philippe.lepert@ifsttar.fr

J. F. de Sousa and R. Rossi (eds.), *Computer-based Modelling and Optimization in Transportation*, Advances in Intelligent Systems and Computing 262,
DOI: 10.1007/978-3-319-04630-3_1, © Springer International Publishing Switzerland 2014

performance of different strategies is an essential step in maintenance management. In this chapter, we present a decision support tool in order to evaluate strategies and choose an adequate one.

A maintenance strategy is a set of rules that define when a maintenance intervention is triggered and the design and technique of this intervention. Strategy rules that trigger interventions are usually based on the extent of the distress and other parameters such as the date of the last intervention. In order to evaluate a strategy, the distress must be computed annually simulating its evolution for the coming years starting from the present date and present condition of the pavement. Whenever a maintenance intervention is triggered by the strategy, the condition of the network is updated accordingly and taken into account for its future evolution. Therefore, the condition of the pavement at a given date in the future is determined by combining the evolution of distress and the effect of maintenance interventions. The cost of maintenance associated to a strategy is the sum of the costs of individual interventions that it triggers.

However, a major difficulty must be addressed in order to simulate a strategy and evaluate it: essential data for this process are missing in real road network databases. As it have been pointed, the distress state of the pavement must be evolved for simulating a strategy and this evolution depends on different factors: pavement design, resistance of materials to cracking, aggressivity of the traffic, etc. It can be expressed by the following formula:

$$\%Ff = \Phi(t\,;\,\alpha\,;\,\beta\,;\,\gamma\,;\,\delta\,;\ldots) \tag{1}$$

where Φ is a complex formula in which t represents the time elapsed since the construction and $\alpha; \beta; \gamma; \delta; \ldots$ represent the factors that influence the evolution. In virtually all road databases describing real networks, the latter variables are missing. There is therefore no possibility of applying an evolution model that takes into account these factors without addressing the missing data problem.

In this chapter, we build a decision support tool for evaluating maintenance strategies. First, the problem of missing data on roads networks is handled by building a VRN that represents the real network, and on which plausible values are allocated to missing or uncertain data. Afterwards, in order to evaluate maintenance strategies, the distress is computed year after year using a distress evolution model, triggering maintenance interventions by strategy rules. The consequences of each strategy are assessed on the future distress state of the network and on the required maintenance budget. Thus, different strategies can be compared.

2 Constructing the Virtual Network (VRN)

A virtual road network is a road database comprising: (1) a list of roads; (2) a reference system and (3) a set of data describing the nature and condition of the roads and their solicitation. As previously stated, the information and data contained in the database of a virtual network are generated according to certain rules and/or by applying certain models in order to accurately represent the real network. Hereafter, the rules followed for creating a virtual network are listed:

- The constitution of the network (number of roads, length) is identical to that of the real network;
- Roads are divided into segments of homogeneous subgrade characteristics. The length of each segment and its characteristics are random samples from the probability distributions of these parameters on the real network (which may be known from the examination of a number of design projects);
- If the traffic on the real network is not known and therefore cannot be assigned directly, each road is divided into segments of homogenous traffic (that are different from the segments mentioned in the previous point). The length of each segment and its traffic are random samples from the probability distributions of these parameters on the real network (obtained by performing a sampling on the latter);
- The pavement design is homogenous per segments that correspond to the original construction segments. Each segment's length is a random sample from a predefined probability distribution that reflects reality. The structure of the segment is chosen according to the characteristics of the subgrade and the traffic that will be supported, as is the case in a conventional design process. These inputs are used to select structures types in a design catalog [1]. The distribution of the types of structures obtained is validated by comparing to the distribution on the real network (obtained by sampling).

In the remainder of this section, we describe the construction of a virtual network. We take the French national road network (FNRN) as a reference real network to illustrate this construction.

2.1 Constitution of a VRN

The VRN is constituted by all the roads of the real network. In order to locate events—such as distress—on the roads, we associate to the network a linear referencing system. Each road is associated with a unique identifier and a number of location points (PR). The position of the various objects on the road is expressed with respect to these points. For practical reasons, roads are segmented into elementary sections of fixed length (typically 200 m). Each section is identified by specifying the "PR + distance" of its start and end points. In the remainder of this chapter we use as building blocks these sections in order to affect different characteristics.

2.2 Attribution of the Bearing Capacity of the Subgrade

The pavement thickness design depends on the bearing capacity (PF) of the platform. This bearing capacity is classified into four groups (PF1, PF2, PF3 and PF4) according to the subgrade's ability to withstand loads. In the process of building the VRN, we assign each section a bearing capacity selected from these four classes. To perform this assignment, two distributions must be defined in advance from sampling performed on the real network:

- The first distribution governs the length of the pavement segments where the bearing capacity is considered homogeneous. The length of each segment is a random sample from a triangular distribution. To reproduce conditions similar to those of the FNRN, the triangular distribution is defined with a minimum of 1 km, a maximum of 5 km and a mode of 2 km.
- The second governs the distribution of bearing capacity classes. Because this distribution wasn't available for the FNRN, the bearing capacity of each homogenous segment is chosen randomly among the four classes. In the case where the distribution between the four classes is known, it should be respected.

2.3 Attribution of the Traffic

The following procedure is applied when the traffic and/or its aggressivity is unknown on all or a part of the network. Like for the bearing capacity, the assignment of traffic on a virtual network is defined in two steps. First, the road network is divided into segments of homogeneous traffic. Then, for each of these segments, the daily vehicle average is defined for each category (cars, trucks…) as random samples from predefined probability distributions. These latter distributions are defined by experts, and reflect the state of the traffic on the real network. In order to reproduce traffic conditions similar to the FNRN the following distributions are used:

- The length of each segment is defined as a random sample from a triangular distribution with a minimum of 5 km, a maximum of 25 km and a mode of 15 km.
- The daily vehicle average is defined as a random sample from a triangular distribution with a minimal traffic of 0 vehicles per day (v/d), maximum traffic of 80000 v/d and a mode 30000 v/d.

In recent design methods, thickness design of pavements depends on the traffic class TCi which takes into account the lifespan of the pavement. The traffic class is determined by the total number of heavy traffic (which is a predefined proportion of the total traffic) that will be supported by the pavement during its lifetime. This lifetime is generally set at 20 years, and is extended to 30 years for pavements supporting high traffic. Traffic classes TC are obtained using the following formula [2]:

$$TC = 365N.\left[d + \tau.d.\left(\frac{d-1}{2}\right).r\right] \qquad (2)$$

where N is the heavy traffic count; τ the linear annual growth rate of traffic rate (2 % by default); d the lifespan in years; and r represents the transversal distribution of heavy traffic. According to the value of TC, the traffic is attributed one of nine classes: TC0, TC1, TC2, TC3, TC4, TC5, TC6, TC7 or TC8.

2.4 Thickness Design

The thickness design of a pavement segment of the VRN is made using the same rules applied for the construction of a real network. It depends on the bearing capacity of the subgrade and the traffic class as described in [2].

The length of segments with a homogenous design results from the length of construction segments. To reproduce the conditions of the FNRN, it is defined as a random sample from a triangular distribution with a minimum of 5 km, a maximum of 25 km and a mode of 15 km. As each construction intervention occurs at a given date, it also defines the age of the pavement.

The pavement design of a segment is given as a function of the bearing capacity of the subgrade and the traffic (Table 1). However, on a single construction segment a multitude of combinations "bearing capacity class/traffic class" are usually identified. Figure 1 shows the diversity of this combination for one construction segment. Given this fact, the design that corresponds to the largest traffic coupled with the smaller PF is chosen. This rule ensures the selection of the stronger design among those that correspond to the various combinations.

3 Evaluating Maintenance Strategies

A maintenance strategy [3–5] is a set of decision rules that determines, mainly based on the condition of a pavement's section, its age, traffic or any other criterion available in the database: the work to be done on this section; the design of the maintenance works; and the priority given to it. Table 2 illustrates a very simple strategy based on the IRI (International Roughness Index). It stipulates that if IRI < 5 then no maintenance intervention is necessary, otherwise an overlay is performed. Usually, strategies are much more sophisticated and rely on many indicators to compute maintenance. Table 3 illustrates a strategy that defines maintenance interventions depending on both the IRI and the percentage of cracks extent.

Table 1 The structure type (GB for base asphalt concrete, GH for cement treated concrete and MX for composite pavement), the base layers thickness and the thickness of the surface layer CS are given as a function of the bearing capacity of the subgrade and the traffic classes [1]

	Plate-forme	Fiche n°	TC2	TC3	TC4	TC5	TC6	TC7	TC8
GB	Risque		30%	18%	10%	5%	2%	1%	1%
	PF2	2 GB3/GB3			CS 11 12	CS 13 13			
	PF3				CS 9 9	CS 10 11	CS 13 13	CS 10 10 11	CS 11 11 12
	PF4				CS 8 8	CS 9 9	CS 11 12	CS 14 14	CS 10 11 11
GH	Risque		12.5%	10%	7.5%	5%	2.5%	1%	1%
	PF2	5 GC4/GC4			CS 18 20	CS 19 20			
	PF3				CS 26	CS 27	CS 18 18	CS 18 18	CS 20 18
	PF4				CS 24	CS 25	CS 18 15	CS 19 15	CS 20 15
MX	Risque (fond.)		50%	35%	20%	10%	3%	2%	1%
	PF2	15 GB3/GC3			CS 7 8 21	CS 7 8 23			
	PF3				CS 12 18	CS 12 20	CS 14 22	CS 8 8 24	CS 8 9 25
	PF4					CS 11 19	CS 13 21	CS 14 22	CS 8 8 24

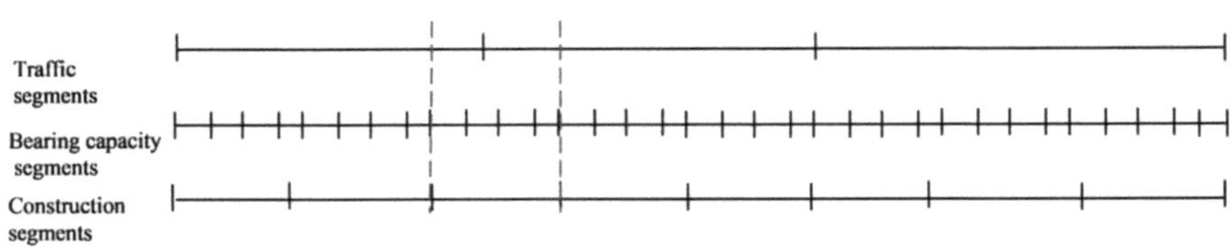

Fig. 1 Combinations "bearing capacity class/traffic class" for one construction segment

Maintenance strategies are evaluated on two criteria:

- Condition of the pavement after a number of years of applying the strategy;
- The overall cost of maintenance it requires.

The condition of a pavement is characterized by a set of information, which we call indicators. An indicator represents de state of a defined distress type (longitudinal cracks, rutting, etc.) on pavement sections. Indicators of different distress types can be combined to form a "functional index". An example of such an index is the "Structural Condition rating" (SC rating) which is used as a global condition indicator of pavements sections on the FNRN [6, 7].

Table 2 A simple maintenance strategy based on the IRI

IRI	
<5	>5
No intervention	Overlay

Table 3 A maintenance strategy where maintenance interventions depend on both the IRI and the percentage of cracks extent

Cracks extent	IRI	
	<5	>5
0–10	No intervention	Reshaping
10–50	Surface dressing	Overlay
50–100	Overlay	Overlay

Simulating as strategy is a step by step process, each step corresponding to one year. It starts at the current date T0 and ends on a date Tend = T0 + x years. At each step of the simulation (every year) two cases may arise for a given section of the network:

- The simulated strategy does not trigger maintenance work, in this case, the distress state is updated (the state is more deteriorated) using an evolution model (described below);
- The simulated strategy triggers maintenance, the state of distress is updated (usually many distress indicators are reinitialized because the pavement is repaired), but also the evolution model to be applied on this section is adapted to take into account the effects of the new layer added by maintenance works.

The diagram in Fig. 2 shows the process of simulating a strategy. Starting from the VRN constructed in Sect. 2 and after the initiation of the network condition, the simulation starts at the current year T0. For each elementary section of the network, the decision rules of the strategy are applied annually to decide if a maintenance intervention is necessary. If no maintenance is triggered, the distress state of the pavement is updated using the evolution model. If maintenance works are triggered, distress indicators are reinitialized and the evolution model applied to the section is updated. At the end of this process, the average cost and average global indicator's value (SC) are calculated for the tested strategy.

In order to initiate the network's condition, distress indicators are set to their original values at current date T0. This can be done straightforwardly by defining an indicator's value on a given section as a random sample of the probability distribution observed on the real network. However, although the distribution of distress obtained on the VRN is identical to that of the real network, this approach has a major drawback because it fails to take into account the dependence between the state of distress on a section and its characteristics such its age, design, the traffic it supports, etc. Therefore, and in order to take into account this dependence, the adopted solution is to simulate the evolution of each section from its construction until T0, while applying the maintenance strategy that has been used on

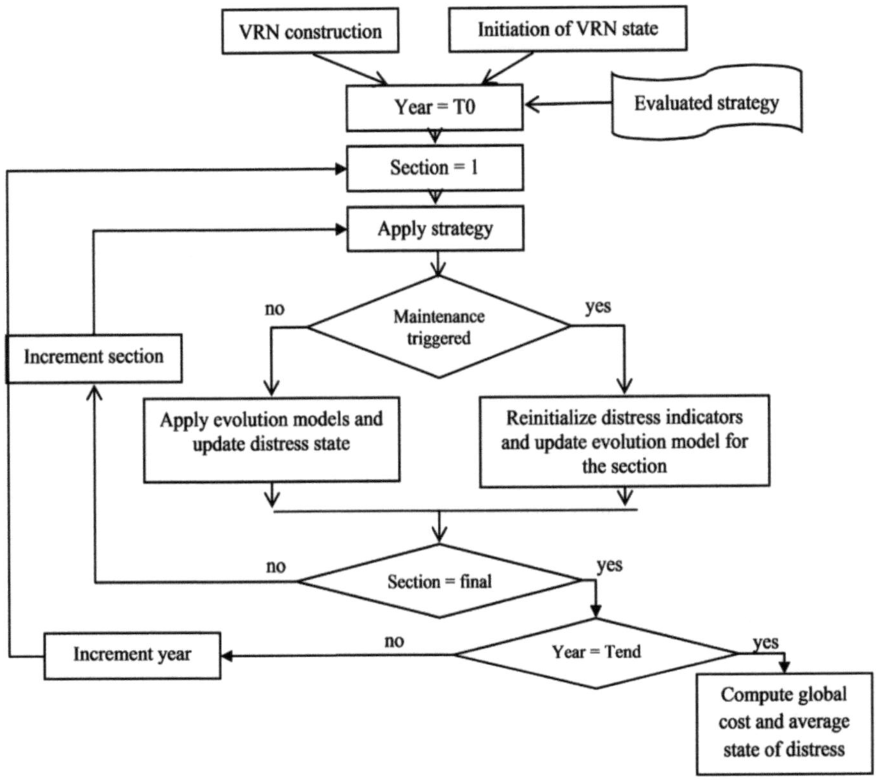

Fig. 2 Diagram that represents the simulation process of maintenance

the real network before T0. The workflow diagram of this process is the same as Fig. 2 (with the exception that distress indicators values start at 0 at the construction date). After the completion of this process, in order to validate the distress state at T0, the corresponding distributions of distress values are checked with the distributions observed on the real network.

In the remainder of this section, we describe some components of the simulation process. First we present an evolution model for distress indicators. Afterwards, we discuss the effects of maintenance works on this model.

3.1 Evolution Model

Distress indicators evolve depending on several factors: layer thickness, modulus and resistance to cracking of materials, traffic, etc. Their evolution can be expressed by the formula given in (1). This equation can be written in a simpler form a follows:

$$\% Ff = \Psi(t; R) \quad (3)$$

In which R represents the contribution of all other parameters, $R(\alpha; \beta; \gamma; \delta\ldots)$. For the evolution of distress, quantified by the length of road affected by the deterioration in proportion to the length of elementary sections, the following model is adopted [8]:

$$V_j(t) = V_d + (V_f - V_d)\, k_j(t) \quad (4)$$

where:

$$\begin{aligned} k_j(t) &= 1 - \frac{R_j}{q_0\, t^{P_0}} & \text{if } t \geq \left(\frac{R_j}{q_0}\right)^{\frac{1}{P_0}} \\ k_j(t) &= 0 & \text{if } t \leq \left(\frac{R_j}{q_0}\right)^{\frac{1}{P_0}} \end{aligned} \quad (5)$$

- V_d and V_f are the min and max values of the evolution of the indicator, usually 0 and 100;
- q_0 is the kick off parameter of the model;
- p_0 is the shape parameter of the model;
- R is the robustness of the law applied to a section. It ranges from 0 to infinity, the value characterizes the deviation of a section from average behavior;
- t is the age of the section.

3.2 *Effects of Maintenance Works*

When a maintenance operation takes place on a section of the network, two changes of the state of the latter should be taken into account:

1. The damage is more or less repaired, the corresponding indicators are then reduced to a predictable value (usually 0);
2. The evolution of distress is altered because the pavement structure has been changed.

Let's assume that, in the evolution model, one of the explanatory variables (contributing to R introduced above) is Odemark equivalent thickness, E_{eq}:

$$R = R(E_{eq};\ b;\ g;\ d;\ldots) \quad (6)$$

This especially occurs when the indicator characterizes the structural fatigue of the pavement. The evolution of this indicator after the addition of a new layer with a modulus E_{ch} and a thickness h_{ch} follows the type of evolution model as before, but E_{eq} is replaced with $E_{eq,ch}$ [9]:

Table 4 Constitution of the network

Road identifier	Name	Length (km)
1	A001	40.5
2	A002	34
…	…	…

Table 5 Pavement data that remains unchanged from year to year

Section number	Road	PR + distance start	PR + distance end	BC	Structure Type	…
1	A001	0 + 000	0 + 200	PF3	GB3/GB3	
2	A001	0 + 200	0 + 400	PF3	GB3/GB3	
…	…	…	…	…	…	
15	A001	2 + 800	3 + 000	PF3	GB3/GB3	
…		…	…	…	…	

$$\mathrm{E}_{\mathrm{eq,ch}} = E_{eq} \cdot \left[\left(\frac{\sum_1^N \left(\mathrm{h_r E_i^{\frac{1}{3}}} \right) + \mathrm{h_{ch} E_{ch}^{\frac{1}{3}}}}{\sum_1^N \left(\mathrm{h_i E_i^{\frac{1}{3}}} \right)} \right) \right]^{-\frac{2p_0}{b}} \tag{7}$$

In this equation, h_i and E_i represent the thickness and the modulus of layer i before maintenance works. In most cases, these parameters (h_i and E_i) are not saved in the database of the real network. Only the average value and the standard deviation may be known. However, the procedure of creating a VRN exposed above will enable to generate these parameters to complete the road database. Then, the effect of overlays on fatigue distress evolution could be assessed with more realism, and correctly taken into account when comparing two strategies.

4 Implementation

Data of a VRN are stored in a relational database. We define three different relations for different types of stored data. The first relation, illustrated in Table 4, records the constitution of the network (IDs, names and lengths of roads). The second relation, illustrated in Table 5, records the unchangeable characteristics of pavements (each elementary section has a single value for each characteristic) such as the bearing capacity of the subgrade, structure types, etc. The last relation, shown in Table 6, records the pavement data that can change from one year to another such as distress and traffic. This last relation specifies the generation (or age) of the data.

The decision support tool described is this chapter has many different modules that interact directly or indirectly through the data created in the VRN. Three different steps are identified where different modules are in action:

Table 6 Pavement data that changes from year to year

Section number	Road	PR + distance start	PR + distance end	Year	Indicator1	…
1	A001	0 + 000	0 + 200	1	0	
1	A001	0 + 000	0 + 200	2	0	
…	…	…	…	…	…	
1	A001	0 + 000	0 + 200	15	12	
…	…	…	…	…	…	
1	A001	0 + 000	0 + 200	30	20	
2	A001	0 + 200	0 + 400	1	0	
…		…	…	…	…	

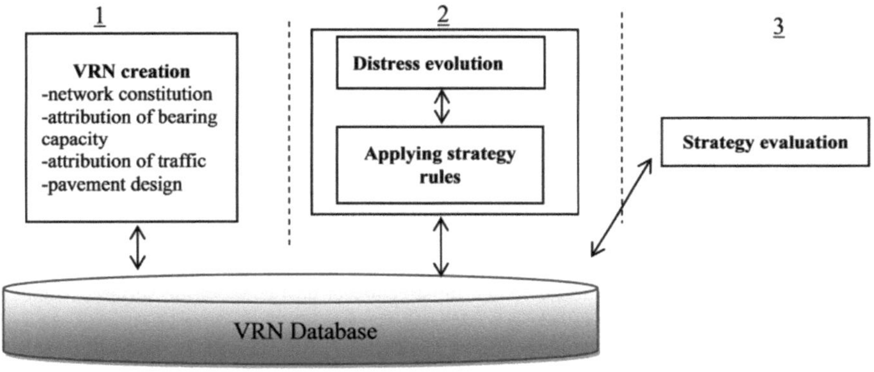

Fig. 3 Architecture

1. Creating the VRN: the dedicated module creates the VRN geometry, affects the subgrade bearing capacity, the traffic and pavement's design.
2. Evolution of the distress and maintenance operations: two interacting modules compute the evolution of distress and the maintenance applied according to the tested strategy.
3. Evaluation of the strategies: the total cost and the value of different indicators (comprising the SC index) are computed.

Figure 3 shows this architecture.

5 Conclusion

In this chapter, we have built a decision support tool for evaluating maintenance strategies. This tool has two major elements: firstly, it handles the problem of essential missing data for the strategy evaluation process. Secondly, it simulates the evolution of distress while applying the rules of a given strategy in order to evaluate it.

The problem of missing data on roads networks is handled by building a VRN that represents the real network, and on which plausible values are allocated to missing or uncertain data. These data are essential for simulating the evolution of the distress of the pavement for the coming years while taking into account important factors such as pavement design, layer thickness, traffic aggressivity, etc.

In order to evaluate maintenance strategies, the distress is computed year after year using a distress evolution model, triggering maintenance interventions by strategy rules. The consequences of each strategy are assessed on two criteria: the resulting distress state of the pavement and on the required maintenance budget. Thus, different strategies can be compared.

References

1. SETRA-LCPC: Catalogue des structures types de chaussées neuves, technical guide, SETRA, Bagneux (2010)
2. SETRA-LCPC: Conception et dimensionnement des structures de chaussée, technical guide, LCPC, Paris (1994)
3. Lepert, P.: Programmation de l'entretien de structure et de l'entretien de surface, technical guide, LCPC, Paris, France (1993)
4. Lepert, P.: Outil d'aide à la programmation d'entretien GiRR : premières applications en site pilote. In: Revue Générale des Routes et Aérodrome (1996)
5. Lepert, P.: Road maintenance policy based on an expert asset management system—concept and case study. In: SURF 2012, Norfolk, USA (2012)
6. Lepert, P.: An evaluation of the French national highway network based on surface damage surveys. In: Proceedings of 3rd International Conference on Managing Pavements, San Antonio (1994)
7. Bertrand, L., Lepert, P.: Relevé des dégradations de surface des chaussées : Méthode d'essai n° 38–2, technical guide, LCPC, Paris, France (1997)
8. Lepert, P., Savard, Y., Leroux, D., Rèche, M.: Méthodes statistiques de prévision de l'évolution d'une chaussée. In: Bulletin des Laboratoires des Ponts et Chaussées, n° 250–251 (2004)
9. Rèche, M.: Effet des travaux d'entretien sur les lois d'évolution des dégradations de chaussées, PhD thesis, University of Clermont-Ferrand (2004)

Models for Train Load Planning Problems in a Container Terminal

Daniela Ambrosino and Silvia Siri

Abstract In this chapter, the train load planning problem for maritime container terminals is dealt with. In the most general case, the optimal assignment of containers to train slots is done considering that it is possible to make reshuffles in the stacking area and to load the train not sequentially; of course, both these types of unproductive movements should be minimized. In the chapter, a general formulation for this problem is provided, as well as other two formulations for the specific cases in which one of these two unproductive operations is not allowed. Then, some experimental results are reported to show the differences among the proposed models.

Keywords Maritime container terminals · Train load planning · Combinatorial optimization

1 Introduction

Container terminals are very complex systems that require the development of optimization methods to support the crucial decisions at the different planning levels, from the strategic to the tactical until the operational one [1]. Some recent surveys on operations research methods applied to container terminals are those provided by Steenken et al. [2] and Stahlbock and Voss [3]. The authors divide the optimization approaches found in the literature according to the different processes

D. Ambrosino
Department of Economics and Business Studies, University of Genova, Genova, Italy
e-mail: daniela.ambrosino@economia.unige.it

S. Siri (✉)
Department of Informatics, Bioengineering, Robotics and Systems Engineering, University of Genova, Genova, Italy
e-mail: silvia.siri@unige.it

J. F. de Sousa and R. Rossi (eds.), *Computer-based Modelling and Optimization in Transportation*, Advances in Intelligent Systems and Computing 262,
DOI: 10.1007/978-3-319-04630-3_2,

in a seaport terminal: ship planning (i.e. berth allocation, stowage planning and crane split), storage and stacking planning, and transport optimization (divided in quayside, landside, and crane movements). With respect to this classification, the present chapter concerns the landside transport planning and presents an optimization approach for the definition of loading plans for trains.

As highlighted in Steenken et al. [2], a loading plan indicates on which wagon a container must be placed; generally speaking, this decision depends on the container destination, type and weight, as well as on the characteristics of the train and wagons. The container location in the stacking area can influence the loading plan as well. In this chapter, we consider the case in which the loading plan is performed by the terminal operator with the aim of optimizing both the pick-up operations in the stacking area where containers are waiting for being loaded on trains and the loading operations of each train.

In the literature few works are specifically devoted to the train loading problem, as it is in our work. Bostel and Dejax [4] deal with rail–rail terminals with rapid transfer yards and propose some models and heuristic methods for container allocation problems on trains. Corry and Kozan [5] consider a terminal where containers are transferred to and from trucks on a platform adjacent to the rail tracks provided with a short-term storage area. They propose several techniques for defining the assignment of containers to slots of a train, minimizing container handling time and optimizing the weight distribution of the train. In that model, only one type of container is considered and the weight restrictions for the wagons are neglected. In a following work, Corry and Kozan [6] treat again the train planning problem, considering more types of containers and different loading patterns and minimizing a weighted sum of number of wagons required and equipment working time. Due to the large number of variables, they propose heuristic algorithms, such as local search and simulated annealing, to solve the problem in practical applications. The load planning problem in intermodal terminals is also studied by Bruns and Knust [7] that consider explicitly the real weight constraints for wagons, as we do in this chapter. They propose three different integer linear programming formulations for solving the problem of loading containers on wagons in order to maximize the utilization of the train and minimize transportation costs for loading containers and set up costs for changing the configuration of wagons. Many types of containers are considered (including also swap bodies) and different types of wagons are treated.

In the present chapter, we develop a mathematical model to optimally plan the train load in order to maximize the train utilization, while minimizing the unproductive activities that can arise both in the stacking area and during the train loading operations executed by the crane. Real weight constraints are explicitly considered, as done by Bruns and Knust [7], and the main novelty of the present approach with respect to the one by Bruns and Knust [7] stands in modeling the reshuffles in the stacking area, since this is a crucial aspect to be dealt with in maritime container terminals. The model proposed in this chapter is an extended version of the one developed by Ambrosino et al. [8] where, again, the train load planning problem was treated but only in the case of sequential loading by the crane. A model similar to the one proposed in this chapter has been considered by Ambrosino et al. [9] to evaluate the impact of various storage policies adopted in the yard on different train loading strategies. In this chapter three different models are validated and compared in order to understand which model is the most suitable for solving real problems in maritime container terminals (i.e. providing good and applicable solutions in an acceptable CPU time).

The chapter is organized as follows. Section 2 is devoted to introduce the problem and the main issues related to it. Section 3 reports the mathematical formulation for the planning problem, both in the general case and in the specific cases of train sequential loading and no-reshuffle policy for the stacking area. Section 4 regards the experimental analysis performed on the three different formulations. Finally, some conclusions are drawn in Sect. 5.

2 Problem Description

The problem studied in this chapter regards the train load planning in seaport terminals. The destination of containers is not taken into account in this load planning problem, since the planning is related to the shuttle trains directed to the inland port (for which the inland terminal is the only common destination). Thus it is assumed that the containers in the stacking area have the same destination. Moreover, the planning problem considers only one train at a time. Anyway, the proposed approach can be easily modified in order to face the loading problem when in the stacking area containers of different destinations are stored.

This work takes inspiration from a real case of an Italian port but it can be easily extended to many other cases. This study refers to a container terminal in which containers that will be loaded on trains are stored in a specific stacking area close to the railway yard. From there, containers are moved near the tracks with trailers; then, a crane loads containers on trains. Generally, the crane starts its work from a wagon and goes on along the train without changing direction (i.e. going forward). Sometimes, during the loading process it can happen that it is not possible to load a container on the train without requiring to the crane to change direction; in this case, for example, the crane has to come back to load a container in a slot of a wagon already visited by the crane itself but remained free (in this way, unproductive movements of the crane are executed).

Containers are stored in the stacking area in stacks of different height. During the loading process, it is not always possible to pick up firstly the containers at the top of the stacks. Sometimes it can be necessary to remove a container from the top of a stack for loading, on the wagon served by the crane, another container that is below it (in this case a reshuffle is executed).

Figure 1 reports a simple example of two different ways for loading, on 2 wagons, 4 containers belonging to the same stack. First of all, in t1 (first operation), for loading container c3 in the first slot of wagon 1, container c4 must be rehandled (container c4 is loaded in t2, i.e. as second operation; obviously we are

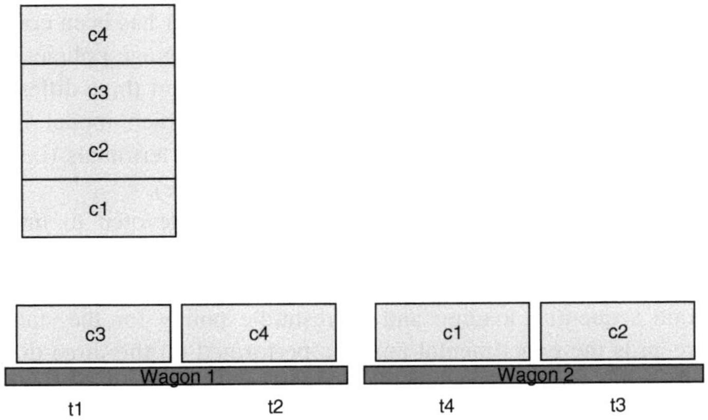

Fig. 1 Sketch of the train loading phase

assuming that c4 cannot be loaded in the first slot e.g. for weight constraints). Then, when loading wagon 2, the crane loads firstly container c2 (third operation) and then goes back for loading container c1; instead of the unproductive crane movement, the same load configuration can be obtained by rehandling container c2 for loading c1 (as happened in wagon 1) and then loading c2.

The assignment of containers to slots is guided by length and weight considerations. One of the characteristics of this problem is the possibility of choosing a particular weight/slot configuration among different ones available for each wagon. These real wagon weight constraints are much stricter than simply considering a maximum weight capacity for each wagon and train. Further details on different wagon configurations can be found in the paper by Bruns and Knust [7] and in the work by Ambrosino et al. [8].

In the problem under investigation the main objective is to plan the train load in order to minimize both the reshuffles in the stacking area and the unproductive movements of the crane loading a train, whilst maximizing the load of the train. As far as the maximization of the load of the train is considered, we have to note that the maximization regards the number of TEUs and the total value of containers loaded instead of the number of containers, since we have to take into account that each container in the stacking area has a different priority to be loaded on a given train. This priority can be directly connected to the due time of the container or to its commercial value.

More formally, given a set of containers with different characteristics (length, weight, and priority) and one train composed by a set of wagons of different types (i.e. with different length, possible configurations and weight constraints), the problem is to choose which containers to load on the train and in which wagon slot. Moreover, the sequence of loading operations must be decided. For this case, a mathematical formulation will be provided. Moreover, other two models can be developed for the specific cases in which either unproductive movements of the crane are not allowed (train sequential loading) or reshuffles are not allowed in the stacking area.

3 Formulation of the Train Load Planning Problem

The mathematical formulation for the train load planning problem is a multi-objective 0/1 Mixed Integer programming model.

3.1 Notation

Let us introduce the notation. First of all, let C denote the number of containers in the stacking area, W the number of wagons of the train to be loaded, S the number of train slots.

For each container $i = 1, \ldots, C$, the weight is denoted as w_i (expressed in tons), the length as λ_i (i.e. 20 or 40 feet), the cost for not being loaded as π_i (it depends on the priority of the container). Moreover, $\gamma_{i,j}, i,j \in \{1, \ldots, C\}, i \neq j$, is related to the position of containers in the stacking area; it is equal to 1 if container i and j are positioned in the same stack and container i is over container j, it is equal to 0 otherwise.

For each wagon $w = 1, \ldots, W, S_\omega$ is the subset of relative slots, B_ω is the subset of weight configurations, ϖ_ω is the weight capacity. Moreover, $B_{s,\omega}$ is the subset of weight configurations for slot s of wagon ω, μ_s is the length of slot s (i.e. 20 or 40 feet), ρ_s is the position of slot s in the train with respect to the first slot of the first wagon (expressed in TEUs), $\delta_{b,s}$ is the weight capacity of slot s in the weight configuration b, $\overline{\Omega}$ is the weight capacity of the train.

Finally, some configuration parameters are the unitary rehandling cost α, the unitary crane movement cost β and the maximum number of possible loading operations on the train T, that corresponds to the TEU capacity of the train.

3.2 General Formulation

In this section let us firstly consider the case in which both reshuffles in the stacking area and unproductive crane movements can be executed. The problem decision variables can be divided in the following sets:

- $x_{i,s,t} \in \{0, 1\}, i = 1, \ldots, C, s = 1, \ldots, S, t = 1, \ldots, T$, equal to 1 if container i is loaded in slot s at operation t, 0 otherwise (these variables are defined only when container i is compatible with slot s in terms of length, i.e. $\lambda_i = \mu_s$);
- $f_{\omega,b} \in \{0, 1\}, \omega = 1, \ldots, W, b \in B_\omega$, equal to 1 if configuration b is chosen for wagon ω, 0 otherwise;
- $y_{i,j} \in \{0, 1\}, i,j \in \{1, \ldots, C\} : \gamma_{i,j} = 1$, equal to 1 if container i is handled to load container j;

- $z_t \geq 0, t = 2, \ldots, T$, unproductive distance traveled by the crane when doing operation t (to compute this variable, it is assumed that the crane is positioned at the beginning over the first wagon on the left and z_t is equal to 0 if the crane, between $t-1$ and t, goes straight, from left to right, whereas it is equal to the covered distance (in TEUs) if the crane goes back, i.e. from right to left);
- $u_t \geq 0, t = 2, \ldots, T$, normally set to 0 except for the operation t such that $t-1$ is the last loading operation by the crane; in that case u_t is positive in order to set $z_t = 0$.

The general formulation is provided in the following:

$$\min \alpha \cdot \sum_{\substack{i,j\in\{1,\ldots,C\}:\\ \gamma_{i,j}=1}} y_{i,j} + \beta \cdot \sum_{t=2}^{T} z_t + \sum_{i=1}^{C} \pi_i \cdot \left(1 - \sum_{s=1}^{S}\sum_{t=1}^{T} x_{i,s,t}\right) \tag{1}$$

$$\text{s.t.} \sum_{s=1}^{S}\sum_{t=1}^{T} x_{i,s,t} \leq 1 \quad i = 1, \ldots, C \tag{2}$$

$$\sum_{i=1}^{C}\sum_{s=1}^{S} x_{i,s,t} \leq 1 \quad t = 1, \ldots, T \tag{3}$$

$$\sum_{i=1}^{C}\sum_{t=1}^{T} x_{i,s,t} \leq 1 \quad s = 1, \ldots, S \tag{4}$$

$$\sum_{b\in B_\omega} f_{w,b} = 1 \quad \omega = 1, \ldots, W \tag{5}$$

$$\sum_{i=1}^{C}\sum_{t=1}^{T} w_i \cdot x_{i,s,t} \leq \sum_{b\in B_{s,\omega}} \delta_{b,s} \cdot f_{\omega,b} \quad \omega = 1, \ldots, W \quad s \in S_\omega \tag{6}$$

$$\sum_{i=1}^{C}\sum_{s\in S_\omega}\sum_{t=1}^{T} w_i \cdot x_{i,s,t} \leq \varpi_\omega \quad \omega = 1, \ldots, W \tag{7}$$

$$\sum_{i=1}^{C}\sum_{s=1}^{S}\sum_{t=1}^{T} w_i \cdot x_{i,s,t} \leq \overline{\Omega} \tag{8}$$

$$\sum_{s=1}^{S}\sum_{t=1}^{T} t \cdot x_{i,s,t} - \sum_{s=1}^{S}\sum_{t=1}^{T} t \cdot x_{j,s,t} \leq T \cdot y_{i,j} + T\left(\sum_{s=1}^{S}\sum_{t=1}^{T} x_{i,s,t} - \sum_{s=1}^{S}\sum_{t=1}^{T} x_{j,s,t}\right) \quad i,j \in \{1,\ldots C\} : \gamma_{i,j} = 1 \tag{9}$$

$$z_t \geq \sum_{i=1}^{C}\sum_{s=1}^{S} \rho_s \cdot x_{i,s,t-1} - \sum_{i=1}^{C}\sum_{s=1}^{S} \rho_s \cdot x_{i,s,t} - u_t \quad t = 2, \ldots, T \tag{10}$$

$$u_t \leq T\left(\sum_{i=1}^{C}\sum_{s=1}^{S}x_{i,s,t-1} - \sum_{i=1}^{C}\sum_{s=1}^{S}x_{i,s,t}\right) \quad t = 2,\ldots,T \tag{11}$$

The objective function (1) minimizes a weighted sum of different cost terms, corresponding to the rehandling cost in the stacking area, the unproductive crane movements, and a penalty paid for containers not loaded on the train. The penalty is higher for containers having a high commercial value (priority).

The first three sets of constraints regard the assignment of containers to train slots: each container can be assigned at most to one slot (2); at most one container-slot assignment is done for each operation (3); and, in each slot, at most one container can be loaded (4). Other constraints regard the weight restrictions. First of all, for each wagon, a given weight configuration must be chosen (5). Moreover, (6), (7) and (8) represent the weight capacity constraints for each slot, each wagon and for the whole train. Constraints (9) ensure that the rehandling variables $y_{i,j}$ are correctly computed; in particular, it is important to remember that container i is rehandled if, when operation t is executed, a container j that is located in the stacking area under i is loaded and container i has not yet been loaded on the train. Finally, constraints (10)–(11) ensure that variables z_t and u_t are correctly computed.

This formulation differs from the one proposed by Ambrosino et al. [9] where the initial position of the crane is not fixed and in the objective function the total distance traveled by the crane is minimized; hence constraints (10) and (11) are different, and in model (1)–(11) variables z_t assume positive values only when the crane goes back to an already visited slot. Moreover, also a different formulation for computing reshuffles is used (i.e. constraints (9) are different in number and size).

3.3 Formulation for the Cases Without Unproductive Crane Movements or Without Reshuffles

In the general formulation (1)–(11), by properly tuning parameters α and β, it is possible to consider a train loading process in which the unproductive crane movements and the reshuffles are weighted in different ways. Hence, by associating a very high value to one of these two parameters, it is possible to represent the specific cases without unproductive crane movements or without reshuffles. However, from the experimental campaign performed, we have realized that it is better (from a computational point of view) to define specific formulations for these particular problems. For the sake of brevity, in the following, these models will be described without reporting the complete formulation, that is straightforward.

As regards the case of train sequential loading (i.e. no unproductive crane movements are allowed), the decision variables to be considered are the following. First of all, the assignment variables are no more indexed with t since, in the case of sequential loading, the order of loading operations is given by the position of the slot where a container is loaded. Then, these variables are $x_{i,s} \in \{0,1\}, i = 1, \ldots, C, s = 1, \ldots, S$, equal to 1 if container i is loaded in slot s, 0 otherwise (again, these variables are defined when $\lambda_i = \mu_s$). Variables $f_{\omega,b} \in \{0,1\}, \omega = 1, \ldots, W, b \in B_\omega$ and $y_{i,j} \in \{0,1\}, i,j \in \{1, \ldots, C\} : \gamma_{i,j} = 1$, are defined exactly as in the general formulation. Then, cost function (1) is rewritten without the second term and the third term is changed considering that now the assignment variables are $x_{i,s}$. Constraints (3), (10), (11) are no more present; constraints (5) remain unchanged, whereas constraints (2), (4), (6)–(9) must be changed according to the new definition of $x_{i,s}$ variables. This formulation differs from the one presented by Ambrosino et al. [8] for the presence of two index assignment variables and for different rehandling constraints (9) .

When instead the stacking policy does not allow reshuffles, the model (1)–(11) must be simplified considering the same variables except $y_{i,j} \in \{0, 1\}$ that are no more present. Then, the new formulation can be obtained from model (1)–(11) by deleting the first term of cost function (1) and constraints (9).

4 Experimental Results

In order to test the effectiveness of the proposed models for the train load planning problem described in Sect. 2 and to compare them, the three models have been implemented in C#; in particular, the 0–1 linear optimization models have been solved using Cplex 12.5 and the IBM ILOG Concert library has been used for building the models from the C# language.

Our experimental analysis is based on 6 groups of instances, whose main characteristics are shown in Table 1. In particular, these 6 groups are characterized by the same number of wagons (i.e. 20), different number of containers present in the stacking area and different number of tiers (maximum number of containers in a stack). For each group, we have randomly generated 5 instances, that differ for the number of $20'$ and $40'$ containers (probabilistically generated), the weight of containers (uniformly distributed between a minimum and a maximum value, specifically defined for $20'$ and $40'$ containers), the priority assigned to containers (again, probabilistically generated, among three priority classes), and the train composition (three different types of wagons are considered, two wagon types have a capacity of 2 TEUs, the third one can carry 3 TEUs). In the last two columns of Table 1 the average capacity of the train T (expressed in TEUs) and the average number of TEUs stored in the stacking area are reported.

These 30 instances have been solved with the 3 models described above, i.e. the general formulation allowing both reshuffles and unproductive crane movements, the formulation for the case without unproductive crane movements and the one for the case without reshuffles. Results are reported in Tables 2, 3 and 4. Each table shows the size of the solved model (i.e. number of variables and constraints), the value of the objective function, the optimality gap expressed in percentage, and

Table 1 Characteristics of the groups of instances

Instances	# containers	# wagons	# tiers	T	TEUs in stacking area
1–5	30	20	4	46.4	39.2
6–10	30	20	6	47.0	43.2
11–15	40	20	4	46.2	56.8
16–20	40	20	6	47.6	58.0
21–25	50	20	4	47.0	77.6
26-30	50	20	6	46.4	74.4

the number of unproductive movements (i.e. number of reshuffles R and number of crane movements M). Please note that the value of the objective function is negative since the constant component of the objective function (1) has not been added in the models in the implementation.

The last 3 columns are useful for understanding the goodness of the obtained solutions in terms of train utilization and "quality" of loaded containers. In particular, L is the percentage ratio between the number of TEUs loaded and the capacity of the train, $\bar{L}$ represents the percentage ratio between the number of TEUs loaded and the TEUs stored in the stacking area, and P is the percentage ratio between the sum of the priority of the loaded containers and the total priority of containers present in the stacking area. The optimality gap is computed as the ratio between (objective function value-lower bound) and (-lower bound). Finally, each row of these tables reports the average data of the five solved instances.

The general model (1)–(11) seems to be very difficult to solve. In 3600 s in some cases the solver is not able to obtain a solution (i.e. one instance of group 16–20, one of group 21–25 and two instances of group 26–30 are not solved). It is worth noting that data reported in Table 2 are obtained by fixing the same weights for reshuffles and unproductive crane movements in the objective function; anyway, the difficulty in solving this model does not change also varying these two weights.

Instead, the case of model without unproductive crane movements is completely different: instances are always solved up to optimality in very few seconds. The related results are shown in Table 3, where also the CPU time in seconds is reported. Except for the instances of the last two groups (21–25 and 26–30) the number of reshuffles is generally very low.

Table 4 shows the results obtained when solving the model without reshuffles with a time limit of 3600 s. The solutions obtained with this model are better than those obtained with the general model, and for the first two groups of instances the solutions are equivalent, in terms of TEUs loaded and priority loaded, to those obtained with the model without unproductive crane movements. The solutions of the remaining groups of instances are quite good in terms of TEUs loaded and priority loaded but are characterized by a very large number of crane movements. Moreover, the optimality gap in the worst case is 16 %.

Table 2 Results obtained with the general model (1)–(11)

Inst.	# var.	# constr.	Obj.	Gap	R	M	L	$\bar{L}$	P
1–5	73166.2	422.4	−1043.0	13	4.4	68.6	76.65	90.45	93.21
6–10	68397.8	461.0	−1225.6	6	5.2	33.2	81.87	88.84	85.09
11–15	85519.6	447.2	−1003.1	41	3.6	83.3	71.07	55.49	62.85
16–20	92797.0	498.6	−682.4	50	10.0	28.6	36.83	28.65	50.11
21–25	99131.6	479.0	−989.1	48	1.0	43.9	48.75	27.52	47.80
26–30	102520.2	520.4	−929.1	30	6.2	40.7	49.35	28.66	39.34

Table 3 Results obtained with the model without unproductive crane movements

Inst.	# var.	# constr.	Obj.	Time	Gap	R	M	L	$\bar{L}$	P
1–5	1837.0	285.2	−1197.0	1.54	0	0.0	–	84.28	99.57	99.85
6–10	1744.2	322.0	−1287.8	1.39	0	2.2	–	85.97	94.12	97.83
11–15	2130.0	310.6	−1621.8	3.88	0	4.2	–	100.00	82.10	93.67
16–20	2273.4	357.8	−1700.0	8.44	0	6.0	–	98.78	83.07	93.16
21–25	2402.8	340.0	−1960.2	2.73	0	9.8	–	100.00	60.96	82.53
26–30	2533.4	383.2	−1883.4	4.46	0	17.6	–	100.00	63.24	85.76

Table 4 Results obtained with the model without reshuffles

Inst.	# var.	# constr.	Obj.	Gap	R	M	L	$\bar{L}$	P
1–5	73123.2	379.4	−1196.10	0	–	0.9	84.28	99.57	99.85
6–10	68322.8	386.0	−1287.70	1	–	2.3	85.97	94.12	97.83
11–15	85459.6	387.2	−1579.20	8	–	28.8	97.45	79.60	95.34
16–20	92701.0	402.6	−1643.70	7	–	48.3	95.92	80.03	92.60
21–25	99058.6	406.0	−1847.30	16	–	66.7	94.50	57.43	80.18
26–30	102399.2	399.4	−1855.60	10	–	43.4	99.61	62.95	85.66

5 Conclusions

In this chapter different models for solving a particular train load planning problem are presented. Results obtained with an extensive experimental campaign show that the general model is very difficult to be solved, whilst the simpler model that enable only reshuffles in the staking area is always solved up to optimality. A constructive heuristic can be used in order to provide a good solution in very few seconds and to avoid expensive unproductive movements. Moreover, it seems that a promising approach is solving the model where only reshuffles are permitted and then applying a local search in order to improve either the load train quality, in cases characterized by a 100 % of TEUs loaded on the train, or the percentage of TEUs loaded in other cases. These ideas will be the focus of a future work.

Acknowledgments We would like to thank our colleague Davide Anghinolfi for his patience and precious support in the implementation phase. The libraries in C# he developed have been very useful and helpful for the experimental tests reported in this work.

References

1. Vis, I.F.A., de Koster, R.: Transshipment of containers at a container terminal: an overview. Eur. J. Oper. Res. **147**, 1–16 (2003)
2. Steenken, D., Voss, S., Stahlbock, R.: Container terminal operation and operations research—a classification and literature review. OR Spectrum **26**, 3–49 (2004)
3. Stahlbock, R., Voss, S.: Operations research at container terminals: a literature update. OR Spectrum **30**, 1–52 (2008)
4. Bostel, N., Dejax, P.: Models and algorithms for container allocation problems on trains in a rapid transshipment shunting yard. Transp. Sci. **32**, 370–379 (1998)
5. Corry, P., Kozan, E.: An assignment model for dynamic load planning of intermodal trains. Comput. Oper. Res. **33**, 1–17 (2006)
6. Corry, P., Kozan, E.: Optimised loading patterns for intermodal trains. OR Spectrum **30**, 721–750 (2008)
7. Bruns, F., Knust, S.: Optimized load planning of trains in intermodal transportation. OR Spectrum **34**, 511–533 (2012)
8. Ambrosino, D., Bramardi, A., Pucciano, M., Sacone, S., Siri, S.: Modeling and solving the train load planning problem in seaport container terminals. In: Proceedings of the 7th annual IEEE Conference on Automation Science and Engineering, pp. 208–213 (2011). doi:10.1109/CASE.2011.6042439
9. Ambrosino, D., Caballini, C., Siri, S.: A mathematical model to evaluate different train loading and stacking policies in a container terminal. Marit. Econ. Logistics **15**(3), 292–308 (2013)

Application of a Rule-Based Decision Support System for Improving Energy Efficiency of Passive Temperature-Controlled Transports

Hans-Dietrich Haasis, Hendrik Wildebrand, Andreas Barz, Guido Kille, Anna Kolmykova, Lydia Schwarz and Axel Wunsch

Abstract A significant proportion of the flow of goods is transported and handled temperature-controlled. Some of these transports are carried out with an active temperature control, while other goods are transported within the scope of a passive temperature control. The project SMITH focuses the issue of passive temperature control using the example of an aluminium producer in Germany which organizes transports of liquid aluminium. The liquid aluminium and the corresponding crucibles need to be heated in a way, which guarantees the customer a delivery in a proper processing temperature. Setting the temperature is currently based on experience. The aim of SMITH is to improve the energy efficiency of passive temperature-controlled logistics. The software predicts the optimum temperature based on factors such as weather conditions. A transfer of the solution to other temperature-controlled transports enables huge energy and CO_2 savings and is an important contribution of the logistics industry to climate protection.

H.-D. Haasis (✉) · A. Barz · G. Kille · A. Kolmykova · L. Schwarz · A. Wunsch
Institute of Shipping Economics and Logistics, Bremen, Germany
e-mail: haasis@isl.org

A. Barz
e-mail: barz@isl.org

G. Kille
e-mail: kille@isl.org

A. Kolmykova
e-mail: kolmykova@isl.org

L. Schwarz
e-mail: schwarz@isl.org

A. Wunsch
e-mail: wunsch@isl.org

H. Wildebrand
Berlin School of Economics and Law, Berlin, Germany
e-mail: hendrik.wildebrand@hwr-berlin.de

J. F. de Sousa and R. Rossi (eds.), *Computer-based Modelling and Optimization in Transportation*, Advances in Intelligent Systems and Computing 262,
DOI: 10.1007/978-3-319-04630-3_3,

Keywords Passive temperature-controlled transport · Expert system · Decision tree · Fuzzy logic · Rule based

1 Introduction

A significant proportion of the national and international flow of goods is transported and handled temperature-controlled. Temperature-controlled goods are frozen or refrigerated foods, pharmaceutical products, chemicals as well as liquid tar or liquid metal in the range of high temperatures. Some of these transports are carried out with an active temperature control, while other goods are transported within the scope of a passive temperature control. The passive temperature control follows without cooling or heating by means of aggregates, the goods must be located within a certain temperature range during the transport. Setting the temperature is currently based on experience, using information like the transport time and the condition of the transport container or weather conditions such as outdoor temperature, wind speed and density of precipitation. The project SMITH at the Institute of Shipping Economics and Logistics (ISL) addresses these temperatures using the example of the transport of liquid aluminium with the aim to improve the energy efficiency of passive temperature-controlled transports. During this project a rule-based expert system is developed that supports shippers and logistics service providers in their decision on the starting temperature of the transported goods. The software predicts the optimum temperature for specific applications based on current factors such as material properties or transport and weather conditions. For the configuration of the expert system and to collect real data from the passive temperature-controlled transports, a multi-sensory tool including data storage and data transmission is developed.

The remainder of this chapter is organized as follows. In §2, we describe the characteristics of temperature-controlled transports. In particular the passive temperature-controlled transport of liquid aluminium and its influencing factors are specified. In §3, we present the developed demonstrator application. In §4, we show the possible reduction of energy consumption and CO_2 emissions. In §5, we offer some concluding comments.

2 Temperature-Controlled Transports

Temperature-controlled logistics and non-temperature-controlled logistics can be distinguished by many features. Beside the differentiated temperature needs, for temperature-critical goods other difficulties like durability, sensitivity, hygiene and security requirements as well as packaging or batch size have to be taken care of. Especially at the transportation of liquid aluminium beside the product-specific features, legal restrictions have to be followed. Based on these reasons, the definition of temperature-controlled logistics written by Truszkiewitz and Vogel [14] which is analogical to the classic definition of logistics can be modified and used in this topic too:

Beside the general objective of logistics to provide goods at the right place, at the right time, in the required quantity and quality, the term temperature-controlled logistics implies the simultaneous observance of all legal restrictions and customer-specific requirements such as production, storage, transportation and distribution of temperature-critical goods.

The Definition refers not only the physical component of the transportation process, but also describes the accompanying information and organization processes. The transportation process is split in two systems: active and passive temperature-controlled transports.

2.1 Active and Passive Temperature Control

The transportation of temperature-critical goods can be done active or passive temperature-controlled. If during the transportation process heaters or cooling units are used, the process is called active temperature-controlled. If isolating packages, containers or carriers are used instead, it is called a passive temperature-controlled transport [6].

Active systems can be used in nearly every form: from packets to containers, almost everything can be cooled or heated. The heating mostly takes place by batteries or external electric sources. The most important advantage of active temperature-controlled transports is the high thermal stability. Disadvantages are beside the high investments and running costs, the lack of flexibility [10].

An alternative for that is the passive temperature-controlled transport. In this case heating or cooling systems are not used to reduce costs. In passive systems the goods are enclosed by isolating charging units, which guarantee that the goods do not get any thermal damages in a defined transport time [4]. But beside a temperature resistant charging unit, this method requires enormous experiences of the employees considering the external influences. The difficulty consists of the right adjustment of the temperature. Influences which could affect the freight temperature like physical, biological, climatic, chemical or abrasive influences or the transport time and speed have to be considered and calculated before departure [6]. Furthermore, the temperature of goods is directly related to its quality, this point is a critical problem of passive temperature-controlled systems.

2.2 Passive Temperature-Controlled Transport of Liquid Aluminium

In the case of liquid aluminium passive temperature-controlled transport takes place in special crucibles, which isolate the aluminium best possibly against external influences. The crucibles have a capacity of 5 to 6 tons and are made of a stable steel case with reinforcing profiles as well as temperature resistant refractory linings for optimum isolating.

Transport Process. The transport process of the liquid aluminium is based on a type of kanban principle. The necessary preparations begin after the smelting of aluminium scrap directly on the furnace. When a vehicle with empty transport units reaches the plant site, new crucibles are heated on a customized preheating station until the temperature of the new transport unit is equivalent to the alloy-specific filling temperature of the liquid aluminium. After the defined preheating temperature is achieved, the preheated crucible is transported to the nearby filling station and the filling of the crucible starts. The filled crucibles are set in exchange for empty ones on the truck and secured with four steel pins [2]. The entire process of loading takes about an hour. Generally a truck transports only one type of alloy per tour to avoid possible confusions at the customer site. Before the loaded trucks leave the factory, they pass a checkweigher. At this point the total weight of the truck and the departure time is recorded. All collected data such as weight, alloy and preheating temperature are given to the customer in form of a protocol. In the best case, the carrier reaches the site of the customer within the desired temperature range and leaves the plant after unloading with empty crucibles and the described process can restart again.

Influencing Factors. Basically, the cooling process is reflected in a digressive falling curve of aluminium temperature along with the transport time, because the effluent heat flow decreases with a decreasing temperature gradient to the outside temperature. The progression of the curve and its pitch are primarily caused by the prevailing parameters of influence. Table 1 shows these parameters of influence that are explained below briefly.

The crucible condition is dependent on the remains of the liquid aluminium on the inside of each crucible. These remains are formed with each transport. They reduce the transporting amounts of liquid aluminium and therefore can have an impact on the cooling process.

Due to the aerodynamic properties of the driver's cab in combination with possible installed air deflectors, the wind load of the crucibles vary depending on the crucibles' position (1 to 3). Thus the position of the transport units has a decisive influence on the course of the cooling process.

The transport time includes the time period in which the liquid aluminium is in the crucible. Therefore it is subject to all kind of influences. The transport time begins at the starting time of filling the aluminium smelter and ends with the discharge at the customer site. The duration of the transport is generally seen in close connection with the transport distance. A longer transport time has a negative effect on the temperature of the smelt.

The temperature gradient between the outdoor temperature and the alloy temperature is primary defined by the outdoor temperature. High temperatures reduce the discharge flow of heat. At lower temperatures a stronger cooling effect is expected.

The precipitation respectively a high density of precipitation leads to decrease of the aluminium temperature. Basically, it can be assumed that raindrops

Table 1 Relevant parameters of influence with entities

Nomination	Entity
Crucible condition	Number of fillings since last cleaning
Crucible position	Position [1, 2, 3]
Transport time	Hours [h]
Outdoor temperature	Degrees celsius [°C]
Density of precipitation	Millimeter per hour [mm/h]
Driving speed	Kilometers per hour [km/h]
Relative humidity	Percent [%]

evaporate by striking the surface of the crucible at a temperature up to 134° C. For the evaporation of a liquid, in other words a phase transition from liquid to gaseous state of aggregation, the heat of evaporation has to be achieved [1]. The necessary energy is withdrawn from the system in form of thermal energy (energy conservation). By that a higher density of precipitation leads to a higher withdrawn thermal energy.

The driving speed causes a turbulent air flow at the surface of the crucibles, which leads to a convective removal of the heat directly around the crucible. The warm air is carried out of the system faster at higher driving speed and will be replaced automatically from the inside of the crucible leading to a greater loss of temperature of the aluminium.

At this time it cannot be estimated which quantitative and qualitative influences are derived on the cooling process of the smelt. But in the past correlations between relative humidity and cooling behavior of the aluminium alloy have been found. Therefore this exogenous factor should be taken into consideration in the progress of work.

Even if the direction of influence is basically clear, precise quantitative statements about the cause-effect relationship of the individual parameters are difficult. However, this would be important and desirable especially for an energy-efficient control of the supply chain.

3 Decision Support System

3.1 Sensor System Architecture

For the recording of weather and crucible data a detection system was developed, which makes it possible to capture the data in real time. The system was installed on a truck‘s semitrailer to do the recordings. It includes a weather station with Integrated Sensor Suits (ISS) and a data logger. The ISS records the actual weather data like wind speed, humidity, precipitation and outdoor temperature during the ride in a 10-min-interval. Further sensors were placed in the walls of the crucibles, which record (also every 10 min) the temperature of the crucibles. The recorded

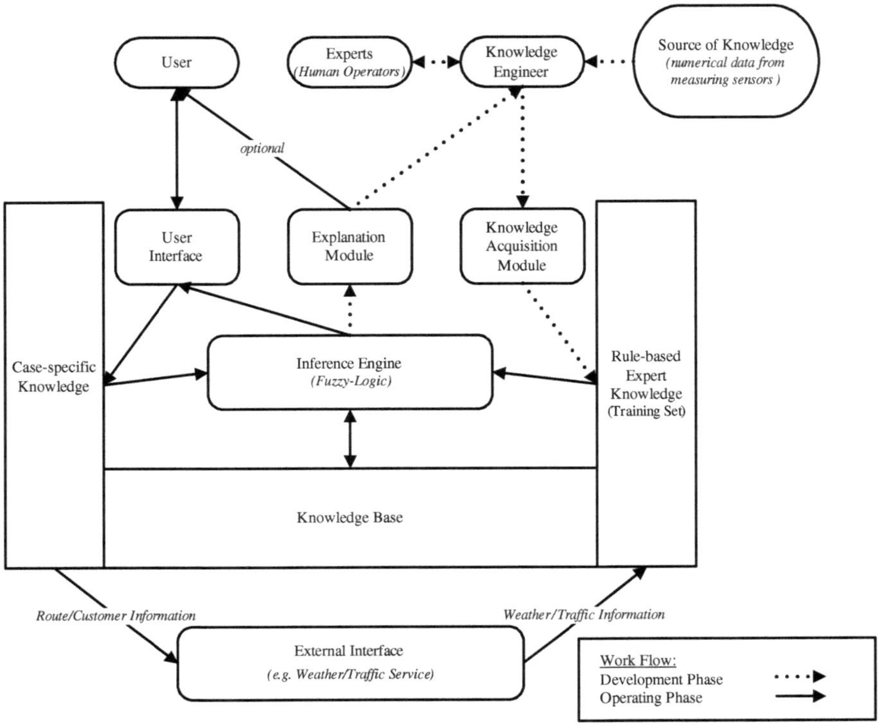

Fig. 1 Expert system architecture

data are transmitted via a cable to the data logger, which saves them in a storage unit. The recorded weather data are being analyzed afterwards.

3.2 Expert System Architecture

The architecture of an expert system, in other words its exterior with different program modules and connections, in general comprises of five components: knowledge base, inference engine, user interface, explanation module and knowledge acquisition module [7]. This basis architecture needs to be modified and extended for this application. Figure 1 shows the resulting schematic structure of the expert system.

The knowledge base is the core and the base of each expert system. It contains the permanent expertise as well as the temporary knowledge of the experts about the individual area of application. In this case the knowledge base distinguishes between rule-based expert knowledge on the one hand and case-specific knowledge on the other hand. While the first one is directly provided by experts, the case-specific knowledge has to be provided by the users. The inference engine

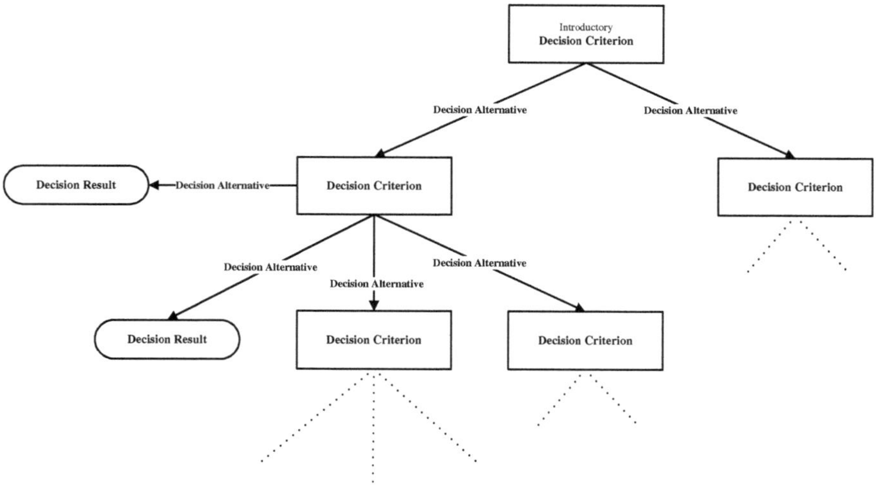

Fig. 2 Decision tree structure

connects and combines both modes of knowledge by using fuzzy logic to draw a conclusion. In addition to this conventional methodology, an external interface uses route or customer information given by user to select route and customer specific weather respectively traffic information.

Another component of the expert system is the user interface, which enables the communication between expert system and user over a graphical user interface (see Sect. 3.5). Besides the result output, the user interface is able to explain the reasoning by showing intermediary results. The knowledge acquisition module is only implemented and used in the development phase for decision-tree-based rule induction. Over and above the human operators, the numerical data from the measuring sensor system (see Sect. 3.1) represents a fundamental source of knowledge.

3.3 Decision-Tree-Based Rule Induction

A decision tree is a decision support tool with nodes, arcs and leaf nodes. It is built up of a hierarchical tree structure where each node contains a branching criterion with associated alternatives for a specific attribute of training set [8]. The outcome of this is the directed graph shown in Fig. 2. The nodes represent the decision criteria, the arcs constitute the possible decision alternatives and the leaf nodes show the closing decision result.

To generate an initial decision tree for this application a set of training cases is obligatory. In this case the current training set comprises about 100 passive temperature-controlled transports of liquid aluminium from supplier to customer. Every transport is listed in the set with its specific attributes or rather its endogenous and exogenous influencing factors. The data acquisition was conducted with the developed sensor system (see Sect. 3.1). In addition to the factor values, the gradient of the liquid aluminium cooling curve is also given.

The identified influence factors (crucible position, crucible condition, driving speed, humidity, density of precipitation, outdoor temperature) and their corresponding measured data are synonymous with the decision criteria respectively decision alternatives. The gradient represents the decision result. In principle, the formed decision tree offers an own competency in solving decision problems, but in this application the decision tree is used as a tool for rule induction. Every decision tree can be translated into an equivalent rule base without any problems [13]. Each decision path, which starts at the introductory criterion and ends with a result, equates one rule. All passed decision criteria in connection with the selected alternatives generate the antecedent (IF). The final decision result is the consequent (THEN) of the rule. Of course, rewriting the decision tree to a collection of rules, one for each path, would not result in anything more simple or flexible than the tree [9]. Therefore and due to the fact that the set of training cases is limited and not able to cover all future possible combination of influence data, the integration of fuzzy logic is necessary.

3.4 Fuzzy Rule-Based Expert System

The fuzzy logic approach helps to formalize human reasoning patterns and to develop high-performance expert systems in contexts where data are uncertain (e.g. "about 10 °C") and/or vagueness (e.g. "very cold").

The use of fuzzy logic combined with the expert system has two central advantages [16]. On the one hand the application of linguistic variables provides the system with elasticity and intuitiveness, and enables to generate the humanlike decisions. On the other hand fuzzy logic helps expert systems to reduce complexity and heterogeneity of their elements [12, 16].

Through the use of fuzzy logic in the expert system, the system is referred to as a fuzzy logic-based expert system or fuzzy expert system. The development of fuzzy logic-based expert systems consists of nine steps: Description of problem and aims, knowledge acquisition, definition of membership functions (linguistic variables and terms), creating the rule base, establishing a weighting factor for each rule, selection of operators, selection of the defuzzification method, testing and fine-tuning respectively optimization of the system [5, 15]. The input and output variables of a fuzzy logic-based expert system can be made available from different information sources, such as from numerical data of measuring sensors or heuristics in the form of linguistic expressions [11].

The knowledge base is a central component of a fuzzy logic-based expert system and contains the knowledge of experts on transport of liquid aluminium as well as collected sensor data (see Sect. 3.2). The representation of this knowledge occurs mostly with if–then-rules. The inference engine component of the expert system includes the fuzzy logic components for fuzzification and defuzzification. In the fuzzification component, the sharp inputs are translated to fuzzy sets with linguistic terms. The processing of the fuzzified inputs with if–then-rules occurs in the inference component of the fuzzy logic-based expert system. Lastly, the defuzzification component calculates a discrete result from the fuzzy sets.

3.5 Demonstrator Application

To support the data analysis process and for later tests in a productive environment, the development of a software demonstrator application has been realized. The composition of the demonstrator can be roughly divided into two parts, namely the calculation logic (containing fuzzy data handling and computations) and the Graphical User Interface (input and output dialogs presented to the user). The demonstrator has been completely developed in Java. As for the calculation logic, a Java-based software library called jFuzzyLogic [3] has been used that supports the Fuzzy Control Language (FCL) for easy import of fuzzy rules and variables.

For the creation of the fuzzy logic rules the method of the decision tree and the fuzzy logic were combined. For a detailed description of these methods see Sects. 3.3 and 3.4. During a defined time period data concerning the different influences on the aluminium temperature during the transports were recorded. Based on the collected data a decision tree was derived. Afterwards the different decision alternatives were compiled and transformed to a set of fuzzified rules. These rules are finally stored in the configuration file of the demonstrator. Required input data for the calculation of the temperature of the liquid aluminium are: crucible condition, crucible position, transport time, outdoor temperature, density of precipitation, humidity and desired temperature of the liquid aluminium at arrival. The cooling curve of the aluminium is approximated with a falling straight line. On the basis of the desired temperature at arrival and the other required data the demonstrator determines the gradient or rather the simple equation. As the travel time is known, the temperature of the liquid aluminium at departure can now be calculated.

The Graphical User Interface (GUI) supports the user regarding the input of all relevant input values for the fuzzy inference system. One of the areas of deployment is the use of the decision support system as a smartphone-application like shown in the following Fig. 3.

Since the needed computing power for the calculations is rather low and the amount of stored data is small, any Android-capable smartphone should be a sufficient platform to handle the execution of the application. When used as a smartphone-application the decision support system grants the user a better availability with less resources.

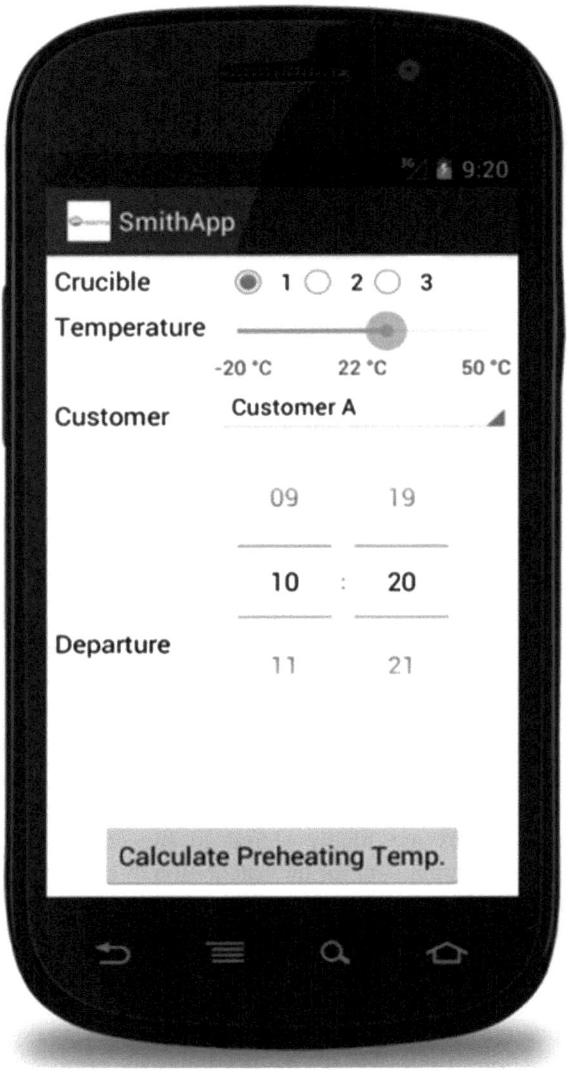

Fig. 3 Screenshot of SMITH android application

4 Improving Energy Efficiency

The use of the SMITH-demonstrator makes it possible to optimally adjust the preheating temperature of the crucible. Because of that less gas is needed for preheating the crucible. This lower gas consumption leads to a lower output of CO_2 during the heating. The calculation of the possible CO_2 savings is done by analyzing the data of the set of training cases. Based on the recorded data the transport is simulated with the use of the SMITH-Software, compared to the data without the use of the SMITH-demonstrator and assessed concerning the possible CO_2 savings. Here, the following assumptions have been made:

Table 2 Symbols, their nomination and units of the equation one

Symbol	Nomination	Entity
CO_{2save}	Possible CO_2 savings	Kilogram [kg]
T_{higher}	Aluminium temperature at arrival above customer requirements (weighted average)	Degrees celsius [°C]
T_{lower}	Aluminium temperature at arrival under customer requirements (weighted average)	Degrees celsius [°C]
$T_{w/S}$	Required temperature of aluminium by customer	Degrees celsius [°C]
W_T	Heating power of gas burner	Kilowatt [kW]
CO_{2aq}	CO_2-conversion factor for burning of methane	–
H_i	Calorific value of methane	Mega joule per kilogram [MJ/kg]
v_T	Heating rate	Degrees celsius per hour [°C/h]

- As the data basis for the comparison is unchanged it is assumed that the temperature of the aluminium at arrival, ceteris paribus, is only adjusted by the preheating temperature of the crucible on the aluminium producer's site.
- The difference of the temperature required by the customer and the actual measured temperature of the aluminium at arrival equates the temperature difference by which the crucible must be preheated more or less before filling.
- The required aluminium temperature at arrival is always maintained when using the SMITH-demonstrator.
- Instead of natural gas pure methane is used for preheating the crucible. In addition, the methane burns completely forming carbon dioxide and water.
- To calculate the possible CO_2 savings CO_{2save} per crucible and tour, taking into account the assumptions mentioned above, the equation one was developed. The equation itself and the results of the calculation will be explained briefly below.

$$CO_{2save} = (T_{higher} + T_{lower} - T_{w/S}) * W_T * CO_{2aq}/(H_i * v_T) \quad (1)$$

The different symbols of the equation, their nomination and their units are shown in Table 2.

T_{lower} and T_{higher} are the actual measured temperatures of the aluminium at arrival on the customer's site. These temperatures are determined by the set of training cases: For all tours with a higher or lower aluminium temperature then the customer requires the weighted average for the measured temperatures is formed. From the sum of the temperatures at arrival the temperature required by the customer $T_{w/S}$ is subtracted. Accordingly to the assumptions above the temperature required by the customer equates the temperature at arrival when using the SMITH-demonstrator. Therefore this difference equates the temperature difference, by which the crucible must be preheated less before filling the aluminium into it. This difference is multiplied by factors for the preheating power W_T and the CO_2-conversion factor CO_{2aq}. Both factors are constant. The preheating power indicates the power of the gas burner. It is 200 kW. The CO_2-conversion factor results from the reaction equation of burning methane. As 1 kg of methane is burned 2,54 kg of CO_2 are generated. In the denominator the heating rate v_T is multiplied with the calorific value of the methane H_i. These factors are constant as well and are 84° C/h respectively 50,013 MJ/kg. By multiplying respectively dividing the different factors the possible CO_2 savings per tour, where the aluminium temperature is too high, can therefore be calculated.

The temperature difference, which is calculated from the higher and lower weighted average temperature of the aluminium at arrival (T_{higher}, T_{lower}) and the temperature required by the customer ($T_{w/S}$), is 6,39 °C. Multiplying the temperature difference with the factors W_T and CO_{2aq} respectively dividing it by the factors vT and Hi results in savings of 3,00 kg CO_2 per crucible and tour, if the SMITH-demonstrator is used. Extrapolated to all crucibles and tours of the German sites there is a reduction of CO_2-emissions in the amount of 124,2 t per year. To get a comparison: these 124,2 t match the annual CO_2-emissions of nearly 84 Volkswagen Golf 1.6 TDI Bluemotion, basing on a mileage of 15.000 km for each car.

5 Conclusions

The method of the passive temperature-controlled transport requires high demands on the employees for setting the right temperature of the goods at departure. To support the employees and to optimize the efficiency of passive temperature-controlled transports, we designed a rule-based expert system. From the point of view of the aluminium producer the use of the SMITH-demonstrator leads to an optimized heating process, time and cost savings. The higher efficiency of temperature controlled transports results in a lower CO_2-output during the preheating of each crucible. As §4 showed a reduction of CO_2-emissions during the preheating of the crucible in the amount of 124,2 t per year is feasible. From the point of view of the customer the use of SMITH-demonstrator leads to an increased delivery reliability and quality.

The development of the software is ongoing. Especially the knowledge base of the expert system is still growing, but the basic function of the system is evident. It is already able to predict a temperature for the liquid aluminium. With a larger knowledge base, the system will be able to create a more accurate prediction.

According to the project SMITH, the demonstrator application respectively the rule-based expert system has been designed to predict the temperature of liquid aluminium. But it is easy to adapt the software to other passive temperature-controlled transports of goods. For example, tar is transported passive-controlled as well. If the solution is transferred to other passive-controlled transports, even higher energy savings and therefore CO_2 savings could be realized.

Acknowledgments This work was funded by the Federal Ministry of Education and Research (BMBF) under the reference number 01LY1104 "SMITH—Improving energy efficiency of passive temperature-controlled transports".

References

1. Baehr, H.D., Stephan, K.: Wärme- und Stoffübertragung. Springer, Heidelberg (2006)
2. Bergrath, J.: Heiße Ware—Werkstofftransport: Flüssig-Aluminium. Lastauto Omnibus—Test Technik Trends **82**(11), 76–78 (2005)
3. Cingolani, P.: Open source fuzzy logic library and FCL language implementation. http://jfuzzylogic.sourceforge.net (2012)
4. Feil, D.: Sichere verpackung für sensible Güter. MM Logistik **5**, 44–45 (2011)
5. Hoenerloh, A.: Unscharfe Simulation in der Betriebswirtschaft: Modellbildung und Simulation auf der Basis der Fuzzy Set-Theorie. Unitext, Goettingen (1997)
6. Lange, V., Hasselmann, G.: Temperaturgeführte transporte. In: Arnold, D., et al. (ed.) Handbuch Logistik, pp. 570–580. Springer, Berlin (2008)
7. Nikolopoulos, C.: Expert Systems: Introduction to First and Second Generation and Hybrid Knowledge Based Systems. Marcel Dekker, New York (1997)
8. Paetz, J.: Soft Computing in der Bioinformatik: Eine Grundlegende Einführung und Übersicht. Springer, Heidelberg (2006)
9. Quinlan, J.R.: C4.5: Programs for Machine Learning. Kaufmann, San Mateo (1993)
10. Roskoss, A.: Temperature-Controlled Packaging Systems—Active or Passive? http://www.iptonline.com/articles/public/Intelsius.pdf. (2011)
11. Shah Hamzei, G.H., Mulvaney, D.J.: Implementation of an intelligent control system using fuzzy ITI. Neural Comput. Appl. **9**(1), 12–18 (2000)
12. Sibbel, R.: Fuzzy-Logik in der Fertigungssteuerung am Beispiel der retrograden Terminierung. Lit Verlag, Muenster (1998)
13. Spreckelsen, C., Spitzer, K.: Wissensbasen und Expertensysteme in der Medizin: KI-Ansätze zwischen klinischer Entscheidungsunterstützung und medizinischem Wissensmanagement. Vieweg und Teubner, Wiesbaden (2008)
14. Truszkiewitz, G., Vogel, S.: Spezielle logistikprozesse—temperaturgeführte logistik. In: Arnold, D., et al. (ed.) Handbuch Logistik, pp. B7-34–B7-46. Springer, Berlin (2004)
15. Zimmermann, H.-J., Angstenberger, J.: Fuzzy Technologien: Prinzipien, Werkzeuge Potenziale. VDI Verlag, Duesseldorf (1993)
16. Zimmermann, H.-J.: Fuzzy Set Theory and Its Applications. Kluwer Academic Publishers, Boston (1996)

An Eco-Traffic Management Tool

Jorge M. Bandeira, Sérgio R. Pereira, Tânia Fontes, Paulo Fernandes, Asad J. Khattak and Margarida C. Coelho

Abstract Drivers routing decisions can be influenced to minimize environmental impacts by using, for instance, dynamic and intelligent road pricing schemes. However, some previous research studies have shown that often different pollutants can dictate different traffic assignment strategies which makes necessary to assign weights to these pollutants so they become comparable. In this chapter, a tool for traffic assignment taking into account eco-routing purposes is presented. The main goal of this work is to identify the best traffic volume distribution that allows a minimization of environmental costs for a given corridor with predetermined different alternative routes. To achieve this, an integrated numerical computing platform was developed by integrating microscopic traffic and emission models. The optimization tool employs non-linear techniques to perform different traffic assignment methods: User Equilibrium (UE), System Optimum (SO) and System Equitable (SE). For each method, different strategies can be assessed considering: (i) individual pollutants and traffic performance criteria; and (ii) all

J. M. Bandeira · S. R. Pereira · T. Fontes (✉) · P. Fernandes · M. C. Coelho
Centre for Mechanical Technology and Automation/Department Mechanical Engineering, Campus Universitário de Santiago, University of Aveiro, 3810-193 Aveiro, Portugal
e-mail: trfontes@ua.pt

J. M. Bandeira
e-mail: jorgebandeira@ua.pt

S. R. Pereira
e-mail: sergiofpereira@ua.pt

P. Fernandes
e-mail: paulo.fernandes@ua.pt

M. C. Coelho
e-mail: margarida.coelho@ua.pt

A. J. Khattak
Civil and Environmental Engineering Department, University of Tennessee, 322 J.D. Tickle Bldg, Knoxville, TN 37996-2010, USA
e-mail: akhattak@utk.edu

J. F. de Sousa and R. Rossi (eds.), *Computer-based Modelling and Optimization in Transportation*, Advances in Intelligent Systems and Computing 262,
DOI: 10.1007/978-3-319-04630-3_4,

pollutants simultaneously. For the latter case, three different optimization approaches can be assessed based on: (i) economic costs of pollutants once released into the air; (ii) human health impacts according to the Eco-Indicator 99; and (iii) real time atmospheric pollutant concentration levels. The model was applied to a simple network, simulating three levels of traffic demand and three different strategies for traffic assignment. The system is developed in Microsoft Excel and offers a user friendly access to optimization algorithms by including a dynamic user interface.

Keywords Eco-routing · Traffic assignment · Microscopic model · Atmospheric emissions

1 Introduction

Although there have been some improvements over recent years, road transport sector is still contributing significantly to nitrogen oxide (NO_X), particulate matter (PM10) and carbon monoxide (CO) emissions in Europe (33, 13 and 27 %, respectively for each pollutant) [1]. Recent studies show that people living near congested roads across Europe are still particularly exposed to air pollution. In 2010, urban traffic air quality stations recorded NO_2 and PM concentrations above legal limits in 44 and 33 % of cases, respectively. These pollutants may have an effect on the cardiovascular system, lungs, liver, spleen and blood [1]. A more efficient management of existing infrastructures has been identified as a key policy with great potential to reduce emissions. These measures may include behavioural changes in the operation of vehicles (eco-driving) as well as route choices with lower emissions impacts associated. In this context, the Eurovignette directive proposes a "user pays" and a "polluter pays" principle for heavy duty vehicles in Europe [2]. In order to encourage the move to transportation patterns with lower environmental impacts, the tolls price could vary according to vehicles' emissions, distance travelled, location and time. Yin and Lawphongpanich [3] demonstrated that there always exists a (non-negative) tolling system that leads to a traffic distribution with minimum emissions. Recently an optimal emission pricing model to reduce emissions in a given transportation network was proposed by Sharma and Mishra [4]. The impact of route selection in terms of emissions has been studied deeply in recent years. Bandeira et al. [5] conducted a detailed revision of the most important studies in this field. Through an empirical study, the relevance of the eco-routing concept has been reinforced [6]. However, in this chapter it was found that the concept of "eco-friendly" should not be strictly confined to CO_2/fuel consumption since a trade-off between CO_2 versus local pollutants minimization has been observed. Barth et al., Ahn and Rakha, Yao and Song [7–9] have conducted important research on environmentally friendly routing based on microscopic emission models (CMEM, VT-micro and VSP respectively). In 2012,

Gazis et al. [10] developed a system of eco-navigation taking into account the relative impact of emissions in terms of cost, life cycle analysis, and current critical air pollutants concentrations. However, in this approach, it was assumed that routing has negligible impact on overall congestion. By modelling two large scales areas, Guo et al. [11] found that a 40 % targeted green routing market penetration yielded 25 % reduction in CO emissions at the expense of 13 % increase in travel time. Instead, Ahn et al. [12] stated that a market penetration of 20 % of eco-routers may produce higher fuel consumption levels.

Some important conclusions of previous research are: (i) an efficient route choice may lead to significant emission reductions; (ii) there may be a conflict in minimizing different pollutants; (iii) the optimization of traffic operations should consider the relative damage impacts of each pollutant; (iv) the implementation of eco-routing systems may lead to unexpected results on large and complex networks.

In a more realistic picture, a more efficient (environmentally) traffic distribution could be performed in certain corridors wherever is possible to implement intelligent toll systems that may lead to a better allocation of traffic. A recent study pointed out that the benefits of dynamic eco-routing and dynamic emissions pricing may lead to 160 billion US dollars of environmental benefits with relative low incremental costs between 2017 and 2055 [13].

This study presents an eco-traffic management tool used to define the most sustainable traffic distribution, given a total demand provided by the user, among n routes linking an Origin/Destination pair (OD). This optimization can be performed using different criteria and assignment methods. In a further step this optimal distribution may be used to estimate optimal emission pricing schemes under different levels of traffic demand. A case study is presented based on a simplified road network linking an OD pair that consists of two arterials and two motorways with different capacities.

2 Methodology

This section presents the methodology for the development of an eco-traffic assignment tool. First, a brief explanation on the development of Volume-Delay functions (VDF) and Volume-Emission functions (VEF) is provided. Then, the optimization methods and the criteria available for eco-friendly traffic assignment strategies are explained.

2.1 Volume-Delay and Volume-Emission Functions

VDF and VEF for each link must be defined before the optimization process is started. These functions use the traffic volume as an independent variable and both

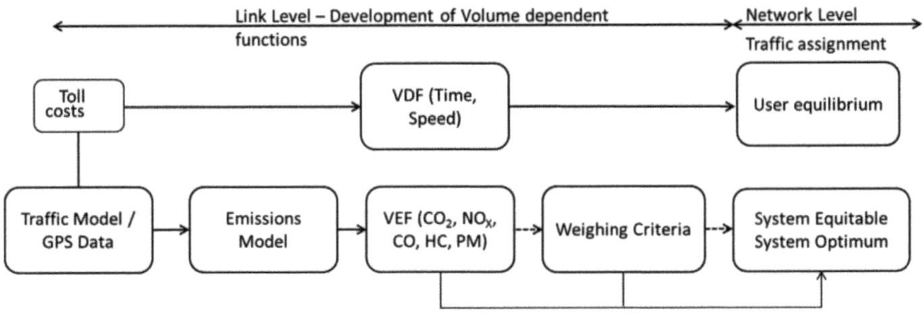

Fig. 1 Overall structure of the optimization platform

travel time (VDF) and emissions (VEF) as dependent variables. Different scenarios of traffic volumes using different links can be performed using commercial microscopic traffic models or real world GPS data using probe vehicles. Microscopic traffic flow models simulate single vehicle-driver units, thus the dynamic variables of the models represent microscopic properties such as the position and second-by-second speed of individual vehicles. Then, emissions can be estimated using instantaneous emission models with the specified level of detail of the road traffic model used previously (Fig. 1).

By conducting a regression analysis, and for a considerable number of routes, a cubic polynomial function (Eq. 1) was shown to be appropriated to interpolate the traffic volume with total pollutant emissions and other traffic parameters (P). Figure 2 exemplifies the VEF for NO_X and CO_2 for 4 different routes (see Sect. 3.1 for further details).

$$\begin{aligned} \mathrm{P} = \mathrm{constant} + \mathrm{b1} \cdot \mathrm{V} + \mathrm{b2} \cdot \mathrm{V}^2 + \mathrm{b3} \cdot \mathrm{V}^3 \\ \mathrm{V} < \mathrm{Capacity}, \quad \mathrm{C} \end{aligned} \tag{1}$$

The likely traffic distribution in the network can be assessed through the traditional volume-delay (or cost) functions and the User Equilibrium (UE) model formulation. To ensure that the assignment converge to a unique solution, the VDF must be strictly increasing. Thus, in this case, a cubic function may not be appropriate, even a high correlation between the observed and predicted values is obtained. In this platform, an additional tool to optimize the most widely used VDF parameters of the BPR [14] and Conical [15] (Eqs. 2 and 3) functions is available. This optimization is conducted by minimizing the Root Mean Square Error between observed/simulated and predicted values of travel time.

$$t = t_0\left(1 + \alpha\left(\frac{V}{C}\right)^{\beta}\right) \tag{2}$$

$$\frac{t}{t_0} = 2 + \sqrt{\left(1 - \frac{V}{C}\right) + \beta^2} - \alpha\left(1 - \frac{V}{C}\right) - \beta \tag{3}$$

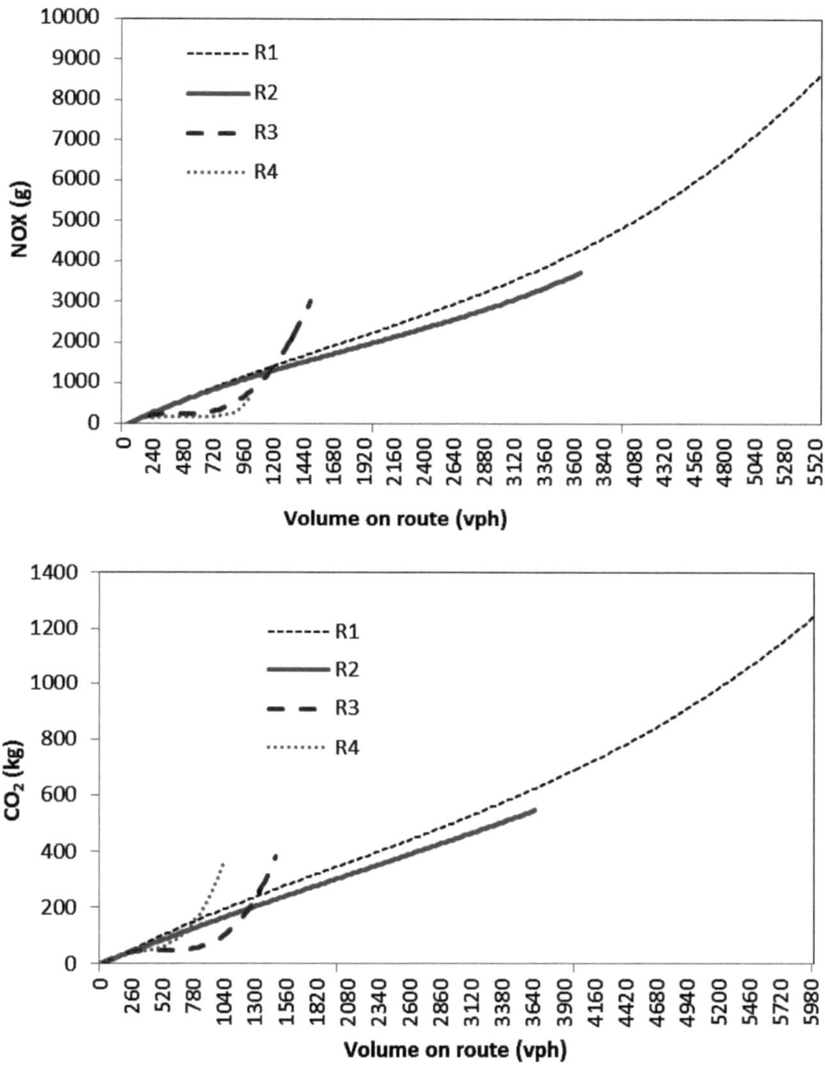

NO_X	r^2	F	sig.
R1	0.996	576	< 0.001
R2	0.999	112495	< 0.001
R3	0.959	55	< 0.001
R4	0.997	762	< 0.001

CO_2	r^2	F	sig.
R1	0.998	1123	< 0.001
R2	0.999	219682	< 0.001
R3	0.951	45	< 0.001
R4	0.997	739	< 0.001

Fig. 2 VEF for NO_X and CO_2 over 4 routes using *cubic polynomial* interpolation

where: t—Travel time for volume V; t0—travel time at free flow; C—Capacity; α; β—dimensionless parameters.

2.2 Criteria

A wide range of criteria is available to optimize the traffic distribution among alternative routes in a corridor, according to criteria and methods selected by the user.

In addition to traffic performance parameters (travel time, speed and, traffic density), individual pollutants (CO, NO_X, HC, PM), and Greenhouse Gases (CO_2), three integrated optimization approaches are available based on: (a) economic cost; (b) health impacts; and (c) air quality levels. The economic cost takes into account the cost of each pollutant once released into the air. Unit benefits can be introduced using published data on the value of reducing emissions or fuel savings. The health impact approach weighs a range of substances according to various damaging effects. Finally, an alternative approach is available if the user wants to consider the real time air quality levels and assign different weights to each pollutant. A more detailed explanation on the methodology for normalization of emissions impacts can be found elsewhere [10]. By default, the parameters for weighing pollutants effects (economic cost and health impacts-indicator) are based on literature [13, 16], and are shown in Table 1. However, since these factors were developed for different contexts, further research is needed to adapt and contextualize these values to particular cases.

2.3 Assignment Method

In addition to the traditional UE approach, two different optimization goals are considered: System Equitable (SE) and System Optimum assignment (SO). In the first case, the traffic distribution between the OD pair is achieved at the same cost for all routes. This concept introduced by Rilett and Benedek [17] has as main goal to distribute equitably the negative effects of traffic among the alternative routes. Moreover, a maximum amount of pollution in the total network or in a specific link can be defined first. This objective is attained by minimizing the standard deviation (among the alternative routes) of the cost associated with the selected criterion. Both the Eqs. 4 and 5 exemplify the optimization process taking into account the criterion "Integrated Economic Cost".

In the second method, the traffic assignment is performed with the aim of maximizing the overall benefit of the whole network (Eq. 5). This objective is attained by minimizing the total environmental costs of the system. This approach indicates a lower bound for the amount of pollution impacts possible, and allows the planners to identify how close to the optimum scenario they are. The constraint functions ensure that the overall and the specific capacity of each link is not exceeded, the non-negativity and the user-defined total demand is met (Eqs. 9–12).

Table 1 Effect of pollutants based on economic costs [13] and eco-Indicator 99 [16]

Pollutant	Cost (USD/g)	Impact (DALYs/kg)
NO_X	0.02480	1.7200
HC	0.00827	0.0248
CO	0.00416	0.0141
CO_2	0.00007	0.00406
PM	0.22920	7.26

$$\min \sqrt{\frac{\sum_{i-1}^{n}(Ec_i - \overline{EC})^2}{n-1}} \tag{4}$$

$$\min \sum_{\mathrm{i}}^{\mathrm{n}} EC_i \tag{5}$$

where:

$$EC = \sum_i^n \sum_j^m P_j CP_j \tag{6}$$

$$P_{ji(vi)} = constant_{ji} + b1_{ji}V_i + b2_{ji}V_i^2 + b3_{ji}V_i^3 \tag{7}$$

$$V_i = VT\ x_i \tag{8}$$

Subject to:

$$\sum_{\mathrm{i}}^{\mathrm{n\ routes}} \mathrm{V_i} = \mathrm{VT} \tag{9}$$

$$\sum_{\mathrm{i}}^{\mathrm{n\ routes}} x_1 = 1 \tag{10}$$

$$x_1 \geq 0 \tag{11}$$

$$\mathrm{V_i} \leq \mathrm{C_i} \tag{12}$$

where:

C_i—Capacity (vph) of route i;
CP_j—Cost of the pollutant j (€/g) released in the air;
EC—Economic cost (€);
m—N° of pollutants considered;
n—N° of alternative routes;
P_{ji}—Total emissions of the pollutant j produced on route i (g);
V_i—Total traffic volume on route i (vph);
VT—Total Demand (vph);
x_i—Relative flow on route i.

Depending on the complexity of the optimization process two optimization strategies can be selected. The Generalized Reduced Gradient (GRG) Nonlinear Solving based on Lasdon and Waren's code [18], selects a basis, determines a search direction, and performs a line search on each major iteration—solving systems of nonlinear equations at each step to maintain feasibility. For more complex problems (non-convex), MS Excel provides an evolutionary algorithm to optimize the relative flows (x_i) that minimize the selected objective function. The use of a population of solutions helps the optimization processes algorithm avoid becoming "trapped" in a local optimum [19].

3 Case-Study

A simple network is presented as a case study for demonstration of some capabilities of this tool. First, the characteristics of the network are described. Then, the process of traffic and emissions simulation is briefly summarized. Finally, the scenarios evaluated are presented.

3.1 Network Characteristics

The case study is based on a stylized road network that consists of four sections of one kilometer of length with different capacities. Four representative scenarios of a Portuguese road network are presented: (a) R1—Motorway with three lanes and with an average toll cost of €0.08/km; (b) R2—Motorway with two lanes and one interchange, with a toll cost of €0.064/km; (c) R3—Highway with one and two lanes sections and one interchange; (d) R4—Arterial in a urban environment with one lane in each direction, five intersections and one traffic light. The main costs perceived by the users, and included in volume-costs functions, are the value of time and the value of tolls. Figure 3 presents the link configuration and Fig. 4 the respective Volume-Costs function, i.e. the user's perceived cost as function of traffic volume in each link.

3.2 Traffic and Emissions Modeling

The evaluation of traffic performance under different traffic demand levels was performed using the VISSIM microsimulation model [20]. Driver behaviour parameters of this model were tested in order to assess their effect on travel times and also speed rates of links with similar characteristics. The calibration parameters can be divided into car-following parameters, lane-change parameters, simulation resolution, desired speed and acceleration distributions and vehicle specific power distribution. A comprehensive explanation on traffic model calibration and evaluation process is available elsewhere [21].

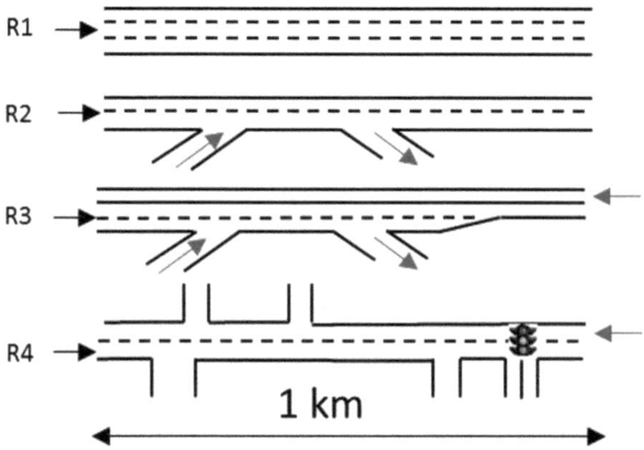

Fig. 3 Layout of alternative routes

To estimate emissions, the Vehicle Specific Power (VSP) method classified in 14 modes was employed. This model allows the estimation of instantaneous emissions based on second-by-second vehicle's dynamics (*speed, (v), acceleration, (a)* and *road grade*) (Eq. 13). The emissions factors used in this study from gasoline and diesel passenger vehicles, as well to buses can be found elsewhere [22–24]. To heavy duty and motorcycle vehicles, there is a lack of VSP emission factors adapted to the European situation. Regarding these specific cases, the CORINAIR methodology [25] was used. This methodology is based on speed, slope and load factor. A C# console application was developed to compute second-by-second vehicle drive cycle data from VISSIM and then calculate emissions based on both mentioned methodologies. A typical Portuguese fleet has been considered [26].

$$VSP = v[1.1a + 9.81\,sin(arctan(grade)) + 0.132] + 0.000302 \times v^3 \qquad (13)$$

3.3 Scenarios

Three traffic assignment scenarios were assessed. The first one simulates the likely traffic distribution using the user equilibrium formulation (UE). In this scenario, each user seeks to minimize his costs without considering environmental issues. The second scenario simulates an optimized traffic distribution scenario (SO) with the aim of minimizing the overall cost of emissions produced on the network. In the third scenario, a SE assignment is performed. Here, the pollution impacts are equally distributed over the various routes. For each scenario, three distinct traffic demands are analyzed: low demand, 1,000 vph; moderate demand, 4,000 vph; and high demand 10,500 vph.

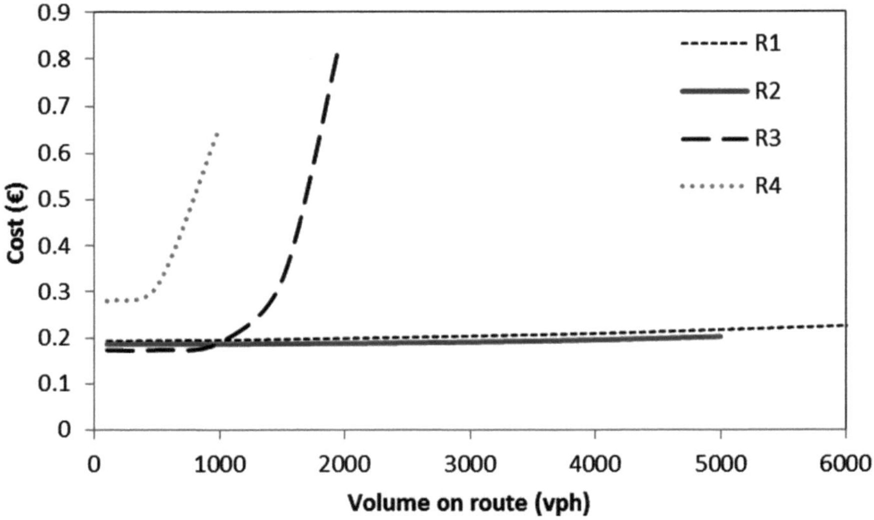

Fig. 4 Volume-cost functions for the alternative routes

4 Results and Discussion

In this section, examples of model outputs are discussed. Firstly, the relative contribution of each pollutant for the total environmental economic costs and the eco-indicator is analyzed. Then, the evaluation of an optimization based on environmental costs is conducted.

4.1 Optimization Parameters

Different approaches were tested to solve the non-linear problem, the GRG method, and genetic algorithms (GA) using a set of recommended settings [27, 28]. Table 2 exemplifies the optimization time and the objective function value. It can be seen that the GRG method is considerably faster than the use of evolutionary algorithms, since this non-linear problem was convex. For non-convex problems, the employ of GA can produce more reliable results and avoid be trapped in a local minimum.

4.2 Relative Impact Of Pollutants Under UE

Figure 5 presents the environmental impact costs and the health impacts (eco-indicator99) related with each pollutant among the alternative routes. In this case

Table 2 Optimization time and objective function result (total integrated cost using different optimization tools

Assignment strategy	System equitable (SE)			System optimum (SO)		
Optimization method	GA_1	GA_2	GRG	GA_1	GA_2	GRG
Optimization time (s)	67	65	2	104	63	9
Relative flow						
R1	44.0 %	43.8 %	44.1 %	53.7 %	52.1 %	52.9 %
R2	25.3 %	25.2 %	25.4 %	26.8 %	29.4 %	29.8 %
R3	17.3 %	17.3 %	17.3 %	10.8 %	10.4 %	10.7 %
R4	13.1 %	13.1 %	13.1 %	6.7 %	6.2 %	6.6 %
Final result (€)	0	0	0	652	653	669

the total impacts were estimated using the traditional UE assignment for a total traffic demand of 4,000 vph. Each bar is an alternative route and each pollutant a different segment of the bar. In terms of mass, all pollutants exhibit very distinct orders of magnitude and CO_2 is by far the most abundant. However, when translated in economic terms, the weight of CO is considerable higher. Considering the health impact, the influence of PM and CO_2 are comparable but in this case NO_X is clearly the pollutant with major impacts. Considering this perspective, HC and CO are negligible. It should again be emphasized that these figures are based on different studies and adopted for different realities. Regional factors such as population densities or land use type influence the environmental costs impact factors, but this is beyond the scope of this study.

4.3 Optimization of Environmental Economic Costs Under Different Levels Of Demand

For reasons of economy of space the optimization analysis will be limited to the criterion integrated cost. Figure 6 shows the traffic distribution for each scenario according different demands. Figure 7 presents the total cost of pollution over similar conditions. Each bar is a different traffic assignment scenario and the contribution of each route is shown in a different colored segment. The relative change (%) of users' total costs between SE or SO assignments and the UE scenario is shown by the black circles.

Regarding the low congestion scenario, an optimized traffic distribution would allow about 18 % reduction of emissions impacts with a marginal impact in the users cost (1 %). This situation occurs because the alternative R3 offers a good alternative in terms of environmental costs without a considerable increase of travel time. Under moderate demand, the SO assignment yields 33 % reduction of pollution costs with 5 % increase in total users cost compared with the UE assignment (Fig. 7). This situation occurs by shifting a considerable amount of traffic from R2 to R1 with higher road capacity but higher toll costs (Fig. 6). The

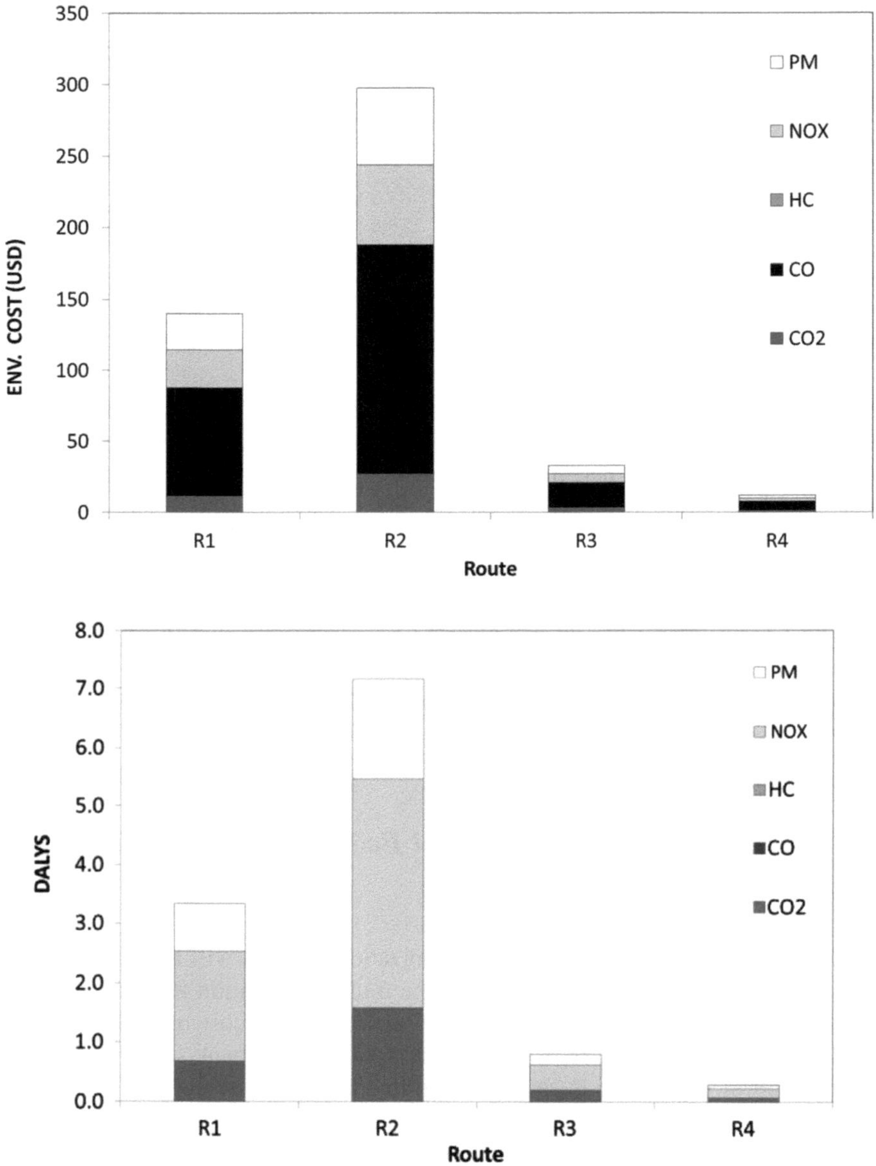

Fig. 5 Example of environmental costs (*top*) and health impact of pollutants over different routes (*Bottom*)

SE scenario would allow a slight reduction in the total environmental costs (compared with UE assignment) but an increase in user costs of 24 %. Considering the higher congestion scenario, there is no significant road capacity to allow considerable improvements in emissions reduction. In this case, the SO assignment

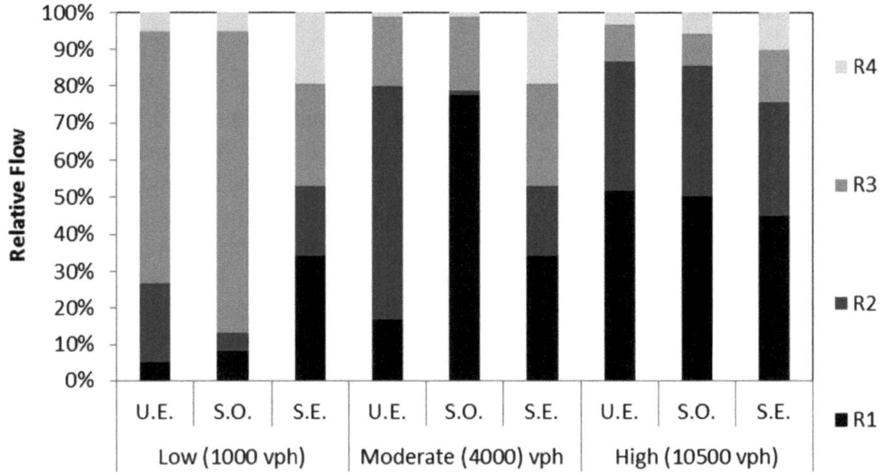

Fig. 6 Top: Flow distribution (%) for each scenario according different demands

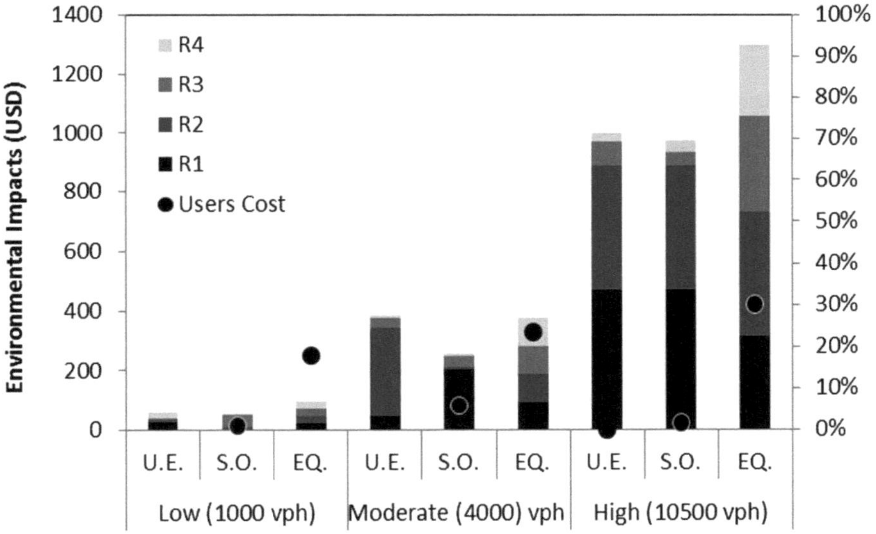

Fig. 7 Environmental costs and relative increase in total users cost in comparison to UE

has a similar distribution with the UE case (Fig. 6). Naturally, the potential of minimizing costs associated with pollution decreases when the V/C ratio for the OD pair is close to 1. In general, an optimization of environmental impacts requires an extra effort in terms of users' cost. Accordingly, we can conclude that the toll rates scheme can be improved with regard to reducing the environmental impact.

5 Conclusions

A tool to help the traffic assignment in a certain corridor more efficiently and environmentally friendly has been developed. The most innovative factor of this tool is the ability to include the impacts of major pollutants in an integrated form according to user's needs. This tool is not intended to replace the traditional traffic assignment models but rather complement them and contribute to a more effective management of traffic. The outputs of this model can be the basis for implementing intelligent traffic management measures. It is common knowledge that SO assignment is an unrealistic scenario since it assumes that drivers will collaborate in making their route choices considering the overall benefit of the complete network, instead of their own benefits. However, new traffic advanced information systems and smart road-pricing schemes may lead to a more efficient allocation of traffic in certain corridors by dynamically change the equilibrium conditions. The case study has demonstrated that it is possible to significantly reduce environmental costs (30 %) by changing the flow distribution along a corridor with 4 alternative routes. Further research is needed to evaluate driver's response to new eco-routing systems. Moreover, it is necessary the development of a methodology to adjust the impact of pollutant emissions according to the characteristics of specific links.

Acknowledgments Bandeira and P. Fernandes acknowledge the support of the Portuguese Science and Technology Foundation—FCT for the Doctoral grants SFRH/BD/66104/2009 and SFRH/BD/87402/2012. This work was partially funded by FEDER Funds through the Operational Program ''Factores de Competitividade COMPETE'' and by National Funds through FCT within the project PTDC/SEN-TRA/115117/2009. The authors also acknowledge the Strategic Project PEst-C/EME/UI0481/2013 and Toyota Caetano Portugal, which allowed the use of vehicles. We are thankful to the US Department of Transportation, Research and Innovative Technology Administration for providing funding through the TranLIVE consortium of university transportation centers." The collaboration between Drs. Coelho and Khattak was under the auspices of the Luso-American Transportations Impacts Study Group (LATIS-G). The contents of this article reflect the views of the authors, who are responsible for the facts and the accuracy of the information presented herein.

References

1. European Environmental Agency—EEA: The contribution of transport to air quality—TERM 2012. EEA European Environment Agency, Copenhagen (2012)
2. European Commission: White paper Roadmap to a Single European Transport Area—Towards a competitive and resource efficient transport system. COM (2011) 144 final, Brussels (2011)
3. Yin, Y., Lawphongpanich, S.: Internalizing emission externality on road networks. Transp. Res. Part D Transp. Environ. **11**, 292–301 (2006)

4. Sharma, S., Mishra, S.: Intelligent transportation systems-enabled optimal emission pricing models for reducing carbon footprints in a bimodal network. J. Intell. Transp. Syst. **17**, 54–64 (2013). doi:10.1080/15472450.2012.708618
5. Bandeira, J.M., Carvalho, D.O., Khattak, A.J., et al.: A comparative empirical analysis of eco-friendly routes during peak and off-peak hours. 91st Annual Meeting of Transportation Research Board (2012)
6. Bandeira, J.M., Almeida, T.G., Khattak, A.J., et al.: Generating emissions information for route selection: experimental monitoring and routes characterization. J. Intell. Transp. Syst. **17**, 3–17 (2013). doi:10.1080/15472450.2012.706197
7. Barth, M., Boriboonsomsin, K., Vu, A: Environmentally-friendly navigation. Conference on Intelligent Transportation Systems, Seattle, WA, USA, pp. 684–689 (2007)
8. Ahn, K.G., Rakha, H.: The effects of route choice decisions on vehicle energy consumption and emissions. Transp. Res. Part D-Transp. Environ. **13**, 151–167 (2008). doi:10.1016/j.trd.2008.01.005
9. Yao, E., Song, Y.: Study on eco-route planning algorithm and environmental impact assessment. J. Intell. Transp. Syst. **17**, 42–53 (2013). doi:10.1080/15472450.2013.747822
10. Gazis, A., Fontes, T., Bandeira, J., et al.: Integrated computational methods for traffic emissions route assessment. Fifth ACM SIGSPATIAL International Workshop on Computational Transportation Science (2012)
11. Guo, L., Huang, S., Sadek, A.W.: An evaluation of environmental benefits of time-dependent green routing in the greater buffalo-niagara region. J. Intell. Transp. Syst. **17**, 18–30 (2013). doi:10.1080/15472450.2012.704336
12. Ahn, K., Rakha, H., Moran, K.: System-wide impacts of eco-routing strategies on large-scale networks. 91st Annual meeting of the transportation research board (2011)
13. DOT US: Applications for the environment: real-time information synthesis (AERIS)—benefit-cost analysis (2012)
14. Bureau of Public Roads: Traffic assignment manual. Washington DC (1964)
15. Spiess, H.: Conical volume-delay functions. Transp. Sci. **24** (1990)
16. Goedkoop, M., Spriensma, R.: The eco-indicator 99—a damage oriented method for life cycle assessment. Methodology Report (2000)
17. Rilett, L., Benedek, C.: Traffic assignment under environmental and equity objectives. Transp. Res. Rec. **1443**, 92–99 (1994)
18. Lasdon, L.S., Waren, A.D., Jain, A., Ratner, M.: Design and testing of a generalized reduced gradient code for nonlinear programming. ACM Trans. Math. Softw. **4**, 34–50 (1978). doi:10.1145/355769.355773
19. Frontline Systems Inc.: Basic solver—nonlinear optimization. http://www.solver.com/content/basic-solver-nonlinear-optimization (2013)
20. PTV.: Vissim 5.30-05 user manual. Planung Transport Verkehr AG, Karlsruhe, German (2011)
21. Fontes, T., Fernandes, P., Rodrigues, H., et al.: Are eco-lanes a sustainable option to reducing emissions in a medium-sized European city? 92nd Transportation Research Board Annual Meeting, Washington DC, January 2013
22. US Environmental Protection Agency—EPA: Methodology for developing modal emission rates for epa's multi-scale motor vehicle and equipment emission system. Prepared by North Carolina State University for US Environmental Protection Agency, Ann Arbor (2002)
23. Coelho, M.C., Frey, H.C., Rouphail, N.M., et al.: Assessing methods for comparing emissions from gasoline and diesel light-duty vehicles based on microscale measurements. Transp. Res. Part D-Transp. Environ. **14**, 91–99 (2009). doi:10.1016/j.trd.2008.11.005
24. Zhai, H., Frey, H.C., Rouphail, N.M.: A vehicle-specific power approach to speed- and facility-specific emissions estimates for diesel transit buses. Environ. Sci. Technol. **42**, 7985–7991 (2008). doi:10.1021/es800208d

25. European Environment Agency—EEA: EMEP EEA Air Pollutant Emission Inventory Guidebook 2010–2011 (2009)
26. ACAP: Automobile Industry Statistics 2010 Edn. (2012)
27. Dejong, K., Spears, W.M.: An analysis of the interacting roles of population size and crossover in genetic function optimization. Proceedings of Parallel Problem Solving from Nature. Berlin, pp. 38–47 (1991)
28. Sadek, A.W., Smith, B.L., Demetsk, M.J.: Dynamic traffic assignment genetic algorithms approach. Transp. Res. Rec.1588 Pap. **1588**, 95–103 (1997)

Evaluating Changes in the Operational Planning of Public Transportation

João Mendes-Moreira and Jorge Freire de Sousa

Abstract Operational planning at public transport companies is a complex process that usually comprises several phases. In the planning phase, schedules are constructed considering that buses arrive and depart as scheduled. Obviously, several disruptions frequently occur, but their impact on the operating conditions is not easy to estimate. This difficulty arises mostly due to the impossibility of testing different solutions under the same conditions. Indeed, typically, the available data are a result of the current plan, while new proposed solutions have not produced real data yet.

Along this chapter we discuss the assessment of the impact of changes in the operational planning on the real operating conditions, before their occurrence. We present a framework for such assessment, which includes two components: the impact on costs, and the impact on revenues. We believe that this framework will be useful in future works on operational planning of public transport companies.

Keywords: Operational planning · Performance evaluation · Public transport

---

J. Mendes-Moreira
Department of Informatics Engineering, Faculty of Engineering, University of Porto, Porto, Portugal

J. F. de Sousa (✉)
Department of Industrial Engineering and Management, Faculty of Engineering, University of Porto, Porto, Portugal
e-mail: jfsousa@fe.up.pt

J. Mendes-Moreira
LIAAD-INESC TEC LA, Porto, Portugal

J. F. de Sousa
UGEI-INESC TEC LA, Porto, Portugal

J. F. de Sousa and R. Rossi (eds.), *Computer-based Modelling and Optimization in Transportation*, Advances in Intelligent Systems and Computing 262,
DOI: 10.1007/978-3-319-04630-3_5, © Springer International Publishing Switzerland 2014

1 Introduction

In the last two/three decades, passenger transport companies have made important investments in information systems, such as Automatic Vehicle Location (AVL), automatic passenger counting, automated ticketing and payment, multi-modal traveler information systems, operations software and data warehouse technology, among others. As a consequence of this effort in Advanced Public Transportation Systems, passenger transport companies have been able to collect massive data. As in other areas of activity, all this information was not proving to be particularly helpful in supporting companies to accomplish their mission in a significantly better way. The quote *we are drowning in information and starving for knowledge* from Rutherford D. Rogers, a librarian from Yale University, summarizes these moments in the company lives.

The existence of these new data has driven to the development of new approaches for the operational planning of public transportation. Some of these approaches imply changes in the initial phases of operational planning. However, the usefulness of such changes from a company point-of-view is of difficult evaluation. Some of the reasons why such difficulty exists are:

- The multi-objective nature of this problem: evaluation of operational costs and; evaluation of clients' satisfaction.
- The dependency on the type of routes in the evaluation of clients' satisfaction: a client has no problems when a bus arrives two minutes in advance when he knows that there is a bus every five minutes; but he will be very upset with such advance if he knows that he must wait thirty minutes by the next bus.
- The impossibility of evaluating different operational plans under the same circumstances: changing the planning when those changes include changes in the trips' offer is very sensitive in terms of public perception. Additionally, due to both the seasonal behavior and high variability of the traffic, it is difficult to identify the factors that can explain the obtained differences using different operational plans.

The main goal of this chapter is to present an evaluation framework in order to measure the impact of changes in the operational planning of public transportation. This chapter starts with the presentation of how operational planning is usually done in public transportation companies. Then, in Sect. 3, a short review on methods for evaluation of operational performance of public transport systems or bus lines is presented. The main issues on such evaluation are presented in Sect. 4. Then, in Sect. 5, a framework for the evaluation of changes in the operational planning is presented. Section 6 presents a case study on the use of travel time predictions for the definition of buses and drivers duties, and Sect. 7 concludes the chapter.

2 Operational Planning

Operational planning at public transport companies is a complex process that is usually divided in a set of successive stages:

1. The network definition: it is, obviously, a planning task for the long/very long term. It comprises the definition of the lines, routes and bus stops. We define route as an ordered sequence of directed road stretches and bus stops. Lines are a set of routes, typically two routes that use roughly the same road stretches but in opposite directions.
2. The trips definition: it is a medium term task, with an horizon much shorter than the network definition. There are typically two different methods for trip definition: (1) headway-based, defining the time between two successive trips on the same route [33]; or (2) schedule-based, defining timetables by explicitly setting the departure time and the time of passage at the main bus stops. The supply of trips is defined by route even if they are articulated between groups of routes/lines [6].
3. The definition of the duties of the drivers and buses: they are medium term (several months) tasks. The goal of both tasks is to define duties. A duty is the work a bus/driver must do in a day. When a duty is defined, in both cases, we do not know which driver or bus will do it. We are just making a logic assignment. The case of bus duties is much simpler than driver duties for obvious reasons: drivers must stop for lunch, cannot drive every day of the week, etc., i.e., they have much more constraints than buses. According to Dias [10], *each driver duty is subject to a set of rules and constraints defined by the government legislation, union agreements and some internal rules of the company*. Typically, bus duties are defined before driver duties.
4. The assignment of duties: it is the task where the driver duties are assigned to drivers and bus duties are assigned to buses. We are now making a physical assignment. Assignment for driver duties is more complex than for bus duties, for similar reasons to the ones explained above. The assignment of driver duties to drivers is called rostering. It can vary significantly from one company to another.

Large and medium sized transport companies usually use computer-aided systems that tackle the vehicle and driver scheduling problems in a sequential basis. Though, the sequential solution approach, that takes the output of the current problem as input of the next one, has no guarantee of always finding a good quality solution to the overall problem. In order to overcome this drawback, solution methods for the integrated vehicle-driver scheduling problem or for the integrated driver scheduling and rostering problems have been proposed in the literature [2, 4, 15, 16, 19, 25].

3 Short Review on Methods for Evaluation of Operational Performance

Since the seventies, the Operational Planning of Mass Public Road Transportation Networks has attracted the attention of many researchers. One of the main reasons has always been the hard but challenging need of balance between two main variables: the operational costs and the passengers' satisfaction.

Increasing concerns with efficiency are shaping the way public transportation systems are designed and operated. These concerns result not only from the enormous financial constraints faced by municipalities, transport authorities and operators but also from a greater awareness of sustainability. The evaluation of performance of transportation systems has attracted widespread attention in the last decade, and the methodological approach has evolved from the analysis of activity ratios and key performance indicators to the use of both parametric statistical methods, such as stochastic frontiers, and non-parametric deterministic methods, such as Data Envelopment Analysis (DEA). Both enable comparisons against best observed performance, with DEA having the advantage of allowing direct comparisons between the units considered (decision making units, DMUs) accounting for multiple dimensions both in terms of the resources used and services produced.

3.1 Analysis of Activity Ratios and Key Performance Indicators

Since the seminal work of Nakanishi [26], where a customer-oriented bus performance indicator program was established for the New York City, containing two schedule adherence indicators—en route on-time performance and service regularity—many authors developed methodologies inspired in bus performance indicators using computer-aided dispatching, and GPS and AVL technologies (e.g. [5, 13, 29]).

Traditionally, there has been an emphasis amongst transport operators and authorities to collect statistics on operational performance that remain popular because the statistics are easy to understand and the data is relatively easy to collect. Many of the studies mentioned in the literature were applied to metropolitan areas or made by entities related to those areas (e.g. [1, 7, 14, 30]).

Despite the significance of reliability indicators only a few studies have examined the merits of alternative measures of reliability performance [22, 31]. None of these studies, however, evaluated both objective and subjective indicators for both frequent and less frequent bus services.

In order to provide a more useful and reliable measurement tool of transit performance, current research about the topic is ever more oriented to consider both objective and subjective service quality measures [11, 32].

More recently, Currie et al. [8] and Eboli and Mazulla [12] try to make both things: the assessment of alternative bus reliability indicators and the junction of objective and subjective indicators.

3.2 Statistical Methods

As previously mentioned, in the transportation sector, the analysis of performance of mass transit companies is a field of study that has attracted widespread attention in recent years. The literature review of De Borger et al. [9] reported 40 published studies on the performance of public transportation companies using frontier methods, with the majority using DEA. Barnum et al. [3] note that since 2002, the number of studies using DEA to compare the efficiency of transportation companies or systems has continuously increased. Whilst these studies are very helpful to compare urban transport organizations or transit systems as a whole [18, 20, 27], they are not effective to help a given organization to evaluate its internal activities.

But the application of DEA to compare subunits within a single public transportation company is growing, even if there are still few papers evaluating the performance of bus routes. In the following paragraphs some examples are presented.

Sheth et al. [28] evaluated the overall performance of bus routes using DEA and goal programming, but used simulated data. The main contribution of this chapter was to identify a set of relevant performance indicators to evaluate bus routes, and to propose the use of a framework that evaluates performance from different perspectives, corresponding to different stakeholders: the service providers, the users and the society in general.

Lin et al. [21] developed a framework for quality control of bus schedule reliability. This framework was based on the use of DEA models and panel data analysis procedures to establish confidence intervals for the DEA scores. This research also recognized the importance of defining appropriate indicators of bus service reliability. The study only used four indicators of on-time performance, leaving aside measures of passenger related activity, resource usage or environmental non-controllable factors such as traffic conditions. Therefore, the DEA model used did not estimate efficiency levels, but it only provided a DEA based composite indicator of schedule adherence.

Barnum et al. [3] used DEA to evaluate the efficiency of bus routes adjusting for the effect of environmental variables. The authors discussed the advantages and disadvantages of existing DEA approaches to account for non-discretionary (ND) factors in efficiency assessments. They proposed a method for correcting the efficiency scores to reflect the ND conditions of bus routes, and reported important managerial insights that a DEA-based performance assessment can provide to bus transit companies.

Hahn et al. [17] applied a DEA approach for evaluating the efficiency of exclusive bus routes in Seoul, considering both the desirable and the undesirable outputs.

4 On the Evaluation of Operational Planning

In many research works, there is data obtained using a given operational plan. The evaluation of new approaches for operational planning typically should be done without having data obtained using the new proposed approaches. This happens because testing new plans have strong implications in terms of operations and, consequently are usually avoided. For that reason, the comparison between the current and the new plan should be done using only data produced using the current plan.

Changes in the operational planning of public transportation companies aim to increase revenue and/or to reduce costs. In this section we discuss the main issues concerning revenues and costs related to the operational management of public transportation.

4.1 Operational Revenues

Operational revenues are obtained from selling trips. To quantify the revenues for a past period is easy. However, to estimate the value of operational changes during the planning phase is very different and will be discussed in Sect. 5. Moreover, changes in the operational planning have, typically, long-term impact on the image of the company that is of difficult evaluation.

4.2 Operational Costs

Two different types of costs can be identified: budgeted costs and non-budgeted costs. The budgeted costs are the ones that were already estimated before the operation's occurrence, i.e., they are, typically, planned costs. The non-budgeted costs are only known when the operations have already occurred and are, typically, consequence of changes to the operational planning. Operational planners can choose between wider or tighter plans. In the first case, budgeted costs will be larger and disruptions will be less frequent. Consequently, non-budgeted costs will be more reduced. Tighter plans have less budgeted costs. However, a larger amount of disruptions will happen and, consequently, non-budgeted costs will be larger.

An example is the definition of trips' travel times and slack times for the schedules. Increasing its values the budgeted costs will increase because the services of the drivers and buses will be larger. In this case the probability of schedule disruption will be reduced comparing against the definition of shorter travel and slack times. Decreasing travel and slack times will reduce budgeted costs but non-budgeted costs will be larger because schedule disruption will be more frequent.

5 A Framework for Evaluating Changes in the Operational Planning

In order to be able to quantify both operational revenues and costs in the planning phase, they should be estimated. Now the question is: how can we estimate the value of the changes done in the operational planning? Or, differently, and assuming that the goal of changing operational planning is to maximize the difference between revenues and costs: which is the plan that maximizes such difference? In practical situations, given two alternative plans, how can we evaluate which plan is the best? The framework we present intends to answer this question.

Considering the stages described in Sect. 2, namely, (1) network definition, (2) trips definition, (3) Duties definition for both drivers and buses, and (4) duties assignment (again for both drivers and buses), only changes in the second (trips definition) and the third (duties definition) stages will be analyzed. Changes in stage 1 are not very frequent and changes in stage 4 are mainly relevant for human resource purposes.

The evaluation of each l version of the planning is done according to the objective function given in Eq. (1).

$$\max_{l} \ \alpha_l \times R - C_l. \quad (1)$$

where α is a value given by the user, R and C are the indicators for, respectively, the revenues and the operational costs, and l identifies each of the versions under evaluation. While α and C can be different for each version of the plan under study, R is fixed. It can be used the value in the last available accounts report for revenues due to trips sold. The value of α_l should reflect the level of change in revenues expected with a given plan l. The value of $\alpha = 1$ represents that the given plan does not have impact on the revenues comparing to the actual plan. As lower limit for α we propose the ratio between the revenues due to pre-purchased passes and the total of revenues due to trips sold, i.e., R. This proposal assumes that the pre-purchased passes represent the less-volatile component of revenues. When the proposed changes are done in stage 3, i.e., in the definition of the buses and/or drivers duties, it is expected that the value of α has minor variations with respect to 1. Indeed, different arrangements in the duties definition hardly changes the number of trips sold. Since the trips and slack times are defined at stage 2, the level of satisfaction perceived by the clients depends much more on the definition of the trips than on the definition of the duties. Consequently, changes in stage 3 affect mostly costs.

5.1 Estimating Costs

Two different types of costs can be identified: budgeted and non-budgeted ones. Additionally, changes in both stages 2 or 3 can affect costs. In stage 2, budgeted costs depends on the used slack times. Largest slack times increase budgeted costs but, potentially, decrease non-budgeted costs. In stage 3, the ability to define buses and drivers duties based on the given cycle times, can vary both budgeted and non-budgeted costs. While budgeted costs can be estimated directly based on unitary costs per time unit, non-budgeted costs cannot be estimated so easily.

The main indicators used to estimate budgeted costs are:

- Buses' costs per time unit (BCTU): it should include all operational costs with buses per unit time.
- Drivers' costs per time unit (DCTU): it should include all operational costs with drivers per unit time.
- Sum of all scheduled cycle times (SCT): since the schedules are usually defined assuming certain slack times, cycle times can be obtained. This indicator can be ameliorated by incorporating a factor for the inclusion of times with non-commercial trips, such as the ones for connection to the bus garage.
- Sum of all durations of buses duties (DBD): since the buses duties are defined, the sum of their durations is direct.
- Sum of all durations of drivers duties (DDD): like DBD it is directly obtained since the drivers duties are defined.

All these five indicators are usually used as performance indicators in public transportation companies. They are used to calculate budgeted costs as presented in Table 1.

Non-budgeted costs result from disruptions between the planned and the real services. Two different indicators should be used to estimate non-budgeted costs: the amount of predicted disruptions and the average cost of disruptions. We discuss next how to estimate these indicators when the changes occur in stages 2 and 3.

We propose the following approach to estimate the amount of predicted disruptions when changes are done in the second stage of operational planning. Assuming an actual schedule AS_l with an scheduled trips defined by their departure times $at_1, at_2, \ldots, at_{an}$ and a new schedule NS_l with nn trips with departure times $nt_1, nt_2, \ldots, nt_{nn}$ for the same line l and a set of effective trips defined by their departure times $RT = rt_1, rt_2, \ldots, rt_{rn}$, where typically $rn >> an$. Considering the middle times between the departure times of each successive trips of NS_l $MDP\text{–}NS = mdp_1, mdp_2, \ldots, mdp_{nn}$, and splitting RT according to the $MDP\text{–}NS$ values, a set of effective trips per each scheduled trip in NS_l is obtained. By comparing the real cycle times per subset against the scheduled cycle time in NS_l. A percentile of

Table 1 Estimating budgeted costs

Changes in:	Budgeted costs
Stage 2	(BCTU + DCTU) × SCT
Stage 3	BCTU × DBD + DCTU × DDD

disruptive trips per scheduled trip in NS_l is obtained. With such percentiles it is possible to obtain a number of disruptive trips per time unit, for instance, per month or per year. The same can be done for AS_l, allowing a comparison in terms of number of disruptive trips between NS_l and AS_l. This approach uses real trips obtained using AS_l to evaluate NS_l, and it is partially described in [23]. This is not, obviously, the ideal situation. However, obtaining effective trips with NS_l is, typically, unfeasible.

The cost of disruptive trips should consider additional costs with drivers and buses. The estimation of the cost per disruptive trip is easily obtained from operational past data.

Changes in stage 3 can result in buses and drivers duties more or less difficult to accomplish than the current buses and drivers duties. If the changes were done for stage 3, all versions would use the same plan for stage 2, i.e., all versions were done using the same cycle times. Consequently, non-budgeted costs are due to: (1) inadequate times to allow drivers to move from one bus to another; or (2) inadequate times for the connection between different lines. The first case, as a consequence on how stage 3 is planned, is negligible. The second case depends on the way noncommercial trips for connection between lines are defined. Such trips typically use paths out of route roads. Hence, typically, there is no data that can be used to evaluate the suitability of the travel times defined for such trips. Anyway, the eventual difficulties derived from the choices made by the company are easily overcome by simple changes of paths or planned travel times. For these reasons non-budgeted costs for changes done in stage 3 may be ignored without significant consequences.

6 A Case Study

Now, we exemplify how the described approach can be used to evaluate the use of long term travel time prediction for the definition of buses and drivers duties instead of the scheduled travel times, as it is usually done.

Let us assume that a mass transit company has the necessary conditions to redefine the planning, namely the duties for the buses and drivers and respective assignment tasks, just three days before the date of the duties. How could the company use Travel Time Predictions (TTP) results 3 days ahead [24]? The first thing to do would be to make a new timetable just for internal planning purposes, i.e., not known to the public. Let us denote the values of this new timetable with the suffix * and let us give to the new scheduled travel time (STT *) the predicted travel time. It is important to note that for planning purposes the only information that is used from this new timetable is the new $STT*$ plus the new Slack Times ($SlT*$). A possible approach to estimate a lower limit for $STT* + SlT*$ would be the use of a decision support system for timetable adjustments, as described in [23] using $STT* + SlT* = p.max$, where $p.max$ is a user defined percentile of disruptive past trips. Another approach would be to define an algorithm to choose one of the acceptable values for this sum. The process of defining $SlT*$ would be eased due to the constraint Cycle Time = Number of Buses in the line × Scheduled Headway [34] which strongly limits the number of acceptable solutions, as discussed in [23]. These new values for $STT* + SlT*$ are the ones needed to define new duties for both buses and drivers.

Summarizing the proposed evaluation of the two approaches:

1. To obtain long term travel time prediction for all trips of a given schedule;
2. To build a new schedule using the predicted travel times;
3. To generate buses and drivers duties using the current and the new schedules;
4. To estimate budgeted costs for the two versions of buses and drivers duties as defined in Table 1;
5. To choose the version with the lowest budgeted cost.

7 Conclusions

The available past data is generally obtained using the current plan. When a new idea emerges in order to improve the current plan it is advisable to demonstrate its validity in order to convince companies to adopt it. How to do it? How to quantify the expected gains that may be obtained with its implementation? This chapter discusses this point. A framework is presented by using data produced with the current plan for the evaluation of the new proposed operational plans. Knowing that the existing plan affects, to some extent, the available data, the validity of the new plans is more difficult to prove using such data than it would be using the data produced by the different plans under appraisal. This circumstance can be seen as a hypothesis test with unknown significance level. Some of the other surveyed approaches, namely DEA, imply the existence of data produced by the different plans under consideration. However, in practical situations, such data typically does not exist and obtaining it is not possible without serious consequences on planning activities and on customers' satisfaction. The proposed framework is, according to the authors' knowledge, a novelty.

Acknowledgements. This work is part-funded by the ERDF - European Regional Development Fund through the COMPETE Programme (operational programme for competitiveness), by the Portuguese Funds through the FCT (Portuguese Foundation for Science and Technology) within project FCOMP - 01-0124-FEDER-022701.

References

1. Achterstraat, P.: Improving the performance of metropolitan bus services nsw transport and infrastructure. In: Nsw Auditor-General's Report Performance Audit, NSW Transport and Infrastructure. Sydney, Australia (2010)
2. Amberg, B., Amberg, B., Kliewer, N.: Increasing delay-tolerance of vehicle and crew schedules in public transport by sequential, partial-integrated and integrated approaches. Procedia Soc. Behav. Sci. **20**, 292–301 (2011)
3. Barnum, D.T., Tandon, S., McNeil, S.: Comparing the performance of bus routes after adjusting for the environment, using data envelopment analysis. J. Transp. Eng. **134**(2), 77–85 (2008)
4. Borndorfer, R., Lobel, A., Weider, S.: A bundle method for integrated multi-depot vehicle and duty scheduling in public transit. In: Hickman, M., Mirchandani, P., VoÃŸ, S. (eds.) Computer-Aided Systems in Public Transport. Lecture Notes in Economics and Mathematical Systems, vol. 600, pp. 3–24. Springer, Berlin (2008)
5. Bullock, P., Jiang, Q., Stopher, P.R.: Using gps technology to measure on-time running of scheduled bus services. J. Public Transp. **8**(1), 21–40 (2005)
6. Ceder, A., Golany, B., Tal, O.: Creating bus timetables with maximal synchronization. Transp. Res. Part A Policy Pract. **35**(10), 913–928 (2001)
7. The Permanent Citizens Advisory Committee to the MTA (PCAC): Minutes matter – a review of performance metrics at the MTA. Research report, MTA - Metropolitan Transportation Authority, New York, USA (2011)
8. Currie, G., Douglas, N., Kearns, I.: An assessment of alternative bus reliability indicators. In: Shaping the future: linking research, policy and outcomes, 35th Australasian Transport Research Forum (STRF), Perth, Australia, 26-29 September (2012)
9. De Borger, B., Kerstens, K., Costa, l.: Public transit performance: What does one learn from frontier studies? Transp. Rev. **22**(1), 1–38 (2002)
10. Dias, T.G.: A new approach to the bus driver scheduling problem using multi-objective genetic algorithms. Phd thesis, Universidade do Porto – Portugal (2005)
11. Eboli, L., Mazzulla, G.: A methodology for evaluating transit service quality based on subjective and objective measures from the passengers point of view. Transp. Policy **18**(1), 172–181 (2011)
12. Eboli, L., Mazzulla, G.: Performance indicators for an objective measure of public transport service quality. In: European Transport - Issue 51, p 3 (2012). ISSN 1825-3997
13. El-Geneidy, A.M., Horning, J., Krizek, K.: Using archived its data to improve transit performance and management. Report, Minnesota Department of Transportation Research Services Section (2007)
14. London for T.: London buses quality of service indicators – route results for London buses services, fourth quarter 2011/12. Technical report, Transport for London (2012)
15. Freling, R., Wagelmans, A., Paixo, J.: An overview of models and techniques for integrating vehicle and crew scheduling. In: Wilson, N. (ed.) Computer-Aided Transit Scheduling. Lecture Notes in Economics and Mathematical Systems, vol. 471, pp. 441–460. Springer, Berlin (1999)
16. Haase, K., Desaulniers, G., Desrosiers, J.: Simultaneous vehicle and crew scheduling in urban mass transit systems. Transp. Sci. **35**(3), 286–303 (2001)
17. Hahn, J.-S., Kim, H.-R., Kho, S.-Y.: Analysis of the efficiency of seoul arterial bus routes and its determinant factors. KSCE J. Civil Eng. **15**(6), 1115–1123 (2011)
18. Hawas, Y.E., Khan, M.B., Basu, N.: Evaluating and enhancing the operational performance of public bus systems using gis-based data envelopment analysis. J. Public Transp. **15**(2), 19–44 (2012)
19. Huisman, D., Freling, R., Wagelmans, A.P.M.: Multiple-depot integrated vehicle and crew scheduling. Transp. Sci. **39**(4), 491–502 (2005)

20. Karlaftis. M.G.: A {DEA} approach for evaluating the efficiency and effectiveness of urban transit systems. Eur. J. Oper. Res. **152**(2), 354–364 (2004). <ce:title>New Technologies in Transportation Systems</ce:title>
21. Lin, J., Wang, P., Barnum, D.T.: A quality control framework for bus schedule reliability. Transp. Res. Part E Logistics Transp. Rev. **44**(6), 1086–1098 (2008)
22. Mazloumi, E., Currie, G., Sarvi, M.: Assessing measures of transit travel time variability and reliability using avl data. In 87th Transportation Research Board Annual Meeting, CDROM (2008)
23. Mendes-Moreira, J., Duarte, E., Belo. O.: A decision support system for timetable adjustments. In 13th EURO Working Group on Transportation Meeting, CDROM (2009). ISBN 978-88-903541-4-4
24. Mendes-Moreira, J., Jorge, A.M., Freire de Sousa, J., Soares, C.: Comparing state-of-the-art regression methods for long term travel time prediction. Intell. Data Anal. **16**(3), 427–449 (2012)
25. Mesquita, M., Moz, M., Paias, A., Pato, M.: A decomposition approach for the integrated vehicle-crew-roster problem with days-off pattern. Eur. J. Oper. Res. **229**(2), 318–331 (2013)
26. Nakanishi, Y.: Bus performance indicators: on-time performance and service regularity. Transp. Res. Rec. **1571**, 3–13 (1997)
27. Sánchez, I.G.: Technical and scale efficiency in spanish urban transport: Estimating with data envelopment analysis. Adv. Oper. Res. **1–16**, 2009 (2009)
28. Sheth, C., Triantis, K., Teodorovi D.: Performance evaluation of bus routes: a provider and passenger perspective. Transp. Res. Part E Logistics Transp. Rev. **43**(4), 453–478 (2007)
29. Strathman, J.G., Dueker, K.J., Kimpel, R.L., Gerhart, R.L., Turner, K., Taylor, P., Callas, S., Griffin, D.: Service reliability impacts of computer-aided dispatching and automatic vehicle location technologies: a tri–met case study. Transp. Q. **54**(3), 85–102 (2000)
30. Suwardo, M.B., Napiah Kamaruddin, I.B.: On-time performance and service regularity of stage buses in mixed traffic. Int. J. bus. Econ. Financ. Man Sci. **3**(7), 942–950 (2009)
31. Trompet, M., Liu, X., Graham, D.J.: Development of key performance indicator to compare regularity of service between urban bus operators. Transp. Res. Rec. J. Transp. Res. Board **2216**(1), 33–41 (2011)
32. Tyrinopoulos, Y., Aifadopoulou, G.: A complete methodology for the quality control of passenger services in the public transport business. Eur. Transport **38**, 1–16 (2008)
33. Vuchic, V.R.: Urban Transit: Operations, Planning, and Economics. Wiley, New Jersey (2005)
34. Zhao, J., Dessouky, M., Bukkapatnam, S.: Optimal slack time for schedule-based transit operations. Transp. Sci. **40**(4), 529–539 (2006)

Part II
Energy and Environmental Impacts (of Transportation)

The complexity of problems and challenges relating to *energy and environmental impacts* often requires interdisciplinary research, covering all aspects of energy conversion and storage, alternative fuel technologies, and the science of environmental impacts and mitigation. In the work by *Chinese et al.*, a case study of limited or locally non-existent market development for CNG in an Italian frontier region is analysed and a mixed-integer non-linear programming model is introduced, to evaluate the effect of incentive measures envisaged by the regional government to foster refueling station development. *Cavadas et al.* propose a method for planning the location of charging stations for electric vehicles in a city in which the aim is to maximize the number of properly charged vehicles if the budget for building stations is already fixed. *Giménez et al.* present a new approach to determine the optimal locations of public charging stations for battery-driven electric vehicles in urban areas, inspired by the low parking time and high rotation rates of the most popular parking places used in pilot studies promoted by many governments. The chapter of *Nocera and Tonin* aims at defining a fair value for the Marginal Social Cost of Carbon to be used within transport planning, and discuss how it is influenced by economic and scientific uncertainty, with the aim of helping researchers, stakeholders and decision-makers to choose among the current range of values supplied by the scientific literature. *Romero et al.* provide an environmental approach based on an optimization-simulation model for planning and managing an urban freight transport system which, using trucks, must serve one or more points of the network which receive and/or generate large volumes of cargo.

A Service Station Location Model to Explore Prospects and Policies for Alternative Transport Fuels: A Case of CNG Distribution in Italy

Damiana Chinese, Piera Patrizio and Monica Bonotto

Abstract CNG is an example of alternative gaseous fuel whose market development requires supply infrastructure (pipelines), refuelling stations and alternative vehicles to exist at the same time, which is known as the "chicken and egg dilemma". In this chapter, a case study of limited or locally nonexistent market development for CNG in an Italian frontier region is analyzed and a mixed integer non linear programming model is introduced to evaluate the effect of incentive measures envisaged by the regional government to foster refuelling station development. It is found that, taking an entrepreneurs' perspective of maximizing profits, even with substantial capital grants investors are more likely to choose higher demand areas, in spite of fiercer competition, rather than areas without stations. Subsidies should be more specifically targeted to critical areas to be efficient.

Keywords CNG filling stations · Compressed natural gas vehicles · Mixed integer non linear programming · Location mode

1 Introduction

The simultaneous existence of fuel supply chains, refuelling stations and alternative vehicles is required for a sustained adoption of alternative transport fuels. In particular, especially the introduction of new gaseous fuels, such as hydrogen, CNG or biogas, faces the challenge of attracting investors in refuelling stations to attain satisfactory refuelling service levels, so that, in turn, more customers find new gaseous fuels an attractive option and market develops [1].

D. Chinese (✉) · P. Patrizio · M. Bonotto
DIEGM, Dipartimento di Ingegneria Elettrica, Gestionale e Meccanica,
University of Udine, 33100 Udine, Italy
e-mail: Damiana.chinese@uniud.it

J. F. de Sousa and R. Rossi (eds.), *Computer-based Modelling and Optimization in Transportation*, Advances in Intelligent Systems and Computing 262,
DOI: 10.1007/978-3-319-04630-3_6, © Springer International Publishing Switzerland 2014

This subject, which is known even in literature as "the chicken and egg dilemma", is investigated in few studies from either a modelling or an empirical perspective. Most empirical [1, 2] or model based [3] studies are performed at a national or international scale, so they give substantial strategic insights but cannot be immediately used at the detailed, local planning level to guide the site and capacity definition of refuelling stations.

On the other hand, the use of operations research models for location planning of service stations is widely spread in literature. Most applications concern future hydrogen based supply chains [4–8], while a single example handling CNG refuelling stations is reported [9].

Upchurch and Kuby [10] present a review of models for optimal location of alternative-fuel stations and summarize three general approaches to locate refuelling stations optimally, i.e.:

- Variants of the p-median model, generally based on census data (about population and car ownership), which tend to and locate stations close to where people live, in harmony with empirical research demonstrating that consumers prefer to refuel near their homes [11].
- Traffic count or VMT methods, based on road traffic data, which tend to locate stations on several adjacent links of high volume freeways.
- Flow intercepting location models, which yield more realistic representations but require a data matrix of traffic flows from origins to destination, which is hardly available at some geographic scales.

For each of these approaches, several variants of objective functions could be conceived, but, to the best of our knowledge, competition factors such as the profitability of single service stations are seldom taken into account. Models focusing on intercepting flow allow to maximize revenues, while a least cost planning philosophy underpins variants of maximum covering algorithms (e.g. Bapna et al. [12]) and strategic planning models at supply chain level [13]. Profitability of service stations is considered implicitly in multicriteria approaches adopted by Frick et al. [9], who use utility models, and by Brey et al. [14], who develop an AHP model. Explicitly, profitability is incorporated in the objective function only by Hugo et al. [15], who deal with the strategic supply chain planning of hydrogen, particularly with refineries location planning, and by Bersani et al. [6], who aim at maximizing net present values of a network of hydrogen fuelling stations.

To overcome the chicken and egg dilemma, the profitability of service stations is, however, a key issue: empirical research has shown that, in cases of successful market penetration of alternative transport fuels, refuelling infrastructure mostly grew through private investment [2]. Therefore, understanding which options for technology, capacity and location planning would be most desirable for potential investors, who aim at maximizing their profits, allows to gain insight on the future evolution of alternative fuel distribution systems and on their chances to thrive or decline.

This especially applies to the case of our concern, that is the development of CNG service stations in Friuli Venezia Giulia (FVG), an Italian region with about one million inhabitants located at the border with Austria and Slovenia. While the market penetration of CNG in Northern Italy is remarkable, reaching a market share of 2 % of total cars statistics [16], and the number of service stations is generally expanding [17] in FVG the market share of CNG cars stops at 0.3 %. And only three refuelling stations exist, located in municipalities marked in black in the upper right miniature in Fig. 1, i.e. mostly in the Western part of the region. Historically, taxes on fuels have been significantly lower in neighbouring countries than in Italy, which makes refuelling abroad the cheapest option, especially for inhabitants living closer to Slovenia. To reduce the resulting flow of refuelling commuters, the regional government of FVG used to finance a system of pricing zones depending on distance to borders, which was modified in 2011 due to objections by the European Union on the grounds of distortion of economic competition between countries. The effectiveness of the discounts was often limited, especially in the first pricing zone (represented in medium gray and marked as F1 in the miniature map in Fig. 1).

This situation is a typical example of a "chicken and egg" dilemma, preventing investors from installing alternative fuel stations, especially in the bordering area. The regional government recently conceived some financial support measures for new CNG distribution stations, which were then stopped as a consequence of national and regional spending reviews. Our aim is to estimate the potential impact of the envisaged subsidies and to evaluate prospects for CNG in the area, by assessing the economical feasibility of expanding the distribution network in the examined region under current and potential circumstances. For this purpose, we analyzed factors affecting actual CNG demand in FVG as reported in Sect. 2 and developed a mixed integer linear programming model for identifying the optimal location, technology and capacity of CNG refuelling stations as shown in Sect. 3. Obtained results are discussed in Sect. 4.

2 Factors Affecting Decisions on the Location of CNG Refueling Stations in Friuli Venezia Giulia

To develop a location model accounting for profits of refuelling stations, potential sales should be estimated. Data on CNG consumption have been collected at regional level in FVG in recent years [18], but they are only available for a restricted time period (from 2007 to 2011) and at a regional aggregation level, so it is not possible to discriminate between sales at different sites. On the other hand, data on the determinants of fuel demand commonly recognized in literature [2, 19] are available at a more local level: gasoline prices between 2007 and 2010 are available at municipal level from studies on the zone tariff mechanism, the share of CNG vehicles is known at regional level since 2006 [16], at province level for the year 2009 (personal communication by Federmetano, 2012)and the number of total

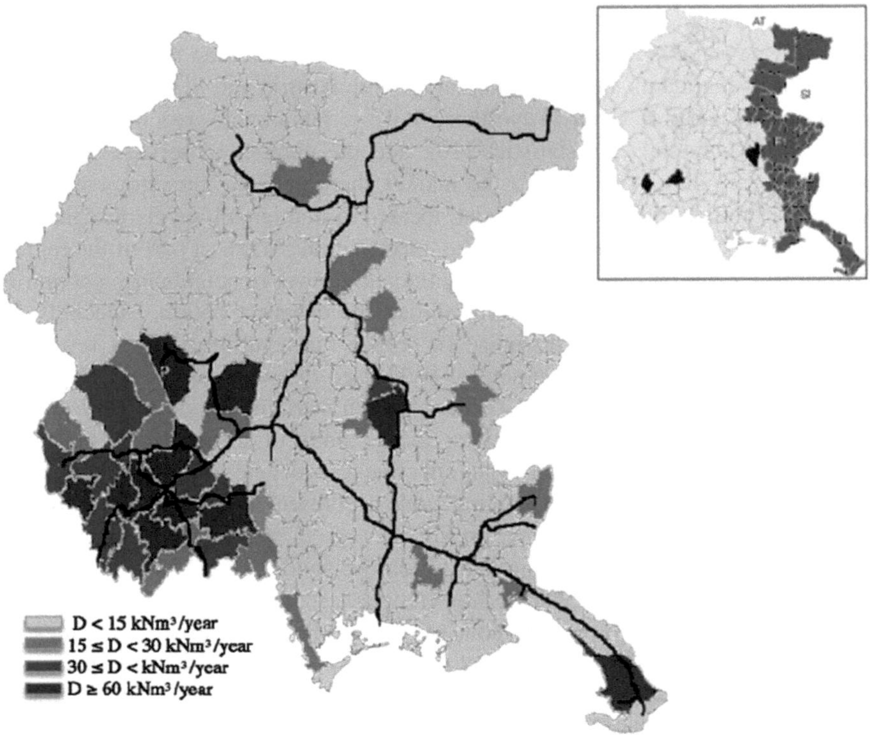

Fig. 1 Distribution of estimated CNG demand and factors affecting the location of CNG stations in FVG

vehicles is known at municipal level for the year 2009 (http://www.comuni-italiani.it/06/statistiche/veicoli.html). For this reason, we used national data to create static econometric models [20], in order to clarify the relation between the involved variables, then applied such models using local data and validated them at regional level by comparing estimated and real regional demand, calculating RMSEs in order to identify best fitting models. In this way, we formulated and tested several alternative models, both logarithmic and linear.

At the end, the best fits were obtained with the simple model expressed by Eq. (1):

$$D_{IT} = 1{,}27\ V_{IT} \tag{1}$$

The obtained model has a coefficient of determination $R^2 = 0.98$ and percent errors between -11 % and +6 % when estimating regional consumption for the years 2007–2011. Thus, we deduce that:

- The model can be used at least at regional level to make reasonable forecasts of demand;

- The specific consumption of CNG per vehicle in FVG is aligned with national data;
- As the model is obtained by regression through the origin, also use at municipal level seems acceptable.

For this purpose, we will use data on the total number of vehicles available at municipal level for the year 2009 and weigh them by the share of CNG vehicles on total vehicles in the same year, available at province level, i.e. at an intermediate aggregation level between regional and municipal ones.

2.1 The Impact of Distance from Slovenia on Demand of Gaseous Vehicle Fuels

While we found that regional specific CNG demand per vehicle is aligned with national values, to apply the model at municipal level it would be desirable to understand how distance from Slovenia may affect CNG demand. While there are no data on CNG, we got data about LPG vehicles and LPG consumption at province level, provided by the Italian Ministry for Economic Development. LPG shares similar features with CNG in that it is a niche market fuel, alternative to gasoline and diesel oil, characterized by tax exemption and consequent lower prices and no zone tariff in FVG. By performing both a general stepwise linear regression and a partial correlation analysis to test the relationship between province LPG demand (D), number of vehicles (V) and province distance (T) from Slovenia, we concluded that factor T will almost disappear when controlling for V. In other words, demand for LPG is affected by distance from Slovenia in that more alternative fuel vehicles are purchased in farthest municipalities from the border, whereas the average consumption per vehicle remains unaffected. We can assume CNG demand to behave similarly, and that, consequently, the coefficient in the model above does not need to be calibrated for the distance from the border, once the number of vehicles at local level is known. We can thus apply Eq. (1) to estimate demand at municipal level based on the total number of vehicles per municipality and the CNG vehicle share at province level, obtaining the main map in Fig. 1.

2.2 How Closeness to Natural Gas Pipelines and Pressure Levels of Natural Gas Supply Affect Costs

Figure 1 also shows the location of the natural gas high-pressure pipeline in FVG, which has a significant impact on effectiveness of CNG stations. In fact, the main element of a CNG station is a compressor plant, which elevates natural gas pressure from municipal distribution (4 bar) or gas pipeline (40 bar) levels to the high-pressure levels required for refueling (220 bar). Both compressor installation and operation costs are higher when connecting to low pressure (LP) infrastructure than to high pressure (HP) infrastructure, but while LP pipeline exists in every municipality considered as eligible location in this study, Fig. 1 shows that HP pipeline is only present in a limited number of municipalities. On the other hand, it should be observed that costs of connecting to LP infrastructure are generally lower than HP pipeline connection costs because distribution pipes are nowadays virtually present under every road, while HP pipeline are usually farther from urban centers. However, rather than incorporating such micro-location issues in an overall optimization model, possibly enhancing its complexity to a great extent, we preferred to preliminarily evaluate the impact of connection costs on annual equivalent costs of CNG stations based on cost data obtained from constructors for various plant capacities and found that connecting to the HP pipeline is the optimal solution when inequality 2 is verified, i.e.:

$$D_{pip}(Q) < 365e^{0.002Q} \tag{2}$$

where D_{pip} is the distance from pipe, D_{be} is break even distance and Q represents the given capacities of the refueling stations in kNm^3/year. It should be noticed that, according to the current market trends, it is unlikely that CNG-dedicated stations shall be built: capital costs considered here refer to the upgrade of existing fuel stations to distribute also CNG.

Given an average surface of municipalities of about 35 km^2, an average number of 2.1 existing fuel stations per municipality and typical capacity ranges of refueling stations between 300 and 1000 kNm^3/year, in our model we will assume that in municipalities served by HP pipeline it will be generally possible to find a fuel station to upgrade to CNG within the economical distance from the pipeline.

2.3 *Subsidies Foreseen by Friuli Venezia Giulia to Overcome the Chicken and Egg Dilemma*

In August 2010, in order to overcome the chicken and egg dilemma, a legislative decree has been issued by the regional government (L.R.14/2010), relating to subsidies' disbursement for CNG fuel stations establishment in the region. Such subsidies, supplied as outright grants, have a maximum value of 50 % of the total construction expenditure, regardless of the location decision.

3 Model Formulation

The main goal of the model is to estimate whether and where entrepreneurs are likely to invest in CNG refuelling stations under current and prospective circumstances, assuming that their rational behavior is directed to maximizing the net present value of their investments. For this reason, we build upon the work by Bersani et al. [6] because they adopt a similar perspective, although for hydrogen distribution. In order to formulate the decision problem for CNG in FVG, following assumptions are introduced:

- Based on previous break-even analysis, we assumed that in the municipalities characterized by the presence of gas pipeline only HP stations should be built;
- The location of the three existing CNG stations is fixed, but their costs are treated from an external viewpoint like the costs of new stations;
- At the moment, it is not realistic to allow the construction of more than one CNG station in each municipality.

3.1 Model Structure

The basic variables of the models are defined as follows:

y_i, $i = 1, \ldots, N$: binary variable associated with the *i*th municipality. Specifically, $y_i = 1$ when a station is located in the ith considered municipality, otherwise $y_i = 0$;
y_{ai}, $i = 1, \ldots, N$: binary variable associated with the *i*th municipality, with $y_{ai} = 1$ if a HP CNG station is located in the considered municipality, otherwise $y_{ai} = 0$;
y_{bi}, $i = 1, \ldots, N$: binary variable associated with the *i*th municipality, with $y_{bi} = 1$ if a LP CNG station is located in the considered municipality, otherwise $y_{bi} = 0$
Q_i: capacity of the *i*th fuel station in kNm^3/year;
P_i: annual equivalent profit of the *i*th station, in €/year
x_{ij}: binary variable representing the fraction of demand associated with the *j*th municipality to be served by a fuel station located in the *i*th municipality.

The parameter D_i represents the CNG demand in each municipality, calculated according to Eq. (1) using the estimated number of CNG vehicles in the *i*th municipality as independent variable. V_i is calculated by multiplying the total number of vehicles in the municipality, which is known for the year 2009, by the share of CNG vehicles on total vehicles in 2009, which is known at province level. Other relevant parameters are the binary parameter p_i, equaling 1 if the *i*th municipality is served by a gas pipeline, 0 otherwise, and the distance t_{ij} between municipalities *i* and *j*. The objective function is to maximize the sum of annual equivalent profits of all stations, as shown in Eq. (3):

$$Max \sum_{i=1}^{N} P_i = \left\{ \sum_{i=1}^{N} \begin{bmatrix} p_{CNG} \cdot Q_i - C_{CNG} Q_i - C_{HR} y_i + \\ -\left(C_{MAIN,HP} + C_{EL,HP}\right) Q_{ai} - \left(C_{MAIN,LP} + C_{EL,LP}\right) Q_{bi} - \\ f \cdot \left(C_{STRF,HP} y_{ai} + C_{STRV,HP} Q_{ai} + C_{STRF,LP} y_{bi} + C_{STRV,HP} Q_{bi}\right) \end{bmatrix} \right\} \quad (3)$$

Where f is the capital recovery factor of a series of uniform amounts, in this case for an interest rate of 7 % for 15 years, while other cost and sale price parameters are summarized in Appendix.

All cost functions are obtained interpolating data obtained by CNG station constructors or managers for at least three different plant capacities.

It should be observed that purchase and sale prices of natural gas obviously do not depend on connection technology, while the cost of human resources for capacities within the technically acceptable range is invariant.

Equations (4–13) represent the main constraints of the model, basically aimed at determining the capacity Q_i of the service station located in the ith municipality according to Eq. (5) as a weighted sum of demand in the municipality of concern and of demand in other municipalities, which can be partially diverted to the ith station depending on attraction factors (Eq. 7) related to distance decay functions (Eq. 8) as indicated in [9] and in [6]. With respect to those references, we do not fix a minimum number of stations, as it is our aim to find it through system optimization. On the other hand, the truncation condition we introduce with Eq. (7) influences the relative distance between stations, in that it imposes that, above a maximum distance t_{max}, the attraction of customers to the fuel station drops to zero.

$$Q_i = \sum_{\substack{j=1 \\ i \neq j}}^{N} x_{ij} D_j + D_i y_i \quad i = 1, \cdots, N \tag{4}$$

$$x_{ij} = \begin{cases} \dfrac{attr_{ji} \cdot y_i \left(1 - y_j\right)}{\sum_{i=1}^{N} attr_{ji} \cdot y_i} & t_{ij} \leq t_{\max} \\ 0 & t_{ij} > t_{\max} \quad i = 1, \cdots, N \quad j = 1, \cdots, N \end{cases} \tag{5}$$

$$attr_{ji} = \frac{1}{t_{ji}} \qquad i = 1, \cdots, N \quad j = 1, \cdots, N \tag{6}$$

$$y_{ai} + y_{bi} \leq y_i \quad i = 1, \cdots, N \tag{7}$$

$$y_{ai} \leq p_i \quad i = 1, \cdots, N \tag{8}$$

$$y_{bi} \leq 1 - p_i \quad i = 1, \cdots, N \tag{9}$$

$$Q_{ai} \leq y_{ai} B_{HP} \quad i = 1, \cdots, N \tag{10}$$

$$Q_{bi} \leq y_{bi} B_{LP} \quad i = 1, \cdots, N \tag{11}$$

$$Q_{ai} + Q_{bi} = Q_i \quad i = 1, \cdots, N \tag{12}$$

$$P_i \geq 0 \quad i = 1, \cdots, N \tag{13}$$

Equations (7–12) deal with HP-LP factors and express logical conditions, requiring that at maximum one station is built in each municipality, either HP or

LP (Eq. 7) and in particular assuring that HP technology is used if we choose to construct stations in municipalities served by HP natural gas pipeline (Eqs. 8 and 9). A maximum technically feasible capacity B equaling 2000 kNm^3/year is imposed through Eqs. (10) and (11) which at the same time force the system to install either HP capacity Q_{ai} or LP capacity Q_{bi}, so that the total capacity calculated with Eq. (12) is actually either equal to Q_{ai} if HP is technically feasible or to Q_{bi} otherwise. Finally, Eq. (13) requires the equivalent annual profit of every single station to be non negative

3.2 Model Implementation

Like similar models in literature, the model is structured as a MINLP problem with binary and continuous decision variables. After a preliminary screening, mainly excluding low population municipalities in the mountain part of the region, 219 eligible locations were identified and distances were calculated and saved in Excel format using RouteBlast (2013). The nature and dimensions of the problem make the identification of global optimum solutions within the branch and bound framework very challenging due to the presence of both the integer variables and the non-convexities. For this reason, we decided to try a genetic solver and, given that our data had been mainly been saved in spreadsheet form, we chose to use the commercial solver Evolver® (2010), with 0.5 crossover rate, a mutation rate automatically determined by the program and a stopping rule entailing a progress of 10 % in the last 1500 trials and a maximum of 15000 trials. Solution times between 10 and 30 h were achieved with these settings and considered acceptable for our purposes.

4 Results and Discussion

In order to asses potential effects of different subsidy schemes, optimal location and capacities were evaluated in four scenarios, i.e.:

- At current demand levels, with no subsidies;
- At current demand levels, with the 50 % capital grant foreseen by the regional government;
- With double demand level in the border area, with no capital grants to stations;
- With double demand level in the border area and 50 % capital grant.

By evaluating these scenarios at different levels of the truncation factor introduced with Eq. (6), we found that such factor has a significant impact on the share of total demand, which is cost-effectively served by stations and on their location and size. The analysis were conducted for two values of $t_{\max}$, namely 20 and

50 km, representing the maximum daily distance for 80 % of European drivers and the maximum daily distance for 70 % of Italian drivers respectively [21].

The analysis with t_{max} at 20 km seem to give a more realistic picture of the current scenario, in that it leads to conclude that only four stations would be sustainable at current conditions with no subsidies, whereas the analysis at $t_{max} = 50$ km tells that even six stations would be viable without incentives. On the other hand, the evaluation of the effect of subsidies seems more realistic with the 50 km analysis, because the other one foresees a proliferation of up to 12 micro-plants with an average capacity of less than 200 kNm^3/year, which does not seem a rational behavior for investors given, in particular, that average sales at national level can be estimated at about 940 kNm^3/year per station. To this respect, from our evaluation with both the 20 and 50 km an average plant capacity of 350 kNm^3/year is already viable at the financial conditions we assume (i.e. 7 % interest rate for 15 years. Post-analysis discussion with constructors who provided cost data pointed out that, based on experience, a minimum size of about 500 kNm^3/year should be economically feasible. This size is smaller than the national sales average, probably due to the more recent practice of upgrading existing gasoline refueling stations—which requires less investment—rather than building standalone stations selling CNG only, which was common practice in the 1990s due to competition and legislation barriers. On the other hand, what we probably underestimated are contingencies, variability in connection costs and the minimum attractive rate of return, which is actually considered by investors to account for those risks. We intend to continue our analysis on these aspects, e.g. by extended sensitivity analysis. Nevertheless, we conclude that at the moment the 50 km scenario is the more realistic and the only one we choose to graphically represent in this chapter (see Fig. 2), for the sake of brevity.

Looking at Fig. 2 we find that a generic 50 % subsidy at current demand conditions would still lead entrepreneurs to choose locations far away from the border, in spite of competition due to relative proximity of existing CNG stations, rather than to invest in the F1 area. A similar pattern was also obtained in the 20 km scenario. As a consequence, the small demand by about 200 vehicles registered in the F1 area, probably in past times of substantial national incentives for CNG vehicle purchase, is not met and at present those vehicles are most likely fuelled with gasoline. Moreover, the more realistic 50 km analysis shows that the 50 % subsidy, which, based on our optimization, would result in an outlay of almost 1 M€ for the regional government, would not substantially change the number of economically viable CNG stations (from 6 to 7), although it would certainly help these investment opportunities to be put into action at these times of difficult access to credit for firms. Still, if the aim of the regional government is to attain a more even distribution of CNG demand in the region, specific measures for the F1 area are needed. For instance, increasing the number of vehicles in the F1 by 100 %, for instance through capital grants for vehicle purchase, would make a fuel station feasible there (although with our probably optimistic minimum capacity). And probably the most effective option would be to invest in both

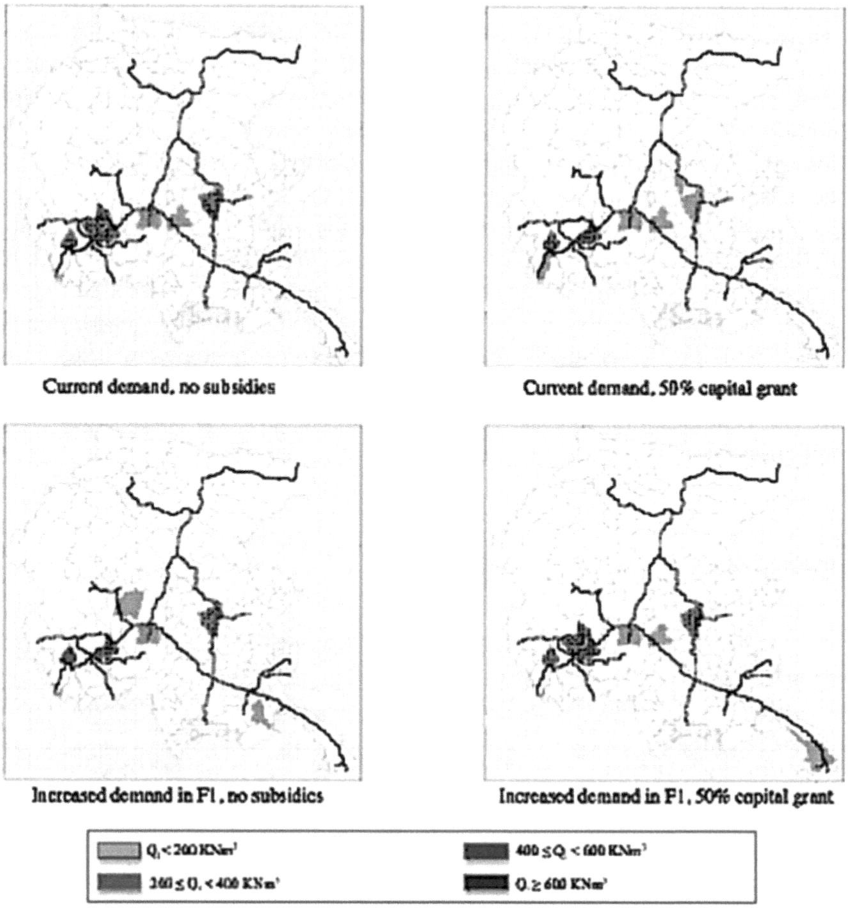

Fig. 2 Optimized capacity and location of CNG refueling station, with $t_{max} = 50$ km

vehicle subsidies and station subsidies (lower right quarter of Fig. 2). In our view, however, in present times of public outlay restriction, incentives should be specifically targeted to current low demand areas, especially F1.

5 Conclusions

Like every model, the presented MINLP optimization model for CNG refueling stations planning in Friuli Venezia Giulia is based on assumptions and simplifications, in part due to computational requirements and in part depending on the features of available data. Collecting further information through empirical research would be a necessary step to increase the validity of the obtained results. In particular, given the demonstrated effect of the attraction function, data collection of actual or stated refueling behavior of CNG vehicle drivers would be needed, given that empirical research on refueling behavior, on which modeling assumptions of this and similar models in literature are based, dates back to the 1980s [11], is focused on gasoline and on the US market. So far, the developed model supported reasonable arguments to rethink the structure of public subsidies making government goals more explicit. Finally, the analysis and statistical modeling of demand data for this case study shows that different policies and tariff structures of neighboring countries impact on consumers' decisions of vehicle purchase, and consequently on alternative fuel demand, which may jeopardize the effects of policies for sustainable transport put into force in single countries. From an European perspective, efforts on policy and infrastructure development could therefore benefit from international coordination, perhaps more than from competition.

Appendix A

Table A.1 Annual LPG demand at province level

Year	P_{lpg} (euro/l)	Province	D(PROV) (l)	$Vehicles_{lpg}$ (PROV)	$Vehicles_{TOT}$ (PROV)
2007	0.626	Gorizia	81810	271	88812
	0.626	Pordenone	2097972	3562	193833
	0.626	Trieste	NA	409	127548
	0.626	Udine	3616002	2679	337664
2008	0.680	Gorizia	136.50	371	8562
	0.680	Pordenone	3950514	4079	196487
	0.680	Trieste	NA	595	127591
	0.680	Udine	4177764	3297	341432
2009	0.563	Gorizia	141.804	724	88598
	0.563	Pordenone	4488642	5781	198013
	0.563	Trieste	NA	1018	127670
	0.563	Udine	4915872	5571	344248
2010	0.661	Gorizia	545400	953	88501
	0.661	Pordenone	6655698	6.903	199270
	0.661	Trieste	NA	1284	127842
	0.661	Udine	7419258	6986	347507
2011	0.755	Gorizia	621756	982	88636
	0.755	Pordenone	7850124	6914	201975
	0.755	Trieste	NA	1321	128006
	0.755	Udine	7010208	7068	351215

Table A.2 Economic coefficients

	High-pressure	Low-pressure
C_{STRV} [€/kNM^{3}year]	287.8	413.5
C_{STRF}[€/year]	249811	255996
C_{EL}[€/kNM^3]	15.21	28.21
C_{MAIN}[€/kNM^{3}year]	$0.402(C_{STRF}y_i + C_{STRV}\ Q_i) - 139930$	$0.269(C_{STRF}y_i + C_{STRV}\ Q_i) - 106840$
C_{CNG}[€/kNM^3]	490	490
C_{HR} [€/year]	35180	35180
P_{CNG}[€/kNM^3]	980	980

References

1. Yeh, S.: An empirical analysis on the adoption of alternative fuel vehicles: the case of natural gas vehicle. Energy Policy (Elsevier) **35**(11), 585–5875 (2007)
2. Collantes, G., Melaina, M.W.: The co-evolution of alternative fuel infrastructure and vehicles: a study of the experience of Argentina with compressed natural gas. Energy policy **39**, 664–665 (2011)
3. Struben, J., Sterman, J.D.: Transition challenges for alternative fuel vehicle and transportation systems. Environ. Plann. B **35**, 1070–1097 (2008)
4. Kuby, M., Lim, S.: The flow-refueling location problem for alternative-fuel vehicles stochastic modeling. Socio Econ. Plann. Sci. **39**(2), 125–145 (2005)
5. Lin, Z., Ogden, J., Fan, Y., Chen, C.W.: The fuel-travel-back approach to hydrogen station siting. Int. J. Hydrogen Energy **33**, 3096–3101 (2008)
6. Bersani, C., Minciardi, R., Sacile, R., Trasforini, E.: Network planning of fuelling service stations in a near-term competitive scenario of the hydrogen economy. Econ. Plann. Sci. **43**, 55–71 (2002)
7. Kuby, M., Lines, L., Schultz, R., Xie, Z., KiM, J.G., Lim, S.: Optimization of hydrogen stations in Florida using the flow-refueling location model. Int. J. Hydrogen Energy **34**, 6045–6064 (2009)
8. Stephens-Romero, S., Brown, T., Kang, J., Recker, W., Samuelsen, G.S.: Systematic planning to optimize investments in hydrogen infrastructure deployment. Int. J. Hydrogen Energy **35**, 4652–4667 (2010)
9. Frick, M., Axhausen, K.W., Carle, G., Wokaun, A.: Optimization of the distribution of compressed natural gas (CNG) refueling station: swiss case studies. Transp. Res. Part D **12**, 10–22 (2007)
10. Upchurch, C., Kuby, M., Lim, S.: A model for location of capacitated alternative-fuel stations. Geogr. Anal. **41**, 133–144 (2009)
11. Kitamura, R., Sperling, D.: Refueling behavior of automobile drivers. Transp. Res. Part A **21**, 235–245 (1987)
12. Bapna, R., Thakur, L.S., Nair, S.K.: Infrastructure development for conversion to environmentally friendly fuel. Eur. J. Oper. Res. **33**, 480–496 (2002)
13. Sabio, N., Gadalla, M., Guillen-Gosalbez, G., Jimenez, L.: Strategic planning with risk control of hydrogen supply chains for vehicle use under uncertainty in operating costs: a case study of Spain. Int. J. Hydrogen Energy **35**, 6836–6852 (2010)
14. Brey, J.J., Carazo, A.F., Brey, R.: Using AHP and binary integer programming to optimize the initial distribution of Hydrogen infrastructures in Andalusia. Hydrogen Energy Publication **37**, 5372–5384 (2011)

15. Hugo, A., Ruttera, P., Pistikopoulosa, S., Amorellib, A., Zoia, G.: Hydrogen infrastructure strategic planning using multi-objective optimization. Int. J. Hydrogen Energy **30**, 1523–1534 (2005)
16. Aci: Annuario statistico dell'Automobile Club d'Italia (2012)
17. GVR (2013), Gas Vehicle Report, **135**(12), 25–27
18. Figisc: Le vendite di benzina e metano in Friuli Venezia Giulia 2007–2011 (analisidefinitiva) (2011)
19. Sterner T., Dahl C.A. (1991).Modeling transport fuel demand. In: Sterner, T. (ed.) International Energy Economics **13**(3), 203–210
20. Dahl, C.A.: Measuring global gasoline and diesel price and income elasticities. Energy Policy **41**, 2–13 (2011)
21. JRC: Driving and parking patterns of European car drivers: a mobility survey (2012)

Electric Vehicles Charging Network Planning

Joana Cavadas, Gonçalo Correia and João Gouveia

Abstract In this chapter we propose a method to plan the location of charging stations for electric vehicles (EV) in a city in which the objective is to maximize the number of satisfied vehicles under a fixed budget for building the stations. We take into consideration the maximum capacity of each possible site for installing a station, in terms of the number of plugs that each one can have, and the distance from that location and each demand point, which is measured in walking time. To be able to apply these models, we develop a charging demand model for based on parking data, considering that the higher the parking time, the greater the probability of charging. We also take in consideration the relation between the demand at different points, e.g., if a vehicle can charge at home, the probability of needing to charge at work will be significantly reduced. We test our mathematical models for the case of the city of Coimbra, where there is already a network of charging stations. We first use an existing mobility survey to extract parking data and establish a demand grid, and then we apply the models that gives us the optimal location for charging stations for the entire city allowing us to compare both.

Keywords Electric vehicle · Charging stations planning · Mixed integer problem · Charging demand

J. Cavadas (✉) · J. Gouveia
Department of Mathematics, University of Coimbra, 3001-454 Coimbra, Portugal
e-mail: joana.cavadas@hotmail.com

G. Correia
Department of Civil Engineering, University of Coimbra, Rua Luis Reis Santos, 3030-788 Coimbra, Portugal

J. F. de Sousa and R. Rossi (eds.), *Computer-based Modelling and Optimization in Transportation*, Advances in Intelligent Systems and Computing 262,
DOI: 10.1007/978-3-319-04630-3_7,

1 Introduction

With the current economic crisis and the growing dependence on fossil fuels for the mobility of people and goods, one of the main causes of the high levels of pollution in our cities and the known greenhouse effects, one observes the relaunch of electric vehicles (EV) as one of the solutions to help mitigate these problems. The electric motors technology was developed in the beginning of the automobile invention (in the end of the nineteenth century and the beginning of the twentieth century) but soon it was abandoned to give place to the combustion engine in the twenties of the twentieth century. This resulted from the discovery of petroleum sources but also it was a consequence of the low autonomy and power of the EV [4]. EVs are more efficient in energy consumption, they are more environmental friendly because they have zero local emissions, they also generate zero noise pollution because they are silent, and even the energy used in the batteries can be obtained through a renewable source, as opposed to the vehicles powered by internal combustion engines that are responsible for 40 % of the CO_2 emissions and 70 % of other greenhouse gas emissions in urban areas [5]. However, EVs still have the same problem that they had before: a low autonomy (between 60 and 160 km); and a new problem: the long time needed to recharge (between 6 and 8 h in normal charging [2]). These limitations hinder the EVs adoption by the automobile market. Webster [14] points out that this low autonomy is more than enough for the majority of the trips done in 1 day. The second limitation can be overcome with the planning of a charging stations network nearby users' main destinations, in order to enable the charge of EVs during parking times.

There have been developed some models to define the best location for charging stations and these have been based different ways of estimating the charging demand. This estimation of demand may take into account different perspectives, namely the traffic flow [6, 8, 9], the charging requirement [1, 12, 13], the parking time spent in the study sites [3] and the number of vehicles and its use [7].

The planning of a charging network has been proposed to be made mainly by optimization models or heuristics. For example: Feng et al. [6] developed a method based in the partition of the network to minimize users' losses on the way to the charging stations and in 2011 [8] proposed a new method based in the Weighted Voronoi Diagram with the same purpose as the model before. Worley et al. [15] formulated the charging stations location problem with a discrete-integer-program whose purpose was to minimize the costs of travelling, charging operations and the charging stations network investment. In Frade et al. [7] the model was based on the p-median problem in order to maximize the satisfied demand; another approach was the use of numerical methods in a multiobjective model in order to minimize the investment, the distance traveled and the cost of the network [13]. In the recent chapter by Chen et al. [3] a mixed-integer program was developed in order to minimize the walking distance between the parking location and the users' destination site. In this chapter we present an approach to the estimation of charging demand which, as in Chen et al. [3], is based on the parking time in each site but it also considers demand transference through the successive trips made between the different parking locations. This lets the model define which site will be more beneficial to receive a charging station without necessary refusing some of the demand of other sites. Given this demand, we develop a mixed-integer optimization model as in Frade et al. [7] with the purpose to maximize the satisfied demand under a fixed budget. We then improve it, taking into account the possibility of transferring demand from sites which have trips between them. We also develop a variation of this model that considers the day split into time intervals, in order to reduce the effects that peak hours may have in changing the solution. We exemplify the application of these models with the case-study city of Coimbra.

The chapter is structured as follows. We start, in Sect. 2, by explaining how to estimate the demand at each discrete point in space and apply in a first mixed-integer optimization model that maximizes the satisfied demand. We proceed, in Sect. 3 by explaining the concept of transferable demand and how to adapt the mathematical programming model in order to consider this possibility. In Sect. 4, we present our final model improvement, which adds time intervals to the optimization formulation, testing the influence that this added realism may bring to the network planning. Finally, in Sect. 5, the three models are applied to Coimbra. The chapter ends with the main conclusions withdrawn from the results.

2 Local Demand Estimation

We will start by estimating the contribution of an EV to the demand for charging vehicles i.e., the time that it is expected to be charging during a day in each public charging station. Our model will be based on two assumptions: first, we will assume that the average number of times an EV charges during the day is constant for all EV's; second we will assume that the probability that a vehicle is charged during one of its stops along the day is proportional to the time it remains parked in that location..

Given an EV's driver m, and a parking location j, let T_j^m be the amount of time that m stays parked at location j, and T^m be the total time that m is parked during the day. By our assumption, and by normalizing the expected number of daily charges of an EV to one, we conclude that the probability that m charges at j is $P_j^m = \frac{T_j^m}{T^m}$. If m charges during its stop, it will charge for the entire time it is parked, so the expected duration of charging will be given by $E_j^m = T_j^m \cdot P_j^m$. To estimate the total potential demand at a particular location j, we then have to sum over all EVs' drivers, and we obtain $D_j = \Omega \cdot \sum_m E_j^m$, where Ω is the average number of daily charges of an EV, which we assumed constant. In practice it will normally be easier to compute this sum over all drivers and multiply by the proportion of EV in the total number of them.

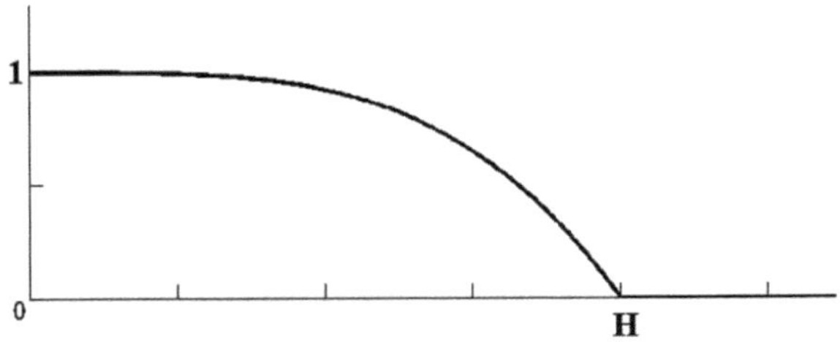

Fig. 1 Example of distance penalty function Γ, where H is the value of the reasonable distance considered

2.1 Basic Charging Station Location Model

We propose a basic mixed integer programming model for determining the optimal location of charging stations using the local demand estimation previously explained. We assume that the demand was estimated at M different locations, and we will represent by D_j the demand at location j. We will also assume that there are N possible locations for placing charging stations and we will represent by C_k and B_k, the capacity of station k (if built) and the cost of building it, respectively. Note that since the demand is given by the number of *cars* · *hour* of occupation, the capacity has to be given also in this unit. Finally, given a demand site j and a potential location k for a charging station, we will define Γ_{jk} as a distance penalty that will be valued one if the two places are very close and will decrease as distance increases, becoming zero if the distance is greater than what we will assume to be a reasonable walking distance. A possible choice for distance penalty function can be seen in Fig. 1.

With this data we can now propose a model for maximizing the demand satisfied, under a fixed budget T as follows:

$$\max \sum_{j=1}^{M} D_j \sum_{k=1}^{N} z_{jk} \cdot \Gamma_{jk} \tag{1}$$

$$\text{subject to} \sum_{j=1}^{M} D_j z_{jk} \leq C_k, \quad k = 1, \ldots, N, \tag{2}$$

$$\sum_{k=1}^{N} z_{jk} \leq 1, \quad j = 1, \ldots, M, \tag{3}$$

$$z_{jk} \leq x_k, \quad j = 1, \ldots, M, \quad k = 1, \ldots, N, \tag{4}$$

$$\sum_{k=1}^{N} B_k x_k < T, \tag{5}$$

$$z_{jk} \in [0, 1], \quad j = 1, \ldots, M, \quad k = 1, \ldots, N, \tag{6}$$

$$x_k \in \{0, 1\}, \quad k = 1, \ldots, N. \tag{7}$$

where:

z_{jk} proportion of demand from j satisfied by the charging station located in k;

x_k 1 or 0 depending on whether a charging station is located in k or not.

The objective function (1) of this mixed-integer optimization model maximizes the satisfied demand, taking into account the distance penalty with the purpose of giving priority to the demand sites closest to the charging station, and avoid forcing users to travel long distances.

Given a demand location, its proportion of satisfied demand cannot exceed 1 (100 %) (3), furthermore the capacity of a charging station cannot be surpassed by the demand it satisfies (2). Finally, only a charging station that is effectively built can satisfy demand (4) and the cost of the built charging stations must be within the budget (5). Expressions (6) and (7) set the domain for the decision variables.

3 Transferable Demand Estimation

The previous model considers demand as a local property, where all sites are independent. However, if an EV's driver stops at several different sites during the day, the presence of a charging station in one of the stops, affects the demand for charging the vehicle in the remaining ones. If all EVs' drivers that could use a particular charging station have other possibilities of charging their vehicle one can save money by not building in that location and instead building stations that would bring users not previously covered by any charging station, typically those that have less stops thus more constrained in their possibilities of charging the EV. The previous model does not consider this possibility, so in order to deal with this we introduce the notion of transferable demand.

Suppose we have an EV's driver m whose daily trips include a journey between two parking locations i and j. We will consider that part of the probability of m charging in location i can be transferred to j and vice versa. Given the demand model previously presented, we conclude that the added demand of m on j, coming from i can be as large as $U_{ij}^{m} = P_{i}^{m} \cdot T_{j}^{m}$. To estimate the potential added demand to j coming from i, we only have to sum over all EVs' drivers that travel from i to j, and we obtain $V_{ij} = \Omega \sum_{m} U_{ij}^{m}$ where, as referred, Ω is the average number of daily charges of an EV. We also have to consider the maximum demand that can be subtracted from i by being transferred to other places. This is the sum of the demand E_{i}^{m} of all the vehicles that go from i to j, which will be denoted as W_{ij}. In short, if all EVs that travel between parking location j and i decide to charge at location j instead of i, the proportion of satisfied demand would decrease by W_{ij} in i and increase by V_{ij} in j.

3.1 Transferable Demand Charging Station Location Model

Using the notion of transferable demand we can now propose an improved version of the previous model.

$$\max \sum_{j=1}^{M} D_j \sum_{k=1}^{N} z_{jk} \cdot \Gamma_{jk} + \sum_{i=1}^{M} \sum_{j=1}^{M} V_{ij} \sum_{k=1}^{N} y_{ijk} \cdot \Gamma_{jk} \tag{8}$$

$$\text{subject to} \sum_{j=1}^{M} D_j z_{jk} + \sum_{i=1}^{M} \sum_{j=1}^{M} V_{ij} y_{ijk} \le C_k, \quad k = 1, \ldots, N, \tag{9}$$

$$\sum_{k=1}^{N} D_i z_{ik} + \sum_{j=1}^{M} \sum_{k=1}^{N} W_{ij} y_{ijk} \le D_i, \quad i = 1, \ldots, M, \tag{10}$$

$$\sum_{k=1}^{N} y_{ijk} \le 1, \quad i,j = 1, \ldots, M, \tag{11}$$

$$z_{jk} \le x_k, \quad j = 1, \ldots, M, \quad k = 1, \ldots, N, \tag{12}$$

$$y_{ijk} \le x_k, \quad i,j = 1, \ldots, M, \quad k = 1, \ldots, N, \tag{13}$$

$$\sum_{k=1}^{N} B_k x_k \le T, \tag{14}$$

$$z_{jk} \in [0, 1], \quad j = 1, \ldots, M, \quad k = 1, \ldots, N, \tag{15}$$

$$y_{ijk} \in [0, 1], \quad i,j = 1, \ldots, M, \quad k = 1, \ldots, N, \tag{16}$$

$$x_k \in \{0, 1\}, \quad k = 1, \ldots, N. \tag{17}$$

In the model, y_{ijk} represents the proportion of the potential demand that can be transferred from i to j that is effectively transferred and satisfied in charging station k.

The objective function was changed from its previous expression in (1)–(8) by taking in consideration not only the satisfied local demand but also the demand that was transferred to each location. This was also taken in consideration in the charging station capacity constrain (9). We also had to replace the demand constraint (3) by taking into account that the total demand satisfied locally plus the total demand lost through transfer cannot be higher than the original demand at each site (10). One last addition to the model was constraint (11) that guarantees that there is no transfer above the transferable demand. Finally, as in the previous model, only a site with a charging station can satisfy a demand location ((12) and (13)) and the cost of the charging station network must be below the budget (14). Expressions (15)–(17) set the domain for the decision variables.

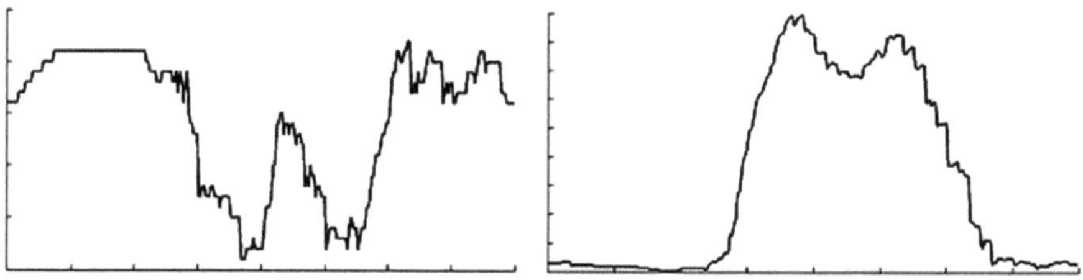

(a) (b)

Fig. 2 Number of vehicles parked at each hour of the day at a residential location (**a**) and at an office/commercial location (**b**) (Mondego region mobility system [10])

4 Considering Time Intervals

It is normally the case that local demand at a site during the day fluctuates widely. For example, one expects that residential locations have most of the demand during the night, while office, industrial or commercial areas would have more people present during working hours (Fig. 2). The previous models do not take these changes into consideration, using instead an average of the charging that the results we consider to be the demand. This induces distortions, since if demand is sharply concentrated at some peak hour, the capacity installed might be inadequate to deal with that peak, and much less of it can be satisfied than what previous models would assume. To have more realistic results we propose a new model where the day is divided in time intervals.

4.1 Estimating Demand at a Time Interval

Given an EV's driver m and a parking location j, we denoted by T_j^m the time that m stays parked at that location. The time the individual remains there might be split between several of the time intervals that we are now considering, so we now consider $T_j^{m,\alpha}$ to be the time that m stays parked at j during the α time interval. In particular, if I is the set of all time intervals considered, $T_j^m = \sum_{\alpha \in I} T_j^{m,\alpha}$. We can now proceed to estimate the demand at a specific time interval in the same way as we did in the previous demand estimation. Each vehicle m will have an expected charging duration at time interval α given by $E_j^{m,\alpha} = P_j^m \cdot T_j^{m,\alpha}$. Therefore total demand at site j at time interval α will be given by $D_j^\alpha = \Omega \sum_m E_j^{m,\alpha}$, where Ω has the same meaning as before. Note that summing D_j^α over all time intervals we recover D_j, illustrating that $D_j^\alpha, \forall \alpha \in I$, is a temporal refinement of local demand.

We now have to deal with the transferable demand, and adapt it to these new settings. Note that the added demand to location j at time β coming from m that traveled from i at time interval α cannot be greater than $U_{ij}^{m,\alpha} = P_i^m \cdot T_j^{m,\beta} \cdot \frac{T_i^{m,\alpha}}{T_i^m}$, where $\frac{T_i^{m,\alpha}}{T_i^m}$ represents the proportion of time that m stays at i during α, so the maximum added demand coming from i at time interval α is just the sum of this figure over all users that travelled from (or to) i and are present at location j at interval β, which we will denote by $V_{ij}^{\alpha\beta}$. The maximum demand that can be subtracted from location i and interval α resulting from transfers to j at β, is then given by the sum over all users present at i during α that will travel to (or come from) j of their demand $E_j^{m,\alpha}$, times the proportion of time that m stays at j during β, $\frac{T_j^{m,\beta}}{T_j^m}$, and we will denote it by $W_{ij}^{\alpha\beta}$. Again, note that summing $V_{ij}^{\alpha\beta}$ and $W_{ij}^{\alpha\beta}$ over all time intervals recovers the previously defined V_{ij} and W_{ij}.

4.2 Improved Location Model

Applying all the previously defined notions we now propose a model that considers demand transference and several time intervals for the demand definition.

$$\max \sum_{\alpha\in I}\left[\sum_{j=1}^{M} D_j^{\alpha}\sum_{k=1}^{N} z_{jk}^{\alpha}\cdot\Gamma_{jk} + \sum_{i=1}^{M}\sum_{j=1}^{M}\sum_{\beta\in I} V_{ij}^{\beta\alpha}\sum_{k=1}^{N} y_{ijk}^{\beta\alpha}\cdot\Gamma_{jk}\right] \tag{18}$$

$$\text{subject to} \sum_{j=1}^{M} D_j^{\beta} z_{jk}^{\beta} + \sum_{i=1}^{M}\sum_{j=1}^{M}\sum_{\alpha\in I} V_{ij}^{\alpha\beta} y_{ijk}^{\alpha\beta} \leq C_k^{\beta}, \quad k=1,\ldots,N, \quad \forall\beta\in I, \tag{19}$$

$$\sum_{k=1}^{N} D_i^{\alpha} z_{ik}^{\alpha} + \sum_{j=1}^{M}\sum_{\beta\in I} W_{ij}^{\alpha\beta}\sum_{k=1}^{N} y_{ijk}^{\alpha\beta} \leq D_i^{\alpha}, \quad i=1,\ldots,M, \quad \forall\alpha\in I, \tag{20}$$

$$\sum_{\alpha\in I}\sum_{k=1}^{N} y_{ijk}^{\alpha\beta} \leq 1, \quad i,j=1,\ldots,M, \forall\beta\in I, \tag{21}$$

$$\sum_{\beta\in I}\sum_{k=1}^{N} y_{ijk}^{\alpha\beta} \leq 1, \quad i,j=1,\ldots,M, \forall\alpha\in I, \tag{22}$$

$$z_{jk}^{\beta} \leq x_k, \quad j=1,\ldots,M, \quad k=1,\ldots,N, \forall\beta\in I, \tag{23}$$

$$y_{ijk}^{\alpha\beta} \leq x_k, \quad i,j=1,\ldots,M, \quad k=1,\ldots,N, \forall\alpha,\beta\in I, \tag{24}$$

$$\sum_{k=1}^{N} B_k x_k \leq T, \tag{25}$$

$$z_{jk}^{\beta} \in [0,1], \quad j=1,\ldots,M, \quad k=1,\ldots,N, \quad \forall\beta\in I, \tag{26}$$

$$y_{ijk}^{\alpha\beta} \in [0,1], \quad i,j = 1,\ldots,M, \quad k = 1,\ldots,N, \quad \forall \alpha,\beta \in I, \tag{27}$$

$$x_k \in \{0,1\}, \quad k = 1,\ldots,N. \tag{28}$$

In this model z_{jk}^{β} is the proportion of the demand at location j in the time interval β that is satisfied by a charging station at k, while $y_{ijk}^{\alpha\beta}$ is the proportion of transferable demand from i at time interval α to j at time interval *beta* that is transferred and satisfied by a charging station at k. The objective function (18) was changed from (8) by summing the satisfied demand over all time intervals.

The capacity constraint (19) and the demand constraint (20) were taken interval-wise. Note that the capacity depends on the length of the time interval, since it is measured in total number of hours of parking for all the cars ($cars \cdot hour$). Constraints (21) and (22) guarantee that the total demand added to j at interval β coming from i at interval α and the total demand subtracted to i at interval α going to j at β do not exceed $V_{ij}^{\alpha\beta}$ and $W_{ij}^{\alpha\beta}$, respectively. Constraints (23)–(25) are analogous to the previous model, while expressions (26)–(28) set the domain for the decision variables.

5 Case Study, City of Coimbra

To illustrate the proposed models we will apply them to the city of Coimbra, in the central region of Portugal. The municipality of Coimbra had a little over 140,000 inhabitants in 2011 (according to the 2011 census), and has an area of 319.41 km^2. For our study we will consider only the main consolidated area of the city and main neighboring suburbs that concentrate most of the population of the municipality. According to the last comprehensive mobility survey done in the city there is a motorization rate of about 530 vehicles/1,000 inhabitants and the private vehicles fleet is of about 55,000 vehicles. In the same study an estimated 70 % of the trips inside the city are done in a private vehicle while only 17 % use the Bus network, the only transit system in the city. Most of the residents do two trips per day (65 %) and the main three motives for a trip are going to work, going to school and shopping [11]. Like in the rest of the country the city has a very small number of EVs in circulation, and an EV charging station network has been installed with nine locations. This network is a result of the Program for Electric Mobility developed in 2009 by the Portuguese government, this program had as its main purpose to create, install and operate a network of charging stations for promoting the use of EV thus reducing the petroleum dependence and reduce pollutant emissions (http://www.mobie.pt/en/homepage). In the first stage the charging network was created in 25 municipalities (Coimbra included). This network was foreseen to grow in a second stage as the adoption of EVs grows. This growth is expected due to the tax benefits given to these vehicles implemented in the plan for stability and growth applied between 2010 and 2013.

Table 1 Permanency profiles given by the mobility survey

Profile number	Number of stops	Stops by duration			% of respondents
		0–2 h	2–8 h	>8 h	
1	1	0	0	1	9.29
2	1	0	1	0	5.32
3	2	0	0	2	28.70
4	2	0	1	1	25.82
5	2	1	0	1	13.78
6	3	1	0	2	2.00
7	4	1	2	1	3.14
Other					11.95

The basis for our study is the geo coded mobility survey that was done in Coimbra, conducted between October of 2008 and March of 2009 [10]. This survey questioned a sample of 10,000 participants, with information about trip origins and destinations and mode choice in a working day. To consolidate our data and make it easier to handle it, the demand was aggregated into sectors using a grid with cells with 800 m sides over the city. The demand was aggregated on each sector at its weighted centre of mass, the squares with negligible demand were eliminated (less than 10 parking) and other squares were subdivided because they had very high demand (higher than 100 parking) resulting in 400 m side squares. At the end of this procedure we obtained an area of 62.88 km^2 divided into 129 sectors, of which 88 have an area of 0.64 km^2 and 41 have an area of 0.16 km^2.

The mobility survey did not ask for the vehicle type driven by the respondent thus there is no distinction between combustion engine vehicles and EVs. To have this in consideration we multiply the contribution of each vehicle on the demand for charging by using a correction factor Λ. This factor depends on the forecasted percentage of EV's in the fleet and has the purpose of representing the estimated average number of charging time of EV's occurring at a public charging station. Given the mobility survey, we are able to define permanency profiles (Table 1) considering the number of stops along the day (1, 2, 3, 4 or greater than 5) by each individual (Fig. 3).

Finally we will use the centroids of each of the squares of our grid as candidate charging station locations. This is just indicative, and should in practice be replaced by a more informed study of physically possible locations in the city (possibly closer to the demand center on each square). In Fig. 4 we can see the resulting grid, the demand points and the candidate sites for placing a charging station.

To estimate the demand from the mobility survey we have to choose values for $\Omega\Lambda$ that represent the number of expected daily charges per vehicle on the road, denote that this factors represents the proportion of EVs given all the vehicles and

Fig. 3 Proportion of parking number

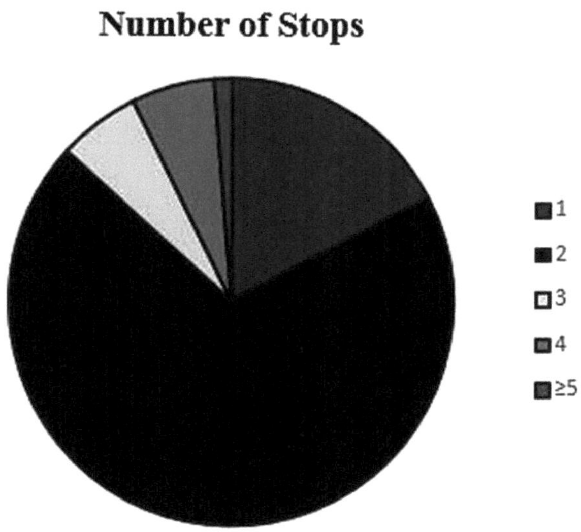

given an EV the probability of its charging occurs in a public charging station. We ran simulations for 1, 1.5 and 2 chargings per 1,000 vehicles. We defined the distance penalty function as follows:

$$\Gamma_{jk} = \begin{cases} \dfrac{\left(-\left(d_{jk}\right)^4 + 20^4\right)}{20^4 \cdot exp\left(\left(\frac{d_{jk}}{40}\right)^3\right)} & \text{se } d_{jk} < 20 \text{ min} \\ 0 & \text{se } d_{jk} > 20 \text{ min,} \end{cases} \quad (29)$$

where d_{jk} is the distance between demand point j and candidate charging station side k measured in minutes of walking time. This assumes that 20 min is the maximum time-distance that the drivers are willing to walk from the parking space where they leave the vehicle charging and their destination (see Fig. 1. with $H = 20$). Since the current charging network installed has nine charging stations, we opted to study the optimal location of a similarly-sized network, so we set the cost of a station to 1 and the total budget to 9. We also assume that each station has four plugs, giving it a total capacity of $C_k = 4 \cdot 24 = 96\ cars \cdot hour$. For the improved model, we will consider the division of the day in four time intervals of 6 h each: $I = \{]2,8],]8,14],]14,20],]20,2]\}$. The capacity in this case will then be $C_k^\beta = 4 \cdot 6 = 24\ cars \cdot hour$.

In Figs. 5 and 6 we present the optimal network obtained for the three models proposed for $\Omega\Lambda \in \left\{\frac{1}{1000}, \frac{1.5}{1000}\right\}$. Table 2 shows the index of coverage, defined by the ratio between the objective function's value and the total local demand, for $\Omega\Lambda \in \left\{\frac{0.5}{1000}, \frac{1}{1000}, \frac{1.5}{1000}, \frac{2}{1000}\right\}$. In Table 2 we also present the index of coverage obtained when applying each model to the network already existing in Coimbra, the MOBI.E network (Fig. 7).

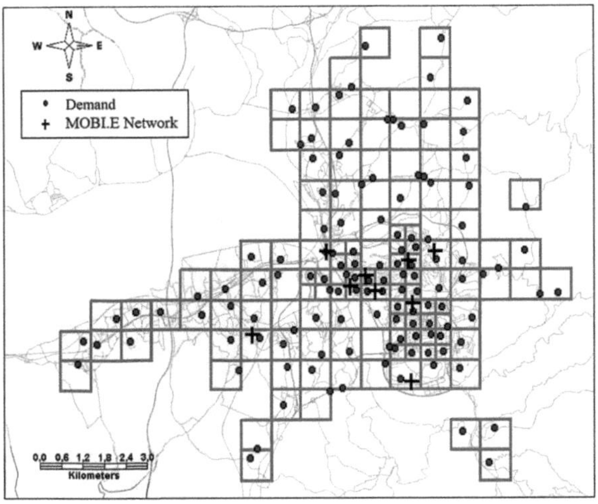

Fig. 4 Dealt area with the resulting grid, the demand points and the candidate sites for placing a charging station

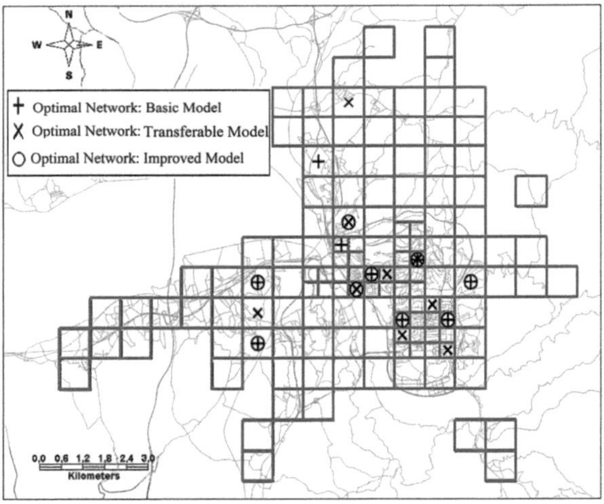

Fig. 5 Optimal network in the three models (basic, transferable and improved) with $\Omega\Lambda = 1/1000$

Note that the addition of transferable demand corrects an underestimated satisfiable demand in the basic model, while adding time intervals, shows that we were dramatically overestimating capacity.

6 Discussion and Conclusions

The main contributions of this chapter are the establishment of a demand model for EV charging based on parking locations and durations and the proposal of a mixed-integer model approach to pick the best locations for deploying charging station network that maximizes the satisfied demand under a fixed budget. Our

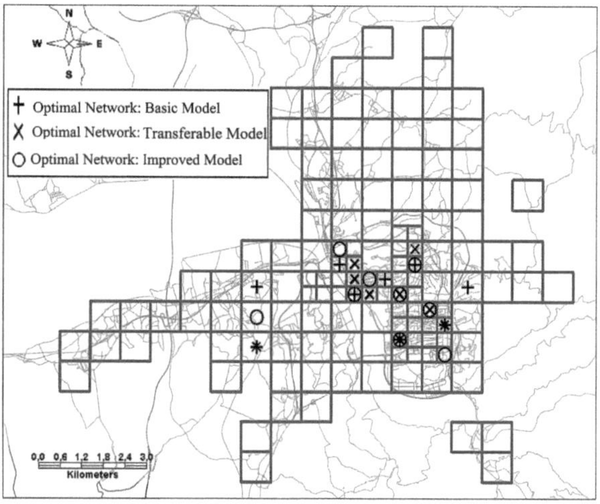

Fig. 6 Optimal network in the three models (basic, transferable and improved) with $\Omega\Lambda = 1.5/1000$

Table 2 Index of coverage for each model and each possible value of $\Omega\Lambda$

$\Omega\Lambda$	0.5/1000	1/1000	1.5/1000	2/1000
Optimal network: basic model	0.5567	0.5463	0.5185	0.4766
Optimal network: transferable model	0.7798	0.7616	0.6752	0.5272
Optimal network: improved model	0.6499	0.6312	0.5820	0.4944
MOBI.E network: basic model	0.4240	0.4240	0.4240	0.4099
MOBI.E network: transferable model	0.6433	0.6435	0.6108	0.5010
MOBI.E network: improved model	0.5168	0.5168	0.5018	0.4486

Fig. 7 MOBI.E network

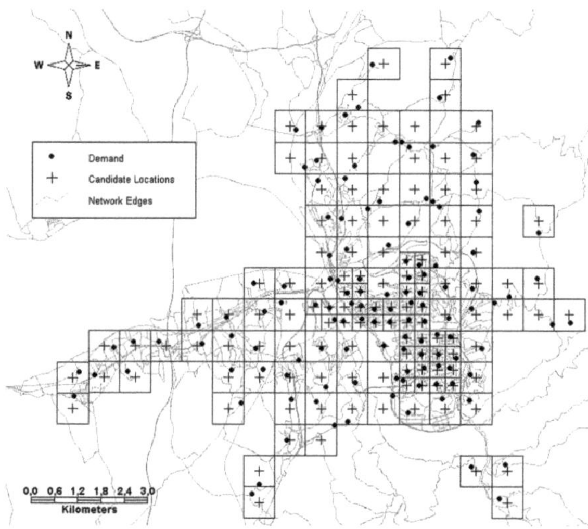

demand model not only deals with each parking location separately (as was previously proposed in [3]) but also uses the daily activities of travelers to link the demand on distinct locations. We were then able to use this data to improve on previously proposed mathematical programming models.

Results obtained from applying our models to travel data for the city of Coimbra, show that in networks where demand is relatively low, the impact of considering demand transference can be very high. Since only a limited number of charging stations is built, we should expect that the users will adjust their behavior accordingly, adding demand to the stations that are actually built. This model seems to be very useful for these situations where lack of demand prevents full-city coverage to be viable. When dividing the day in time intervals, the results obtained show again a very big impact on satisfied demand. This means that disregarding peak-hour effect, averaging demand over the whole day, leads to a significant overestimation of capacity. Comparing the optimal and the MOBI.E networks, we can conclude that our solutions are not very different from the existing network, but the values of covered demand are lower is this network.

Note that several variations of these models can easily be adopted and studied. We can use the time intervals to model differences between weekday/weekend demand patterns; We could also make demand transfer harder, by limiting it to some fraction of its theoretical value, modeling demand inertia; We can separate demand points into residential and industry/commercial areas and use this data to further refine the demand model; Another possible adaptation is to consider the number of plugs that must be installed in each location instead of charging stations with a constant number of plugs. In fact, one of the main strengths of both the estimation and the optimization models is that they are theoretically simple, and therefore easy to adapt and refine. After a solution is obtained, it might also be possible to use some local numerical optimization methods to slightly perturb the proposed charging station locations and improve the quality of the solution.

The proposed models present a good performance in our case study in terms of Coimbra's dimension that is, considering the number of possible locations for charging stations and demand locations. However, this may not happen for a larger scale problem, due to the increasing of computation complexity. In the particular case of the improved model, the way the time is discretized also contributes to increase the complexity of the problem. The more we increase the number of time intervals in order to better fit the drivers' profiles in the demand estimation, the greater the number of decision variables. These scaling limitations can potentially be circumvented by the use of heuristics, preprocessing the data to reduce the number of potential charging stations or dividing a large problem into subproblems.

The main difficulties encountered in the development of the study have to do with access to some of the useful real data, namely, good estimates of future penetration rate of EVs and viable locations to charging stations, which would require further research before application of these models. Another aspect that should be taken in consideration is the actual behavior of EVs' drivers, which is the determining factor for the average number of daily charges, and for which no reliable data seems to be available at least for the Portuguese reality.

Despite this, we believe that the methodology developed in this chapter can provide a good planning for a charging station network and offers a valid contribution to the growing field of electric mobility.

Acknowledgments Support for this work was provided by Project mobiOS—mobility Operating System, financed by COMPETE/QREN—Agência de Inovação. The authors would like to thank Metro Mondego for making available the data used in the case-study and Intergraph for providing the software Geomedia Professional 6. Gouveia was also supported by the Centre for Mathematics at the University of Coimbra and Fundação para a Ciência e a Tecnologia, through the European program COMPETE/FEDER.

References

1. Bae, S., Kwasinski, A.: Spatial and temporal model of electric vehicle charging demand. IEEE Trans. Smart Grid **3**, 394–403 (2011)
2. Bostford, C., Szczepanek, A.: Fast charging vs. slow charging: pros and cons for the new age of electric vehicles. In: EVS 24 International Battery, Hybrid and Fuel Cell Electric Vehicle Symposium. Stavanger, Norway (2009)
3. Chen, T., Khan, M., Kockelman, K.: The electric vehicle charging station location problem: a parking-based assignment method for Seattle. In: 92nd Annual Meeting of the Transportation Research Board. Washington DC, USA (2013)
4. Cowan, R., Hultén, S.: Escaping lock-in: the case of the electric vehicle. Technol. Forecast. Soc. Chang. **53**, 61–79 (1996)
5. European Commission. A sustainable Future for Transport. Publications Office of the European Union (2009)
6. Feng, L., Ge, S., Lui, H.: Electric vehicle charging station planning based on weighted voronoi diagram. In: 2011 International Conference on Transportation, Mechanical and Electrical Engineering (TMEE). Changchun, China (2011)
7. Frade, I., Ribeiro, A., Gonçalves, G., Antunes, A.P.: Optimal location of charging stations for electric vehicles in a neighborhood in Lisbon, Portugal. Transp. Res. Rec. J. Transp. Res. Board **2252**, 91–98 (2011)
8. Ge, S., Feng, L., Liu, H.: The planning of electric vehicle charging station based on grid partition method. In: IEEE Electrical and Control Engineering Conference. Yichang, China (2011)
9. Ip, A., Fong, S., Liu, E.: Optimization for allocating BEV recharging stations in urban areas by using hierarchical clustering. In: The 2nd International Conference on Data Mining and Intelligent Information Technology Applications (ICMIA 2010). Seoul, Korea (2010)
10. Mondego Region Mobility System. Inquérito à Mobilidade da Região do Mondego. Database Metadata. Coimbra, Portugal (2009)
11. Mondego Region Mobility System. Transportation Planning Model of the Mondego Region Mobility System. Report Volume 1—Design and Results, Coimbra, Portugal (2011)
12. Qian, K., Zhou, C., Allan, M., Yuan, Y.: Load model for prediction of electric vehicle charging demand. In: 2010 International Conference on Power System Technology. Hangzhou, China (2010)

13. Wang, H., Huang, Q., Zhang, C., Xia, A.: A novel approach for the layout of electric vehicle charging station. In: Apperceiving Computing and Intelligence Analysis Conference (2010)
14. Webster, R.: Can the electricity distribution network cope with an influx of electric vehicles? J. Power Sources **80**, 217–225 (1999)
15. Worley, O., Klabjan, D., Sweda, T.: Simultaneous vehicle routing and charging station siting for commercial electric vehicles. In: IEEE International Electric Vehicle Conference 2012. Greenville, NC (2012)

Charging-Stations for Electrical Vehicles: Analysis and Model to Identify the Most Convenient Locations

Diego-Alejandro Giménez, Anabela Ribeiro, Javier Gutiérrez-Puebla and Antonio Pais-Antunes

Abstract Most of the plans to deploy public pilot networks of slow charging-stations in urban areas are choosing locations at popular parking places, such as city centers, shopping areas, train stations, and university campuses. The low parking time and high rotation rates often observed there could deliver an inadequate solution. This chapter presents a new approach to determine the optimal locations of public charging stations in urban areas. This approach relies on three different parts: an analysis of the specific battery electrical vehicles (BEV) charging characteristics obtained from a literature and market overview; the application of a location optimization model which maximizes the population access to these stations and in consequence the potential use of them; and the use of a target market index. Some of the specific characteristics of the charging problem addressed by this approach are: the required time of charge, access distance, and charging opportunities.

Keywords Electric vehicles · Charging stations · Optimal location problem

D.-A. Giménez (✉) · A. Ribeiro · A. Pais-Antunes
Department of Civil Engineering, FCTUC, Coimbra, Portugal
e-mail: diego@dec.uc.pt

A. Ribeiro
e-mail: anabela@dec.uc.pt

A. Pais-Antunes
e-mail: antunes@dec.uc.pt

J. Gutiérrez-Puebla
Department of Human Geography, UCM, Madrid, Spain
e-mail: javiergutierrez@ghis.ucm.es

J. F. de Sousa and R. Rossi (eds.), *Computer-based Modelling and Optimization in Transportation*, Advances in Intelligent Systems and Computing 262,
DOI: 10.1007/978-3-319-04630-3_8,

1 Introduction

There are many pilot programs for installing charging-stations in urban areas but most of them lack of a comprehensive analysis for the location of the stations. In the majority of the cases, the plans are choosing locations at popular parking places, such as city centers, shopping areas, train stations, and university campuses. These places are highly visible, however, the low parking time and high rotation rates often observed there could deliver an inadequate solution for the daily charging needs of the users. The higher refueling times and lower autonomy ranges of BEV makes previous refueling station location models not suitable for these conditions.

This chapter presents a new approach to determine the most convenient locations of public charging stations in urban areas and therefore, to optimize its distribution. This approach describes the particular characteristics of the BEV recharging process and valuates the expected effects of cover a determined location with a charging station. Indexes as the number of residents, workers or daily shoppers are used to measure the importance of each location. These indexes are also complemented by the time spent on each activity, as also by the level of the target market.

This approach relies on three different parts: an analysis of the specific BEV charging characteristics obtained from a literature and market overview, which allows evaluating the daily charging needs; the application of a location-optimization model which maximizes the population access to these stations and in consequence the potential use of them; and an estimation of the target market, by the adoption of an index related to socio-economic characteristics of the user. After these parts, a maximal gradual covering location model is used to maximize the benefits of the public charging network and decide the optimal location of a fixed amount of stations. This model maximizes simultaneously the benefits for the users, by enlarging the opportunities to have access to charging stations, and also the benefits for the operator, by increasing the potential use of their stations.

Covering models such as the one underlying the proposed approach were introduced in Church and Velle [8] and recently surveyed in Snyder [25]. Models of this type have been recently applied to alternative fuel charging station location problems by e.g. Wang and Wang [28] and Frade et al. [13].

2 Charging Technologies

The main challenge of electric vehicles lies on the energy storage system. The research goal is to find the best balance between storage capacity (energy density), performance (charge time, power and life-span), and cost. There are commonly four types of energy storage systems considered for electrical vehicles: electrochemical batteries, ultra-capacitors, fuel cells and flywheels. Detailed analysis can be found comparing the basic technical characteristics of different storage technologies in Burke [5] and Chan [6, 7], with more updated and more general comparison in Boulanger et al. [2], Eberle and von Helmolt [9] and Van Bree [26]. According with these studies, it is argued that electro-chemical battery electrical vehicles (BEV) are the most convenient mid-term technology to implement the electrical mobility. This is mainly because electro-chemical batteries have the most balanced characteristics, even considering the long charging times, which in the case of urban uses is feasible to be handled if the charging process is smartly managed. Instead, some fast charge alternatives as the ultra-capacitors and flywheels have a high cost and low energy density; fuel cells is a promising alternative with higher density, but is still a non-mature technology with very high costs and the requirement of a dedicated considerable infrastructure.

Considering the electro-chemical battery vehicles, there are mainly three types of commercially available recharging technologies: slow charge, fast charge and exchange of batteries. Some comparison between these charge methods can be found in Botsford and Szczepanek [1] or Boulanger et al. [2]. The slow charge is the preferred and recommended system by auto manufacturers. This type of charge presents a series of advantages over the rest: it is the best condition to maintain the battery life-span; requires a simple installation that could be made at home; it has better efficiencies; it does not have important impacts in the grid and could use electricity directly from it. The alternative of fast charge has negative effects on the battery life-span an on the electricity network. The exchange of batteries it is an interesting alternative to solve charging times, but require additional batteries (the most expensive component) and their standardization, which is not likely to be easily accepted by the manufacturing industry and the users. The main limitation of slow charge, it is precisely the considerable needed time to charge the batteries. An efficient public slow charging network can easily supply the energy needs of the commute and urban trips of BEV.

Most of the market available BEVs have autonomies from 100 to 200 km (Table 1) in ideal conditions. For common urban uses these autonomies are reduced to approximately 80 to 160 km. A full charge of these vehicles varies from 3.5 to 9 h of slow charge, depending on the size of the batteries (which is directly related to the autonomy range) and the capacity of the charger. In the case of a public charger this time should not exceed 8 h.

According to a study from Eurostat, most of the Europeans make on average 3 trips per day and travel between 30 and 40 km per day [12]. Passenger car transportation accounts on average for about a 70 % of the total passenger transport, and the average commute distance is from 6 to 8 km, which is less than a quarter than the total average daily distance. The average day distance includes the long distance trips and other modes different from car, suggesting that in the specific case of the use of cars in urban areas this distance will be significantly lower. Therefore, it can be assumed that an autonomy range of 40 km will cover the great majority of BEV urban trips and the energy required for this range can be usually supplied from 2 to 3 h. Hence, the average charging process of a BEV will require 3 h a day, and a maximum of no more than 8 h for a full charge.

Table 1 Characteristics of the typical market available BEV

Model	Release year	Autonomy (Km)	Battery (Kwh)	Time of charge (h at 240 V)
Mitsubishi i MiEV (Citroen C-Zerp/ Peugeot iOn)	2010	160	16	7.0
Nissan Leaf	2010	175	24	8.0
Renault Fluence ZE	2012	185	22	9.0
Renault Twizy	2012	100	7	3.5
Ford Focus Electric	2013	122	23	4.0
Smart ED	2013	110	17.6	6.0

3 Charging Process

Given the considerable required charging time, the common refuelling process taking place in some minutes at the middle of a trip is no longer feasible or convenient for the BEV technology. Instead, the recharge process can take place during the parking time. A private urban car is usually park the great majority of the day near the usual activities of the driver. These activity locations will represent then the potential locations for the charging process. The charging coverage of the users will be measured by the access to charging stations near these places and weighted by the usual time spent on each one.

3.1 Access Distance

The instant coverage concept measures whether a given location is within an acceptable walking distance of the nearest public charging station (through the shortest possible network path, as recommended in Gutiérrez and García-Palomares [15]). If this condition is achieved, then the location is considered to be instantly covered. To the best of our knowledge, there is no (published) work on acceptable walking distances to charging stations. However, several studies have addressed the same topic in relation to bus (transit) stops. This includes, for instance, Van Nes and Bovy [27] and Furth and Rahbee [14]. Distances between bus stops worldwide are of the order of 400 m, being usually a little larger in Europe than in North America. In line with the results of Furth and Rahbee [14], Murray [21] and Horner and Murray [17] using the same distance as the radius for the coverage area of a bus stop in their studies. However, as pointed out in Gutiérrez et al. [16] based on data from Madrid, Spain, binary coverage functions such as the ones underlying these studies do not capture the decay pattern observed empirically: coverage starts to decay significantly for walking distances above 200 or 300 m, is relatively small after 400 or 500 m, but is not null until distances are as large as 1,400 m (a detailed discussion of distance-decay functions is available in Martínez and Viegas [19]).

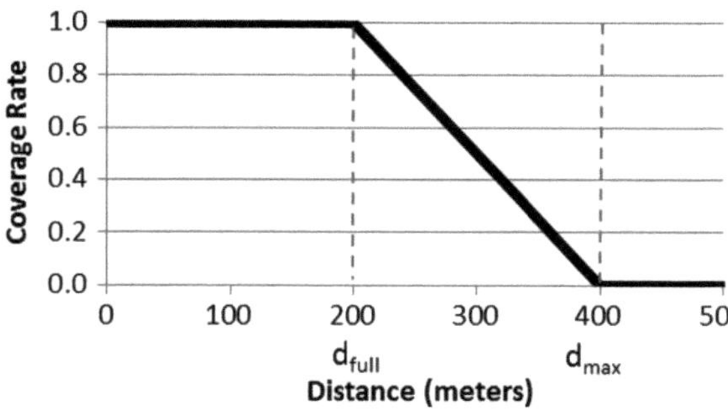

Fig. 1 Instant coverage function

In order to represent the observed empirical behaviour, a coverage function of three segments is considered to evaluate instant coverage: a first segment with a coverage rate of 1 (100 %) for distances below the (maximal) full coverage distance (d_{full}); a second segment with a linear coverage rate decay, from 1 to 0, between the full coverage distance and the maximal (partial) coverage distance (d_{max}); and finally, no coverage for distances higher than the maximal coverage distance. That is:

$$c_{ij} = \begin{cases} 1 & \Leftarrow d_{ij} \leq d_{full} \\ \frac{d_{max} - d_{ij}}{d_{max} - d_{full}} & \Leftarrow d_{full} \leq d_{ij} \leq d_{max} \\ 0 & \Leftarrow d_{ij} > d_{max} \end{cases}, \forall i \in \mathbf{N},\ j \in \mathbf{K} \tag{1}$$

where

N set of zones
K set of possible charging station locations
c_{ij} coverage rate for location i with a charging station at k
d_{ij} distance from location i to charging station k.

In accord with the empirical studies mentioned previously, plausible values for the full coverage distance and the maximum coverage distance are 200 and 400 m, respectively (Fig. 1).

The instant coverage of location i (v_i) is equal to the coverage rate obtained by the nearest charging station:

$$v_i = \max_{j \in \mathbf{K}} c_{ij}, \forall i \in \mathbf{N} \tag{2}$$

3.2 Parking Locations and Times

The average time distribution of population activities in Europe is summarized in Fig. 2 [11]. It can be observed there that, in general, home and work are the only places where BEV drivers can completely fulfill their average daily charging

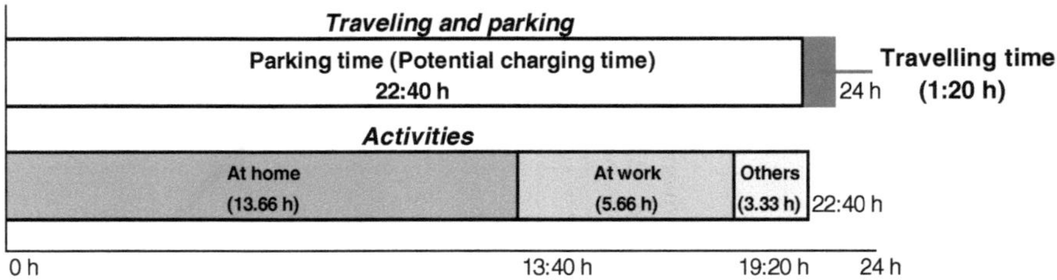

Fig. 2 Daily distribution of parking time in Europe

needs, as the time spent on average in other locations (mainly because of shopping and leisure activities) is much shorter. This observation is backed by the analysis of some enquiries, as for example those mentioned in Skippon and Garwood [24], where it was found that the preferred places to charge the electric vehicles are at home and at work.

Given this usual proportion of the daily activities it is possible to identify the potential charging locations and the average parking time spend in each one. The relation between the proportions of time represents the relative importance of having access on each location. For example, to have access to charging stations at home, represents 2.48 (13.66 h/5.66 h) more time access than in work.

3.3 Home Charging Access

The use of private chargers at home is a comfortable alternative to the use of public charging stations, and BEV users who can pay their cost (approximately 1000 euros) and, most importantly, have a private garage (with a proper electric connection), will most certainly install them. The latter requirement is commonly met in North America but not in Europe, where most people live in apartment buildings and private parking lots are the dominant home parking solutions. Such parking lots are not ideal to install private chargers, because this involves modifications on buildings that require a relatively large expenditure and may be difficult to approve by the building's community of property owners. For BEV users who can charge their vehicles at home, public charging stations normally will only provide coverage when they are working or doing other activities.

3.4 Charging Coverage

The charging coverage measures the sum of hours of access to charging station of the users in a determined location. For each location, three types of users will be considered: the residents; the workers; and the shoppers or leisure visitors. As it was stated on the previous subsection, the residents with access to a private garage will not be considered as users of charging stations (will use a private home charger), and therefore, just the residents without a private garage are considered. The covered time provided to each user corresponds with the usual average time spent on the activity. The expression for the charging coverage of the location i is as it follows:

$$w_i = (R_i \times f_R + E_i \times f_E + S_i \times f_s) \times v_i \tag{3}$$

where

R_{i} amount of residents (without access to a private garage) on the location i
E_{i} amount of workers (employments) on the location i
S_{i} amount of daily shoppers of leisure visitors on the location i
f_{R} average time spent at home
f_{E} average time spent at work
f_{S} average time spent in other activities (shopping and recreation)
v_{i} instant coverage at location i.

4 Target Market

The BEV technology is an attractive alternative to some small market segments with certain characteristics. These segments—the potential BEV buyers—are the target users for charging stations.

To address the target market, an index of the level of potential electric vehicle adoption is going to be created. This index can be assessed by the relation with socio-economic characteristics of the population that are commonly related to the adoption of electrical vehicles. This relation should be calibrated for the particular location of application, mainly by revealed or state preference studies.

Some of the characteristics that usually are related with the adoption of electrical vehicles or similar type of vehicles are education, income, car ownership and commuting distance. A group of studies in which these characteristics and other ones are evaluated are: Kurani et al. [18], McCarthy and Tay [20], Brownstone and Train [3], Brownstone et al. [4], Williams and Kurani [29], Potoglou and Kanaroglou [22], Sangkapichai and Saphores [23], Erdem et al. [10], Skippon and Garwood [24] and Zhang et al. [30]. Most of the studies found the same factors and also similar impacts on the willingness to adopt electrical vehicles or similar types of vehicles.

The final index should measure the relative difference between the segments of the population and is applied to different locations distinguished also by the type of activity of the users in that location.

5 Location Model

Once defined the charging coverage and the adoption potential the objective function of our location model is easily to be defined. The charging coverage weighted by the respective potential adoption index of the users of a determined activity in a determined location, would represent the formulation of the objective function. The maximization of this objective will look for the greatest amount of hours of access to charging station of the users, giving more focus to those who are more willing to adopt an electric vehicle. Most of the parts of the charging coverage are, in fact, weight factors of a determined location, as it is also the case of the adoption potential. The overall configuration of the objective function corresponds to maximal gradual covering model.

The model formulation is expressed as it follows:

$$\max Z = \sum_{i \in \mathbf{N}} \sum_{j \in \mathbf{K}} (a_i^R \times R_i \times f^R + a_i^E \times E_i \times f^E + a_i^S \times S_i \times f^S) \times c_{ij} \times x_{ij} \quad (4)$$

Subject to:

$$x_{ij} \leq y_j\,, \ \forall i \in \mathbf{N},\ j \in \mathbf{K} \quad (5)$$

$$\sum_{j \in \mathbf{N}} x_{ij} \leq 1\,, \ \forall i \in \mathbf{N} \quad (6)$$

$$\sum_{j \in \mathbf{K}} y_j = p \quad (7)$$

Decision variables:

y_j decision to locate a station at j (binary)
x_j decision to cover location i from station at j

where:

N set of zones
K set of possible charging station locations
c_{ij} coverage rate for location i with a charging station at j
a_i^R adoption potential of residents (without access to a private garage) of location i
a_i^E adoption potential of workers of location i
a_i^S adoption potential of shoppers and leisure activity visitors of location i
R_i amount of residents (without access to a private garage) on the location i
E_i amount of workers (employments) on the location i
S_i amount of daily shoppers of leisure visitors on the location i
f_R average time spent at home
f_E average time spent at work
f_S average time spent in other activities (shopping and recreation)
p amount of stations.

First, in expression (5), the decision to cover the location i from the location j, requires to have a station at j. Secondly, the expression (6), determines that only one station can provide service to each location. Finally, expression (7), limits the total amount of station to be equal to p.

6 Conclusions

This chapter presents a new approach to determine the optimal locations of public charging-stations in urban areas. A common maximal gradual covering model is used to find the best location for a limited amount of stations, providing the best solution for the users (increasing the user access to stations) and for the operator (increasing the uses of the stations).

The new approach relies in the particular definition of two components of the gradual maximal covering model in accordance with the BEV charging requirements, the access to charging stations and the weight of each location. In regards of the access a simplified two step function is used based on empirical information. In regards of the weight, a composition between the level of activities of the location; the usual time spend on each activity, and the particular characteristics of the users in relation with the target market characteristics are used.

The data requirements of the model are the amount of residents without access to a private garage, amount of employments, amount of daily shoppers or leisure activity visitors and some socio-economic characteristics of those users. The computational efforts to solve the model can take from one second for areas of 50 sectors to few minutes in areas of 1000 sectors in optimization software like FICO Xpress (FICO 2012). The new formulation provides a better understanding of the BEV charging problem, identifying the best locations based on previous market research.

A further work to improve the model is to introduce the capacity issues of the stations and also expand the coverage analysis from a general point of view to the specific distribution between the users. The introduction of uncertainties on the characteristics of the market is also under consideration. Finally, hypothetical and real case studies are also under development.

Acknowledgments This chapter has been supported by the Energy and Mobility for Sustainable Regions—EMSURE—Project (CENTRO-07-0224-FEDER-002004).

References

1. Botsford, C., Szczepanek, A.: Fast charging vs. slow charging: pros and cons for the new age of electric vehicles. In: EVS24 - International Battery, Hybrid and Fuel Cell Electric Vehicle Symposium, Stavanger, Norway, pp. 1–9, 13–16 May 2009
2. Boulanger, a G., Chu, a C., Maxx, S., Waltz, D.L.: Vehicle electrification: status and issues. Proc. IEEE **99**(6), 1116–1138 (2011)

3. Brownstone, D., Bunch, D.S., Train, K.: Joint mixed logit models of stated and revealed preferences for alternative-fuel vehicles. Transp. Res. Part B: Methodol. **34**(5), 315–338 (2000)
4. Brownstone, D., Train, K.: Forecasting new product penetration with flexible substitution patterns. J. Econom. **89**(1–2), 109–129 (1998)
5. Burke, A.F.: Batteries and ultracapacitors for electric, hybrid, and fuel cell vehicles. Proc. IEEE **95**(4), 806–820 (2007)
6. Chan, C.C.: The state of the art of electric and hybrid vehicles. Proc. IEEE **90**(2), 247–275 (2002)
7. Chan, C.C.: The state of the art of electric, hybrid, and fuel cell vehicles. Proc. IEEE **95**(4), 704–718 (2007)
8. Church, R., Velle, C.R.: The maximal covering location problem. Pap. Reg. Sci. **32**(1), 101–118 (1974)
9. Eberle, D.U., von Helmolt, D.R.: Sustainable transportation based on electric vehicle concepts: a brief overview. Energy Environ. Sci. **3**(6), 689–699 (2010)
10. Erdem, C., Şentürk, İ., Şimşek, T.: Identifying the factors affecting the willingness to pay for fuel-efficient vehicles in Turkey: a case of hybrids. Energy Policy **38**(6), 3038–3043 (2010)
11. Eurostat: European time use survey: how is the time of Europeans distributed? http://epp.eurostat.ec.europa.eu/cache/ITY_PUBLIC/3-27072004-AP/EN/3-27072004-AP-EN.HTML (2004)
12. Eurostat: Passenger mobility in Europe. Office for official publications of the European communities, Eurostat, European Commission. Luxembourg (2007)
13. Frade, I., Ribeiro, A., Gonçalves, G., Pais Antunes, A.: Optimal location of charging stations for electric vehicles in a neighborhood in Lisbon, Portugal. Transp. Res. Rec. J. Transp. Res. Board. **2252**(1), 91–98 (2011)
14. Furth, P.G., Rahbee, A.B.: Optimal bus stop spacing through dynamic programming and geographic modeling. Transp. Res. Rec.: J. Transp. Res. Board **1731**(1), 15–22 (2000)
15. Gutiérrez, J., García-Palomares, J.C.: Distance-measure impacts on the calculation of transport service areas using GIS. Environ. Plann. B: Plann. Des. **35**(3), 480–503 (2008)
16. Gutiérrez, J., Cardozo, O.D., García-Palomares, J.C.: Transit ridership forecasting at station level: an approach based on distance-decay weighted regression. J. Transp. Geogr. **19**(6), 1081–1092 (2011)
17. Horner, M.W., Murray, A.T.: Spatial representation and scale impacts in transit service assessment. Environ. Plann. B: Plann. Des. **31**(5), 785–797 (2004)
18. Kurani, S., Turrentine, T., Sperling, D.: Testing electric vehicle demand in "Hybrid Households" using a reflexive survey. Transp. Res. Part D: Transp. Environ. **1**(2), 131–150 (1996)
19. Martínez, L.M., Viegas, J.M.: A new approach to modelling distance-decay functions for accessibility assessment in transport studies. J. Transp. Geogr. **26**, 87–96 (2013)
20. McCarthy, P.S., Tay, R.S.: New vehicle consumption and fuel efficiency: a nested logit approach. Transp. Res. Part E: Logistics Transp. Rev. **34**(1), 39–51 (1998)
21. Murray, A.T.: Strategic analysis of public transport coverage. Socio-Econ. Plann. Sci. **35**(3), 175–188 (2001)
22. Potoglou, D., Kanaroglou, P.S.: Household demand and willingness to pay for clean vehicles. Transp. Res. Part D: Transp. Environ. **12**(4), 264–274 (2007)
23. Sangkapichai, M., Saphores, J.-D.: Why are Californians interested in hybrid cars? J. Environ. Plann. Manag. **52**(1), 79–96 (2009)
24. Skippon, S., Garwood, M.: Responses to battery electric vehicles: UK consumer attitudes and attributions of symbolic meaning following direct experience to reduce psychological distance. Transp. Res. Part D: Transp. Environ. **16**(7), 525–531 (2011)
25. Snyder, L.V.: Covering problems. In: Eiselt, H.A., Marianov, V. (eds.) Foundations of Location Analysis, pp. 109–135. Springer, London (2011)

26. van Bree, B., Verbong, G.P.J., Kramer, G.J.: A multi-level perspective on the introduction of hydrogen and battery-electric vehicles. Technol. Forecast. Soc. Chang. **77**(4), 529–540 (2010)
27. van Nes, R., Bovy, P.H.L.: Importance of objectives in urban transit-network design. Transp. Res. Rec.: J. Transp. Res. Board **1735**(1), 25–34 (2000)
28. Wang, Y.-W., Wang, C.-R.: Locating passenger vehicle refueling stations. Transp. Res. Part E: Logistics Transp. Rev. **46**(5), 791–801 (2010)
29. Williams, B.D., Kurani, K.S.: Estimating the early household market for light-duty hydrogen-fuel-cell vehicles and other "Mobile Energy" innovations in California: a constraints analysis. J. Power Sources **160**(1), 446–453 (2006)
30. Zhang, Y., Yu, Y., Zou, B.: Analyzing public awareness and acceptance of alternative fuel vehicles in China: the case of EV. Energy Policy **39**(11), 7015–7024 (2011)

A Joint Probability Density Function for Reducing the Uncertainty of Marginal Social Cost of Carbon Evaluation in Transport Planning

Silvio Nocera and Stefania Tonin

Abstract This chapter aims at defining a fair value for the Marginal Social Cost of Carbon (MSCC) to be used within transport planning, briefly discussing how it is influenced by economic and scientific uncertainty, with the scope of helping researchers, stakeholders and decision makers to choose among the current range of values of four orders of magnitude provided from the scientific literature. The method here proposed estimates a joint probability density function for MSCC using a database of almost 600 available estimates, and then defines a subsample of 80 to be used for the evaluation of transport planning policies and projects, so that the variability of MSCC decreases significantly to a single order of magnitude.

Keywords Transport policy · Carbon dioxide estimation · Marginal social cost of carbon (MSCC)

1 Introduction

Providing a reliable value for carbon dioxide (CO_2) emissions is a fundamental task if our society wants to develop a sustainable transport industry and protect the environment from the consequences of global warming. Technological advances in recent years have been considerable and play a noticeable role towards this goal. Nonetheless, transport is the end-use sector that has seen by far the most rapid increase in emissions over the last 20 years: some of the issues regarding freight mobility seem still open [1], and CO_2 emissions in transport increased by 2.2 Gt from 1991 to 2011, or by almost 50 % [2]. Some previous researches [3–6] seem to identify an issue in the quality perceived of specific modes, it seems

S. Nocera (✉) · S. Tonin
IUAV University of Venice, Santa Croce 191 Tolentini, 30135 Venice, Italy
e-mail: nocera@iuav.it

J. F. de Sousa and R. Rossi (eds.), *Computer-based Modelling and Optimization in Transportation*, Advances in Intelligent Systems and Computing 262,
DOI: 10.1007/978-3-319-04630-3_9,

however that this complicated challenge cannot be tackled by mere technology but should be sustained from adequate policy interventions instead [7]. Reducing emissions in transport thus forms a crucial element for any comprehensive strategy to reduce global CO_2 emissions.

Setting a fair value for the cost of climatic change is not a mere economic valuation exercise, but a more complex procedure that attempts to quantify the negative externalities resulting from a rise in temperature levels. Due to the cumulative nature of such consequences, greater and greater impacts are expected from the progressive increase in carbon concentrations [8]. The real challenge seems to predict with a certain accuracy how CO_2 will affect future climatic change and impact the physical environment. Even if theoretically rigorous, this logic suffers from a number of economic and scientific uncertainties that are endemic to all these estimates [9]. The process is based on five main steps: firstly, the estimation of future CO_2 emission levels; secondly, the determination of a link between emissions and atmospheric concentration; thirdly, the assessment of CO_2 consequences on climatic change; fourthly, the measurement of the physical impacts of climate change; and finally their economic appraisal.

Methods for calculating CO_2 production [10] and specific integrated assessment models [11] have been developed recently to determine the externalities of endogenous greenhouse warming and to monetize them. Scientific literature also describes other methods to quantify the CO_2 economic cost such as the Avoidance Cost [12] or market-based prices, i.e., carbon trading cost [13, 14] and carbon tax [15].

However, the current relative abundance of estimates, somehow very divergent, in turn increases the uncertainties for researchers. This chapter aims particularly at developing a method for reducing such uncertainty, with the aim of helping transport researchers and decision makers to orient themselves among the vast range of values provided by the scientific literature. The array of such values derives from the selection of the main integrated assessment models (IAMs) of climate change through a damage function that reflects the interaction between climate variation and the impacts in the economy and society.

This chapter will focus on the assessment of an efficient economic value for the so-called "Marginal Social Cost of Carbon" (MSCC)[1]—i.e. the net present value of the incremental damage due to a small increase in CO_2 emissions [17]. Technically, MSCC represents the marginal cost of global damage from climate change or, conversely, the value of damages avoided for a small emission reduction (i.e. the benefit of a CO_2 reduction).

A fair estimate of MSCC can be used by policy-makers to infer about the carbon efficiency of a certain transport action (such as new light-duty vehicles with low CO_2 emission, better transport infrastructure plans, new policies or regulations expected to cut CO_2 emissions, etc.). It can either be adopted in terms of

[1] MSCC is also defined as "Marginal Climate Change Damage Cost" or "Social cost of Carbon" [16].

Cost-Benefit Analysis (CBA) to value the benefits of climate policy or to set a Pigouvian tax imposed by a benevolent social planner, or in a Multi Criteria Evaluation (MCE), where an efficiently valued MSCC can stand for the impact of greenhouse gases of a certain action. The possibility of using both these methods jointly, as it often happens, also holds.

For this reason, quantifying carbon impacts fairly, and reducing the variation range for MSCC have significant practical importance in transport planning, as policy makers may—by purpose or unconsciously—tend to choose the pathway that best addresses their objectives. This approach basically overcomes the risk profile of the outcomes on the way to maximize the chances of hitting a certain target, potentially disregarding some assumptions and input uncertainties. This is especially dangerous when decisions and setting standards that shape the future over several decades are at stake, because of the notable uncertainties in technology and economic fluctuations in the mid to long term. The prospect to reduce the uncertainty of MSCC, and consequently to set a correct market price signal, allows implementing the most efficient and lowest costs action (be it a project, a plan, or an investment), for reducing CO_2 impacts in transport sector. Greater certainty of the value of MSCC in the long term policy will also encourage long-run technological innovation and greater investment in more energy-efficient and reduced CO_2 emission, and capital equipment in the transport sector [18].

A method for estimating a probability density function will be therefore suggested in this chapter, based on a database made available by Tol[2] [17], which collected almost 600 different estimates of MSCC. Each of them is the result of some structural characteristics and specific assumptions underlying the IAMs, which provide estimates of the monetary impacts of CO_2 emissions. It also includes different ways to convert the total economic cost of climate change into marginal costs of current emissions, such as the different discount rates, the regional economic inequalities, the growth rate of per capita consumption, etc. The analysis performed in this chapter will be based on a subsample of 80 MSCC values related to emission scenarios equal or later than 2015, so as to capture the more recent studies, and consequently the new technological innovation in climate change modeling.

This chapter first estimates the main determinants of the MSCC through a regression model. Then, it infers a probability density function with the aim of highlighting some of the assumptions used in the carbon assessment, focusing on the uncertainty of the different estimates currently available. The aim is to provide a realistic estimation for one of the most worrisome environmental impacts of transport.

The structure of the chapter is divided into five sections. This current section aimed at introducing the theme and provided a motivation for conducting such kind of analysis. Section 2 deals with the economic impacts of climate change and

[2] The database on the marginal damage costs of CO_2 emissions can be found at: http://www.sussex.ac.uk/Users/rt220/marginaldamagecost.xlsx.

the related uncertainty. Section 3 proposes a regression model to determine the main factors that produce such a high uncertainty in the MSCC valuation, and Sect. 4 tests the values by adopting a two-parameter Gumbel distribution. Finally, Sect. 5 discusses the results and concludes the chapter.

2 Economic Impacts of Climate Change

According to Tol [17], there are four different ways to estimate economic impacts of climate change. It is for instance possible to interview a limited number of experts [19]. It is also possible to multiply estimates of the "physical effects" of climate change with estimates of their price [20–27]. Alternatively, Bosello et al. [28] use similar estimates of the physical impacts but compute the general equilibrium effects on welfare. Finally, other methods may consist in using observed variations (across space) in prices and expenditures to discern the effect of climate [29–32], or in drafting self-reported well-being [33, 34].

These methods link the variation of GDP to the increase of temperatures (Table 1): the forecast impacts on the national economies are very different, varying from a considerable decrease (−11.5 %: [33]) to a slight increase of the GDP (+2.3 %: [26]).

A shared agreement between these studies is that the uncertainty is vast and right-skewed. This means that undesirable surprises are more likely than desirable ones. For instance, it is hard to make an upper bound limit to the value of the climate sensitivity. Estimates stop conventionally at 3 °C of global warming, but climate change may well go beyond that. The uncertainties about the impacts are compounded by extrapolation [35]. Moreover, impacts change drastically in the various countries, as the poorest tend to be more vulnerable to climate change. They have a large share of their economic activity in sectors such as agriculture, that are directly exposed to the weather. Furthermore, they also tend to be worse at adaptation, lacking resources and capacity [36].

3 A Regression Model for the Estimates of MSCC

As previously shown (Table 1), a significant amount of studies has calculated the economic damage of climate change in terms of total cost of carbon for a benchmark scenario. This value allows to determine also the MSCC, which is defined as the additional damage caused by an additional tonne of carbon emissions. The process goes as follows: firstly, the total cost of carbon is measured in terms of loss of world GDP with respect to a doubling of atmospheric CO_2 concentrations from pre-industrial levels. These values are generally obtained running a main impact assessment model (IAM) such as PAGE [11]. Secondly, once estimated the total cost of climate change, varying different assumptions to take

Table 1 Estimates of the welfare loss due to climate change (as equivalent income loss in percent)

Study	Warming (Celsius degrees)	Impact (% GDP)
[19]	3.0	−4.8 (−30.0 to 0.0)
[23]	3.0	−1.3
[21]	2.5	−1.4
[25]	2.5	−1.9
[49] [a]	2.5	−1.7
[50] [a]	2.5	−2.5 (−0.5 to −11.4)
[30] [a,b,c]	2.5	0.0b 0.1b
[51]	2.5	−1.5
[26]	1.0	2.3 (1.0)
[29] [a,d]	2.5	−0.1
[34] [a,c]	1.0	−0.4
[11] [a,e]	2.5	0.9 (−0.2 to 2.7)
[32]	2.5	−0.9 (0.1)
[24]	3.0	−2.5
[33] [a]	3.2	−11.5
[28]	1.9	−0.5

[a] The global results were aggregated by the author
[b] The top estimate is for the "experimental" model, the bottom estimate for the "cross-sectional" model
[c] Mendelsohn et al. [30] only include market impacts
[d] Maddison [29] only considers non-market impacts on households
[e] The numbers used by Hope are averages of previous estimates by Fankhauser [21] and Tol [26]; Stern et al. [44] adopted the work of Hope
Estimates of the uncertainty are given in bracket as standard deviations or 95 % confidence intervals
Source [17]

into account several physical climate and key economic variables, it is possible to estimate the social cost of carbon (that is, the additional damage caused by the emissions of an additional tonne of carbon)[3]. Recently, Tol [17] made available a dataset of 588 values coming from 75 different studies of the social cost of carbon[4]. The database gives information about the authors' names, the publication year, the currency used, the years of emission, and a set of other variables created by the author for taking other effects into account. It includes for instance, the growth rate of the social cost of carbon, the welfare loss of the impacts of climate change expressed as equivalent income losses, the pure rate of time preference used in the different studies, if the equity-weighting was applied, and many more (for further information see [17]).

[3] Scientific literature reasons substances emitted in tonnes of carbon (tC) or tonnes of carbon-dioxide (tCO_2) generally, the equivalence relation 1tC=3.664 tCO_2 [37] holds.

[4] Only some of the total cost estimates have been used for estimating the marginal social cost of carbon (such as [29], [30], [32], [34]).

The range of those estimates, expressed in 2010 U.S. dollars and related to emissions in the year 2010 goes from \$−10.4/tC to \$7,243.7/tC. The negative values mean that climate change can initially have positive impacts. Tol asserts that this is partially explained by the fact that "the global economy is concentrated in the temperate zone, where a bit of warming may well be welcomed because of reductions in heating costs and cold-related health problems" [38], or even because "the higher ambient concentration of carbon dioxide would reduce water stress in plants and may make them grow faster" [39].

It is also well-known that the MSCC is influenced by different variables, such as the Pure Rate of Time Preference (PRTP)[5], the growth rate of per capita consumption, the total welfare impact of climate change, the elasticity of marginal utility of consumption, the projections of CO_2 emissions, the carbon cycle, the rate of warming, the economic scenarios, and some others [17, 40]. Any of these factors is a potential source of uncertainty.

Tol's database is used in this chapter to estimate the main determinants of the value of the MSCC through a robust Ordinary Least Square (OLS) regression analysis:

$$\ln\text{MSCC} = \beta \text{X} + \varepsilon \tag{1}$$

where X is a vector of independent variables, and ε is the error term.

Through the previous regression analysis, we establish a relation between the dependent variable MSCC and the set of independent variables X, examining which of the latter has more influence on the value of the former. We chose to express the equation in the form of a natural log to take the right skewedness of the estimates of MSCC into account. The log transformation also allows to interpret the coefficients of the regression model easily, as one percentage change in the independent value leads to a $\beta(100)$ % change in the dependent one.

Table 2 refers to the whole sample of 588 cases available, analyzing only those which present every information needed. It summarizes the main variables selected for estimating the OLS model, the estimates of their coefficients, and the value of t-student. It shows that all the explanatory variables are significant at the 5 % level, except the coefficient of the dummy variable accounting for the possibility of the study to adopt an independent impact assessment for the estimation of the total costs of climate change.

The model indicates a moderate, statistically significant association with the R-squared value of 47 %, which is comparable with many meta-analysis studies in the literature [41]. Results of the OLS regression show that higher PRTP implies lower value of MSCC, all else the same. For instance, the increase of one point in the average discount rate would result in a decrease of 63.7 % change in the average value of MSCC. This result confirms that the choice of the discount rate is central to any assessment of climate change policy, especially those that will have

[5] The PRTP is the rate at which time is discounted, it is also defined as the "Rate of Impatience": people would prefer consume now than in the future [42].

Table 2 Specification and definition of the variables considered and results of the regression model

Dependent variable	Definition		
lnMSCC	Natural logarithm of the Marginal Social Cost of Carbon in $2010 and pertaining to emission 2010		
Intercept and independent variables	Definition	Coefficient estimates	t-student
Intercept		76.11**	3.16
Emission scenarios	Time horizon of emissions	−0.013**	2.65
Publ year	Year of study publication	−0.022**	2.33
PRTP	Pure Rate of Time Preference chosen in each study	−0.637**	12.92
Ew	Dummy variable equal to 1 if equity weighting is adopted, 0 otherwise	0.480**	3.50
Peer-reviewed study	Dummy variable equal to 1 if the study is peer-reviewed, 0 otherwise	−0.654**	4.65
Impact independently estimated	Dummy variable equal to 1 if the study is based on an independent impact assessment and does not borrow total cost estimates of the total costs of climate change, 0 otherwise	0.174	1.02
Method for marginal impacts	Dummy variable equal to 1 if the study uses the incremental or marginal calculus to estimate MSCC, 0 otherwise	0.721**	2.29
Dynamic impact model	Dummy variable equal to 1 if the study is based on dynamic climate change scenarios, 0 otherwise	−0.813**	5.48

N. obs: 402 $F(8,393) = 44.00$** R-squared = 0.47

Note **significant at the 5 % level

the major effects on a distant future such as generally those related to transport sector. Given a certain fixed amount of costs, this implies that higher discount rates tend to reduce the present value of benefits, hence weakening the implementation of current strong action. Similarly, more recent publications and a longer time horizon for emissions also reduce the values of MSCC variable. The dummy variables chosen to describe the quality of the different studies considered in this chapter so far indicate that peer-reviewed studies and the adoption of integrated assessment models that used a dynamic approach decrease the MSCC [43]. On the contrary, studies that compute the marginal damage costs from total damages increases the value of the MSCC. The value of MSCC is also predicted to increase when the practice to weighting impacts in different regions is applied.

As transport policy options range here from comprehensive legislation to targeted regulations to reduce carbon emissions and improve the efficiency of vehicles, a false estimation of the economic damage—in any direction—leads to an error in the determination of the priority of the actions to undertake and to a consequent penalization for the community. This is the reason for which we tried to specialize the argument developed so far to transport policy, by choosing in the next section the estimates which could be influent to future transport actions.

4 A Joint Probability Density Function for the MSCC Estimates

In this part of this chapter, we calibrate an appropriate probability density function with the aim to reducing part of the uncertainty of the MSCC estimates currently available in literature and to narrow down the results for future strategies of transport policy. Focusing primarily on feasibility analysis, we restrict our attention on those studies in which the emission scenarios are referred to the year 2015 or later (henceforth, emission scenarios $\geq$2015). The aim is to estimate CO_2 impacts of future transport actions and strategies. The different emissions projections used in the integrated assessment models are based on specific socio-economic assumptions in terms of GDP, population growth, and technological change that might modify the emissions pathway and the resulting damage. The subsample considered accounts for 80 observations, and it is statistically described in Table 3.

Table 3 shows that for the overall sample the mean estimate of the MSCC is \$173.32/tC. This high value is determined by some extreme values present in the sample, as further indicated by the other statistics reported in the table (median, standard deviation, mode, and 95th percentile). If the analysis is restricted to the emission scenarios of year 2015 or later, the mean estimate of the MSCC drops to \$109.91/tC, while the mode unexpectedly increases to \$25.57/tC.

To investigate the high variability a little further, the different PRTPs have been carefully considered, since the choice of the appropriate discount rate for climate change policy is considerably debated in economic literatures [44, 45]. We have

Table 3 Main descriptive statistics of the MSCC for the whole sample and the subsample of values pertaining to emissions after 2015

	Whole sample	MSCC ≥ 2015 ($/tC)	MSCC ≥ 2015, PRTP 3 % ($/tC)	MSCC ≥ 2015, PRTP 1 % ($/tC)
Mean	173.32	109.91	27.50	73.55
Median	37.56	32.74	25.93	38.56
Std dev	494.69	345.58	24.62	88.78
Mode	10.51	25.58	25.57	37.93
95 %	711.00	305.10	88.68	241.82
N. obs.	588	80	38	28

Source Tol database [17], elaborated

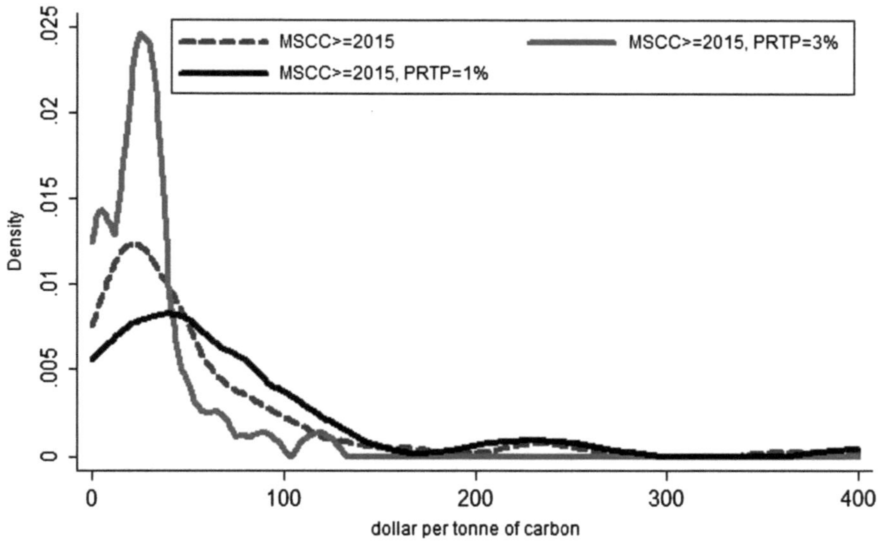

Fig. 1 Kernel density of different distributions with target emission after 2015

already highlighted in the previous section that a higher discount rate implies that the costs incurred in the future for climate change adaptation and mitigation policies will have lower present values.

The estimated mean for the subsample related to a PRTP equal to 3 % is lower than the one of the subsample where the PRTP is equal to 1 % (Fig. 1), confirming what should have been expected, but also partially explaining the variation in the estimates. Moreover, the distribution of the different estimates of MSCC is right skewed and Fig. 1 shows the kernel density functions of the MSCCs for all observations of the sample considered in this study, for those equal to a 3 and a 1 % PRTP.

To account for the skewedness of the distribution, and to illustrate the right tail distribution of the estimates related to the MSCC, a two-parameter Gumbel

Table 4 Results of the two-parameter Gumbel distribution used

SCC	Coefficient	z-value
β	80.89 (8.42)	9.53 (<0.000)
μ	40.62 (9.12)	4.45 (<0.000)
Log Likelihood	−499.98	

Note standard errors in parentheses

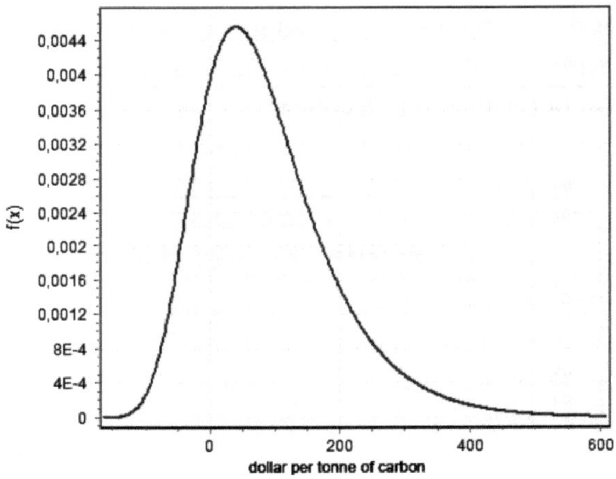

Fig. 2 Probability density function of sample data MSCC ≥ 2015

Table 5 Sample statistics and characteristics of the Gumbel distribution

	MSCC ≥ 2015 ($/tC)	MSCC ≥ 2015, PRTP 3 % ($/tC)	MSCC ≥ 2015, PRTP 1 % ($/tC)
Mean	87.31	26.99	67.87
Median	70.27	23.46	58.34
Std. dev.	102.60	24.61	57.95
Mode	40.62	17.31	41,79
95 %	280.59	67.18	174.66
N	80	38	28

distribution is applied to build up an overall distribution of the estimates and their uncertainties [46]. Using Stata®, the scale and location parameters were inferred by maximum likelihood methods.

It is well-known that the Gumbel two-parameter (μ, β) density belongs to the Generalized Extreme Value distributions [47]. Its probability density function is the following:

$$f(x, \mu, \beta) = \frac{z}{\beta} \cdot e^{-z} \tag{2}$$

where μ is the location parameter of the distribution and β is the scale parameter, and

$$z = e^{-\left(\frac{x-\mu}{\beta}\right)} \text{ and } \beta > 0 \tag{3}$$

As interesting properties of the Gumbel distribution, the mode is equal to the location parameter μ, the mean (μ_1) is equal to:

$$\mu_1 = \mu + \beta \cdot \gamma \tag{4}$$

where γ is the Euler's constant and it is equal to 0.5772.

Finally the median (μ_2) can be expressed as:

$$\mu_2 = \mu - \beta \cdot \ln(\ln 2). \tag{5}$$

5 Results and Conclusions

The subsample of 80 estimates of the MSCC, as previously defined, has been used to fit the two-parameter Gumbel distribution, and Table 4 shows the main results. The best-fit Gumbel distribution has a scale parameter of 80.89 and location parameter of 40.62, and the two parameters are significantly different from zero (as shown from the z-values).

Figure 2 displays the probability density function with the parameters of Table 4, when only the value of MSCC related to emissions after 2015 is considered.

Table 5 reports the sample statistics and characteristics of the Gumbel distribution fitted to the observations when emissions are later than 2015, and two alternative ways to split the sample, when PRTP is equal to 3 and 1 %.

The mean of the probability density function for the restricted sample is $87.31/tC, but the mode is only $40.62/tC. This large difference seems to suggest that the mean estimate is still driven by some large estimates of MSCC, as also shown from the estimated MSCC at the 95th percentile ($280.59/tC). Once again the higher rate of time preference implies that the costs of climate change incurred in the future will have a lower present value. The mean estimate of the MSCC for the studies with a 3 % rate is in fact $26.99/tC, while it is $67.87/tC for studies that choose a 1 % PRTP.

Furthermore, the results of the regression analysis show that it is possible to distinguish among the different MSCC estimates: the most recent studies reflect some innovation on climate change research, at the same time yielding lower estimates with smaller uncertainties than the first pioneering studies did. Finally, it must be noted that the choice of discount rate and the aggregation over countries (equity weighting) may considerably change the final estimate, and assume hence important policy implications. Even if we managed to reduce it consistently through the regression proposed in Sect. 3, our results show that the uncertainty surrounding the estimates of MSCC is still high, and that a deeper investigation is necessary to reduce it any further. Additional efforts should be devoted to disentangle the quantitative effects of the different explanatory variables on the dependent variable such as the climate policy regime adopted to meet the targets (international or national trading of emission permits, carbon tax, etc.), geographic information, the sectors considered in the models, and others. This was not yet available in the current database.

As transport emissions are a significant contributor to climate change, their economical estimation is extremely important and should not be left out from feasibility studies. Particularly, MSCC may be a key input in Cost Benefit Analysis and in other valuation approaches that can help in the decision making process: hence, reducing the uncertainty surrounding its calculation is vital to obtain sound results. It must be considered that climate changing affects transport planning in at least two ways: firstly, the increasing environmental threats will reduce the resilience and performance of surface transport systems, especially in some countries, forcing the transport system to adapt to these changes and carrying to potential waste of economic resources. Secondly, since transport sector is one of the largest emitters of carbon dioxide, better strategies to mitigate the impact of transportation emissions are essential. In both cases, climate change considerations have to be integrated into the transport planning process already in its first steps.

Through a top-down approach, the evaluation of carbon emissions through the method described in this chapter allows transport policy analysts to identify some possible risks, and in case of necessity to carry out a number of steps for reducing uncertainty in carbon emission evaluation, thus efficiently monitoring their effects. At the same time, a bottom-up approach should allow a concrete estimation of the impacts on some parts of the complex transport system that cause CO_2 emissions. Provided that travel demand, modal share and elasticity could be estimated with some reliability [48] this would allow an efficient planning of technical and policy measures for hitting transport emissions and consequences, including an estimation of specific investments in new systems (for instance, driverless subways) or technologies (e.g., ITS, hybrid or electric vehicles).

References

1. Cappelli, A., Nocera, S.: Freight modal split models: data base, calibration problem and urban application. In: Brebbia, C.A., Dolezel, V. (eds.): Urban Transport XII—Urban Transport and the Environment in 21st Century. The Wit Press, Southampton. ISBN: 1-84564-179-5, ISSN: 1746-4498 (2006)
2. International Energy Agency [IEA]: Redrawing the energy-climate map. http://www.worldenergyoutlook.org/energyclimatemap/ (2012)
3. De Oña, J., De Oña, R., Eboli, L., Mazzulla, G.: Perceived service quality in bus transit service: a structural equation approach. Transp. Policy **29**, 219–226 (2013)

4. Nocera, S.: Un approccio operativo per la valutazione della qualità nei servizi di trasporto pubblico/An operational approach for quality evaluation in public transport services. Ingegneria Ferroviaria **65**(4), 363–383 (2010)
5. Nocera, S.: The key role of quality assessment in public transport policy. Traffic Eng. Control **52**(9), 394–398 (2011)
6. Eboli, L., Mazzulla, G.: L'influenza dei fattori di qualità del servizio nelle preferenze d'uso tra auto e bus/The influence of service quality factors in the preferences concerning the use of car and bus. Scienze Regionali **11**(3), 75–92 (2012)
7. Nocera, S., Cavallaro, F.: Policy effectiveness for containing CO_2 emissions in transportation. Procedia Soc. Behav. Sci. **20**, 703–713 (2011)
8. Nocera, S., Cavallaro, F.: Economical evaluation of future carbon impacts on the Italian highways. Procedia Soc. Behav. Sci. **54**, 1360–1369 (2012)
9. Clarkson, R., Deyes, K.: Estimating the Social Cost of Carbon Emissions. Department of Environment, Food and Rural Affairs, London (2002)
10. Nocera, S., Maino, F., Cavallaro, F.: A heuristic method for evaluating CO_2 efficiency in transport planning. Eur. Transp. Res. Rev. **4**, 91–106 (2012)
11. Hope, C.W.: The marginal impact of CO_2 from PAGE2002: an integrated assessment model incorporating the IPCC's five reasons for concern. Integr. Assess. J. **6**(1), 19–56 (2006)
12. Kuik, O., Brander, L., Tol, R.S.J.: Marginal abatement costs of carbon-dioxide emissions: a meta-analysis. Energy Policy **37**(4), 1395–1403 (2008)
13. European Union (EU): Analysis of options to move beyond 20% greenhouse gas emission reductions and assessing the risk of carbon leakage. http://eur-lex.europa.eu/LexUriServ/LexUriServ.do?uri=COM:2010:0265:FIN:EN:PDF (2012a). Accessed 14 Feb 2012
14. European Union (EU): EU action against climate change. http://ec.europa.eu/climateaction/eu_action/index_en.htm (2012b). Accessed 14 Feb 2012
15. Santos, G., Behrendt, H., Maconi, L., Shirvani, T., Teytelboym, A.: Externalities and economic policies in road transport. Res. Transp. Econ. **28**, 2–45 (2010)
16. Kuik, O.J., Buchner, B.K., Catenacci, M., Goria, A., Karakaya, E., Tol, R.S.J.: Methodological aspects of recent climate change damage cost studies. Integr. Assess. J. **8**(1), 19–40 (2008)
17. Tol, R.S.J.: Targets for global climate policy: an overview. J. Econ. Dyn. Control **37**(5), 911–928 (2013)
18. US Department of Transportation: Transportation's Role in Reducing U.S. Greenhouse Gas Emissions: Volume 1. http://ntl.bts.gov/lib/32000/32700/32779/DOT_Climate_Change_Report_-_April_2010_-_Volume_1_and_2.pdf (2010)
19. Nordhaus, W.D.: Expert opinion on climate change. Am. Sci. **82**(1), 45–51 (1994)
20. Fankhauser, S.: The economic costs of global warming damage: a survey. Glob. Environ. Change **4**(4), 301–309 (1994)
21. Fankhauser, S.: Valuing Climate Change: The Economics of the Greenhouse, 1st edn. EarthScan, London (1995)
22. Nocera, S., Cavallaro, F.: A methodological framework for the economic evaluation of CO_2 emissions from transport. J. Adv. Transp. (2013). doi:10.1002/atr.1249
23. Nordhaus, W.D.: Managing the Global Commons: The Economics of Climate Change. The MIT Press, Cambridge (1994)
24. Nordhaus, W.D.: A Question of Balance: Weighing the Options on Global Warming Policies. Yale University Press, New Haven (2008)
25. Tol, R.S.J.: The damage costs of climate change toward more comprehensive calculations. Environ. Resour. Econ. **5**(4), 353–374 (1995)
26. Tol, R.S.J.: Estimates of the damage costs of climate change: part 1: benchmark estimates. Environ. Resour. Econ. **21**(1), 47–73 (2002)
27. Tol, R.S.J.: Estimates of the damage costs of climate change: part 2: dynamic estimates. Environ. Resour. Econ. **21**(2), 135–160 (2002)
28. Bosello, F., Eboli, F., Pierfederici, R.: Assessing the economic impacts of climate change. Rev. Environ. Energy Econ. (2012). doi:10.7711/feemre3.2012.02.002

29. Maddison, D.J.: The amenity value of the climate: the household production function approach. Resour. Energy Econ. **25**(2), 155–175 (2003)
30. Mendelsohn, R.O., Morrison, W.N., Schlesinger, M.E., Andronova, N.G.: Country-specific market impacts of climate change. Clim. Change **45**(3–4), 553–569 (2000)
31. Mendelsohn, R.O., Schlesinger, M.E., Williams, L.J.: Comparing impacts across climate models. Integr. Assess. **1**(1), 37–48 (2000)
32. Nordhaus, W.D.: Geography and macroeconomics: new data and new findings. Proc. Nat. Acad. Sci. **103**(10), 3510–3517 (2006)
33. Maddison, D., Rehdanz, K.: The impact of climate on life satisfaction. Ecol. Econ. **70**(12), 2437–2445 (2011)
34. Rehdanz, K., Maddison, D.J.: Climate and happiness. Ecol. Econ. **52**(1), 111–125 (2005)
35. Tol, R.S.J.: The uncertainty about the total economic impact of climate change. Environ. Resour. Econ. **53**, 97–116 (2012)
36. Yohe, G.W., Tol, R.S.J.: Indicators for social and economic coping capacity: moving towards a working definition of adaptive capacity. Glob. Environ. Change **12**(1), 25–40 (2002)
37. Intergovernmental Panel on Climate Change (IPCC): Climate Change 1995. Economic and social dimensions of climate change. Contribution of Working Group I to the Second Assessment Report of the Intergovernmental Panel on Climate Change (1995)
38. Tol, R.S.J.: Why worry about climate change? a research agenda. Environ. Values **17**(4), 437–470 (2008)
39. Long, S.P., Ainsworth, E.A., Leakey, A.D.B., Noesberger, J., Ort, D.R.: Food for thoughts: lower-than-expected crop yield stimulation with rising CO_2 concentrations. Science **312**(5811), 1918–1921 (2006)
40. Tol, R.S.J.: The economic effects of climate change. J. Econ. Perspect. **23**(2), 29–51 (2009)
41. OECD: Mortality Risk Valuation in Environment, Health and Transport Policies, OECD Publishing. http://dx.doi.org/10.1787/9789264130807-en (2012)
42. Olsson, M., Bailey, M.J.: Positive time preference. J. Polit. Econ. **89**(1), 1–25 (1981)
43. Tol, R.S.J.: The economic impact of climate change. Perspektiven der Wirtschaftspolitik **11**(s1), 13–37 (2010)
44. Stern, N., Peters, S., Bakhshi, V., Bowen, A., Cameron, C., Catovsky, S., Crane, D., Cruickshank, S., Dietz, S., Edmonson, N., Garbett, S.-L., Hamid, L., Hoffman, G., Ingram, D., Jones, B., Patmore, N., Radcliffe, H., Sathiyarajah, R., Stock, M., Taylor, C., Vernon, T., Wanjie, H., Zenghelis, D.: Stern Review: The Economics of Climate Change. HM Treasury, London (2006)
45. Weitzman, M.L.: A review of the stern review on the economics of climate change. J. Econ. Lit. **45**(3), 703–724 (2007)
46. Tol, R.S.J.: The social cost of carbon: trends, outliers and catostrophes. Economics-The Open Access Open Assess. E-J. **2**(25), 1–24 (2008)
47. Forbes, C., Evans, M., Hastings, N., Peacock, B.: Statistical Distributions. Wiley, Hoboken (2011)
48. Libardo, A., Nocera, S.: Transportation elasticity for the analysis of Italian transportation demand on a regional scale. Traffic Eng. Control **49**(5), 187–192 (2008)
49. Nordhaus, W.D., Yang, Z.: RICE: a regional dynamic general equilibrium model of optimal climate-change policy. Am. Econ. Rev. **86**(4), 741–765 (1996)
50. Plamberk, E.L., Hope, C.W.: PAGE 95: an updated valuation of the impacts of global warming. Energy Policy **24**(9), 783–793 (1996)
51. Nordhaus, W.D., Boyer, J.G.: Warming the World: Economic Models of Global Warming. The MIT Press, Cambridge (2000)

An Environmental Approach to Optimize Urban Freight Transport Systems

Juan P. Romero, Juan Benavente, Jose L. Moura, Angel Ibeas and Borja Alonso

Abstract This chapter proposes an optimization-simulation model for planning and managing an urban freight transport system, which has to serve one or more points of the network that receive and/or generate a great volume of cargo, using trucks. This type of transport has special characteristics and generates significant impacts: increased traffic congestion, due to the presence of large vehicles which take up much space and are very slow; and air pollution caused by the extra traffic volume and the extra congestion. Therefore, the purpose of the model is to minimize these negative effects on the environment and on the users of the local road network. To achieve this goal, the authors propose and solve an optimization problem to minimize the total system cost (operating costs of the suppliers, costs supported by private vehicle users and public transport users, operating costs of the public transport, etc.). The proposed optimization problem is a bi-level mathematical programming model, where the upper level defines the total cost of the system, and the lower level defines the behaviour of private and public users, assuming that each of them chooses the route that minimizes his total journey cost. Then, this model is applied to the real case in the city of Santander (Northern Spain) obtaining a series of interesting conclusions from the corresponding sensitivity analysis.

J. P. Romero · J. Benavente · J. L. Moura (✉) · A. Ibeas · B. Alonso
G.I.S.T. Group of Research in Transport Systems, University of Cantabria, Cantabria, Spain
e-mail: mourajl@unican.es

J. P. Romero
e-mail: romerojunquerajp@gmail.com

J. Benavente
e-mail: juan_benavente_ponce@yahoo.es

A. Ibeas
e-mail: ibeasa@unican.es

B. Alonso
e-mail: alonsobo@unican.es

J. F. de Sousa and R. Rossi (eds.), *Computer-based Modelling and Optimization in Transportation*, Advances in Intelligent Systems and Computing 262,
DOI: 10.1007/978-3-319-04630-3_10,

Keywords Urban freight · Optimization · Environment

1 Introduction

On the subject of urban freight transport, the situation in which one or more points of the network require large amounts of supplies, and/or generate a significant volume of waste material, usually construction and demolition debris, has not been sufficiently considered in the literature. Due to the characteristics of the vehicles used to move the cargo, and how traffic flow is affected by their presence, this type of transport has a significant impact on the urban environment: increased traffic congestion; more air pollution; and, due to longer journey times, a raise in private transport costs.

This problem can be approached as a typical supply chain problem; where materials need to be delivered, in predetermined quantities, to a point, following a schedule. There are many studies where supply chain modeling and simulation have been applied to predict the behavior and optimize the design of many kinds of industry. One example is [1], who modeled and designed the supply chain structure for a food company. With the same aim in mind, other types of tools and techniques have been developed to study urban goods movement in supply chains: simulation techniques to study production, accounting and distribution policies, as in the work of [2]; the Goodtrip model by Boerkamps and Binsbergen [3]; microscopic-level models for mode choice and vehicle routing, as in the work of [4], who use adaptive stated preferences for designing a freight mode choice model; and the freight routing model of time-definite delivery by Lin [5].

As previously stated, none of the references above mention the particular case of transporting large amounts of cargo to or from one or more points of an urban transport network, a subject which has hardly been studied; although some work does exist, such as [6], who designed an integrated model that combines concrete production scheduling with its transport by trucks. Their objective was to minimize the operator costs only, thus social and environmental impacts were not taken into account.

Most of the studies that examine social and environmental impacts have mainly concentrated on the development of rules, regulations, measurement and legislation in order to minimize the impact of goods transport in urban areas. The work of [7] stands out in this field, discussing measures taken and the effects they produced in large European cities; and identifying three characteristics of the urban mobility of goods: the movement of goods is not affected by the internal structure of the city; urban policies regarding freight mobility are inefficient; and the provision of adequate logistic services is growing slower than the need for them in urban areas. From a social point of view, the work of [8] proposes a model for the movement of containers using trucks with time constraints at origins and destinations, guaranteeing that the drivers will not work more than a certain number of hours per shift.

Therefore, this chapter presents a model to optimize freight transport to and/or from one point of the urban network, based on the minimization of the overall costs of the system. Apart from quantifying the costs associated with transport planning, the proposed model considers the emission of pollutants throughout the study area.

This section has presented the Introduction and State of the Art. In Sect. 2, we describe our methodology; Sect. 3 provides specific details of the case study; and finally our main conclusions are shown in Sect. 4.

2 Methodology

We present a model to optimize the planning and management of a system that uses large vehicles (trucks) to supply and/or retrieve great amounts of supplies/waste materials from one point of an urban network. This model considers a number of potential routes, and determines the optimal way to distribute truck trips among them from an economic, social, and environmental point of view. To achieve this goal, a network with car, bus, and truck modes has been modelled and then calibrated; using the modal split and the trip assignment to the network steps to implement the interactions between modes. Therefore, any variation in the characteristics of the freight transport system affects both car and bus modes, as it can lead to modal shifts and changes in the routes chosen by drivers, or lines selected by bus users.

The optimization model is based on the minimization of the total system cost, which is a social cost function composed of car and bus user costs, and bus and truck operating costs [9–11]. Bi-level mathematical programming has been applied to find the best alternative: the urban network model on the lower level returns the data (flows, access times, waiting times, travel times, etc.) needed by the upper level to calculate the total system cost.

Social Cost $= Cu + Cop$

$$\begin{aligned} Cu_{\mathrm{T}} &= Cu_{\mathrm{C}} + Cu_{\mathrm{B}} \\ Cu_{\mathrm{C}} &= \varphi_{Viaje,\mathrm{C}} \cdot T_{Viaje,\mathrm{C}} \\ Cu_{\mathrm{B}} &= \varphi_{Acc,\mathrm{B}} \cdot T_{Acc,\mathrm{B}} + \varphi_{Egr,\mathrm{B}} \cdot T_{Egr,\mathrm{B}} \\ &\quad + \varphi_{Esp,\mathrm{B}} \cdot T_{Esp,\mathrm{B}} + \varphi_{Travel,\mathrm{B}} \cdot T_{Travel,\mathrm{B}} + \varphi_{Tra,\mathrm{B}} \cdot T_{Tra,\mathrm{B}} \end{aligned} \tag{1}$$

where:

Cu_{T}	Total users cost
Cu_{C}	Car users cost
Cu_{B}	Bus users cost
$T_{Travel,C}$	Car travel time
$\varphi_{Travel,C}$	Car travel time worth
$T_{Acc,B}$	Bus access time

$\varphi_{Acc,B}$ Bus access time worth
$T_{Egr,B}$ Bus egress time
$\varphi_{Egr,B}$ Bus egress time worth
$T_{Esp,B}$ Bus waiting time
$\varphi_{Esp,B}$ Bus waiting time worth
$T_{Travel,B}$ Bus travel time
$\varphi_{Travel,B}$ Bus travel time worth
$T_{Tra,B}$ Bus transfer time
$\varphi_{Tra,B}$ Bus transfer time worth.

Operating costs are calculated using the following formulation:

$$\begin{aligned} Cop_T &= Cop_B + Cop_{Tr} \\ Cop_B &= CR + CP + CF \\ CR &= \varphi_{CR} \cdot Total\,Km. \\ CP &= \varphi_{CP} \cdot Person\,hours \\ CF &= \varphi_{CF} \cdot N^{\circ}Buses \\ Cop_{Tr} &= \sum_i T_i \cdot f_i \cdot C_u \\ T &= T_{outward} + T_{return} + T_{loading} + T_{unloading} \end{aligned} \tag{2}$$

where:
Cop_T Total operating costs
Cop_B Bus operating cost
Cop_{Tr} Truck operating cost.

Bus operating costs (Cop_B) is made up of three factors: Cost proportional to travelled distance (*CR*), personnel costs (*CP*), and fixed costs (*CF*).
Total cost due to the distance travelled by the buses is equal to:

$$CR = \varphi_{CR} \cdot \ Total\ Km. \tag{3}$$

where:
φ_{CR} Unit cost per kilometer covered by bus

$$Total\ Km. = \sum_i L_i \cdot f_i$$

where:
L_i Length of route i
f_i Frequency of route i.

Employee costs are calculated considering only the personnel who are really working on the buses:

$$CP = \varphi_{CP} \cdot Person\ hours \tag{4}$$

where:
φ_{CP} The hourly employee cost (€ per hour)

$$Man - Hours = \sum_i \frac{tc_i}{h_i}$$

where:
tc_i Time of a round trip (min)
h_i Headway on route i (min).

Fixed costs are calculated with the following formula that only considers the buses that are really circulating:

$$CF = \varphi_{CF} \cdot N^{\circ} buses \tag{5}$$

where:
φ_{CF} Fixed cost per hour of bus (€ per hour)

$$N^{\circ} buses = \sum_i \frac{tc_i}{h_i}$$

where:
tc_i Time of a round trip (min)
h_i Headway on route i (min).

Truck operating cost (Cop_{Tr}) is estimated as:

$$\begin{gathered} Cop_{Tr} = \sum_i T_i \cdot f_i \cdot C_u \\ T = T_{outward} + T_{return} + T_{loading} + T_{unloading} \end{gathered} \tag{6}$$

where:
$T_{outward}$ Truck outward time
T_{return} Truck return time
$T_{loading}$ Truck loading time
$T_{unloading}$ Truck unloading time
C_u Cost per hour of truck use
f_i Truck flow.

To gauge the environmental impact of the different alternatives, the emissions of 5 types of pollutants have been calculated (CO, NOx, NMVOC, CH_4 and PM). Each transport mode's fuel consumption depends on the total distances travelled

Table 1 Vehicle's consumption rates (litres/Km) and vehicle's emission rates (g of pollutant/Kg of fuel)

	Emissions (g of pollutant / Kg of fuel)					Consumption (l/Km)		
	CO	NOx	NMVOC	CH_4	PM	Congested	Uncongested	Kg/l
Gasoline cars	75.99	10.89	13.44	1.19	0.03	0.08	0.06	0.680
Diesel cars	3.77	11.12	0.61	0.07	0.80	0.07	0.05	0.850
Buses	6.62	32.67	0.99	0.24	0.81	0.34	0.26	0.850
Trucks	9.82	34.84	3.06	0.38	1.34	0.34	0.26	0.850

by vehicles of that mode through congested and uncongested roads [12, 13]. Then, the emissions produced by these consumptions can be estimated [14].

Table 1 shows the different fuel consumption rates for the different kinds of vehicles in our model, depending on if the road is congested or not, and each kind of vehicle's emission rates (g of pollutant/Kg of fuel):

To solve the optimization problem, due to the size of the case study in relation to the number of variables, an exhaustive search algorithm will be applied. It will return all possible solutions, allowing us to analyze how the system behaves.

3 Case of Study

The methodology described above is applied to a real case: the city of Santander (Spain). It is a medium-sized city, with approximately 180,000 inhabitants, located on the north coast of the Iberian Peninsula.

A large construction project in the southeast of Santander will require a flow of 20 trucks per hour. The size and speed of these trucks create a substantial negative impact, increasing air pollution and traffic congestion.

The three alternatives to supply materials to the construction site are shown in Fig. 1. R1 route passes for the most part through a 2-lane urban road, except in the section closest to the construction site, where it goes through a tunnel of 800 m with a single lane in each direction. The route R2 has a initial leg in common with route R1, passing in its final stage to a single lane road in each direction, going around housing areas instead of through the tunnel to get to the construction. Finally, Route R3, even though runs through 2-lane and 3-lane urban roads in each direction, passes through areas of the city with high traffic density.

Applying the methodology previously described, we determine the social cost (user and operating costs), and pollutant emissions of all the different ways to distribute 20 trucks between the three routes.

Also, we perform a sensitivity analysis, studying how different values of the maximum speed for the trucks (20, 15, and 10 km/h) affect social cost and emissions in the city of Santander. The results are shown in Fig. 2.

Moreover, we represent the social cost of all simulated cases, ordering these from lowest to highest social cost. See Fig. 3.

Fig. 1 Considered routes

Analyzing Figs. 2 and 3, it can be seen that, as expected, the lower the truck speed becomes, the further to the right the center of mass of each cloud of points is located; because slower trucks increase the negative influence of the construction project in the urban system. Furthermore, lower truck speeds have the consequence of a wider range of possible social costs (the points are arranged closer to a straight line): from 470 units in the case of a truck speed of 20 km/h, to 670 units in the case of a truck speed of 15 km/h, and finally 1161 units in the case of a truck speed of 10 km/h.

It can also be seen in Fig. 2 that slower trucks means that the cloud of points will resemble a straight line more closely.

Regarding emissions, their overall value hardly changes at all, because we are working with mean fuel consumption rates, instead of considering fuel consumption as function that depends on the vehicle's speed. It would be necessary a detailed analysis at this point. See Table 2.

Regardless of the chosen truck speed, we can minimize the social cost, the emissions, or choose an intermediate solution. Thus, if we want to minimize the social cost, will have to move along the Ox axis ($\alpha = 0°$) until the perpendicular from our position touches the curve shown in black in Fig. 4 (Pareto boundary). In the same way, if we want to minimize emissions, we will travel along the Oy axis ($\alpha = 90°$). If the planner wants an intermediate optimal solution, he should use: $\alpha \mid 0° < \alpha < 90°$. As an example, we represent in Fig. 4 the Pareto optimal for $\alpha = 45°$.

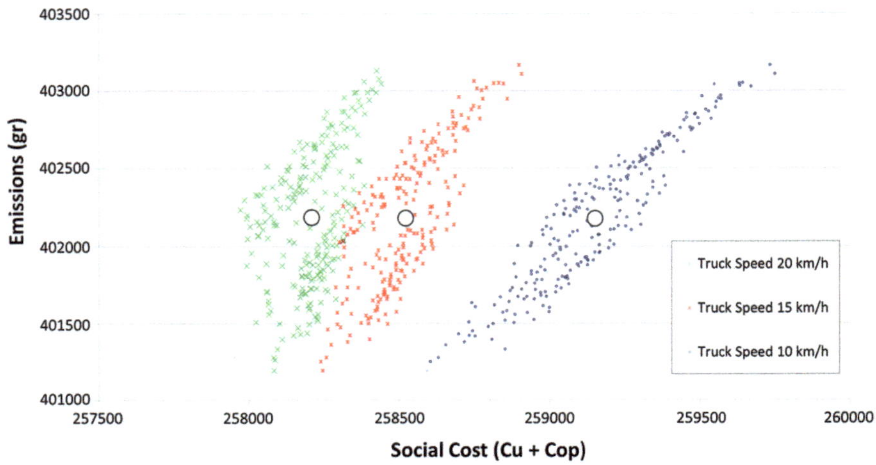

Fig. 2 Social cost versus emissions

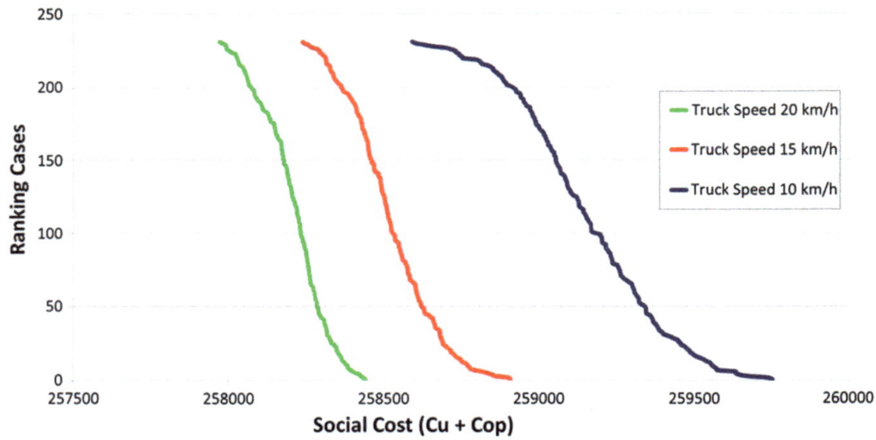

Fig. 3 Social cost versus ranking cases for different truck speeds

Table 2 Social cost and emissions values

Truck speed	Social cost			Emissions
	20	15	10	
Maximum	258442	258909	259753	403166
Minimum	257972	258239	258592	401190
Centre of mass	258209	258521	259153	402181

This way we can obtain many different solutions, according to the truck speed, and the chosen objective: social cost minimization, emission minimization, or a combination of both.

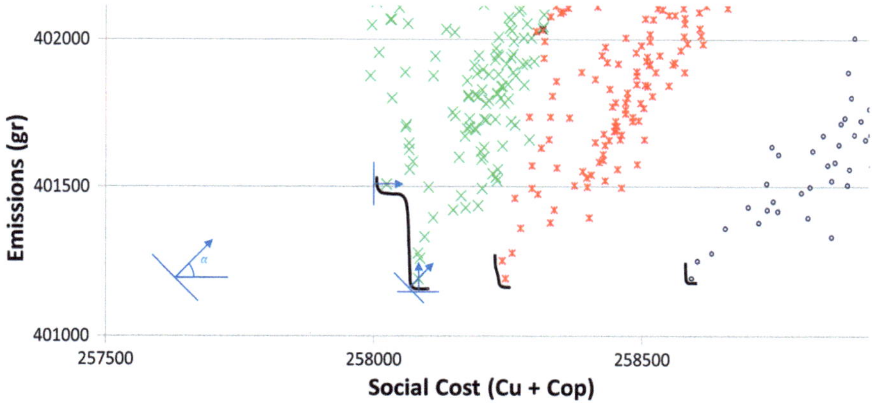

Fig. 4 Detail of social cost versus emissions

4 Conclusions

This chapter proposes a model to distribute truck trips along different routes in an urban environment, in a way that makes possible to analyze the emissions and the cost of the alternatives. In this way, we can propose policies to minimize the negative consequences, from a completely environmental, purely social, or intermediate point of view.

Due to the special characteristics of the case study, we opted for an exhaustive search algorithm, which yielded plentiful data, which was examined and used to perform a sensitivity analysis to determine how variations in the speed of the trucks influence the model's output.

Considering a family of solutions as the points that represent, for a certain value of the speed of the trucks, the social cost and emissions consequence of all possible ways to distribute the 20 trucks between the three routes; we observe that, as expected, the social cost of the center of mass of a family of solutions increases as the speed of the trucks decreases. For instance, if we compare the centers of masses of the families corresponding to truck speeds of 20 and 15 km/h, the latter's social cost is 0.12 % times greater. Analogously, studying 20 and 10 km/h families reveals a 0.37 % increase in the social cost of the center of mass. Also, the greater a family of solutions' truck speed, the wider its range of social cost values: 20 km/h family has a social cost range of values 42 % greater than the 15 km/h family; and 147 % greater than the 10 km/h family. It is also worth mentioning that as truck speed decreases, a family of solutions' outline becomes less steep, longer, and thinner.

References

1. Reiner, G., Traka, T.: Customized supply chain design: problems and alternatives for a production company in the food industry. Int. J. Prod. Econ. **89**(2), 217–229 (2004)
2. Siprelle, A.J., Parsons, D.J., Clark, R.J.: Benefits of using a supply chain simulation tool to study inventory allocation. In: Proceedings of the 2003 Winter Simulation Conference (2003)
3. Boerkamps, J., Binsbergen, A.V.: GoodTrip a new approach for modelling and evaluation of urban goods distribution. In: Proceedings 1st International Conference on City Logistics: City Logistics, Australia (1999)
4. Shinghal, N., Fowkes, T.: Freight mode choice and adaptive stated preferences. Transp. Res. Part E Logist. Transp. Rev. **38**(5), 367–378 (2002)
5. Lin, C.: The freight routing problem of time-definitive freight delivery common carriers. Transp. Res. Part B Methodol. **35**(6), 525–547 (2001)
6. Shangyao, Y., Weishen, L., Maonan, C.: Production scheduling and truck dispatching of ready mixed concrete. Transp. Res. Part E **44**, 164–179 (2008)
7. Dablanc, L.: Goods transport in large European cities: difficult to organize, difficult to modernize. Transp. Res. Part A **41**, 280–285 (2007)
8. Jula, H., Dessouky, M., Ioannou, P., Chassiakos, A.: Container movement by trucks in metropolitan networks: modelling and optimization. Transp. Res. Part E **41**, 235–259 (2005)
9. Moura, J.L., Ibeas, A., dell'Olio, L.: Optimization-simulation model for planning supply transport to large infrastructure public works located in congested urban areas. Netw. Spat. Econ. **10**(4), 487–507 (2008)
10. Romero, J.P., Moura, J.L., Ibeas, A., Benavente, J.: Car-bicycle combined model for planning bicycle sharing systems. Transp. Res. Board. 91st Annual Meeting No. 12-3062 (2012a)
11. Romero, J.P., Ibeas, A., Moura, J.L., Benavente, J., Alonso, A.: A simulation-optimization approach to design efficient systems of bike sharing. Procedia—Soc. Behav. Sci. **54**, 646–655 (2012b)
12. IDAE: Guía para la estimación del Combustible en las Flotas de Transporte por Carretera (2006)
13. IDAE: Guía de Vehículos Turismo de venta en España, con indicación de consumos y emisiones de CO_2 (2013)
14. Ntziachristos, L.: 1.A.3.b. Road Transport TFEIP endorsed draft. Emision Inventory Guidebook (2009)

Part III
(Urban) Network Design

Many transportation applications such as capital investment decision-making, parking planning, or traffic-light signal setting involve some form of *network design*. *Cipriani et al.* present and test a method to support the design of urban road transport systems, highlighting the sustainability of the proposed measures. *Polimeni and Vitetta* design jointly the road network and transit routes in an urban area. *Ceylan et al.* develop a simulation/optimization model for solving the problem of determining on-street parking places in urban road networks.

A Road Network Design Model for Large-Scale Urban Network

Ernesto Cipriani, Andrea Gemma and Marialisa Nigro

Abstract The aim of this work is to propose and test a methodology to support the design of urban road transport systems, highlighting the sustainability of the proposed measures. Finally, a tool for public administration support is provided. The problem is formulated as a road network design problem (NDP) with fixed demand, with design variables representing street direction and lane addition on links of the road network; the proposed methodology is based on two main phases: (1) a first phase aiming at reducing the solution's search space; (2) a second phase concerning the optimization procedure. The latter consists in a heuristic method based on a genetic algorithm. The procedure has been initially tested on a sub-network of the city of Rome (Eur network), and subsequently applied to the city of Brindisi (Southern Italy).

Keywords Road network design problem · Sustainability · Noise pollution

1 Introduction

In recent years, metropolitan areas are converting into the "automobile cities" [1]: the private demand increases more and more and, due to the lack of available spaces and to the high construction costs, it is not always possible to supply to this

E. Cipriani · A. Gemma · M. Nigro (✉)
Department of Engineering, Roma Tre University, Via Vito Volterra 62,
00146 Rome, Italy
e-mail: marialisa.nigro@uniroma3.it

E. Cipriani
e-mail: ernesto.cipriani@uniroma3.it

A. Gemma
e-mail: andrea.gemma@gmail.com

J. F. de Sousa and R. Rossi (eds.), *Computer-based Modelling and Optimization in Transportation*, Advances in Intelligent Systems and Computing 262,
DOI: 10.1007/978-3-319-04630-3_11,

increase with new infrastructures. Moreover, the "automobile city" is not consistent with sustainable issues, implying high social costs in terms of safety and health problems. In such a context, the reorganization of the current supply configuration is one of the available method to use existing resources efficiently and in a sustainable way.

This concept can be expressed mathematically as a Road Network Design Problem (RNDP): it is usually differentiated between the pure Network Design Problem (NDP), that consists in the optimal definition of link directions and capacities [2], and the Traffic Signal Setting (TSS) problem, i.e. to define the optimal signal setting at each junction [3, 4].

A lot of literature exists about the RNDP, however the basic classification of the different adopted approaches is based on: (1) the type of variables analyzed, that can be integer (topology) and/or continuous (signal settings), (2) the way the demand is dealt with (elastic or fixed, mono or multi-modal), (3) the optimization criteria (minimization of total costs or maximization of the reserve capacity etc. in combination with a descriptive or normative user behavior), (4) the solution algorithms.

Cantarella and Vitetta [2] deal with both the variables types, considering an elastic demand with respect to mode choice and different optimization criteria. Cipriani et al. [5] face with the Transit Network Design Problem (TNDP). Miandoabchi et al. [6] address a bi-modal multi-objective NDP. Drezner and Wesolowsky [7] put as objective the minimization of the total construction and transportation costs, while Ziyou and Yifan [8] the maximization of the reverse capacity of the road network, that is the maximum possible increase in traffic demand accepted by a given network structure.

About the solution algorithms, a lot of metaheuristic approaches have been explored: Simulated Annealing [9], Tabu search, Genetic Algorithm [7], Hill Climbing, Path Relinking [10], Descent Algorithm and Scatter Search algorithm [11], but also Hybrid Genetic Algorithm and Evolutionary Simulated Annealing [12].

The aim of this work is to propose and test a methodology to support the design of urban road transport systems, highlighting the sustainability of the proposed measures.

The problem is formulated as a road network design problem (NDP) with fixed demand, with design variables representing street direction and lane addition on links of the road network; specifically: (1) to modify the number of lanes for each direction, possibly increasing the capacity of the links entering new lanes if there are not planning constraints and enough space is available (2) to find the optimal direction on one-way traffic links.

The proposed methodology is based on two main phases: (1) a first phase in order to reduce the solution's search space; (2) a second phase in order to optimize the solution.

During the first phase a hierarchical set of sequences is created: the definition of the sequences is one of the main novelty of the study and a sequence consists in a set of contiguous links (in a given direction of traffic) with similar geometric characteristics representing a main road. This is based on the consideration that modern large urban networks are usually very detailed in the representation of main roads split in many short links that it is reasonable to be consistent with each other. This phase permits to reduce the search space, as it lowers the number of possible combination of values that links belonging to the same main direction can assume: the sequences, and no single links, are considered as variable and thus optimized, so increasing the performances of the solution methods, as reported by Cantarella et al. [10] and suggested by Russo and Vitetta [13].

In order to solve the optimization expected in the second phase of the methodology, an heuristic method based on a genetic algorithm has been implemented and calibrated. The choice of the genetic algorithm derives from the promising results obtained by this type of algorithm in literature [6, 7].

The procedure has been tested on a sub-network of the city of Rome (Eur network) initially with the objective of minimizing the total travel time of the network and then applied to the city of Brindisi (Southern Italy); in this last case a penalty term has been introduced in the objective function of the NDP in order to take into account the noise pollution.

2 The Methodology

The problem is formulated as a road network design problem (NDP) with fixed demand, with design variables representing street direction and lane addition on links of the road network; specifically: (1) to modify the number of lanes for each direction, possibly increasing the capacity of the links entering new lanes if there are not planning constraints and enough space is available (2) to find the optimal direction on one-way traffic links.

The proposed methodology is based on two main phases: (1) a first phase in order to reduce the solution's research space; (2) a second phase in order to optimize the solution.

2.1 First Phase: Reducing the Search Space

The first phase consists of five steps:

1. associating lane constrains to each link in terms of maximum number of lanes for the roadway (Cmax) and minimum number of lanes for any one-way road section (Cmin);
2. identifying the invariants links, such as main corridors of high priority or links belonging to the local road network, which will not be subjected to the next optimization phase;

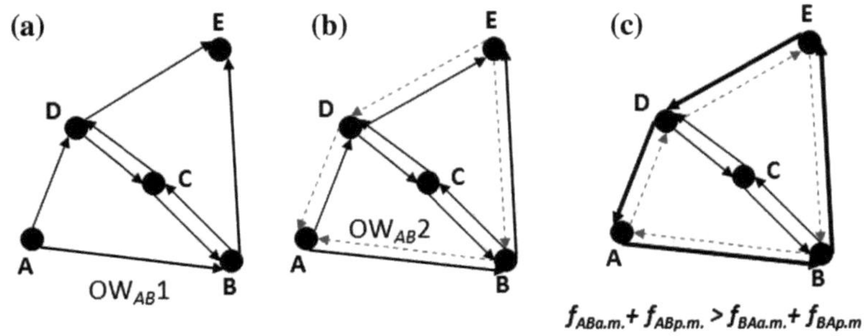

Fig. 1 One way optimal direction definition

3. identifying the links necessarily at one-way ride (links relative to roadway section with one lane only);
4. defining an optimal one-way direction for links of step 4 (strictly one-way links);
5. defining a hierarchical set of sequences, where the sequence is a set of contiguous links (in a given direction of traffic) with similar geometric characteristics representing a main road.

In the step 1, the attribute Cmax is a positive integer representing the maximum number of lanes for the whole roadway section; it can be obtained in the design phase only according to physical constraints. Instead, the attribute Cmin could be also equal to zero, when dealing with one-way links. Steps 3 and 4 proceed with the definition of the optimal direction for the links necessarily at one-way ride. These links will not be taken into account during the second phase of the process (optimization phase).

One way optimal direction definition.

The procedure adopted to define the optimal one-way direction derives from the procedure firstly proposed by Montella [14]:

1. for each one-way link (OW_l1, Fig. 1a), the other way round link is created (OW_l2, Fig. 1b);
2. the morning peak hour demand is assigned to the network with an equilibrium assignment model and the derived link flows are collected ($\mathbf{f}_{AM}$);
3. the afternoon peak hour demand is assigned to the network with an equilibrium assignment model and the derived link flows are collected ($\mathbf{f}_{PM}$);
4. for each OW_l1 and OW_l2 the total link flow is computed as the sum of $f_{l,AM}$ and $f_{l,PM}$;
5. if the total link flow on OW_l1 is greater than the total link flow on OW_l2, the last one is deleted or vice versa (Fig. 1c).

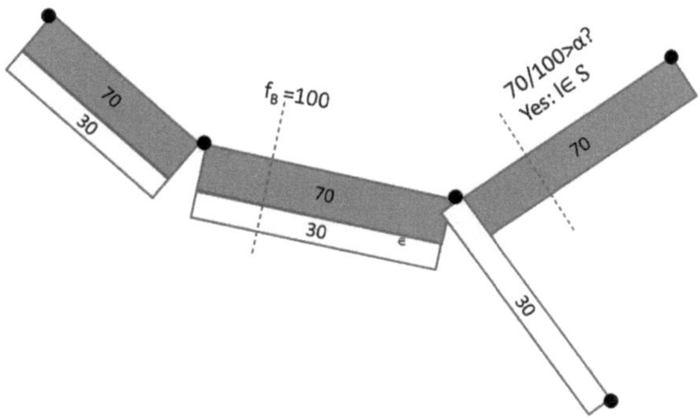

Fig. 2 Sequences definition

As usually in urban traffic networks the traffic is differently oriented between the morning and the afternoon period, the procedure considers both peaks demand matrices.

Sequences definition.

In order to avoid the occurrence of discontinuity phenomena, a procedure that automatically assigns to each link an attribute indicating the belonging to a sequence has been implemented and applied preliminarily in the optimization phase.

A sequence is a set of contiguous links (in a given direction of traffic) with similar geometric characteristics representing a main road. The sequences have a hierarchy of relevance and they are the decisional variables entering in the optimization phase (second phase).

The construction of the hierarchical set of sequences occurs as follows (Fig. 2):

1. a list containing the candidate links of the network is initialized;
2. the link with the higher flow value f_B is selected (link B) and removed from the above list;
3. link B defines the (first) sequence S1; any other link R of the list belongs to the sequence S1 if:

   a. using a select link analysis procedure, the path flow vector $\mathbf{F}_B$ passing on the link B is derived;
   b. for each link R crossed by a path flow $F \in \mathbf{F}_B$ the following condition is checked: $F/f_B > \alpha$, where α is a threshold; if the inequality is satisfied the link R is considered belonging to the sequence S1, and removed from the list;

The procedure restarts from point 1, considering the next remaining link in the list with the higher flow value in order to construct sequence S2 and so on until the list is empty. The construction of the sequences is applied to both the design scenarios of morning and afternoon peak hours, obtaining two sets of hierarchical sequences: the procedure is suited to deal with the occurrence that a same sequence can have a different relevance in morning and afternoon hours, as can be detected in real networks, where roads play different roles in the morning and in the afternoon periods, being affected by different types of flows.

First sequences created by the procedure (for both the design scenarios of morning and afternoon peak hours) are the most hierarchically important (sequences firstly created are characterized by higher flows): this hierarchy will be fully exploited during the implementation of the optimization phase.

2.2 Second Phase: Optimization

The optimization phase works only on links l belonging to the set of the sequences S, as constructed in the first phase of the procedure; the optimal solution is searched in order to minimize a mono criteria objective function containing the total travel time (TTT) on the network for both the rush hours (morning and afternoon):

$$\mathrm{O.F.} = \mathrm{TTT}^{\mathrm{a.m.}} + \mathrm{TTT}^{\mathrm{p.m.}} = \left(\sum\nolimits_{l \in S} f_l^{\mathrm{a.m.}} t_l^{\mathrm{a.m.}}\right) + \left(\sum\nolimits_{l \in S} f_l^{\mathrm{p.m.}} t_l^{\mathrm{p.m.}}\right) \quad (1)$$

In order to solve the optimization, a heuristic method based on a genetic algorithm has been implemented considering the promising results obtained by this type of algorithm in literature (see for instance Cipriani et al. [5]).

The implemented genetic algorithm is based on the following common structure [15]:

1. starting population generation: each member (chromosome) of the starting population has a genome composed by elements equal to the number of sequences S. At first iteration, a number of chromosomes (NC) are randomly created considering for each element of the genome a number of lanes in the constraint interval [A,B], where: A = $max_{l\ \mathrm{Si}}$(Cmin); B = $max_{l\ \mathrm{Si}}$(Cmax);
2. solutions evaluation: each chromosome is evaluated, assigning the supply characteristics contained in the genome to the sequences, performing an assignment of the a.m. and p.m. traffic demand and computing the O.F. reported in (1);
3. elitism: at each iteration a percentage value p_{el} of the chromosomes belonging to the population of the previous iteration is sent directly to the population of the next iteration without changes. The elitism starts from the second iteration of the algorithm and the chromosomes candidates to this operation are those with the best values of the objective function;
4. mutation: the mutation is carried out on a percentage value p_m of the chromosomes in all iterations; the chromosomes to be mutated are randomly extracted from those remaining as a result of the elitism. Each element of the genome of the chromosomes entering in the mutation process has a probability (mutation rate, *mr*) to be changed;
5. selection: using a Roulette wheel procedure, chromosomes are selected in pairs;

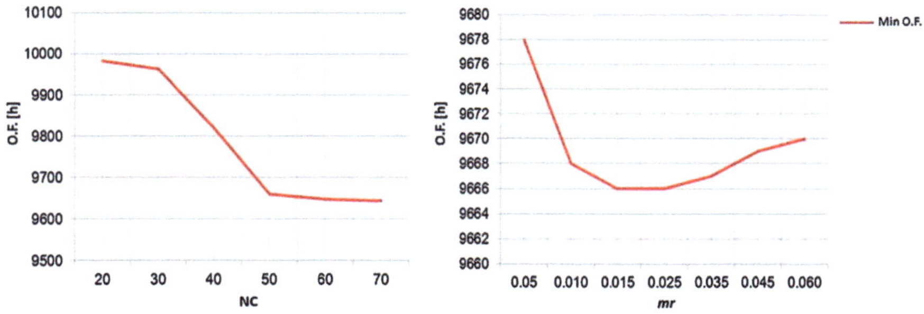

Fig. 3 NC and *mr* parameters calibration

6. crossover: a multipoint crossover is applied between each pair of chromosomes derived from the selection phase, in order to generate a number of new chromosomes equal to percentage value p_{cross} of the starting population.

The solutions, in the form of number of lanes for each sequence, are transferred to the level of individual links (number of lanes for each link), verifying the constraints on the local maximum and minimum number of lanes: once made, these choices are considered definitive, thus the number of lanes relative to the links belonging to sequences of lower relevance is conditioned by the number of lanes already identified for the links belonging to sequences of higher relevance.

The algorithm is stopped when, for the current population, the minimum value of the objective function is very close to the average value of the values of all the objective functions. This means that the population is composed of very similar chromosomes. The best solution (effective solution of the optimization problem) is represented by the chromosome with the lowest value of the objective function inside the population.

3 Calibration of the NDP: Eur Network

A first calibration of the parameters related to the number of chromosomes of the starting population (NC) and the mutation rate (*mr*) has been performed using a real test network: the Eur district located in the Southern of Rome. It is composed by 21 centroids, 49 regular nodes, 176 links and a demand in the morning peak hour of 30,000 vehicles.

The values of p_{el}, p_m and p_{cross} have been set to respectively 10, 15 and 75 %.

It can be observed (Fig. 3, NC parameter) that the goodness of the obtained solutions, in terms of objective function value, improves as the number of chromosomes of the population increases and, in particular, this improvement is very marked up to the case of 50 chromosomes; after, the improvement is much less evident. As a consequence, a number of chromosomes equal to 50 represents the best compromise between goodness of the solutions and computational times.

Table 1 Results of NDP on Eur network

	TTTa.m.	TTTp.m.	Global O.F.
Current state (h)	5,060	5,180	10,240
After optimization (h)	4,819	4,840	9,659
Percentage difference (%)	−4.7	−6.5	−5.6

Fixed the NC parameter to 50, the calibration of the mutation rate (*mr*) has been performed (Fig. 3, *mc* parameter): in this case the goodness of the solutions obtained is observed for values of *mr* equal to 0.015 and 0.025.

Setting finally the NC parameter to 50 and the *mr* parameter to 0.015, the NDP has been applied on the Eur network, obtaining the results reported in Table 1:

All the traffic assignments required by the model have been performed using the EMME simulator (INRO).

4 Application of the NDP: Brindisi Network

The NDP procedure, appropriately tested and calibrated, has been applied to a network greater than the previous one. In this way it is possible to underline the real potential of the procedure, while in the test network the configurations to be evaluated are limited and the O.F. convergence needs few iterations.

The analyzed network is the Brindisi network, a city of about 90,000 inhabitants located in the Southern Italy: it is composed by 43 centroids, 368 regular nodes and 885 links.

The NDP procedure has been applied for two different scenarios:

1. a first scenario, where only one way links optimal direction and optimal lane layout has been derived;
2. a second scenario, where not only one way links optimal direction and optimal lane layout has been derived, but also the optimal capacity increment.

For both the scenarios, the noise impact has been introduced within the NDP procedure as a penalty in (1); this penalty is calculated as:

$$P = C \cdot N_{VR} \tag{2}$$

with

C = a constant to be calibrated;

N_{VR} = residual number of violations of the acoustic limit as a result of a reduction of the link free speed of 5 km/h.

Results for both the scenarios are reported in Table 2: in the first case, scenario 1, it is possible to work only on the current structure of the network, without extension of it; in such a case, the model shows its capability to reduce both the total travel time and the number of acoustic violations. In particular, the acoustic

Table 2 Results of NDP on Brindisi network

		TTTa.m. (h)	TTTp.m. (h)	Global O.F. (h)	N_{VR}	Saturation degree > 1.5 (n°links)
	Current state	14,326	9,609	23,935	12	25
Scenario 1	After optimization	13,113	9,483	22,596	0	18
	Percentage difference (%)	−8.50	−1.30	−5.30	–	−28.00
Scenario 2	After optimization	10,142	7,548	17,690	2	13
	Percentage difference (%)	−27.10	−18.10	−23.50	−83.33	−48.00

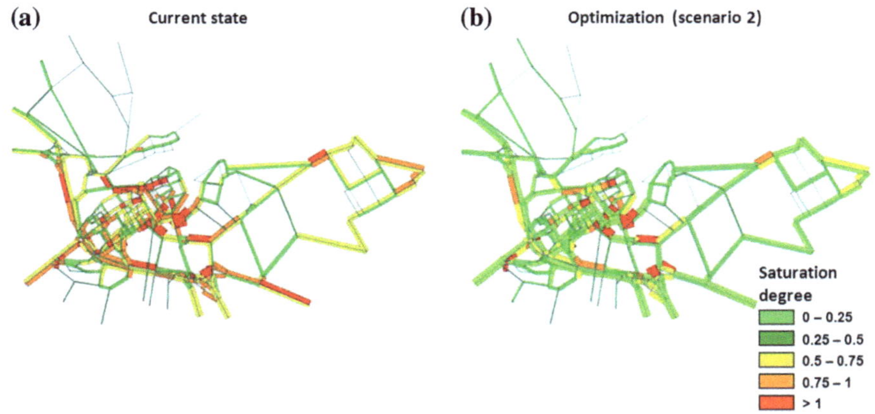

Fig. 4 Saturation degree ante (**a**), and post (**b**) optimization—scenario 2

violations are totally cleared although the reduction of the total travel time is only equal to 5.30 % respect to the current state.

In the second case, scenario 2, where it is possible to add new capacity to the current structure of the network, the total travel time reduction reaches the 23.50 % and it is translated in a high reduction of congestion (Fig. 4, the links with a saturation degree greater than 1.5 are reduced of 48 %, respect to the 28 % of scenario 1). The supply increase obtained optimizing scenario 2 is equal to the 17 % of the current state (moving from 450 linear kilometers of lanes to 530 linear kilometers).

5 Conclusions

The study proposes and tests a methodology to support the design of the road urban transport systems, considering as design variables street direction and lane addition on links of the road network. The problem is formulated as a road network design problem (NDP) with fixed demand. The proposed methodology is based on two main phases: (1) a first phase in order to reduce the solution's search space; (2) a second phase in order to optimize the solution.

The definition of the sequences during the first phase of the procedure is one of the main novelty of the study and it permits to reduce the search space, as the sequences and no single links are optimized.

Results of the conducted applications demonstrate the reliability of the procedure in terms of both travel times and noise impacts, as the possibility to adopt the procedure for public administration support.

Future developments will deal with: (1) more elaborated multi-objective functions, for example introducing also building costs, (2) introducing the free speed as a variable (integer variable—speed classes), (3) building dynamically the sequences, (4) other metaheuristic algorithms, (5) the elasticity of demand, (6) the join of the Road network design problem with the Transit network design problem.

References

1. Gori, S., Nigro, M., Petrelli, M.: The impact of land use characteristics for sustainable mobility: the case study of Rome. Eur. Transp. Res. Rev. 1–14 (2012). ISSN: 1867-0717. doi: 10.1007/s12544-012-0077-6
2. Cantarella, G.E., Vitetta, A.: The multi-criteria road network design problem in an urban area. Transportation **33**, 567–588 (2006)
3. Cipriani, E., Fusco, G.: Combined signal setting design and traffic assignment problem. Eur. J. Oper. Res. **155**, 569–583 (2004). ISSN: 0377-2217
4. Adacher, L., Cipriani, E.: A surrogate approach for the global optimization of signal settings and traffic assignment problem. In: IEEE Conference on Intelligent Transportation Systems, 13th IEEE ITSC, pp. 60–65. Funchal, Madeira Island, Portugal (2010). (ISSN: 2153-0009), ISBN: 978-1-4244-7657-2. doi: 10.1109/ITSC.2010.5625295
5. Cipriani, E., Gori, S., Petrelli, M.: Transit network design: a procedure and an application to a large urban area. Transp. Res. Part C: Emerg. Technol. **20**(1), 3–14 (2012)
6. Miandoabchi, E., Farahani, R.Z., Dullaert, W., Szeto, W.Y.: Hybrid evolutionary metaheuristics for concurrent multi-objective design of urban road and public transit networks. Netw. Spat. Econ. **12,** 441–480 (2012)
7. Drezner, Z., Wesolowsky, G.O.: Network design: selection and design of links and facility location. Transp. Res. Part A **37**(2003), 241–256 (2003)
8. Ziyou, G., Yifan, S.: A reserve capacity model of optimal signal control with user-equilibrium route choice. Transp. Res. Part B **36**(2002), 313–323 (2002)
9. Meng, Q., Yang, H.: Benefit distribution and equity in road network design. Transp. Res. Part B **36**(2002), 19–35 (2002)
10. Cantarella, G.E., Pavone, G., Vitetta, A.: Heuristics for urban road network design: lane layout and signal settings. Eur. J. Oper. Res. **175**(2006), 1682–1695 (2006)
11. Gallo, M., D'Acierno, L., Montella, B.: A meta-heuristic approach for solving the Urban Network Design Problem. Eur. J. Oper. Res. **201**(2010), 144–157 (2010)
12. Miandoabchi, E., Farahani, R.Z.: Optimizing reserve capacity of urban road networks in a discrete Network Design Problem. Adv. Eng. Softw. **42**(2011), 1041–1050 (2011)

13. Russo, F., Vitetta, A.:A topological method to choose optimal solutions after solving the multi-criteria urban road network design problem. Transportation **33**, 347–370 (2006)
14. Montella, B.: Pianificazione e controllo del traffico urbano. Modelli e metodi. CUEN (1996)
15. Michell, M.: An Introduction to Genetic Algorithms. MIT Press, Cambridge, MA (1998)

A Method for Topological Transit Network Design in Urban Area

Antonio Polimeni and Antonino Vitetta

Abstract The goal of this chapter is to design jointly the road network and the transit routes in an urban area. Generally, the transit route design is made without evaluating the possible changes in the path due to the road network layout design. In the problem, two main aspects can be considered: it is necessary to design, in a joint model, road network for cars and buses; it is necessary to design route for buses integrated with the optimized road network. Starting from a rigid road supply and an elastic demand, the road and the transit network are designed in accordance with one or more objectives (minimum travel time, maximum users satisfaction). The problem is formulated as a discrete problem. The proposed algorithm implemented is heuristic, based on genetic algorithm. To test the proposed procedure, an application to a main transit line of Reggio Calabria is reported.

Keywords Bus transit line · Genetic algorithm · Elastic demand · Optimization problem

1 Introduction

In the Network Design Problem (NDP) the main aim is to optimize the network configuration, optimizing a set of criteria related to a set of objectives. Relating to the problem formulation, a mono or multi-objective approach (user, public

A. Polimeni (✉) · A. Vitetta
DIIES—Dipartimento di Ingegneria dell'Informazione, delle Infrastrutture e dell'Energia Sostenibile, Università degli Studi Mediterranea di Reggio Calabria, Feo di Vito, 89060 Reggio Calabria, Italy
e-mail: antonio.polimeni@unirc.it

A. Vitetta
e-mail: vitetta@unirc.it

J. F. de Sousa and R. Rossi (eds.), *Computer-based Modelling and Optimization in Transportation*, Advances in Intelligent Systems and Computing 262,
DOI: 10.1007/978-3-319-04630-3_12,

manager and community) can be followed. Relating to the problem solution, its complexity does not allow using exact algorithms (at least for real problems).

In an urban area, a design problem should consider: (i) the road network design problem (RNDP), related to link directions and signals setting at the junctions for all traffic components (car, bus, …); (ii) the transit network design problem (TNDP), related to the transit routes and frequencies. In most cases, in literature, the two problems are studied separately.

A possible RNDP classification can be made considering the variables involved in the problem. So, three set of problems can be identified: problems with discrete variables [1–7], problems with continuous variables [8–17], problems with mixed variables [18–21]. Discrete variables dealing with the road layout; continuous variables dealing with junction regulation and price (road pricing and park pricing).

The transit design consists of three main aspects: the routes, the frequencies and the scheduling. In [22–24], first indications on the routes and frequencies design are supplied, proposing heuristic approaches to generate the routes. The frequency design is based on flow maximization on the transit lines, considering also the cost management. In [25, 26], first indications on the routes and frequencies design considering the user behaviour are supplied. The scheduling design considers, generally, two topics: the vehicles and the crew scheduling. The two topics can be considered separately or joint in the design model (one of the papers relative to this topic is [27].

In this chapter, a formulation trying to link in a joint method the RNDP and the TNDP is proposed. The problem is formulated considering the route cost for transit system in the objective function of the RNDP. The design variables are, in the RNDP the road layout (topology) and the junctions regulation (link capacity); in the TNDP are the transit routes (topology) and the frequency (route capacity). In term of algorithms, two main stages can be considered: a *first* stage, in which the road network and the transit routes topology are designed; a *second* stage, in which the buses frequencies and scheduling are designed (Fig. 1).

The highlights of this chapter can be summarized in: joint road and transit network design, a genetic algorithm codified to design reserved lanes for transit vehicles, comparing transit design in optimized and non-optimized network, testing the application of the method in a real network and comparing the result with a real transit line.

The chapter is structured as follows. In Sect. 2 the general model is proposed, in Sect. 3 a solution algorithm is presented. In Sect. 4 some numerical examples relative to the application of the model and algorithm for real system are discusses. Some conclusions and future developments are reported in Sect. 5.

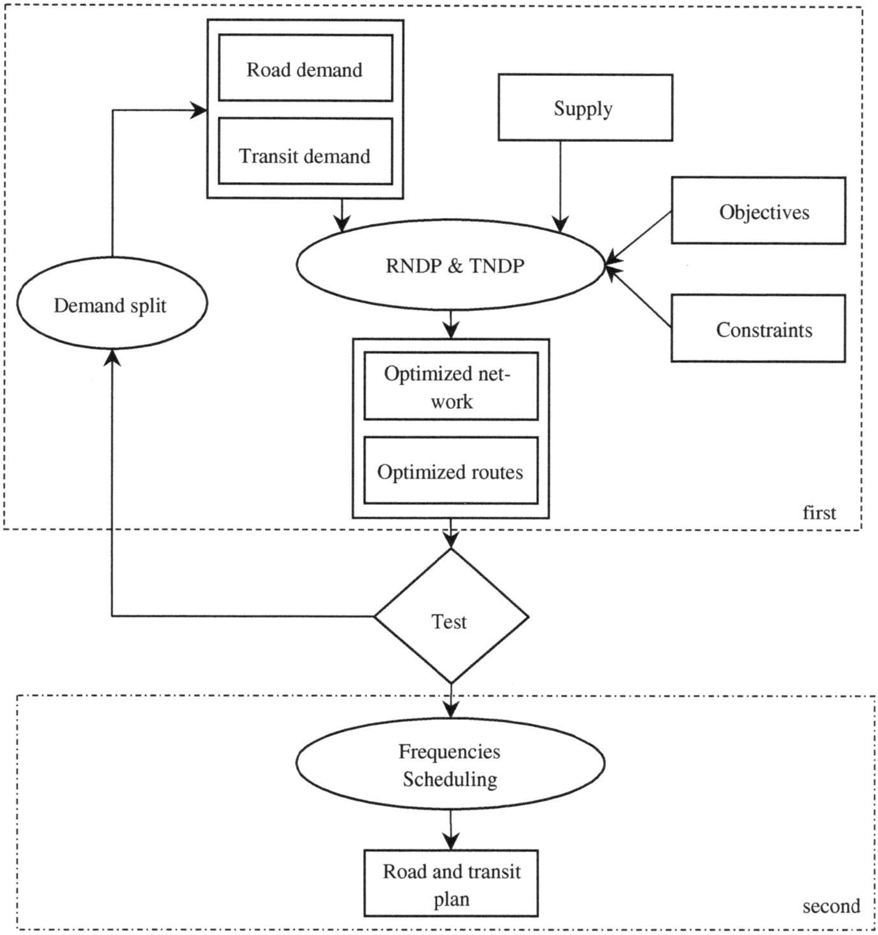

Fig. 1 Joint road network and transit design problem

2 Problem Formulation

In the problem (Fig. 1) the inputs are the supply and the demand, distinct into road demand (potential users that move using your vehicle) and the transit demand (potential users that move using the transit system). The design module (RNDP and TNDP) allows configuring the network both for private vehicles (cars, motorcycles, …) and buses. The outputs of this module are an optimized network in terms of topology and junction regulation and the optimized routes for buses. A test is performed to evaluate the results: if it is passed, a module allows designing frequencies and scheduling; else a demand split procedure is applied to evaluate the variation in the demand due the new network configuration. The output is the optimal road and transit plan.

Generally, a network design model is structured considering:

1. the objective function;
2. the design variables;
3. the set of constraints.

The objective function is responsible for explaining the objectives of the problem, both on mono-objective case (only an objective) and in multi-objective case (some objectives to achieve). Example of objectives are the minimum travel time, the maximum safety, the minimum management costs.

The design variables are responsible to formalize the problem, describing the aspects that can be optimized in the problem.

The set of constraints considers can be split in some subset: technical (network connection, signals, number of lanes, number of bus), economic (budget), normative (i.e. maximum CO emission) and the behavioural constraint that simulates the demand–supply interaction.

A general formulation of the design problem, considering the mono-objective case and the minimization approach is reported in Eqs. (1–4):

$$\text{Objective:} \quad \min_{f,y} z(\mathbf{f}, \mathbf{y}) \tag{1}$$

$$\text{Design variables:} \quad \mathbf{y} \in S_y \tag{2}$$

$$\text{Constraints:} \quad \mathbf{f} = \mathbf{f}_{\mathrm{SNL}}(\mathbf{c}(\mathbf{f}, \mathbf{y})) \tag{3}$$

$$f \in S_f \tag{4}$$

where z is the objective function, **f** is the link flow vector in the multimodal (road and transit) network and **y** is the configuration vector (design variables).

The vector **f** has two parts:

- a link flow vector $\mathbf{f}^{(r)}$ for road;
- a link vector $\mathbf{f}^{(t)}$ for transit.

The vector **y** has three parts:

- a sub-vector $\mathbf{y}^{(1)}$ for junction setting;
- a sub-vector $\mathbf{y}^{(2)}$ for link layout;
- a sub-vector $\mathbf{y}^{(3)}$ for transit routes layout.

The flow vector **f** is function of the link cost vector functions **c** with stochastic network loading function $\mathbf{f}_{\mathrm{SNL}}$.

The vector **y** and the vector **f** belong respectively to:

S_y the set of admissibility of design variables;
S_f the set of admissibility of link flow.

A possible specification for the objective function is:

$$z = z_1(\mathbf{f}^{(r)}, \mathbf{y}^{(1)}, \mathbf{y}^{(2)}) + \beta \times z_2(\mathbf{f}^{(r)}, \mathbf{f}^{(t)}, \mathbf{y}^{(1)}, \mathbf{y}^{(2)}, \mathbf{y}^{(3)}) \quad (5)$$

where

- β is a weight coefficient;
- z_1 is the cost for the road users;
- z_2 is the cost for the transit users.

In Eq. (5) the component z_1 is related to network design (RNDP) while the component z_2 is related to the transit routes design (TNDP and RNDP).

3 Algorithms

In this section, two classes of algorithms are considered: an algorithm for TNDP, it allows designing the transit routes; an algorithm for RNDP integrated with TNDP, it allows designing the road network. For simplicity sake, in the follows (Sects. 3.1 and 3.2) the two algorithms are treated separately, but they are correlated considering that the TNDP algorithm work on the designed road network with a feedback. To highlight this correlation, Fig. 2 shows the logical flows of the topological stage reported in Fig. 1. As mentioned in Sect. 1, the input date are the supply and the demand (distinguishing road demand and transit demand). The RNDP procedure allow obtaining an optimized road network configuration; starting from it a feasible transit network is extracted as input for TNDP, that applies a search procedure to finding an optimal topology for transit routes and evaluate the routes cost. A test is performed to evaluate the current solution, if it is passed the solution is accepted, else the demand is split between road and transit and a new iteration is performed.

3.1 Algorithms: RNDP

The output of RNDP are the best road link layout (optimizing the network layout), junctions regulation (optimizing the network in term of signals at junctions) and reserved transit lanes (optimizing the transit routes).

The algorithm implemented to solve the link layout and lane allocation (including the lane reservation) is based on genetic algorithm proposed in [18].

The genetic algorithm allows evolving an initial set of solutions (population) until achieve a goal (in our case, minimize the objective function).

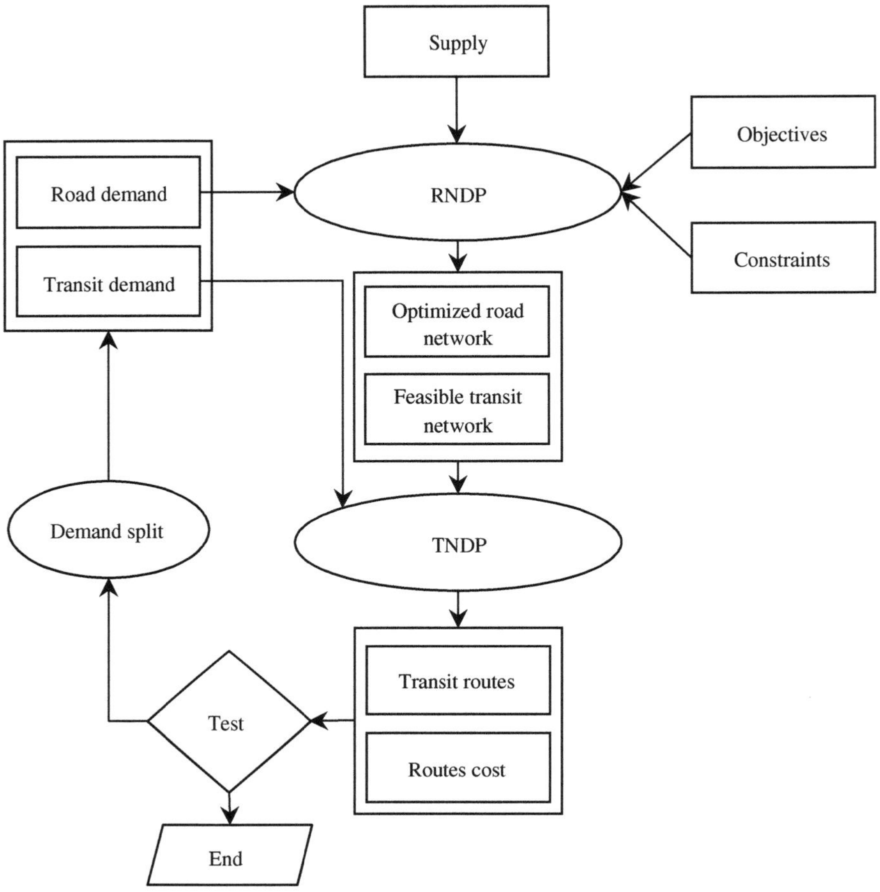

Fig. 2 Proposed algorithms for RNDP and TNDP

Basic elements of the genetic algorithm are:

- the selection: is the process of choosing solutions from the population (i.e. an approach is the roulette tournament);
- the crossover: is the process of taking two solutions and producing from them one or more other solutions;
- mutation: is the process of create a perturbation in one or more solutions.

The junctions can be solved as in [18], by using a projected gradient algorithm. Known the network layout by the solution of integer variables problem, an assignment is effected in order to calculate the objective function with the current signals setting variables, the procedure is iterated until convergence.

Another simplified way is to optimize the signals at junctions by using an approximate method (such as Webster method).

3.2 Algorithms: TNDP

Designing the transit routes, two phases are considered [28]: design of the potential routes and extraction of the final set of routes.

The design of the potential routes implies: place the last stations; individuate the feasible network (from designed road network); build the transit routes.

The last stations are placed in the neighborhood of other transport modes (like train stations or sea stations) to maximize the inter-modality and the accessibility.

The movements of a transit vehicle (a bus) are more constrained than car movements (i.e. for the vehicle shape or the vehicle length). For this reason, it is necessary to extract in the RNDP level a feasible network starting from the designed road network, eliminating some links with specific criteria (low width, low curve radius, high slope). If in the RNDP are designed reserved links for transit, the feasible network contains the reserved links.

The transit routes design is based on a constructive heuristic, implementing a greedy search. At each iteration three links type (waiting, boarding, alighting) are added, optimizing the component z_2 of Eq. (2). The link addition procedure change if in the current solution there are or there are not partial transit routes. In first case, to add the links, only the forward star of the initial node of the routes or the backward star of the final node is considered. In second case, to add the links, all the links in the feasible network are considered.

The final solution is a set of routes extracted considering the features of designed lines.

4 Application

The test application is performed on the city of Reggio Calabria (Italy).

The aim of this demo application is to highlight how change the transit design joint with an optimized network. Traditional applications in transit design field design the lines considering the demand levels (i.e. drawing the line for the more loads network links) but not considering the variations in road network as optimal lanes allocation and junctions' regulation.

An existing transit line is considered, crossing the city from south (Airport) to north (University) and vice versa. In this test, the design of the transit line between Airport and University is considered to compare the existing and the designed transit line. For simplicity sake, the comparison is made considering the direction Airport-University; the same analysis could be done in the direction University-Airport. In all the network configurations proposed below, the transit line always passes through some interchange points (i.e. train station, sea station) that do not change in design phase.

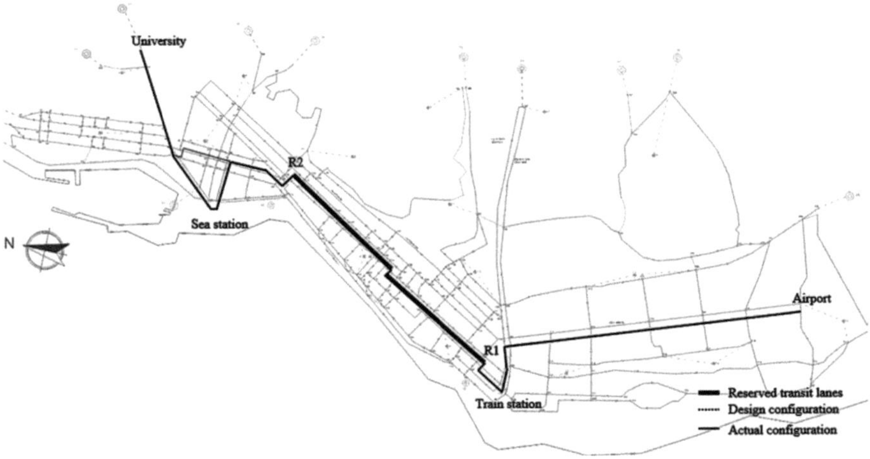

Fig. 3 Actual transit line configuration (—I) versus designed transit line in non-optimized network (···II)

(i) The first configuration of the examined transit line is the present configuration, designed by the transit manager, with the aim to connect some interchange points in the city. In this context, the line touches, among others, the airport, the train station, the sea station. In Fig. 3, the continuous line shows the actual transit route and some of the main stops. A reserved lane for buses and a lane (in the same direction) for other vehicles characterize the way between the point R1 and the point R2 (Fig. 3, the reserved lanes are indicated with a bold line).

(ii) A second configuration is due solving the TNDP in a non-optimized road network (as in consolidate literature). The assignment on the feasible network allows identifying, at each iteration, the link with greater flow, generating the transit route between Airport and University. The generated route (dotted line in Fig. 3) differ from the actual route in the links near to the train and the sea station. The reason is that actual route wants to encourage the accessibility stations while the model for transit route design select links belonging on city centre with higher demand.

(iii) The third configuration is the solution of the joint problem RNDP and TNDP: in this case, the transit line is designed together with the road network optimization, and the transit line configuration influences the road network design. Two cases are considered: designing reserved lanes (in addition to the existing ones) or not. In first case (III.a), the line designed in optimized network overlap the line designed in non-optimized network. In the second case (III.b), some modifications are introduced in the system, as in Fig. 4. The dotted line is the transit line in non-optimized network, whereas the dashed line is the transit line in optimized network. The main differences is the change in the transit line topology, which after the train station diverges

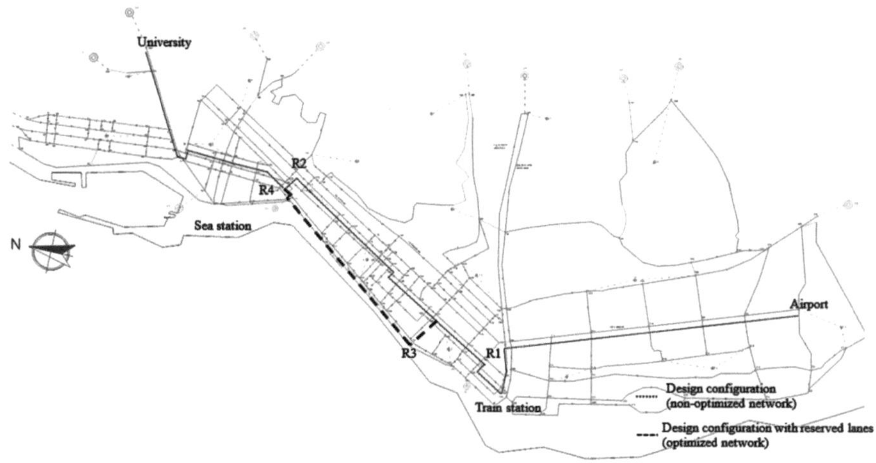

Fig. 4 Designed transit line in optimized network (- - - III.a) versus designed transit line in non-optimized network (···III.b)

following the reserved lanes (way between R3 and R4 in Fig. 4). Another change is in the way between R1 and R2, now dedicated only at the others vehicles (two lanes in direction north–south). The solution with reserved lanes offers a gain, in term of cost function (2), about of 5 %.

5 Conclusion and Future Developments

In this chapter, a method for joint road and transit network design in urban area is proposed. The method consists of two main procedures: first, a road network design to optimize the link layout; second the transit routes generation in the optimized network. The consolidated literature gives the transit route generation considering only the demand and addressing the transit route on the more load links with the road network fixed. The proposed procedure want to take into account both the demand and the road network design, with a loop that ties the transit route with the road network design. In fact, the proposed problem formulation, considers jointly the road users and the transit line users and the whole objective function evaluates the road users and the transit users costs and the general management costs. In this formulation, the infrastructures in the supply is assumed rigid whereas the demand is elastic (to consider the variations in the users choice). The proposed procedure to solve the whole problem, considers two algorithms: a genetic algorithm to solve the road network design problem, a greedy algorithm to solve the transit route design.

A preliminary test in a real road network is performed, considering two different cases for network design: reserving some lanes to transit lines or not.

Is emerged that reserving some lanes to the transit lanes, in this experiments, a gain for all users in the objective function is possible considering that also the road system is optimized.

The results have to be considered preliminary and future developments concern the problem extension, considering more than one transit line and updating the search algorithm.

Acknowledgments Partially supported by national MIUR under PRIN2009 grants n. 2009EP3S42_001.

References

1. Billheimer, J.W., Gray, P.: Network design with fixed and variable cost elements. Transp. Sci. **7**, 49–74 (1973)
2. Chen, M., Alfa, A.S.: A network design algorithm using a stochastic incremental traffic assignment approach. Transp. Sci. **25**(3), 215–224 (1991)
3. Foulds, L.R.: A multi-commodity flow network design problem. Transp. Res. Part B **15**, 273–283 (1981)
4. Gao, Z., Wu, J., Sun, H.: Solution algorithm for the bi-level discrete network design problem. Transp. Res. B **39**, 479–495 (2005)
5. Poorzahedy, H., Abulghasemi, F.: Application of ant system to network design problem. Transportation **32**, 251–273 (2005)
6. Herrmann, J.W., Ioannou, G., Minis, I., Proth, J.M.: A dual ascent approach to the fixed-charge capacitated network design problem. Eur. J. Oper. Res. **95**, 476–490 (1996)
7. Kalafatas, G., Peeta, S.: Planning for evacuation: insights from an efficient network design model. J. Infrastruct. Syst. **15**(1), 21–30 (2009)
8. Webster, F.W.: Traffic signal settings. Road Research Technical Paper no. 39 (1958)
9. Webster, F.V., Cobbe, B.M.: Traffic signals. Road Research Laboratory Technical Paper 56, London, UK (1966)
10. Allsop, R.E.: SIGCAP: a computer program for assessing the traffic capacity of signal-controlled road junctions. Traffic Eng. Control **17**, 338–341 (1976)
11. Gartner, N.H.: Area traffic control and network equilibrium. In: Florian, M. (ed.) Traffic Equilibrium Methods, vol. 118, pp. 274–297. Lecture Notes in Economics and Mathematical SystemsSpringer, Berlin (1976)
12. Smith, M.J.: The existence, uniqueness and stability of traffic equilibria. Transp. Res. B **13**(4), 295–304 (1979)
13. Sheffi, Y., Powell, W.B.: Optimal signal setting over transportation networks. Transp. Eng. **109**(6), 824–839 (1983)
14. Meneguzzer, C.: An equilibrium route choice model with explicit treatment of the effect of intersections. Transp. Res. B **29**, 329–356 (1995)
15. Chiou, S.W.: Optimal design of signal-controlled road network. Appl. Math. Comput. **189**, 1–8 (2007)
16. Marcianò, F.A., Musolino, G., Vitetta, A.: Signal setting design on a road network: application of a system of models in evacuation conditions. WIT Transactions on Information and Communication Technologies **43**(part I), 443–454 (2010)
17. Cantarella, G.E., Velonà, P., Vitetta A.: Day-to-day dynamic network modeling and optimization. In: IEEE International Intelligent Transportation Systems Conference, pp. 2086–2092 (2011)

18. Cantarella, G.E., Pavone, G., Vitetta, A.: Heuristics for urban road network design: lane layout and signal settings. Eur. J. Oper. Res. **175**, 1682–1695 (2006)
19. Russo, F., Vitetta, A.: A topological method to choose optimal solutions after solving the multi-criteria urban road network design problem. Transportation **33**, 347–370 (2006)
20. Poorzahedy, H., Rouhani, O.M.: Hybrid meta-heuristic algorithms for solving network design problem. Eur. J. Oper. Res. **182**(2), 578–596 (2007)
21. Polimeni, A., Vitetta, A.: A procedure for an integrated network and vehicle routing optimisation problem. Procedia Soc. Behav. Sci. **54**, 65–74 (2012)
22. Baaj, H.M., Mahmassani, H.S.: Hybrid route generation heuristic algorithm for the design of transit networks. Transp. Res. C **3**(1), 31–50 (1995)
23. Ceder, A., Wilson, N.H.M.: Bus network design. Transp. Res. B **20**(4), 331–344 (1986)
24. Ceder, A.: Designing public transport network and routes. In: Lam, W.H.K., Bell, M.G.H. (eds.) Advanced Modeling For Transit Operations and Service Planning, Chapter 3, pp. 59–91. Emerald Group Publishing Limited (2003)
25. Nuzzolo, A., Russo, F., Crisalli, U.: Transit Network Modelling. The Schedule-Based Dynamic Approach. FrancoAngeli, Milano (2003)
26. Fusco, G., Gori, S., Petrelli, M.: An heuristic transit network design algorithm for medium size towns. In: Proceedings of the 13th Mini-EURO Conference, Bari (2002)
27. Mesquita, M., Paias, A.: Set partitioning/covering-based approaches for the integrated vehicle and crew scheduling problem. Comput. Oper. Res. **35**, 1562–1575 (2008)
28. Russo F.: Metodi per la progettazione dei sistemi di trasporto collettivo. Quaderno di dipartimento, QD-SD 1/10 (2010)

Determining On-Street Parking Places in Urban Road Networks Using Meta-Heuristic Harmony Search Algorithm

Huseyin Ceylan, Ozgur Baskan, Cenk Ozan and Gorkem Gulhan

Abstract This study aims to develop a simulation/optimization model for the solution to the problem of determining on-street parking places in urban road networks. The problem is dealt within the Discrete Network Design (DND) context due to the binary decision variables and the bi-level programming technique is used for the solution of the problem. The upper level represents the determination of on-street parking places while the reaction of drivers' to the design is handled in user equilibrium manner in the lower level. The upper level problem is formulized as a non-linear mixed integer programming problem and the meta-heuristic Harmony Search (HS) optimization technique is employed for the solution. In the proposed model, VISUM traffic analysis software is utilized as the simulation tool for solving the lower level problem. The performance of the proposed model is tested on Sioux-Falls road network which has widely been used on DND studies in the previous works. Results show that determining optimal or near-optimal on-street parking places may be achieved by using the proposed model.

Keywords On-street parking · Discrete network design problem · Harmony search algorithm · Traffic management · VISUM traffic model

H. Ceylan (✉) · O. Baskan · C. Ozan
Department of Civil Engineering, Pamukkale University, Denizli, Turkey
e-mail: hceylan@pau.edu.tr

O. Baskan
e-mail: obaskan@pau.edu.tr

C. Ozan
e-mail: cozan@pau.edu.tr

G. Gulhan
Department of Urban and Regional Planning, Pamukkale University, Denizli, Turkey
e-mail: ggulhan@pau.edu.tr

J. F. de Sousa and R. Rossi (eds.), *Computer-based Modelling and Optimization in Transportation*, Advances in Intelligent Systems and Computing 262,
DOI: 10.1007/978-3-319-04630-3_13,

1 Introduction

Mobility demand of people living in urban and metropolitan areas has continuously been growing due to the increasing socio-economical needs which lead varied activities. Hence, people tend to use individual motorized transport modes in order to satisfy this ever-changing mobility demand. Increasing trend of modal shift in favor of the private car leads parking problems in urban road networks. Inadequacy of parking facilities comes with serious problems about the urban economics and travel quality of the citizens. Drivers tend to solve their parking problems, which arise from inadequate parking spaces, by on-street parking while the local authorities seek for solutions to develop parking policies that would not decrease the capacity of the road network.

Shoup [1] stated that a significant number of drivers cruse in search of an available parking space in congested traffic. Free/underpriced and unplanned on-street parking may lead to serious problems in terms of the network capacity. On the other hand, properly planned and market-clearing charged on-street parking places may provide an ideal source of local public revenue. Yousif and Purnawan [2] investigated the effects of the design of the parking spaces on maneuver time and the gap acceptance to merge into the traffic stream when leaving a parking space. Portilla et al. [3] quantified the influence of on-street parking maneuvers and badly parked vehicles on average link travel times. It was stated that badly parked vehicles and parking maneuvers have a significant impact on link travel times and roadway capacity. Guo et al. [4] proposed a proportional hazard-based duration model to investigate the influence of on-street parking places on travel time. It was stated that the occupancy, number of parking maneuvers and effective lane width have a significant impact on travel time. According to our current knowledge, researchers have focused on the influence of on-street parking on link travel times and capacities. On the other hand, there is a need for a model that could be used to determine the network-wide available lanes for on-street parking purpose.

Due to the binary variables, which represent whether a lane is allocated to the on-street parking space, this problem may be handled in Discrete Network Design (DND) context. Numerous researchers have discussed on DND and developed solution methods for different problems in the literature. Bruynooghe [5] proposed an integer programming model for determining optimal investment strategies to improve the performance of road networks. LeBlanc [6] implemented the Branch and Bound (BB) method to the solution of the DND problem in which the link flows were calculated in the User Equilibrium (UE) manner. Poorzahedy and Turnquist [7] developed a bi-level programming model for solution of DND problem. In their model, minimization of the total network travel time was achieved with a BB-based heuristic algorithm. The BB method, which has been used to solve DND problems, has some disadvantages such as high memory and long computation time requirement for the problems including large number of decision variables [8, 9]. Gao et al. [10] introduced a solution algorithm based on a support function concept to describe the relationship between flows and the new additional links in the network. The bi-level problem is discontinuous due to the use of binary variables to represent lane allocations and capacity improvements and it necessitates heuristic solution methods [11]. Ceylan and Ceylan [12] formulated the bi-level DND problem as mixed integer programming problem and employed the meta-heuristic Harmony Search (HS) algorithm for the solution. It was stated that the HS based solution method gave remarkable results for the solution of capacity improvement and lane allocation problems.

In this study, a bi-level mixed integer programming model is proposed to determine the optimal on-street parking places on urban road networks. On the upper level, lane allocations to the on-street parking space are carried out using meta-heuristic HS algorithm. The reactions of drivers are taken into account in the UE manner and the traffic assignment problem is solved using VISUM traffic simulation tool in the lower level. Note that the delays and congestions arise from the parking maneuvers are out of scope of this chapter.

The rest of this chapter is organized as follows. The problem formulation is summarized in Sect. 2. In Sect. 3, basics of the HS algorithm and model development are presented. A numerical application is presented in Sect. 4. The chapter ends with some conclusions in Sect. 5.

2 Problem Formulation

The nomenclature used in the formulation is given in the table below.

Nomenclature	
A	Set of links $(\forall a \in A)$
M	Set of candidate links for on-street parking $(\forall m \in M)$
K_{rs}	Set of paths between Origin-Destination (O-D) pair rs $(\forall r \in R)$ $(\forall s \in S)$
R	Set of origins $(\forall r \in R)$
S	Set of destinations $(\forall s \in S)$
c_a	Capacity of link a $(\forall a \in A)$
D_{rs}	Travel demand between origin r $(\forall r \in R)$and destination s $(\forall s \in S)$
f_k^{rs}	Traffic volume on path k between O-D pair rs $(\forall r \in R)$ $(\forall s \in S)$
l_a	Length of link a $(\forall a \in A)$
t_a^0	Free flow travel time on link a $(\forall a \in A)$
$t_a(v_a)$	Travel time on link a $(\forall a \in A)$
v_a	Traffic volume on link a $(\forall a \in A)$
α	Arbitrary penalty factor
$\delta_{a,k}^{rs}$	Element of the link/path incidence matrix. $\delta_{a,k}^{rs} = 1$ if route k uses link a, and $\delta_{a,k}^{rs} = 0$ otherwise

The problem of determining the on-street parking places may be expressed as the maximization of total length of available lanes which are allocated to on-street parking as follows:

$$\max \sum_{i=1}^{m} u_i l_i \tag{1}$$

s.t.

$$\frac{v_m}{c_m} \leq 1 \quad (\forall m \in M) \tag{2}$$

where u_i is a binary variable which may be expressed as follows:

$$u_m = \begin{cases} 1 & \text{if one lane of link } m \text{ is allocated to on-street parking} \\ 0 & \text{otherwise} \end{cases} \tag{3}$$

The constraint, which is given in Eq. (2), ensures that the UE flows would not exceed the capacity of the candidate links. The UE link flows are calculated by the solution of the following convex optimization problem:

$$\min_x z = \sum_{a \in A} \int_0^{v_a} t_a(x) dx \tag{4}$$

s.t.

$$\sum_{k \in K} f_k^{rs} = D_{rs} \qquad \forall r \in R, \quad s \in S, \quad k \in K_{rs} \tag{5}$$

$$v_a = \sum_{rs} \sum_{k \in K_{rs}} f_k^{rs} \delta_{a,k}^{rs} \qquad \forall r \in R, \quad s \in S, \quad a \in A, \, k \in K_{rs} \tag{6}$$

$$f_k^{rs} \geq 0 \qquad \forall r \in R, \quad s \in S, \quad k \in K_{rs} \tag{7}$$

3 Harmony Search Algorithm and Model Development

3.1 Meta-Heuristic Harmony Search Algorithm

HS is a heuristic optimization technique that has been created by getting inspired from a musical improvisation process of an orchestra by Geem et al. [13]. In an orchestra, musicians literally seek for a perfect harmony by improvising successive melodies while the global or a near-global optimum solution is investigated throughout iterations in an optimization process. In this analogy, decision variables and their values may represent the musicians and the notes performed by the orchestra, respectively. In the recent years, the HS algorithm has been applied to the solution of many engineering optimization problems including transport energy demand modeling, continuous network design and area traffic control problems [14–16]. The HS algorithm has five steps:

Step 1: Initialization of the problem and HS parameters

Consider an optimization problem as follows:

$$\max f(\vec{x}) \tag{8}$$

s.t.

$$g_i(\vec{x}) \geq 0;\ i = 1, 2, \ldots, P \tag{9}$$

$$x_h \in \left[x_{h,\min},\ x_{h,\max}\right];\ h = 1, 2, \ldots, Q \tag{10}$$

where $f(\vec{x})$ is the objective function to be maximized, $g_i(\vec{x})$ is the inequality constraint $(i = 1, 2, \ldots, P)$, $\vec{x} = [x_1, x_2, \ldots, x_Q]^{\mathrm{T}}$ is the set of decision variables, Q is the number of decision variables, P is the number of inequality constraints, $x_{h,\min}$ and $x_{h,\max}$ are the lower and upper bounds for decision variables $(h = 1, 2, \ldots, Q)$.

Three parameters, which are used in the HS progress, are set at this step. The first one is Harmony Memory Size (HMS) which represents the number of solution vectors in the Harmony Memory (HM). Harmony Memory Considering Rate (HMCR), which is the second one, is the probability of assigning the values to the variables from the HM. The third one is Pitch Adjusting Rate (PAR) that is the probability of slightly adjusting by moving to neighboring values of a value selected from the HM.

Step 2: Initialization of the HM

The HM, which includes randomly generated solution vectors and their corresponding fitness values, is initialized as given in Eq. (11).

$$\left[\begin{array}{ccccc|c} x_1^1 & x_2^1 & \cdots & x_{Q-1}^1 & x_Q^1 & f(x^1) \\ x_1^2 & x_2^2 & \cdots & x_{Q-1}^2 & x_Q^2 & f(x^2) \\ \vdots & \vdots & \vdots & \vdots & \vdots & \vdots \\ x_1^{HMS-1} & x_2^{HMS-1} & \cdots & x_{Q-1}^{HMS-1} & x_Q^{HMS-1} & f(x^{HMS-1}) \\ x_1^{HMS} & x_2^{HMS} & \cdots & x_{Q-1}^{HMS} & x_Q^{HMS} & f(x^{HMS}) \end{array}\right] \tag{11}$$

Step 3: Improvisation of a new harmony

Improvising a new harmony represents generation of a new solution vector. First of all, it is decided whether the value of a decision variable is selected from the HM $\left(\text{e.g.}\ x_1' \in \left[x_1^1 \ldots x_1^{HMS}\right]\right)$ based on the HMCR or not, as follows:

$$x_h' = \begin{cases} x_h' \in \left\{x_h^1, x_h^2, x_h^3, \ldots, x_h^{HMS}\right\} & \text{with probability HMCR} \\ x_h' \in \mathbf{X}_h & \text{with probability } (1 - \text{HMCR}) \end{cases} \tag{12}$$

where $\mathbf{X}_h$ is the possible random range for each decision variable $(h = 1, 2, \ldots, Q)$. In Eq. (12), value of the h-th decision variable is selected from HM with the probability of HMCR or it is selected from the possible range with the probability of (1−HMCR). If the value of a decision variable is selected from the HM then it is decided whether the pitch adjusting process will be performed based on the PAR or not, as follows:

$$x'_h = \begin{cases} x'_h \pm \text{Rand}(0,1) \times \text{bw} & \text{with probability PAR} \\ x'_h & \text{with probability } (1 - \text{PAR}) \end{cases} \tag{13}$$

where *bw* is an arbitrary bandwidth, $\text{Rand}(0,1)$ is a uniform random number between 0 and 1. Note that the values of the all decision variables $\left(x'_2, x'_3, x'_4, \ldots, x'_Q\right)$ are selected in the same manner.

Step 4: Updating the HM

At this step, corresponding objective function values of the newly generated vector and the worst solution vector in the HM are compared. If the new solution vector provides a better objective function value then it is replaced with worst one in the HM.

Step 5: Checking the termination criterion

If a preset termination criterion is met then the algorithm is terminated. Otherwise, the computation is continued by iterating Steps 3–5.

3.2 Model Development

Flow chart of the proposed model is given in Fig. 1.

It can be seen in Fig. 1 that the characteristics of the road network, candidate lanes, O-D demands, HS parameters and the termination criterion are initialized at Step 1. At Step 2, the HM is filled with initial solution vectors which include randomly generated binary variables. Then the UE link flows are calculated for each solution vector in the HM based on Eq. (4). In this study, the UE assignment problem is solved by VISUM traffic simulation tool [17]. In order to calculate the total length of the lanes, which are allocated to the on-street parking, the objective function, which was given in Eq. (1), may be modified as follows:

$$\max\left(\sum_{i=1}^{m} u_i l_i - G_i(v_i)\right) \tag{14}$$

where $G_i(v_i)$ is the static penalty function, which may be used to handle the inequality constraints given in Eq. (2), and it has the following form:

$$G_i(v_i) = \begin{cases} \alpha & \text{if } \frac{v_i}{c_i} > 1 \quad (\forall i \in M) \\ 0 & \text{otherwise} \end{cases} \tag{15}$$

At Step 3, a new solution vector is generated based on the HM and it should be adjusted by the probability of PAR. However, considering the binary space, in which the value of each decision variable in the HM is bounded to be “0” or “1”,

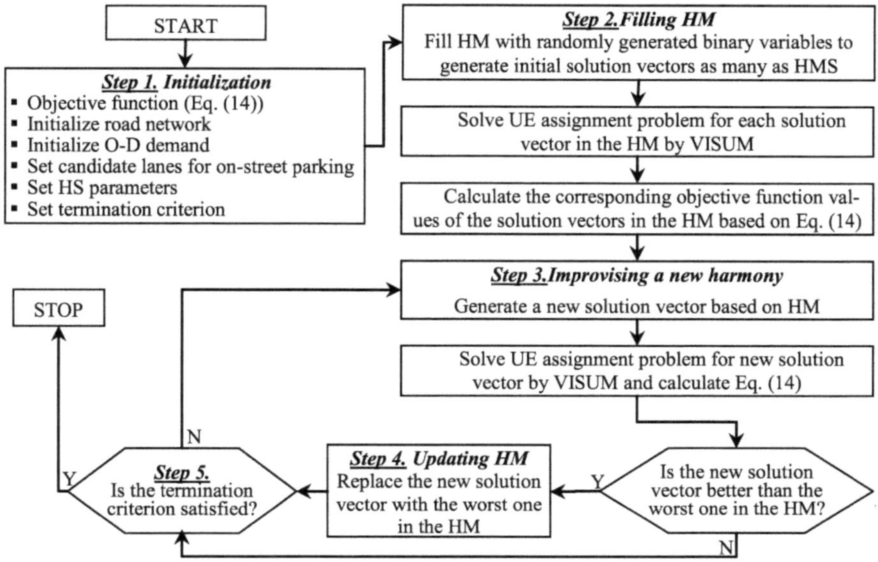

Fig. 1 Flow chart of the proposed model

traditional pitch adjusting process may become non-functional. Wang et al. [18] have modified Eq. (13) in order to improve the local search performance of the HS algorithm as follows:

$$u_h' = \begin{cases} u_h^1 & \text{if } \mathrm{Rand}(0,1) \le \mathrm{PAR} \\ u_h' & \text{otherwise} \end{cases} \tag{16}$$

where u_h^1 is the h-th element of the first, so-called best, solution vector in the HM. Therefore, the algorithm performs a better local search based on the current and best available solutions. Newly generated solution vector represents a new on-street parking space design and then the UE assignment problem is solved by VISUM traffic simulation tool, similarly to the previous step. Then, the corresponding objective function value is calculated based on Eq. (14). At Step 4, new solution vector is compared with the worst one in the HM in terms of their corresponding objective function values. If the new solution vector provides a better objective function value then it is included to the HM while the other is excluded. A preset termination criterion is checked at Step 5. If it is not satisfied then the computation is continued by iterating Steps 3–5.

4 Numerical Application

In this section, a numerical example is given to prove the performance of the proposed model. The test network, which represents the road network of the city of Sioux Falls, South Dakota, is given in Fig. 2.

Fig. 2 Layout of the test network

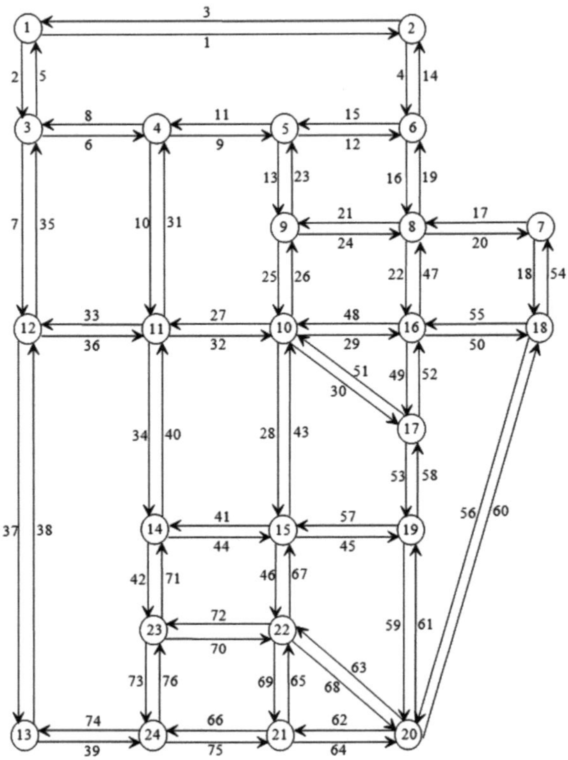

As can be seen in Fig. 2 that the test network consists of 76 links, 24 nodes and 552 O-D pairs. It is considered that each link consists of two traffic lanes. Link capacities and free flow travel times are taken from [19] and the original O-D travel demands, which were given in [6], are adjusted from daily to peak-hour figures by a factor of 0.05 for this study.

For the solution of the test problem, HMS, HMCR and PAR parameters are set as 40, 0.95 and 0.4, respectively and the model run is terminated after the 10000-th iteration. The arbitrary penalty factor α is set as 4 kms for each link with a volume to capacity ratio greater than "1". In this study, link travel times are calculated based on Bureau of Public Roads (BPR) travel time function, which has the following form [20]:

$$t_i = t_i^0\left[1 + 0.15(v_i/c_i)^4\right] \tag{17}$$

In Table 1, link lengths of Sioux Falls network are given. In this study, 26 links, for which the volume to capacity ratio is greater than "1", are excluded from the on-street parking candidates list. Thus, 50 links, one lane of which may be allocated to on-street parking, are candidates with total length of 112 km. The convergence history of the proposed model is presented in Fig. 3.

Table 1 Link lengths

Links	Length (km)
9, 11, 16, 18, 19, 49, 52, 53, 54, 58, 65, 69, 73, 76	1.0
17, 20, 25, 26, 37, 38, 45, 46, 50, 55, 57, 66, 67, 75	1.5
2, 5, 6, 7, 8, 12, 15, 29, 34, 35, 39, 40, 42, 48, 56, 59, 60, 61, 70, 71, 72	2.0
4, 13, 14, 22, 23, 27, 32, 41, 44, 47, 63, 68	2.5
1, 3, 10, 28, 31, 33, 36, 43, 62, 64	3.0
30, 51	4.0
21, 24	5.0

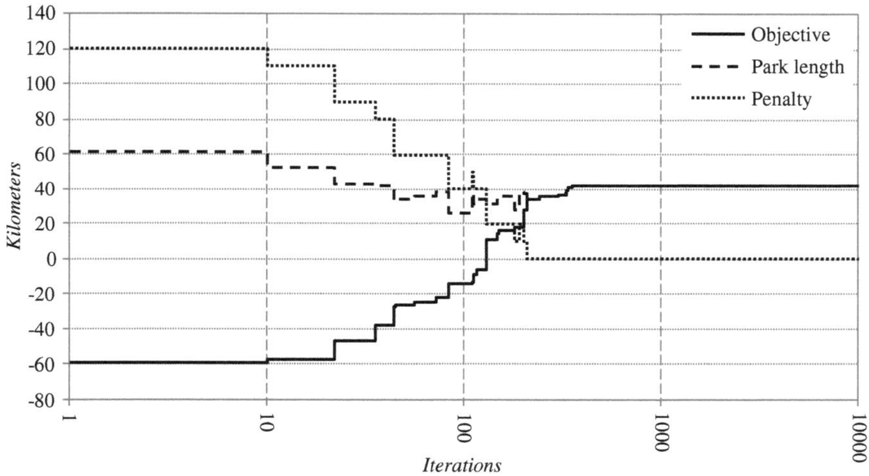

Fig. 3 Convergence history of the proposed model

Table 2 The results of the proposed model

Links allocated to on-street parking	Total length (km)
1, 2, 3, 5, 6, 7, 8, 9, 11, 18, 21, 24, 35, 37, 38, 50, 54, 55, 56, 60	42.0

It can be seen from Fig. 3 that the convergence is achieved after about 350 iterations. The results and new layout of the network are given in Table 2 and Fig. 4, respectively.

As can be seen in Fig. 4 that one lane of each 20 links are allocated to on-street parking and 42 km long parking space is provided. Changes of total travel time and average volume to capacity ratio through the road network are given in Table 3.

Table 3 shows that the total travel time on network changes below 1 % while average capacity usage changes about 9 % through the network after the application of the proposed model. In order to show the effect of different HS

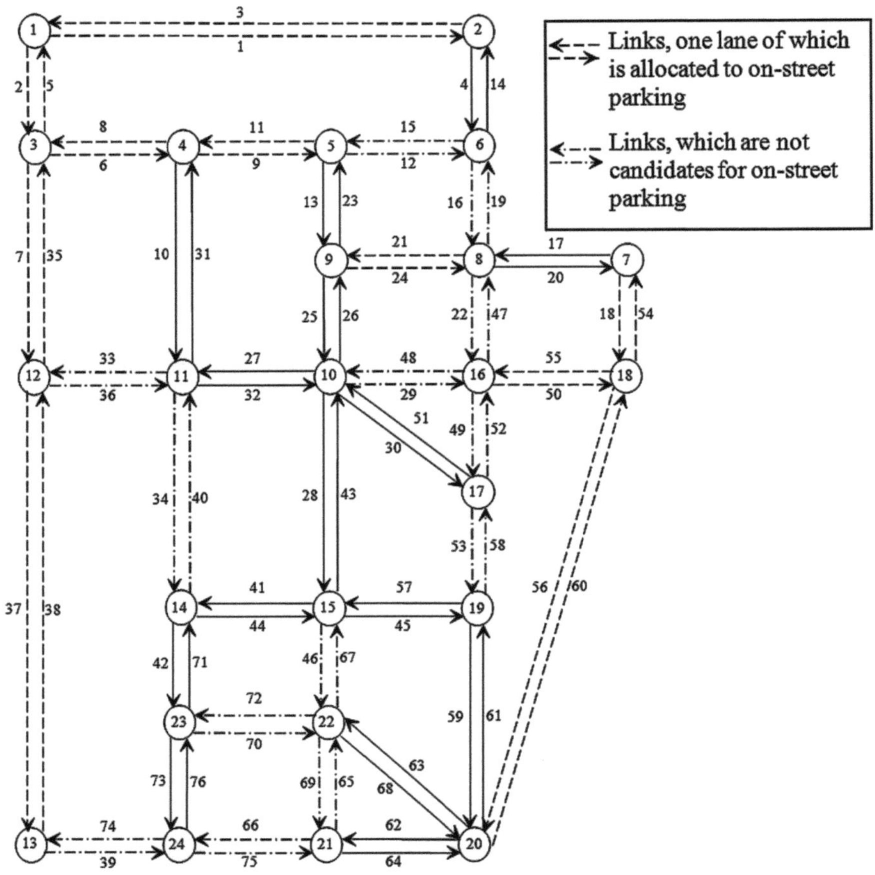

Fig. 4 Layout of the network after model application

Table 3 Total travel times and average volume to capacity ratios for the network

Case No	Original	Model results	Change
Total travel time (veh-h)	18705.88	18784.08	0.42
Average volume/capacity ratio	0.76	0.83	9.21

parameters on the solution accuracy of the model, the problem has been solved with different sets of HS parameters. Data sets and obtained results are given in Table 4.

It can be seen in Table 4 that the optimal/near-optimal solution has been achieved for all cases. Note that the solution has been reached after 539 and 2658 improvisations as the minimum and maximum number of iterations by the cases 7 and 3, respectively.

Table 4 Results of the proposed model for different sets of HS parameters

Case no	HMS	HMCR	PAR	Total length of parking places (km)	Iterations
1	40	0.85	0.20	42	1143
2	40	0.85	0.30	42	1654
3	40	0.85	0.40	42	2658
4	40	0.90	0.20	42	989
5	40	0.90	0.30	42	1114
6	40	0.90	0.40	42	1289
7	40	0.95	0.20	42	539
8	40	0.95	0.30	42	547

5 Conclusions

In this chapter, the problem of optimizing on-street parking places in urban road networks has been dealt within the DND context. For this purpose, a bi-level simulation/optimization model has been developed. In this model, the upper level represents the designer's strategy, which includes the decision of the lanes allocated to on-street parking, while the drivers' response are modeled in the lower level within the UE manner. Note that the traffic assignment problem is solved using VISUM traffic simulation tool while the upper level decisions are made by meta-heuristic HS optimization algorithm. Proposed model has been applied to a well-known test network and encouraging results have been obtained. In the future studies, delays and congestions arise from the parking maneuvers should be taken into account to improve the applicability of the model to the more realistic problems.

Acknowledgments Scientific Research Foundation of Pamukkale University with the Project No. 2013BSP009 is acknowledged.

References

1. Shoup, D.C.: The ideal source of public revenue. Reg. Sci. Urban Econ. **34**, 753–784 (2004)
2. Yousif, S., Purnawan.: Traffic operations at on-street parking facilities. P. I. Civil Eng.-Transp. 157, 189–194 (2004)
3. Portilla, A.I., Orena, B.A., Berodia, J.L.M., Diaz, F.J.R.: Using M/M/∞ queuing model in on-street parking maneuvers. J. Transp. Eng.-Asce. **135**(8), 527–535 (2009)
4. Guo, H., Gao, Z., Yang, X., Zhao, X., Wang, W.: Modeling travel time under the influence of on-street parking. J Transp. Eng.-Asce. **138**(2), 229–235 (2012)
5. Bruynooghe, M.: An optimal method of choice of investments in a transport network. Presentation, Planning and Transport Research and Computation Seminars on Urban Traffic Model Reasearch (1972)
6. LeBlanc, L.J.: An algorithm for the discrete network design problem. Transp. Sci. **9**(3), 183–199 (1975)

7. Poorzahedy, H., Turnquist, M.A.: Approximate algorithm for the discrete network design problem. Transp. Res. B.-Meth. 16(1), 45–55 (1982)
8. Heragu, S.S.: Facilities Design, p. 647. PWS Publishing Company, Boston (1997)
9. Pinedo, M.L.: Scheduling Theory. Algorithms and Systems, 3rd edn, p. 647. Springer, LLC (2008)
10. Gao, Z.Y., Wu, J.J., Sun, H.J.: Solution algorithm for the bi-level discrete network design problem. Transp. Res. B.-Meth. **39**, 479–495 (2005)
11. Duthie, J., Waller, S.T.: Incorporating environmental justice measures into equilibrium-based network design. J. Trans. Res. B. **2089**, 58–65 (2008)
12. Ceylan, H., Ceylan, H.: Discrete design of urban road networks with meta-heuristic harmony search algorithm. Tek Dergi **24**(1), 6211–6231 (2013)
13. Geem, Z.W., Kim, J.-H., Loganathan, G.V.: A new heuristic optimization algorithm: harmony search. Simulation **76**(2), 60–68 (2001)
14. Ceylan, H., Ceylan, H., Haldenbilen, S., Baskan, O.: Transport energy modeling with meta-heuristic harmony search algorithm, an application to Turkey. Energy Policy **36**, 2527–2535 (2008)
15. Baskan, O.: Harmony search algorithm for continuous network design problem with link capacity expansions. KSCE J. Civ. Eng. (2013). doi:10.1007/s12205-013-0122-6
16. Ceylan, H., Ceylan, H.: A hybrid harmony search and TRANSYT hill climbing algorithm for signalized stochastic equilibrium transportation networks. Transp. Res. C.-Emer. **25**, 152–167 (2012)
17. PTV AG.: VISUM 12.5 Fundemantals. Karlsruhe, Germany (2012)
18. Wang, L., Mao, Y., Niu, Q., Fei, M.: A multi-objective binary harmony search algorithm. Lect. Notes Comput. Sc. **6729**, 74–81 (2011)
19. Suwansirikul, C., Friesz, T.L., Tobin, R.L.: Equilibrium Decomposed Optimization: A Heuristic for the Continuous Equilibrium Network Design Problem. Transp. Sci. **21**(4), 254–263 (1987)
20. Bureau of Public Roads: Traffic Assignment Manual. U.S. Department of Commerce, Washington (1964)

Part IV
Optimization and Simulation (in Transportation)

Many of the problems which arise in transportation are solved by *optimization* approaches and *simulation* models. *Gallo et al.* study the Global Optimization of Signal Settings (GOSS) problem and propose a meta-heuristic algorithm for its solution. *Castelli et al.* present two bid price-based heuristic approaches to tackle stochastic price-oriented demand for air cargo. *Maia and Couto* propose a freight network optimization model, developed as a support tool for planning and policy decisions involved in improving rail networks at regional and national levels. *Gomes et al.* have developed a Decision Support System (DSS), integrating simulation and optimization, to help design and operate Demand Responsive Transportation services, minimizing operating costs and maximizing service quality. *Dell'Orco et al.* propose a microscopic model of crowd evacuation which incorporates the fuzzy perception and anxiety inherent in human reasoning. *Reyes et al.* formulate the O/D matrix adjustment problem based upon traffic counts as a bi-level optimization problem in which the Traffic Assignment Problem (TAP) is the lower level, and use new TAP methods in order to accelerate the convergence of the TAP and reduce the computational cost of the process.

Global Optimisation of Signal Settings: Meta-Heuristic Algorithms for Solving Real-Scale Problems

Mariano Gallo, Luca D'Acierno and Bruno Montella

Abstract In this chapter the *Global Optimisation of Signal Settings* (GOSS) problem is studied and a meta-heuristic algorithm is proposed for its solution. The GOSS problem arises when the parameters of all (or some) signalised intersections of a network are jointly optimised so as to minimise the value of an objective function (such as total travel time). This problem has been widely studied elsewhere and several algorithms have been proposed, mainly based on descent methods. These algorithms require high computing times for real-scale problems and usually lead to a local optimum since the objective function is hardly ever convex. The high computing times are due to the need to perform traffic assignment to determine the objective function at any iteration. In this chapter we propose a multi-start method based on a *Feasible Descent Direction Algorithm* (FDDA) for solving this problem. The algorithm is able to search for a local optimal solution and requires lower computing times at any iteration. The proposed algorithm is tested on a real-scale network, also under different demand levels, by adopting different assignment algorithms proposed in the literature. Initial results show that the proposed algorithms perform well and that computing times are compatible with planning purposes also for real-scale networks.

Keywords Signal settings · Network design · Metaheuristic algorithms

M. Gallo (✉)
Department of Engineering, University of Sannio, Benevento, Italy
e-mail: gallo@unisannio.it

L. D'Acierno · B. Montella
Department of Civil, Architectural and Environmental Engineering,
'Federico II' University of Naples, Naples, Italy
e-mail: luca.dacierno@unina.it

B. Montella
e-mail: bruno.montella@unina.it

J. F. de Sousa and R. Rossi (eds.), *Computer-based Modelling and Optimization in Transportation*, Advances in Intelligent Systems and Computing 262,
DOI: 10.1007/978-3-319-04630-3_14,

1 Introduction

In this chapter we consider the problem of optimising the signal settings of all (or some) signalised intersections of an urban network, assuming its physical configuration (topology and link dimensions) as fixed and invariable. This problem is a particular case of the more general *Equilibrium Network Design Problem* (ENDP), where signal settings assume the role of decision variables. This problem, also known as the *Signal Setting Design Problem* (SSDP), can be solved by following two different approaches [1]: a global approach (GOSS—*Global Optimisation of Signal Settings*) and a local approach (LOSS—*Local Optimisation of Signal Settings*). In the first case the signal settings of the network are designed so as to minimise total user travel time and the problem can be formulated with an optimisation model. In the second case, instead, the signal settings of each junction are designed so as to minimise only the total delay at that given junction, according to a specific local control policy, leading to a fixed-point model.

In this chapter we study the GOSS problem; the LOSS problem has been treated by the authors elsewhere in previous papers [2, 3]. Vis-á-vis the consolidated literature, we propose some metaheuristic algorithms for solving the problem that are able to reduce the computing times significantly when compared with other algorithms, and to search for several local optima since, except for simple and particular cases, the objective function is usually not convex.

The general formulation of the SSDP and the distinction between the global and local approach can be found in [1, 4–8]. The GOSS problem was studied in a static environment by [1, 9–15]. References [16–18] also considered the offsets as decision variables. References [19, 20] proposed group-based methods, while [21] studied the joint optimisation of signal settings and road pricing. A dynamic approach was proposed in [22] while Artificial Neural Network solution methods were proposed by [23, 24].

This chapter is organised as follows: Sect. 2 formulates the optimisation model; the proposed solution algorithms are described in Sect. 3; numerical results on a real-scale network are summarised in Sects. 4 and 5 concludes the chapter.

2 Optimisation Model

The GOSS problem consists in optimising jointly the values of the parameters of all (or some) signalised intersections of an urban network in order to optimise a given objective function. Since the objective function usually represents the total travel time on the network, the problem is one of minimisation. In this problem, the signal settings (effective green times, cycle lengths, etc.) take the roles of decision variables. In general, the GOSS problem can be formulated as follows:

$$\hat{\boldsymbol{g}} = \operatorname{Arg}\min_{\boldsymbol{g}} \; w(\boldsymbol{g}, \boldsymbol{f}) \tag{1}$$

subject to:

$$f = f^* \tag{2}$$

$$g \in G \tag{3}$$

where $\boldsymbol{g}$ is the signal settings vector; $\boldsymbol{f}$ is the link flow vector; $\hat{\boldsymbol{g}}$ is the optimal solution; $\boldsymbol{f}^*$ represents the equilibrium link flow vector (obtained by solving an equilibrium assignment problem on the network); $w(\cdot)$ represents the objective function; $\boldsymbol{f} = \boldsymbol{f}^*$ represents the assignment constraint; and G represents the feasible set for the signal settings, that is all constraints other than that of assignment.

In this chapter we consider the following assumptions: (a) transportation demand is rigid; (b) the cycle lengths are not considered variables of the problem; (c) all signalised intersections have only two phases; (d) the objective function is the total travel time on the network; (e) the route choice model is stochastic.

Assumptions (b) and (c) can be removed, although in this case the variables of the problem increase, as do the computing times for reaching a solution. Assumptions (b) and (c) allow the number of decision variables to be reduced to one for each signalised intersection. Indeed, if C_i stands for the effective cycle for intersection i, with g_i^A the effective green time for phase A of intersection i, the effective green time of the other phase is simply obtained as $g_i^B = C_i - g_i^A$.

Assumption (a) should be easily removed, complicating only the assignment procedure that should consider an elastic demand assignment model. Since the optimisation of traffic signals does not generally produce significant effects on user choices differing in path choice, such as destination or mode, the removal of this assumption is not useful. As for assumption (d), it is possible to choose another objective function (average travel speed, total delay, air pollution emissions, etc.) without problems; we choose this objective function because it is the one that best represents the overall operation of the network.

Finally, also assumption (e) could be easily removed, but on urban networks stochastic route choice models are better than their deterministic counterparts, since the alternative routes on an OD pair are numerous and their costs are sometimes close.

Under these assumptions, the GOSS model proposed in this chapter can be formulated as follows:

$$\hat{g} = \underset{g}{\operatorname{Arg\,min}} \; [\boldsymbol{c}(\boldsymbol{g}, \boldsymbol{f}^*)]^{\mathrm{T}} \boldsymbol{f}^* \tag{4}$$

with:

$$g^{\mathrm{T}} = \left[g_1^A, g_2^A, \ldots, g_i^A, \ldots, g_n^A\right] \tag{5}$$

subject to:

$$\boldsymbol{f}^* = \boldsymbol{\Delta P}\left(\boldsymbol{\Delta}^{\mathrm{T}} \boldsymbol{c}(\boldsymbol{g}, \boldsymbol{f}^*)\right)\boldsymbol{d} \tag{6}$$

$$g_{min} \leq g_i^A \leq C_i - g_{min} \qquad \forall i \tag{7}$$

$$g_i^B = C_i - g_i^A \qquad \forall i \tag{8}$$

where, over the already defined terms, $\boldsymbol{c}(\cdot)$ is the link cost vector; $\boldsymbol{P}(\cdot)$ is the route choice probability matrix; $\boldsymbol{\Delta}$ is the link-route incidence matrix; d is the demand vector; g_{min} is the minimum value of effective greens (for instance 15 s).

This model is a non-linear constrained optimisation model; the objective function is not linear, the assignment constraint is not linear and not in a closed form and the variables could be assumed either continuous or discrete. This model can be seen as a bi-level model where the upper level is the optimisation one and the lower level is represented by the solution of the assignment problem.

Under some quite mild assumptions on link cost functions, to each feasible solution, $\boldsymbol{g}$, corresponds one and only one configuration of equilibrium traffic flows, $\boldsymbol{f}^*$; the relation between signal settings and equilibrium traffic flows is an application: once $\boldsymbol{g}$ is fixed, $\boldsymbol{f}^*$ is exactly determined. Therefore, the constraint (6) can be formally expressed as:

$$\boldsymbol{f}^* = \boldsymbol{f}^*(\boldsymbol{g}) \tag{9}$$

The problem consists in finding the signal settings, $\hat{\boldsymbol{g}}$, to which equilibrium traffic flows, $\boldsymbol{f}^*(\hat{\boldsymbol{g}})$, correspond, which minimise the value of the objective function. Inserting constraint (9) inside the objective function, the optimisation model is simplified as follows:

$$\hat{\boldsymbol{g}} = \operatorname{Arg}\min_{\boldsymbol{g}} \; [\boldsymbol{c}(\boldsymbol{g}, \boldsymbol{f}^*(\boldsymbol{g}))]^{\mathrm{T}} \boldsymbol{f}^*(\boldsymbol{g}) \tag{10}$$

subject to:

$$g_{min} \leq g_i^A \leq C_i - g_{min} \qquad \forall i \tag{11}$$

$$g_i^B = C_i - g_i^A \qquad \forall i \tag{12}$$

The solution of this model requires calculation of equilibrium traffic flows, $\boldsymbol{f}^*$, at each objective function evaluation, solving the fixed point problem (6). Therefore, it is very important, in order to reduce computing times, to minimise the time for calculating the equilibrium traffic flows, whatever the algorithm adopted for solving the optimisation problem (10–12) is.

3 Solution Algorithms

For solving the optimisation problem described by equations (10–12) we propose to use a multi-start method based on a *Feasible Descent Direction Algorithm* (FDDA). The multi-start approach is necessary since the objective function is not convex, except for simple cases, and looking for more local optimal solutions can improve the final results, even if it requires higher computing times. In particular, in the proposed algorithm we will assume that decision variables (i.e. $\boldsymbol{g}$) are discrete and express the duration in seconds of the effective green times.

The algorithm considers different solutions as starting points for the descent direction local search, leading to several local optima. The different starting points can be generated in different ways. Hence, we propose the following starting points: (a) all variables g_i^A equal to 50 % of the cycle; (b) all variables g_i^A equal to g_{min}; (c) all variables g_i^A equal to $C_i - g_{min}$; (d) all variables g_i^A equal to g_i^{A*}, where g_i^{A*} represents the solution of the LOSS problem, i.e. the values of signal settings that are congruent with equilibrium traffic flows and local control policy (see [1, 8]); (e) random values for all variables g_i^A. Obviously, several combinations of these methods can be adopted. In what follows, the variable g_i represents the variable g_i^A.

Concerning the FDDA, in order to minimise computing times we do not use a numerical gradient to generate descent directions since it requires that derivative values of the objective function be numerically calculated, which implies the solution of an equilibrium assignment problem for each partial derivative. Thus the proposed algorithm chooses a variable, g_i, and searches for the optimal value of the variable assuming the other variables as fixed. In particular, the search phase of the FDDA uses a predetermined step of some seconds, added with or subtracted from the current value of the analysed variable. Moreover, the step is reduced when the objective function stops decreasing, until a minimum step value (for instance, 1 s) along the direction is reached. The algorithm then chooses another direction and operates in the same way, stopping when no steps in any direction are able to reduce the objective function further.

Various algorithms can be proposed according to the approach used to choose the direction to follow (i.e. the variable to minimise). In this chapter we propose to analyse two approaches:

- a *Steepest Descent Method* (SDM), where the descent direction in terms of decisional variable (i.e. vector component) to be analysed (and related increase or decrease) is chosen so that it produces the best reduction of the objective function;
- a *Random Descent Method* (RDM), where the descent direction (i.e. the vector component to be modified and its increase or decrease) is chosen randomly.

Although the SDM approach seems to be the better one since it allows us to identify the direction which potentially minimises the objective function to a greater extent, it requires that an equilibrium assignment problem be solved for each direction (i.e. 2 $\times$ the number of variables) prior to choosing the descent direction; the latter method, instead, does not require these preliminary assignments. Moreover, several applications of a neighbourhood search algorithm in the case of large-scale networks (see, for instance, [25, 26]) have shown that an RDM approach provides good results in reasonable calculation times when compared to the SDM.

3.1 Assignment Algorithms

To solve the fixed-point problem (6), we analyse the use of three kinds of algorithms based on an MSA framework ([27, 28]). Algorithms based on the MSA (*Method of Successive Averages*) framework are widely used for solving traffic assignment problems formulated as fixed-point problems. The main MSA algorithms are the MSA-FA (*Flow Averaging*), which was proposed by [28]; the MSA-CA (*Cost Averaging*) devised by [29]; and the MSA-ACO (*Ant Colony Optimisation*), which was proposed by [30]. Even if these algorithms could also be used under the assumption of deterministic route choice models, in the following we will refer only to stochastic route choice models, that work better for simulating user behaviour on urban networks.

All these algorithms are based on the calculation of a sequence of network loadings (i.e. assignment for uncongested networks) and stop when the link traffic flows are equal (in practice, a stop threshold is used) to the uncongested network loading traffic flows.

The MSA-FA averages at each iteration the uncongested link flows, $\boldsymbol{f}^k_{UNL}$, with the results of the previous iteration $\boldsymbol{f}^{k-1}$, as follows:

$$\begin{aligned}
&k = k + 1\\
&\boldsymbol{c}^k = \boldsymbol{c}\left(\boldsymbol{f}^{k-1}\right)\\
&\boldsymbol{f}^k_{UNL} = \boldsymbol{f}_{UNL}\left(\boldsymbol{c}^k\right)\\
&\boldsymbol{f}^k = \boldsymbol{f}^{k-1} + 1/k\left(\boldsymbol{f}^k_{UNL} - \boldsymbol{f}^{k-1}\right)
\end{aligned}$$

The MSA-CA is based on the same general framework of the MSA-FA but the costs, instead of flows, are averaged, as follows:

$$\begin{aligned}
&k = k + 1\\
&\boldsymbol{f}^k = \boldsymbol{f}_{UNL}\left(\boldsymbol{c}^{k-1}\right)\\
&\boldsymbol{y}^k = \boldsymbol{c}\left(\boldsymbol{f}^k\right)\\
&\boldsymbol{c}^k = \boldsymbol{c}^{k-1} + 1/k\left(\boldsymbol{y}^k - \boldsymbol{c}^{k-1}\right)
\end{aligned}$$

Finally, the MSA-ACO algorithm, based on Ant Colony Optimisation ([31, 32]), uses the general framework of the MSA as follows:

$$\begin{aligned}
&k = k + 1\\
&\boldsymbol{c}^k = \boldsymbol{c}\left(\boldsymbol{f}^{k-1}\right)\\
&\Delta\boldsymbol{\tau}^k = \boldsymbol{\tau}\left(\boldsymbol{c}^k\right)\\
&\boldsymbol{\tau}^k = \boldsymbol{\tau}^{k-1} + 1/k\left(\Delta\boldsymbol{\tau}^k - \boldsymbol{\tau}^{k-1}\right)\\
&\boldsymbol{f}^k = \boldsymbol{f}_{UNL}\left(\boldsymbol{\tau}^k\right)
\end{aligned}$$

where $\boldsymbol{\tau}^k$ and $\Delta\boldsymbol{\tau}^k$ represent, respectively, the pheromone trail and the related increase at iteration k.

Generally, the above algorithms adopt as a starting condition a null flow vector (i.e. $\boldsymbol{f}^0 = 0$). Moreover, MSA-CA and MSA-ACO need initial network loading in order to have, respectively, a link cost vector $(\boldsymbol{c}^0)$ and a pheromone trail vector $(\boldsymbol{\tau}^0)$ consistent with the network topology and travel demand.

Since in the implementation of the GOSS problem we assume travel demand as rigid and the network topology as invariant, we may implement assignment algorithms according to two different approaches: (a) the starting flow vector at each iteration of the FDDA algorithm is always the null vector (i.e. $\boldsymbol{f}^0 = 0$); (b) the starting flow vector at each iteration of the FDDA algorithm, except for the first iteration, is equal to the equilibrium flow vector of the previous iteration. Indeed, in the latter case the initial flow vector, albeit not an equilibrium according to the new signal setting configuration, is closer than the null vector to the equilibrium solution. Moreover, in the case of MSA-CA and MSA-ACO it is possible to avoid the pre-loading phase since equilibrium vectors of the previous iterations are always consistent with network topology and travel demand.

Hence, in the numerical application we indicate as MSA-xx-F0 the algorithms based on the first approach and MSA-xx-UE the algorithms based on the latter approach.

4 Numerical Results

The proposed algorithms were tested on the urban network of Benevento, a town of about 61,000 inhabitants in the south of Italy. The transportation model (i.e. the supply network and travel demand) was built when the town's Urban Traffic Plan was drawn up. The network graph consists of 1,577 oriented links (corresponding to 216 kms of roads) and 678 nodes (66 internal zone centroids, 14 external zone centroids and 598 generic nodes). The travel demand matrix was estimated by a system of random utility models (see [33]) calibrated for other urban networks and adapted to the specific case by means of traffic data collected at 139 count sections. The network has only eight signalised intersections whose operation can be modelled by means of nine two-phase intersections, since one of them can be split into two different signalised intersections. Figure 1 shows the network graph where the grey and black nodes represent respectively the 80 centroids and 9 signalised intersections.

Initial tests were implemented to compare the SDM approach with that of the RDM, by applying all assignment algorithms in the case of the starting point (a), i.e. when all design variables (effective green times) are equal to 50 % of the effective cycle.

In particular, Figs. 2, 3 and Table 1 show that, consistently with the literature (such as [25, 26]), the RDM almost always requires less computational effort both

Fig. 1 Supply model

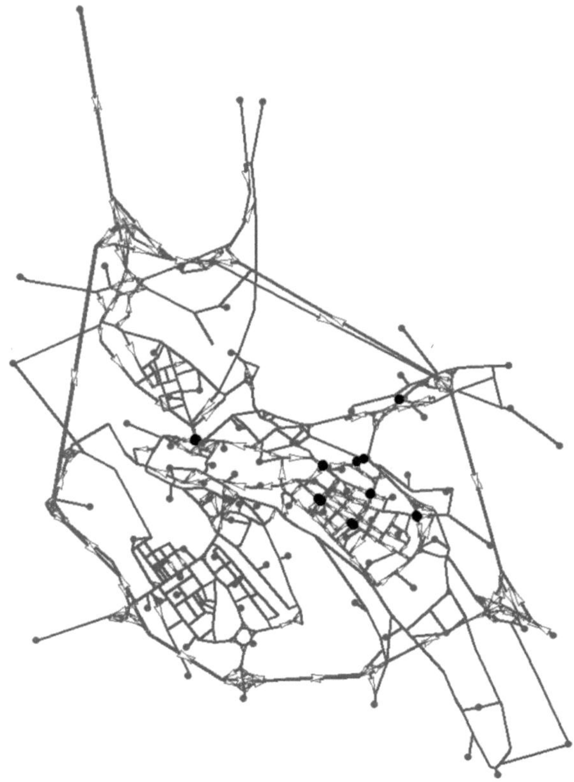

in terms of calculation times and algorithmic steps. In particular, as shown below, the quantity which best describes the computational effort is the number of *Uncongested Network Loadings* (UNLs) rather than the number of *Algorithm Iterations* (AIs).

Likewise, in terms of assignment algorithms, the MSA-ACO provides similar results to the other algorithms with almost always a lower number of UNLs and hence the calculation time is lower too. In particular, the approach which uses equilibrium flows from the previous assignment (i.e. the MSA-ACO-UE), allows a considerable reduction in the number of UNLs and hence elaboration time.

However, it is worth noting that, as shown elsewhere (see, for instance, [30]), although assignment models have the same theoretical solution numerically due to the stop threshold, they provide slightly different solutions. This difference in terms of equilibrium flow estimation obviously yields a difference in objective function values related to a generic $\boldsymbol{g}$ vector, since the objective function depends on equilibrium flow values. With these considerations, a generic assignment algorithm based on the first assignment approach (i.e. initial flow always null) will always provide the same result in terms of equilibrium flow and hence in terms of objective function value. Likewise, assignment algorithms based on the second

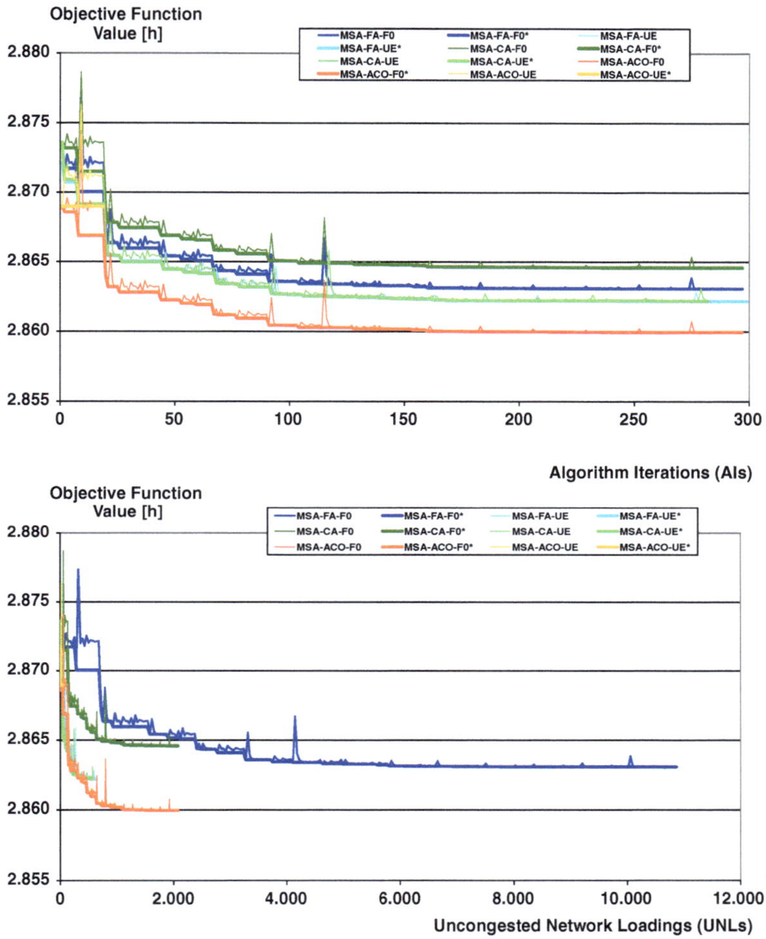

Fig. 2 SDM approach in the case of starting point (*a*)

assignment approach (i.e. initial flows equal to the equilibrium flows of the previous iteration) will provide equilibrium flows, and hence an objective function estimation, depending on the initial flow vector. Hence, the calculation of the objective function related to a generic $\boldsymbol{g}$ vector, once the assignment algorithm is fixed, could not be unable to provide always the same result.

Further applications were implemented to test the usefulness of applying the FDDA algorithm by means of an RDM approach in the case of MSA-ACO-UE as an assignment algorithm. In particular, confirming the above considerations, we analysed:

- the implementation of the RDM approach, by increasing travel demand by a factor of 1.6, in order to verify whether the positive performance of assignment algorithms is maintained (see Table 2);

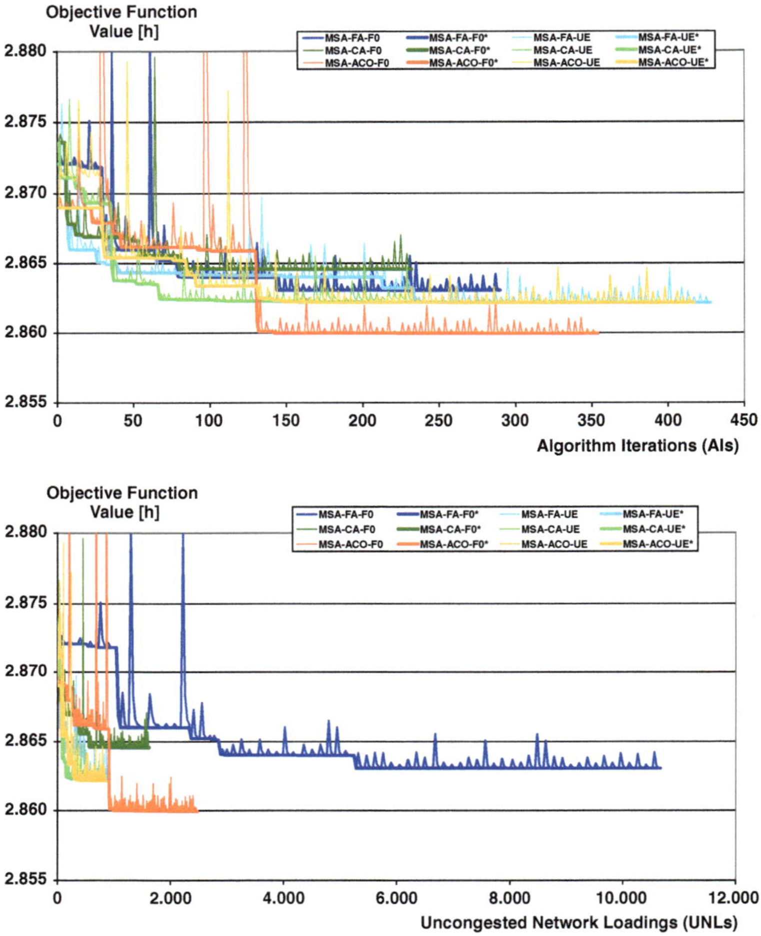

Fig. 3 RDM approach in the case of starting point (*a*)

- the implementation of the RDM approach with an MSA-ACO-UE by adopting different random sequences for identifying descent directions (see Table 3).

Finally, in order to apply the multi-start approach with the FDDA-RDM, we applied it in the following cases (see Table 4):

- 1 initial solution where all variables are equal to g_{min} (i.e. starting point *b*);
- 1 initial solution where all variables are equal to g_{max} (i.e. starting point *c*);
- 1 initial solution where all variables are equal to the solution of the LOSS problem (i.e. starting point *d*);
- 5 initial solutions where any variable is obtained by means of consecutive random draws (i.e. starting point *e*).

Table 1 Comparison between SDM and RDM approaches

FDDA approach	Assignment algorithm	Decisional variables									Optimal objective function value	Algorithm iterations (AIs)	UNLs	UNLs/AIs	Calculation times [min]
		g_1	g_2	g_3	g_4	g_5	g_6	g_7	g_8	g_9					
Initial values: starting point (a)		45	45	45	45	45	45	45	45	45					
SDM	MSA-FA-F0	55	47	55	65	52	58	53	43	45	2863.078	297	10870	36.599	145.62
	MSA-FA-UE	55	45	55	65	50	58	53	43	45	2862.182	299	638	2.134	8.85
	MSA-CA-F0	55	47	55	65	52	58	53	43	45	2864.588	297	2079	7.000	26.30
	MSA-CA-UE	55	49	55	66	50	58	55	42	45	2862.212	282	574	2.035	7.96
	MSA-ACO-F0	55	47	55	65	52	58	53	43	45	2859.963	297	2079	7.000	27.90
	MSA-ACO-UE	45	45	45	45	45	45	45	45	45	2868.986	19	43	2.263	0.61
RDM	MSA-FA-F0	55	47	55	65	51	58	53	43	45	2863.071	290	10668	36.786	133.29
	MSA-FA-UE	55	47	55	65	51	58	53	42	45	2862.165	428	919	2.147	12.31
	MSA-CA-F0	55	47	55	65	51	58	53	43	45	2864.581	232	1624	7.000	21.50
	MSA-CA-UE	51	45	56	65	49	58	49	45	45	2862.216	232	483	2.082	6.53
	MSA-ACO-F0	55	47	55	65	51	58	53	43	45	2859.956	354	2478	7.000	31.29
	MSA-ACO-UE	55	47	55	65	51	58	53	43	45	2862.164	417	854	2.048	10.93

Table 2 Implementation of the FDDA-RDM in the case of travel demand multiplied by a factor of 1.6

FDDA approach	Assignment algorithm	Decisional variables									Optimal objective function value	Algorithm iterations (AIs)	UNLs	UNLs/AIs	Calculation times [min]
		g_1	g_2	g_3	g_4	g_5	g_6	g_7	g_8	g_9					
Initial values: starting point (a)		45	45	45	45	45	45	45	45	45					
RDM	MSA-FA-F0	56	49	51	66	49	60	56	42	54	10827.231	385	94719	246.023	1199.48
	MSA-FA-UE	56	49	52	66	50	60	59	42	54	10824.643	311	1781	5.727	23.07
	MSA-CA-F0	56	49	51	66	50	60	56	42	54	10825.164	410	4490	10.951	54.80
	MSA-CA-UE	56	49	52	66	49	60	56	43	54	10824.569	472	1992	4.220	24.37
	MSA-ACO-F0	56	49	51	66	49	60	56	42	54	10824.553	345	4830	14.000	61.44
	MSA-ACO-UE	56	49	51	66	49	60	57	41	54	10824.635	296	990	3.345	12.63

Table 3 Implementation of the FDDA-RDM with an MSA-ACO-UE with different random sequences

	Decisional variables									Optimal objective function value	Algorithm iterations (AIs)	UNLs	UNLs/AIs	Calculation times [min]
	g_1	g_2	g_3	g_4	g_5	g_6	g_7	g_8	g_9					
Initial values: starting point (a)	45	45	45	45	45	45	45	45	45					
Random sequence no. 1	55	47	55	65	51	58	53	43	45	2862.164	417	854	2.048	10.93
Random sequence no. 2	57	45	55	66	50	60	55	43	45	2862.225	160	336	2.100	4.33
Random sequence no. 3	56	46	55	65	51	58	55	43	45	2862.167	531	1080	2.034	13.99
Random sequence no. 4	55	47	55	65	51	58	53	43	46	2862.164	545	1115	2.046	14.18
Random sequence no. 5	55	47	55	65	51	58	53	43	45	2862.164	364	739	2.030	9.61
Random sequence no. 6	55	47	55	65	52	58	53	42	45	2862.137	483	993	2.056	13.07

Table 4 Implementation of the multi-start approach in the case of FDDA-RDM

	Decisional variables									Optimal objective function value	Algorithm iterations (AIs)	UNLs	UNLs/AIs	Calculation times [min]
	g_1	g_2	g_3	g_4	g_5	g_6	g_7	g_8	g_9					
Initial values: starting point (*b*)	15	15	15	15	15	15	15	15	15					
Algorithm implementation	54	48	55	65	51	58	53	42	45	2862.163	446	1344	3.013	17.58
Initial values: starting point (*c*)	75	75	75	75	75	75	75	75	75					
Algorithm implementation	55	47	55	65	51	58	53	42	45	2862.164	578	1211	2.095	15.02
Initial values: starting point (*d*)	59	51	50	69	53	59	58	40	59					
Algorithm implementation	56	49	51	66	49	60	56	41	55	2862.165	287	942	3.282	11.57
Initial values #1: starting point (*e*)	56	65	54	34	62	44	55	39	73					
Algorithm implementation	56	45	55	65	52	58	53	44	45	2862.180	287	600	2.091	7.87
Initial values #2: starting point (*e*)	48	37	36	65	27	16	57	68	35					
Algorithm implementation	54	47	56	65	51	58	53	44	45	2862.157	270	567	2.100	7.44
Initial values #3: starting point (*e*)	21	18	46	25	55	38	73	51	24					
Algorithm implementation	56	46	55	65	51	58	53	43	45	2862.158	566	1234	2.180	16.19
Initial values #4: starting point (*e*)	28	49	73	61	65	65	50	40	19					
Algorithm implementation	56	47	55	65	50	58	55	43	46	2862.175	309	680	2.201	8.93
Initial values #5: starting point (*e*)	44	34	27	49	52	40	44	37	52					
Algorithm implementation	56	47	55	65	51	58	53	43	47	2862.168	559	1154	2.064	15.16

5 Conclusions and Research Prospects

This chapter focused on the *Global Optimisation of Signal Settings* (GOSS) problem that arises when we assume that the signal settings of each junction are designed jointly so as to minimise total travel time on the network. This problem can be formulated as a bilevel constrained optimisation problem where the lower level consists in implementing an equilibrium traffic assignment problem. In this chapter we applied a multi-start method based on a *Feasible Descent Direction Algorithm* (FDDA). In particular, we defined the descent direction according to two different approaches: a *Steepest Descent Method* (SDM) and a *Random Descent Method* (RDM). Finally, we proposed to solve the assignment problem by means of three algorithms (MSA-FA, MSA-CA and MSA-ACO) formulated with two different approaches according to initial network flows.

The first major result is that, according to the literature (see, for instance, [30]), in the case of real-scale networks, the MSA-ACO allows the assignment problem to be solved in a lower calculation time but with the same accuracy in determining equilibrium solutions compared with other traditional MSA algorithms (i.e. MSA-FA and MSA-CA). Likewise, the adoption of an RDM approach in the FDDA allows local optima to be determined in fewer iterations compared to the SDM approach. Moreover, due to the non-convexity of the objective function, it is not possible to state a priori that an RDM approach provides a worse solution than an SDM approach (indeed, in some cases, the RDM has provided a better result).

Numerical applications have shown that the use of different draw sequences for implementing the RDM provides different solutions both in terms of objective function value and in terms of decisional variable optimal values.

Finally, the use of initial flows equal to the equilibrium flows in the previous iteration allows the assignment problem to be solved in a shorter calculation time. However, it is worth noting that since MSA algorithms stop when a stopping criterion (which consists in a threshold of the difference between initial flows and uncongested network flows) is satisfied, the same MSA-xx-UE algorithm, in the case of different initial flows, could provide different equilibrium flows which all verify the stopping criterion but provide different values of the objective function. Hence, if the stopping criterion threshold of the assignment algorithm is comparable in order of magnitude with the objective function variation in a neighbourhood of the local optimum, the stopping criterion could lead to different optimal solutions.

Future research could fruitfully focus on the search for other algorithms to solve the GOSS problem, further reducing computing times. MSA-ACO-F0 and MSA-ACO-UE could be combined to obtain a good compromise between the reduction in computation time and the uniqueness of the objective function values. Finally, the solution of the GOSS problem could be used within methods and algorithms for solving the (topological) Urban Network Design Problem.

Acknowledgments Partially supported by the Italian MIUR under PRIN2009 grant no. 2009EP3S42_002.

References

1. Cascetta, E., Gallo, M., Montella, B.: Models and algorithms for the optimization of signal settings on urban networks with stochastic assignment. Ann. Oper. Res. **144**, 301–328 (2006)
2. D'Acierno, L., Gallo, M., Montella, B.: An ant colony optimisation algorithm for solving the asymmetric traffic assignment problem. Eur. J. Oper. Res. S **217,** 459–469 (2012)
3. Gallo, M., D'Acierno, L.: Comparing algorithms for solving the local optimisation of the signal settings (loss) problem under different supply and demand configurations. Procedia Soc. Behav. Sci. **87**, 147–162 (2013)
4. Marcotte, P.: Network optimization with continuous control parameters. Transport. Sci. **17**, 181–197 (1983)
5. Fisk, C.S.: Game theory and transportation systems modelling. Transport. Res. B-Meth. **18**, 301–313 (1984)
6. Cantarella, G.E., Improta, G., Sforza, A.: Road Network Signal Setting: Equilibrium Conditions. In: Papageorgiou, M. (ed.) Concise encyclopedia of traffic and transportation systems, pp. 366–371. Pergamon Press, Amsterdam (1991)
7. Cantarella, G.E., Sforza, A.: Network design models and methods for urban traffic management. In: Gartner, N.H., Improta, G. (eds.) Urban traffic networks–dynamic flow modeling and control, pp. 123–153. Springer, Berlin (1995)
8. Cascetta, E., Gallo, M., Montella, B.: An asymmetric sue model for the combined assignment-control problem. In: Selected proceedings of 8th WCTR, vol. 2, pp. 189–202. Pergamon Press, The Netherlands (1999)
9. Sheffi, Y., Powell, W.B.: Optimal signal settings over transportation networks. J. Transp. Eng.-ASCE **109,** 824–839 (1983)
10. Heydecker, B.G., Khoo, T.K.: The equilibrium network design problem. In: Proceedings of AIRO '90, Conference on Models and Methods for Decision Support, pp. 587–602. Sorrento, Italy (1990)
11. Yang, H., Yagar, S.: Traffic assignment and signal control in saturated road networks. Transport. Res. A-Pol. **29**, 125–139 (1995)
12. Oda, T., Otokita, T., Tsugui, T., Mashiyama, Y.: Application of simulated annealing to optimization of traffic signal timings. In: Preprints of the 8th IFAC Symposium on Transportation Systems. Chania, Greece (1997)
13. Wong, S.C., Yang, H.: Reserve capacity of a signal-controlled road network. Transport. Res. B-Meth. **31**, 397–402 (1997)
14. Chiou, S.-W.: Optimization of area traffic control for equilibrium network flows. Transport. Sci. **33**, 279–289 (1999)
15. Ziyou, G., Yifan, S.: A reserve capacity model of optimal signal control with user-equilibrium route choice. Transport. Res. B-Meth. **36**, 313–323 (2002)
16. Heydecker, B.G.: A decomposition approach for signal optimisation in road networks. Transport. Res. B-Meth. **30**, 99–114 (1996)
17. Pillai, R.S., Rathi, A.K., Cohen, S.L.: A restricted branch and bound approach for generating maximum bandwidth signal timing plans for traffic networks. Transport. Res. B-Meth. **32**, 517–529 (1998)
18. Wey, W.-M.: Model formulation and solution algorithm of traffic signal control in an urban network. Comput. Environ. Urban Syst. **24**, 355–377 (2000)
19. Wong, S.C.: Group-based optimisation of signal timings using parallel computing. Transport. Res. C-Emer. **5**, 123–139 (1997)
20. Wong, S.C., Wong, W.T., Leung, C.M., Tong, C.O.: Group-based optimization of a time-dependent TRANSYT traffic model for area traffic control. Transport. Res. B-Meth. **36**, 291–312 (2002)
21. Smith, M.J., Xiang, Y., Yarrow, R.: Bilevel optimisation of signal timings and road prices on urban road networks. In: Preprints of the 8th IFAC Symposium on Transportation Systems, Chania, Greece (1997)

22. Abu-Lebdeh, G., Benekohal, R.F.: Design and evaluation of dynamic traffic management strategies for congested conditions. Transport. Res. A-Pol. **37**, 109–127 (2003)
23. Saraf, R.K.: Adaptive traffic control using neural networks. PhD Thesis, Vanderbilt University, Nashville (TN), USA (1994)
24. Spall, J.C., Chin, D.C.: Traffic-responsive signal timing for system-wide traffic control. Transport. Res. C-Emer. **5**, 153–163 (1997)
25. Gallo, M., D'Acierno, L., Montella, B.: A meta-heuristic approach for solving the urban network design problem. Eur. J. Oper. Res. **201**, 144–157 (2010)
26. Gallo, M., Montella, B., D'Acierno, L.: The transit network design problem with elastic demand and internalisation of external costs: an application to rail frequency optimisation. Transport. Res. C-Emer. **19**, 1276–1305 (2011)
27. Powell, W.B., Sheffi, Y.: The convergence of equilibrium algorithms with predetermined step sizes. Transport. Sci. **6**, 45–55 (1982)
28. Sheffi, Y., Powell, W.B.: An algorithm for the traffic assignment problem with random link costs. Networks **12**, 191–207 (1982)
29. Cantarella, G.E.: A general fixed-point approach to multimodal multi-user equilibrium assignment with elastic demand. Transport. Sci. **31**, 107–128 (1997)
30. D'Acierno, L., Montella, B., De Lucia, F.: A stochastic traffic assignment algorithm based on ant colony optimisation. Lect. Notes Comput. Sci. **4150**, 25–36 (2006)
31. Dorigo, M.: Optimization, learning and natural algorithms (in Italian). PhD Thesis, Department of Electronics, Polytechnic of Milan, Italy (1992)
32. Dorigo, M., Stützle, T.: Ant Colony Optimization. The MIT Press, Cambridge (2004)
33. Cascetta, E.: Transportation Systems Analysis: Models and applications. Springer, New York (2009)

Bid-Price Heuristics for Unrestricted Fare Structures in Cargo Revenue Management

Lorenzo Castelli, Raffaele Pesenti and Desirée Rigonat

Abstract In the present work we propose two bid-price based heuristic approaches to tackle a stochastic price-oriented demand of air cargo transportation. We assume fares are non-decreasing over time: the earlier the booking, the cheaper the fare. We consider a single-leg flight without overbooking practices or no-show customers. The proposed framework is suited for air cargo carriers providing a unique product to all its price-oriented customers. The business sustainability relies on a significant reduction in fares that would outperform other benefits, an earlier time of delivery above all. Nevertheless, our modelling framework may be easily extended to other modes of cargo transportation, such as maritime, where a given shipment receives the same service regardless the paid fare, which, in turn, only depends on the time the booking request is made.

Keywords Heuristics • Revenue management • Capacity management • Air cargo • Dynamic programming • Bid-price

1 Introduction

Airlines traditionally prevent high fare paying passengers from buying down into lower fare classes by associating restrictions to each fare level. They make different fares correspond to different products whose characteristics fit the needs

L. Castelli (✉) · D. Rigonat
Dipartimento di Ingegneria e Architettura, Università degli Studi di Trieste,
Via A. Valerio 10, 34127 Trieste, Italy
e-mail: castelli@units.it

D. Rigonat
e-mail: desiree.rigonat@phd.units.it

R. Pesenti
Dipartimento di Management, Università Ca' Foscari di Venezia,
Fondamenta San Giobbe 837, 30121 Cannaregio, VE, Italy
e-mail: pesenti@unive.it

J. F. de Sousa and R. Rossi (eds.), *Computer-based Modelling and Optimization in Transportation*, Advances in Intelligent Systems and Computing 262,
DOI: 10.1007/978-3-319-04630-3_15,

of only one class of customers. This fare policy is reasonable in presence, for each class, of a product-oriented demand, interested to a specific fare product and independent of the availability of cheaper services [25]. The emergence of low-cost carriers shows that this assumption of independent demand segments is becoming more and more unrealistic. Indeed, low-cost carriers offer a single type of product to price-oriented passengers that ignore ticketing restrictions and purchase solely on price [6]. Also, these passengers typically exploit the potentiality of the Internet-based distribution channels to compare the fares of several different airlines.

Precise modelling of customer choice behaviour has been a subject of growing interest in recent years [7, 16]. In fact, the application of pricing algorithms that assume independent demand to a non-segmented market gives rise to the spiral-down effect: customers willing to pay a higher fare but accepting a lower one if available are recorded as lower fare demand when the cheaper product is available. Then, forecasts built on these cheaper product sales underestimate demand for higher fare levels and more low fare products than necessary are made available and, consequently, revenues spiral down [10]. To contrast such an effect, the recent literature proposes pricing policies based on Revenue Management (RM) approaches that segment passengers according to their willingness to pay instead of their compliance to restrictions [27].

In the present work we study the application of these RM techniques to the air cargo industry. Specifically, we propose and analyse the performances of two bid-price based heuristic approaches to tackle a stochastic price-oriented demand of cargo shipments.

The RM approaches are essential tools for cargo shipment as the demand in this industry is in general price-oriented, although some segmentation of the demand by offered product still exists. The same shipment may be charged differently depending on the guaranteed delivery time (express delivery vs. standard delivery). In addition, some airlines, e.g., Lufthansa and American Airlines, offer further optional product features that include boarding priority, pick-up time and location at destination.

The air cargo industry accounts for tens of millions of dollars a year in revenue and, according to the International Air Transport Association (IATA), has stabilised and even shown some weak signs of reprise in certain markets (http://www.iata.org/pressroom) after the recession following the 2008 crisis.

Price oriented demand has been a much explored field of study in RM in recent years. In particular, Westermann [27] describes how to integrate revenue management and dynamic pricing concepts based on willingness to pay at airlines with different fare structures. Hopperstad and Belobaba [12] introduce seat inventory control schemas in the single-leg case when demand is not independent from fare class. They forecast the total demand at the lowest fare and repartition it to the different higher fare classes by taking into account the passengers' willingness to pay higher fares. Thus they are in the position to compute the booking limits using traditional algorithms. Fiig et al. [11] address the coexistence of restricted and unrestricted fare structures in markets sharing the same leg(s) on a network. Using simulations, Cléaz-Savoyen [9] shows that the simultaneous application of the approaches described in Fiig et al. [11] and Hopperstad and Belobaba [12] allows a partial mitigation of the spiral down effect in certain markets. Unfortunately, RM approaches used for passenger flights cannot be directly applied to cargo flights. Indeed, Kasilingam [14], Billings et al. [5] and Slager and Kapteijns [22] point out that the structure of demand and services in the two industries present many differences. For example, each passenger requires just one seat, while each cargo shipment consumes capacity in terms of both weight and volume. Passenger demand presents seasonality patterns while the cargo one is usually more erratic, hence the former is easier to forecast than the latter. The number of passenger customers is usually greater than cargo customers. However, the latter ones make larger bookings, so the behaviour of few of them can significantly influence the prices paid by other customers, a condition which is generally not true for passengers.

In bid-price RM policies, threshold, or "bid", prices are set for each unit of resource. This kind of price setting was first introduced for airlines' seat booking by Smith and Penn [23] and Simpson [21]. Since then, they have become widely used due to their conceptual simplicity and easiness of implementation. Talluri and van Ryzin [24] give a comprehensive overview of bid-price techniques pointing out the difficulties arising in determining the right bid. Generally speaking determining the optimal bids may be computational cumbersome, even because they may change dynamically as the flight departure times approach. For this reason, Adelman [1] proposes to compute dynamic bid-prices through a Linear Programming (LP) approximate model. A drawback of this approach is that the number of variables grows exponentially. Bijvank et al. [4] aim at improving robustness towards uncertainty in the demand. To this end, they propose three heuristics that exploit scenario-based stochastic programming methods. Pricing policies for a price-oriented demand are also determined through dynamic programming approaches. As an example, Zhang [28] introduces a dynamic programming decomposition approach and shows that it outperforms static bid-fares one even when bids are frequently recomputed. Popescu et al. [19] use a dynamic programming approach to determine the bid-prices in presence of large shipments. Differently, in presence of small shipments they use a bid-price approach whose bids are obtained by approximating the booking requests with passenger arrival models.

Different authors discuss the consequences of imprecise demand models or incomplete demand data in the air cargo industry. Totamane et al. [26] point out that imprecise demand forecasting causes most cargo airlines to operate at an average ratio between the utilized capacity and the total capacity, the so-called load factor, of 50–70 %. To overcome this limit, they propose a learning algorithm, based on a producer/consumer model, which is able to deliver a 9 % revenue improvement. Luo et al. [17] address the problem of defining overbooking policies that take into account that most booking reservation systems do not keep track of unfulfilled requests. In this framework, they develop an overbooking model that, under appropriate assumptions, they prove providing the optimal overbooking limits. Amaruchkul et al. [3] study the role of asymmetrical information on customers' demand, operating cost, margin and reservation profit. They investigate under which conditions the maximum combined profit of the involved agents can be obtained in the presence of such an asymmetry. The same authors, in a previous work [2], address uncertainty in package volume, whereas Huang and Hsu [13] and Chew et al. [8] deal with uncertainty in the supplied capacity.

The remainder of this paper is organised as follows. In Sect. 2, we formulate the problem of interest using dynamic programming. In Sect. 3, we introduce two heuristic bid price-based approaches. In Sect. 4, we describe the experiments we carried out to compare the performances of the algorithms and we address the analysis of the results. Finally, in Sect. 5, we draw some conclusions.

2 Dynamic Programming Formulation

In this section, we formulate the problem that a revenue manager faces when he tries to maximise a cargo flight expected revenues as a dynamic programming problem.

The revenue manager works within a single user-class framework, i.e., demand is not segmented or restricted. We assume that there exists a finite set F of M fares $F = \{f_1, \ldots, f_M\}$ which are unknown to the customer. The revenue manager has to decide which of the M fares to propose to the customer, and this decision is taken a posteriori, i.e., after having known the size of the shipment, and within the limits set by the following marketing rules that we assume to hold:

2.1 Assumptions

1. A shipment can be accepted only if the flight has residual capacity in terms of volume and weight to accommodate it.
2. A shipment must be charged proportionally to its weight, so that the revenue for a shipment is computed as its weight times the paid fare.
3. The company discourages last minute opportunistic behaviour, hence the succession of fares displayed to the users must be non-decreasing as time approaches flight departures.
4. No price negotiation is allowed.
5. Customers are served one at a time.

We observe that Assumption 2 may sound quite simplistic. However, the results that we present in this work generalize trivially when more complex cost functions are considered.

We also stress the following consequences of the above assumptions. Assumptions 1 and 4 imply that a customer is lost if the displayed fare does not satisfy his willingness to pay or if there is not enough weight and volume capacity to accommodate his shipment. Also, the revenue manager will always display no more than one fare at a time, since customers are price-oriented and hence always purchase the necessary capacity for their shipments at the lowest available fare that is not higher than their willingness to pay.

Hereafter, we denote by:

- $\{0, 1, \ldots, t, t+1, \ldots, T\}$ the set of the time instants at which customers can arrive and be served, 0 is flight reservation opening and T is the closing time;
- $t_k \in \{0, \ldots, T\}$ the arrival time of the k customer;
- ϕ_t the probability a customer shows up at time t;
- f the generic fare;
- C_w and C_v the flight storage capacities in weight and volume, respectively;
- (ω, υ) the size of the generic shipment, where ω and υ are its weight and its volume, respectively; both are integer values less than or equal to (C_w, C_v);
- p_{km} the customer k willingness to pay toward a fare $f_m \in M$, that is the probability that a customer arriving at time t is willing to pay $f_m\omega$ to send a package of size (ω, υ);
- $q_{\omega\upsilon}$ the probability that a shipment has size (ω, υ);
- $J_m(t, w, v)$ the function that returns the expected optimal revenues from time t on under the hypothesis that the revenue manager displays in t a fare f_m and there are residual capacities w in weight and v in volume.

At each time t, being w and v the residual capacities, the revenue manager faces the following dynamic programming problem:

$$J_m(t, w, v) = \left[(1-\phi_t) + \phi_t \sum\nolimits_{(\omega\upsilon) > (w,v)} q_{\omega\upsilon}\right] J_m(t+1, w, v) + \phi_t \sum\nolimits_{(\omega\upsilon) \leq (w,v)} q_{\omega\upsilon} \max\nolimits_{j>m} \cdot \{p_{kj}(f_j\omega + J_j(t+1, w-\omega, v-\upsilon)) + (1-p_{kj}) J_j(t+1, w, v)\} \tag{1}$$

with final conditions: $J_m(T, w, v) = J_m(T, 0, v) = J_m(T, w, 0) = 0$ for all $0 \leq w \leq C_w$ and $0 \leq v \leq C_v$.

The first term of Eq. (1) r.h.s. states that the expected revenues from t on coincide with the expected revenues from $t + 1$, when no customer shows up in t, or if the size of the shipment exceeds the available capacity, when a customer shows up in t. Differently, the second term states that, when a customer shows up, the revenue manager must choose the fare to display after having seen the size of the shipment. In this situation, the expected revenues in t are given by the sum of the revenues from the current customer (*stage revenues*) and the expected revenues from $t + 1$ on (*revenues to go*). In presence of a customer and being displayed a fare f_j, two situations may occur:

- with probability p_{kj} the current customer accepts f_j, then the stage revenues are equal to $f_j\omega$ and the next customer finds $(w - \omega)$ and $(v - \upsilon)$ as available capacities;

- with probability $(1 - p_{kj})$ the current customer refuses f_j, then stage revenues are 0 and the next customer finds w and v as available capacities.

We conclude this section observing that the difference $J_m(t, w, v)$–$J_m(t, w - \omega, v - \upsilon)$ represents the opportunity cost at fare f_m of a shipment of size (ω, υ) making a request at time t when capacities w in weight and v in volume are still available, i.e., it is the expected loss in future revenue from using the capacity now rather than reserving it for future use [25]. Accordingly, if $J_m(t, w, v)$ is differentiable then $\partial J_m(t, w, v)/\partial w$ and $\partial J_m(t, w, v)/\partial v$ are the weight marginal opportunity cost and volume marginal opportunity cost, respectively.

3 Bid-Price Heuristics

Solving problem (1) is, in general, impractical from a computational point of view. For this reason, in this section, we present two heuristics based on threshold values called bid-prices. The rationale behind these heuristics is the following. An optimal policy, solution of (1), accepts a shipment if and only if the revenue it generates is larger or equal to its opportunity cost and bid-prices can be fixed as approximate estimations of the marginal opportunity costs. On the base of this property, bid-price heuristic policies accept a shipment if its revenue is greater or equal to the estimation of its opportunity cost that can be derived from the bid-prices [25]. More formally, we denote by $\pi_w(w, t)$ and $\pi_v(v, t)$ the bid-prices for weight and volume, respectively, when capacities w in weight and v in volume are still available at time t. Then, the opportunity costs of such a request are approximated by $\pi_w(w, t)\omega + \pi_v(v, t)\upsilon$. Hence a booking request generating revenues r is accepted if and only if [18]:

$$r = f\omega \geq \pi_w(w,t)\omega + \pi_v(v,t)\upsilon, \tag{2}$$

where f is the fare applied.

3.1 Static BP Heuristic (SBP)

We define bid-prices as static if they are fixed at the beginning of the booking period, i.e., they do not change over time and do not depend on the remaining capacity: $\pi_w(w, t) = \pi_w$ and $\pi_v(v, t) = \pi_v$.

Once the bid-prices are chosen, our heuristic fixes a same fare f for all customers. Specifically, f is set equal to the first accepted shipment's minimum available fare such that the acceptance rule (2) is almost surely respected; that is

$$f = \min\{f_j : f_j\omega_h \geq \pi_w\omega_h + \pi_v\upsilon_h \text{ and } p_{hj} = 1\}$$

where, h denotes the first customer for which a fare $f_j \in F$ satisfying the above condition exists. The uniqueness of f trivially guarantees that the displayed fares do not decrease over time. However, an evident drawback of this choice is that actual generated revenues depend on the willingness to pay of the first accepted customer k. Heuristic SBP may perform poorly when demand is inverse, especially if bid-price values are low. This performance problem holds true for static bid-price approaches in general since they accept a request as long as its revenues are higher than the computed threshold, without taking into account that the marginal value of the remaining capacity increases over time. A common solution to this is to recalculate bid-prices at pre-defined time intervals (which become smaller as flight booking closure grows closer) in order to take into account the increased marginal value of remaining capacity.

To fix the values of the static bid-prices our heuristic averages the optimal static bid-prices of a set of training instances (possibly based on historical data). Specifically, paralleling the work in Pak and Dekker [18], we combine the findings in Lenstra et al. [15] and in Rinnooy Kan et al. [20], and we observe that, if the demand is known in advance, the choice of accepting shipments of size (ω_k, υ_k) generating (*potential*) *revenue* $r_k = f^* \omega_k$, where f^* is the maximum fare that customer k is willing to pay, can be made by solving a multi-dimensional knapsack problem. Then, we compute the bid-prices for each training instance by solving the associated knapsack problem through the following greedy algorithm.

Greedy algorithm. For each customer k we define the ratio

$$\delta_k = r_k/(\lambda\omega_k + \mu\upsilon_k)$$

where λ and μ are appropriate positive multipliers. The ratio δ_k is a measure of the revenue for unit of the overall resources used by shipment k. The multipliers λ and μ are necessary to express volume and weight capacity requirements as single resource consumption. Items are then ordered by decreasing values of δ_k and accordingly inserted into the knapsack as long as there is available capacity. Clearly, the choice of the multipliers may affect the sequence of shipments entering the knapsack and thus the associated final revenues. However, Rinnooy Kan et al. [20] prove that there exists a pair of multipliers λ^* and μ^* maximising the revenues that can be computed in $O(n^3 \log n)$, where n is the number of shipments.

Let δ^* be the ratio value associated to the last shipment inserted in the knapsack when the multipliers are λ^* and μ^*. We define the instance optimal static bid-prices as:

$$\pi_w = \delta^*\lambda^* \quad \pi_v = \delta^*\mu^*.$$

Indeed, is trivial to observe that, in the above procedure, the shipment of a customer k is inserted into the knapsack if and only if $\delta_k \geq \delta^*$ and there is enough capacity. Under these circumstances, condition (2) is met, as $r_k \geq \delta^* (\lambda^* \omega_k + \mu^* \upsilon_k) = \pi_w \omega_k + \pi_v \upsilon_k$. □

3.2 Dynamic BP Heuristic (DBP)

In our dynamic bid-price heuristic bid-prices are updated as the requests arrive, in order to capture and exploit the increasing willingness to pay of the demand. Let L_k and M_k be the weight and volume (dynamic) bid-price values, respectively. Hence the acceptance rule at time t_k is $r_k \geq L_k\ \omega_k + M_k\ \upsilon_k$.

After each accepted request, static bid-prices are calculated by running SBP on the remaining capacity and demand. Let $\underline{\pi}_w(w, t)$ and $\underline{\pi}_v(v, t)$ be the weight and volume bid-prices respectively, when SBP is run on $N - k + 1$ customers with remaining capacities w and v.

Initially, we set: $L_1 = \underline{\pi}_w(C_w, t_1)$, $M_1 = \underline{\pi}_v(C_v, t_1)$. Then we update the dynamic bid-prices according to the following policy:

- $L_{k+1} = L_k$, $M_{k+1} = M_k$: if request k is rejected or if one or both static bid-prices turn out to be lower than or equal to current values of L_k or M_k, that is, either $\underline{\pi}_w(w, t_{k+1}) \leq L_k$ or $\underline{\pi}_v(v, t_{k+1}) \leq M_k$.
- $L_{k+1} = \underline{\pi}_w(w, t_{k+1})$, $M_{k+1} = \underline{\pi}_v(v, t_{k+1})$ if request k is accepted and static bid-prices turn out to be greater than or equal to current values of L_k or M_k, that is $\underline{\pi}_w(w, t_{k+1}) > L_k$ and $\underline{\pi}_v(v, t_{k+1}) > M_k$.

Finally, we choose the fare to display to customer k as $f = \min\{f_j\colon f_j\omega_k \geq L_k\omega_k + M_k\upsilon_k$ and $p_{kj} = 1\}$. If no fare satisfies this last condition, customer k is rejected.

The rationale behind the choice of updating dynamic bid-prices only when both SBP bid-prices simultaneously increase is three-fold. First, in this way, at each update, the threshold given by the pair (L_{k+1}, M_{k+1}) is the optimal one for the remaining capacity and demand, according to the SBP algorithm. Second, we cannot update both the bid-prices to lower SBP bid-prices values as otherwise we could not guarantee that the displayed fare does not decrease over time. Third, we cannot update a single bid-price to a higher value when only one SBP bid-price is greater than the current dynamic bid-prices as otherwise we obtain a too selective policy. Indeed, the acceptance threshold increases very rapidly over time since it is updated whenever at least one bid-price augments.

We finally point out that the choice to consider updating dynamic bid-prices only after a request is accepted, and not after any request, is due to reducing computational time.

4 Experimental Results

In this section, we assess the quality of the heuristics introduced in the previous section. The revenues obtained with the two heuristics are then compared also with the ones obtained solving problem (1) under the assumption that the number of customers and their characteristics, in terms of both arrival times and the shipment sizes, are known in advance. In fact, under this (relaxed although unrealistic) assumption, problem (1) becomes computational tractable and provides an upper bound for the optimal expected revenues $J_1(0, C_w, C_v)$. Hereafter, we indicate this last kind of revenues as obtained through dynamic programming (DYM).

4.1 Experiment Design

Each shipment is characterised by three attributes: weight, volume and willingness to pay of its owner; they are expressed in kilos (Kg), cubic meters (m^3) and US dollars ($), respectively. As it is common practice within the air cargo industry, we distinguish between small and large shipments: small when its weight is between 2 and 45 kg and large when it is between 46 and 500 kg. This distinction is generally applied because of the different weight and volume capacities reserved on the aircraft and the dissimilar fares applied by the carriers. In principle, fares per unit of weight decrease as the weight increases and roughly depend on the distance to be flown. Since in this work all the customers book for the same single-leg flight, we can neglect the dependence on the distance.

Hence, fares are only related to the shipment weight category and to the time of the booking request. We considered four different fares ($f_1 < f_2 < f_3 < f_4$) for both small and large shipments. As anticipated, fares are non-decreasing over time and only one fare is available at each time instant.

In order to deal with uncertainty regarding shipment volume [2, 22], it is usual practice within airlines to associate an average volume for a given weight, i.e., a shipment k of weight ω_k has an average volume $\gamma\omega_k$ where γ is a constant. We set $\gamma = 0.00581$ as in Pak and Dekker [18]. To add variability to the volume, its value υ_k is randomly chosen in an interval centred in $\gamma\omega_k$ whose length becomes larger as the weight of the shipment increases.

Tests are run with willingness to pay of the demand either random or inverse. In the former case, the willingness to pay is stationary over the time. Indeed the willingness to pay of customer k is chosen randomly in the interval $[f_1\omega_k - l_m, f_4\omega_k + l_M]$ with $l_m = 0.92$ and $l_M = 2.26$ for small shipments and $l_m = 16.56$ and $l_M = 21.26$ for large shipments. In the latter case (i.e., inverse demand), we introduce a dummy fare $f_0 < f_1$ and we divide the N potential customers in four intervals of approximately equal length. Each customer in interval i, ($i = 1, 2, 3, 4$) randomly chooses between fare f_{i-1} and fare f_i with equal probability $p = 0.5$.

By averaging the values presented by airlines on their websites, we fixed the relevant data as reported in Table 1.

Table 1 Experiment parameters

Small shipments	
Weight capacity C_w	4000 kg
Volume capacity C_v	26.7 m^3
Fares	$f_1 = 4.34\$$ $f_2 = 5.31\$$ $f_3 = 6.04\$$ $f_4 = 7.72\$$
Shipment weight ω_k	$2 \leq \omega_k \leq 45$ kg (integer random)
Shipment volume υ_k	υ_k is a random number between: $[\gamma\omega_k - 0.0051, \gamma\omega_k + 0.0051]$ when $\omega_k \leq 9$ $[\gamma\omega_k - 0.0101, \gamma\omega_k + 0.0101]$ when $10 \leq \omega_k \leq 45$
Demand amount N	750
Large shipments	
Weight capacity C_w	15500 kg
Volume capacity C_v	80.25 m^3
Fares	$f_1 = 1.97\$$ $f_2 = 2.67\$$ $f_3 = 3.67\$$ $f_4 = 4.03\$$
Shipment weight ω_k	$46 \leq \omega_k \leq 500$ kg (integer random)
Shipment volume υ_k	υ_k is a random number in: $[\gamma\omega_k - 0.0201, \gamma\omega_k + 0.0601]$ when $46 \leq \omega_k \leq 99$ $[\gamma\omega_k - 0.0151, \gamma\omega_k + 0.171]$ when $100 \leq \omega_k \leq 299$ $[\gamma\omega_k - 0.0101, \gamma\omega_k + 0.451]$ when $300 \leq \omega_k \leq 500$
Demand amount N	450

4.2 Computational Results

We distinguish four different scenarios in accordance with the four different sets of instances used as input demand. Tables 2 and 3 present the average values obtained over the same set of Λ instances by the different heuristics for each scenario.

Average execution time refers to the training instances for SBP and DBP. The testing instances simply need to check at each booking request whether the threshold is respected or not, which is a very quick operation.

On the other hand, the DYM algorithm does not require a training phase, so the testing instances are optimally solved through dynamic programming and average execution time refers to these latter.

The results provided shows that, at least for the instances considered, both the heuristics are reasonable. Here, we recall that the DYM results are upper bounds on the optimal ones that a revenue manager could obtain solving (1) without a priori deterministically knowing the demand. Static bid-price heuristic DBP, as expected performs better than SBP with inverse demand. Differently, SBP outperforms DBP when the demand is random. This latter observation is not surprising as, given the non-decreasing assumption, the DBP tries to increase the

Table 2 Results for random demand (average values)

	Small shipments, random demand		
	DYM	SBP	DBP
Revenues	2,458,003	2,283,183	2,082,226
Weight LF	0.999	0.964	0.849
Volume LF	0.869	0.839	0.738
Accepted requests (No.)	178	166	152
Running time (s)	552	<1	624
	Large shipments, random demand		
	DYM	SBP	DBP
Revenues	4,752,162	4,641,544	4,421,279
Weight LF	0.820	0.816	0.770
Volume LF	0.997	0.992	0.935
Accepted requests (num.)	51	49	48
Running time (sec.)	64	<1	2

Table 3 Results for inverse demand (average values)

	Small shipments, inverse demand		
	DYM	SBP	DBP
Revenues	2,712,138	2,053,969	2,121,717
Weight LF	0.999	1.000	1.000
Volume LF	0.870	0.870	0.870
Accepted requests (num.)	171	173	174
Running time (sec.)	594	<1	122
	Large shipments, inverse demand		
	DYM	SBP	DBP
Revenues	5,110,171	4,680,050	4,736,291
Weight LF	0.821	0.823	0.824
Volume LF	0.990	0.999	0.998
Accepted requests (num.)	47	48	49
Running time (sec.)	79	<1	5

revenues becoming more and more selective. However, in the case of random demand, the willingness to pay of the customers does not increase over time and then the DBP may reject too many customers. Indeed, in presence of a stationary demand, it may be reasonable to become less selective when we approach the flight closing time. Unfortunately, the implementation of a similar policy could induce an opportunistic behaviour in the demand and modify its statistics.

5 Conclusions

In the present work we introduced the problems that can arise in RM from an imprecise forecast of customer demand. In this context, we focused on customer's *willingness to pay* as a proven robust measure on which to build capacity management algorithms. Referring to a bi-dimensional capacity scenario (i.e., weight and volume, which is the case of air cargo) and considering demand as deterministic at the time of booking, we first introduced an optimal Dynamic Programming model. Then we proposed two bid-prices based algorithms, one, static and one dynamic, and showed through computational tests their performances in terms of average revenues and computational times.

Future development of this work may address further improvements to the BP policies and comparison to DP based heuristics, another solving approach that has drawn a wide interest in literature.

References

1. Adelman, D.: Dynamic bid prices in revenue management. Oper. Res. **55**(4), 647–661 (2007)
2. Amaruchkul, D., Cooper, W., Gupta, D.: Single-leg air-cargo revenue management. Technical report, University of Minnesota (2005)
3. Amaruchkul, D., Cooper, W., Gupta, D.: Air cargo capacity contracts under asymmetric information. Working paper, supply chain and operations research laboratory, University of Minnesota
4. Bijvank, M., Haensel, A., Ecuyer, P.L., Marcotte, P.: Time-dependent bid prices for multi-period network revenue management problems. Working paper, (2012)
5. Billings, J., Diener, A., Yuen, B.: Cargo revenue optimisation. J. Pricing Revenue Manag. **2**(1), 69–79 (2003)
6. Boyd, E., Kallesen, R.: The science of revenue management when passengers purchase the lowest available fare. J. Pricing Revenue Manag. **3**(2), 171–177 (2004)
7. Bront, J., Méndez-Díaz, I., Vulcano G.: A column generation algorithm for choice-based network revenue management. Oper. Res. **57**(3), 769–784 (2007)
8. Chew, E., Huang, H., Johnson, E., Nemhauser, G., Sokol, J., Leong, C.: Short-term booking of air cargo space. Eur. J. Oper. Res. **174**, 1979–1990 (2006)
9. Cléaz-Savoyen, R.: (2005). Airline revenue management methods for less restricted fare structure. MS thesis, Massachusetts Institute of Technology, Cambridge
10. Cooper, W., Homem-de Mello, T., Kleywegt, A.: Models of the spiral- down effect in revenue management. Oper. Res. **54**(5), 968–987 (2006)
11. Fiig, T., Isler, K., Hopperstad, H., Cléaz-Svaoyen, R.: Davn-MR: A unified theory of O&D optimization in a mixed network with restricted and unrestricted fare products. In: AGIFORS Reservation and Yield Management Meeting. Cape Town, South Africa, 15–19 May 2005
12. Hopperstad, C., Belobaba, P.: Alternative RM algorithms for unrestricted fare structures. In: AGIFORS Reservation and Yield Management Meeting. Aukland, New Zealand, 28–31 Mar 2004
13. Huang, K., Hsu, W.: Revenue management for air cargo space with supply uncertainty. Proc. East. Asia Soc. Transp. Stud. **5**, 570–580 (2005)
14. Kasilingam, R.: Air cargo revenue management: Characteristics and complexities. Eur. J. Oper. Res. **96**, 36–44 (1996)

15. Lenstra, A., Lenstra, J., Rinnooy Kan, A., Wansbeek, T.: Two lines least squares. Ann. Discret. Math. **16**, 201–211 (1982)
16. Liu, Q., van Ryzin, G.: On the choice-based linear programming model for network revenue management. Manufact. Serv. Oper. Manag. **10**(2), 288–310 (2008)
17. Luo, S., Çakanyıldırıma, M., Kasilingam, R.: Two-dimensional cargo overbooking models. Eur. J. Oper. Res. **197**(3), 862–883 (2009)
18. Pak, K., Dekker, R.: Cargo revenue management: Bid-prices for a 0-1 multi knapsack problem. Technical report, Erasmus University Rotterdam, The Netherlands (2004)
19. Popescu, A., Barnes, E., Johnson, E., Keskinocak, P.: Bid prices when demand is a mix of individual and batch bookings. Transp. Sci. **47** (2), 198–213 (2013)
20. Rinnooy Kan, A., Stougie, L., Vercellis, C.: A class of generalized greedy algorithms for the multi-knapsack problem. Discret. Appl. Math. **42**, 279–290 (1993)
21. Simpson, R. W.: Using network flow techniques to find shadow prices for market and seat inventory control. Technical report, MIT Flight Transportation Laboratory Memorandum M89-1, Massachusetts Institute of Technology, Cambridge (1989)
22. Slager, B., Kapteijns, L.: Implementation of cargo management at KLM. J. Pricing Revenue Manag. **3**(1), 80–90 (2004)
23. Smith, B.C., Penn, C.W.: Analysis of alternate origin-destination control strategies. In: Proceedings of AGIFORS 28th Annual Symposium, pp. 123–144. New Seabury (1988)
24. Talluri, K., van Ryzin, G.: An analysis of bid-price controls for network revenue management. Manag. Sci. **44**(1), 1577–1593 (1998)
25. Talluri, K., van Ryzin, G.: The Theory and Practice of Revenue Management. Springer, New York (2004)
26. Totamane R., Dasgupta A., Rao S.: Air cargo demand modeling and prediction. Syst. J. IEEE forthcoming (2012). doi:10.1109/JSYST.2012.2218511
27. Westermann, D.: (Realtime) dynamic pricing in an integrated revenue management and pricing environment: An approach to handling undifferentiated fare structures in low-fare markets. J. Pricing Revenue Manag. **4**(4), 395–405 (2006)
28. Zhang, D.: An improved dynamic programming decomposition approach for network revenue management. Manufact. Serv. Oper. Manag. **13**(1), 35–52 (2010)

A Rail Network Optimization Model Designed for Freight Traffic

Luís Couto Maia and António Fidalgo do Couto

Abstract The freight network optimization model presented in this chapter was developed as a support tool for planning and policy decisions involved in the improvement of rail networks on a regional and national level. It is based on a strategic traffic assignment model designed to model macro networks with a high aggregation level, being exclusively designed for freight traffic. The model contemplates road and rail transport modes, and considers two different types of cargo: intermodal cargo, which is generally transported in containers and is easily interchanged between different modes at intermodal terminals; and general cargo, which represents all the remaining cargo. The optimization process is based on a local search heuristic which delivers good solutions in a reasonable computing time, with the quality of each network improvement solution being assessed based on the reduction of the total generalized costs and CO_2 emissions. This freight network optimization model is innovative in the fact that it is not limited, allowing for both the improvement of existing links as well as the construction of new ones, and not having a limit on the number or variety of network improvement possibilities. Its adaptability to different conditions is emphasized when the model is applied to a network under two different investment scenarios, by delivering considerably different solutions adapted to the conditions of each scenario.

Keywords Network optimization · Freight transportation · Traffic assignment

L. C. Maia (✉) · A. F. do Couto
Faculty of Engineering, University of Porto, Porto, Portugal
e-mail: luis.maia@fe.up.pt

A. F. do Couto
e-mail: fcouto@fe.up.pt

J. F. de Sousa and R. Rossi (eds.), *Computer-based Modelling and Optimization in Transportation*, Advances in Intelligent Systems and Computing 262,
DOI: 10.1007/978-3-319-04630-3_16,

1 Introduction

While freight transportation is an activity that plays a crucial role in the everyday life of any modern economy, being critical to a large part of the economy, it usually gets less attention in the academia than its passenger counterpart. This is probably justified by the fact that it is not as appealing to policy makers and the general public as passenger transportation, but also because it is a considerably more complex subject, due to the multiplicity of goods transported, the complexity of the freight supply chain and the difficulty in getting the needed data. While it may be less appealing and more complex than passenger transportation, it is important to study freight transportation using models specifically made for it, in order to account for its distinct characteristics and for the fact that the network investments needed to improve freight transportation can be considerably different from those aimed at improving passenger transportation. Due to that, the network optimization model and the associated traffic assignment model that are presented in this chapter have been developed specifically for this type of transportation, although they may be combined with passenger models in the future, in order to create a model for the whole transportation system.

The presented model, developed in the scope of a broader project [10], uses a strategic planning traffic assignment model [5] designed to model macro networks with a high aggregation level. This assignment model does not require very detailed data inputs, with the outcome of its application being the estimation of the movement of freight at a regional, national, or international scale. It considers road and rail transport modes, being intended to simulate medium and long distance flows of inland transportation. The model contemplates two different types of cargo, namely general cargo and intermodal cargo, in order to make a distinction between the cargo that may be easily interchanged between different modes at intermodal terminals, which is generally transported in containers, and the rest of the cargo. The above characteristics make this traffic assignment model particularly suited for the planning and policy decisions that are going to be performed by the network optimization model [15].

As for the optimization process in itself, it is quite flexible and innovative, allowing for both upgrades in the quality of existing rail and intermodal terminal links as well as the construction of new ones, not having a limit on the number or variety of improvement solutions. This is achieved by defining a set of possible link levels for each link type, according to the users' preferences, including the mere possibility of building a link. As for the quality of each network improvement solution, it is assessed based on the reduction of the total generalized costs and CO_2 emissions, with the weight given to each of those parameters being defined by the user according to its preferences. The optimization model is based on a local search heuristic and tries to meet a balance between efficiency and effectiveness, by delivering good solutions in a reasonable computing time.

This chapter is structured in six sections. After the introduction, there is a background section, containing a brief literature review on the subject of freight traffic assignment and network optimization models. The third section is dedicated to the traffic assignment model, while the fourth section is devoted to the network optimization process. The fifth section describes an application of the network optimization model, with the sixth and last section being dedicated to the final conclusions.

2 Background

The different traffic assignment techniques that are presented in the literature can be divided in four big groups: All-or-nothing (AoN), Equilibrium, Stochastic-multi-flow and Stochastic-equilibrium [8]. The two factors whose usage determines to which of the four groups a model belongs are the existence of capacity constraints imposed by congestion (Equilibrium and Stochastic-equilibrium models) and the use of a variable perception of costs (Stochastic-multi-flow and Stochastic-equilibrium models). Although there are many different models present in the literature, with many being created for just one specific work, there are two major freight traffic assignment models that are worth mentioning, due to their importance and extensive use. Those are STAN, which was developed in 1990 in Canada [4, 7], making use of an Equilibrium assignment technique, and the NODUS software, which was developed in Belgium a few years later [2, 8, 9] and that has been employed using all the three most common assignment techniques: AoN, Equilibrium and Stochastic-multi-flow. As for the method used to perform the distribution of traffic, the Logit formulation has consistently been chosen to address this problem, due to its versatility and convenience [8, 12, 14].

Most of the research found in the literature on the subject of network optimization was performed using two types of models: the discrete network design problem (DNDP) [1, 13, 16] and the continuous network design problem (CNDP) [17]. While the former tends to concentrate on the addition of new links, and the latter on the (continuous) improvement of existing links, it is also possible to use a discrete approach allowing for both the addition of new links and the improvement of existing links [13]. Due to the considerable complexity of the transportation networks and to the discrete nature of most models, there is no practical analytical solution for this problem, which leads to the adoption of heuristic techniques. Several techniques have been successfully used to address this kind of problems, predominantly metaheuristics such as tabu search, simulated annealing and genetic algorithms [1, 3, 13, 16].

3 Freight Traffic Assignment Model

3.1 *Model's General Attributes and Computation of the Shortest Paths*

The presented model is a strategic freight traffic assignment model, being based on a previous model developed by the authors [11]. It considers two different types of cargo: general cargo and intermodal cargo. Each link has a set of attributes, with some of them being inherent to the link, namely its length and capacity, while others depend on the vehicles that use those links, which may be different for each type of cargo. Those are the average speed, vehicle capacity, cost per distance and CO_2 emissions of the vehicles, and the value of time for each type of cargo. The calculation of the generalized cost per unit of cargo is constituted by a vehicle cost component and a time cost component, and is given by Eq. 1:

$$Generalized\,cost\,per\,unit\,of\,cargo = \frac{Length * Vehicle\,cost\,per\,distance}{Vehicle\,capacity} + \frac{Length}{Average\,speed} * Value\,of\,time \tag{1}$$

Based on the defined generalized costs it is possible to calculate the shortest path (with the least generalized costs) between any given pair of nodes for each type of cargo. The shortest path algorithm that is employed in the model is the Floyd-Warshall algorithm [6] with path reconstruction, which computes the value of the shortest paths between all the nodes, as well as the path in itself (the links used in each shortest path).

3.2 *Assignment Process*

With the model's basic features properly defined, it is now possible to describe the assignment process, whose main features are resumed in Fig. 1.

As it can be seen in Fig. 1, there are different assignment techniques for the two different types of cargo, mainly due to the fact that only intermodal cargo is allowed to use intermodal terminals, which is a solution already applied to other studies in this area [2]. This means that, while intermodal cargo may use more than one mode of transport per trip, general cargo is limited to using the same transport mode in each trip, which is reflected in the traffic distribution techniques that are used for each type of cargo. In the case of general cargo, as each trip may only use one mode of transport, there is a clear mode choice decision between the least costly paths using road and rail transport. This allows for the distribution of traffic between the two modes, using a Stochastic-multi-flow technique, which is implemented using a Logit function [14] that gives the percentage of traffic using each mode. As for intermodal cargo, its traffic is assigned to the least costly path

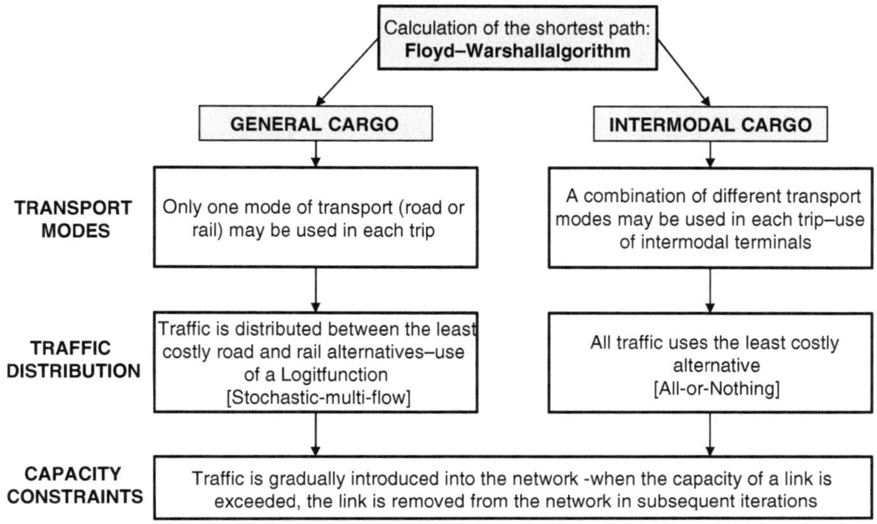

Fig. 1 Assignment process

between the origin and the destination, which may include intermodal terminals. This is done using an AoN technique, which allows for road and rail links to be used indifferently in every trip, as long as they are part of the absolute least expensive path.

The model does not consider capacity limits on road links, due to the fact that congestion is mostly observed in and around urban areas [8] and not on intercity routes, but it considers it on rail links and intermodal terminals. This is justified by the fact that the capacity of rail links and intermodal terminals is relatively rigid, with empirical evidence showing that rail links are more likely to be used to capacity than their modern intercity road links counterparts. Given that most of the capacity problems in rail networks are due to specific point in the network, such as congested rail junctions, the model allows for the introduction of congested rail nodes with a limited capacity. As it can be seen in Fig. 2, which displays the technique used to model rail nodes [4], the model considers a virtual link that represents the total capacity of the rail node.

It is important to notice that what is defined in the network is the physical capacity of each link, which means that the capacity that is left for freight trains is the total capacity minus the flow of passenger trains. That flow is obtained by assigning an origin/destination (O/D) matrix of passenger trains to the rail network, using the shortest distance path.

The total freight flow is gradually inserted into the network, with the user defining in how many interactions is the traffic flow introduced into the network. If any new link has reached its capacity after each iteration, it is removed from the network and the shortest paths and traffic distributions are recalculated. This process continues until all the traffic is assigned to the network.

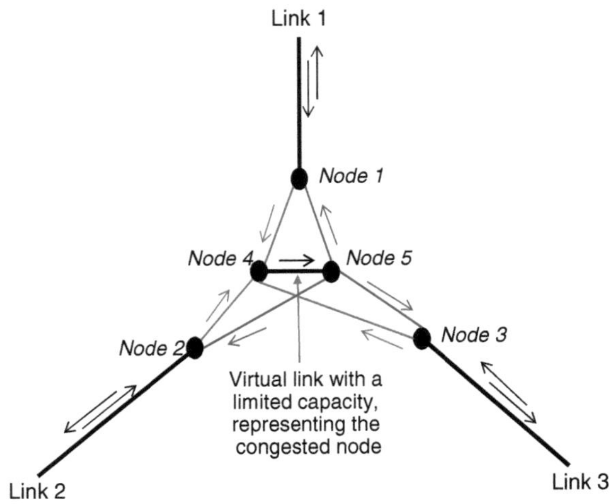

Fig. 2 Congested rail nodes scheme (example of a congested rail node with 3 rail links converging on it)

4 Network Optimization Model

The first step in the development of a network optimization process is the definition of the adopted network structure, defining all the possible link levels and network improvement possibilities. The solution employed in this model is a network structure where the links have a limited number of discrete quality levels, which each level corresponding to a different link type. Link's quality levels can vary from zero, which corresponds to the mere possibility of building a link, to the highest level, corresponding to the best possible link quality. Each link level has an associated set of characteristics for each type of cargo, which may be freely defined by the user. This network structure allows for both the improvement and the construction of new links, permitting unlimited improvement possibilities. The model allows for the improvement of rail, intermodal terminal and virtual links, meaning that all the links which are related to rail transport may be improved, in order to meet the goal of the model, which is the optimization of rail networks. All the possible improvement operations that are defined by the user have to have an associated cost, in order to quantify the money that is spent in each improvement scenario.

The factors that are considered for the assessment of the quality of each network improvement solution are the total generalized costs, and the total emissions of CO_2. The total generalized cost reflects the economic costs that are supported by the freight carriers, and according to which they make their transportation decisions. As for the total emissions of CO_2, they quantify the total CO_2 emitted by all the vehicles transporting freight, serving as a measure of the environmental impact caused by freight transportation. The quality of each improvement solution is measured based on the minimization of both of this parameters. Given the existence of more than one optimization parameter, it is necessary to define the weight that is given to each one of them, which is something that is defined by the user according to each case's planning priorities.

Fig. 3 Optimization process

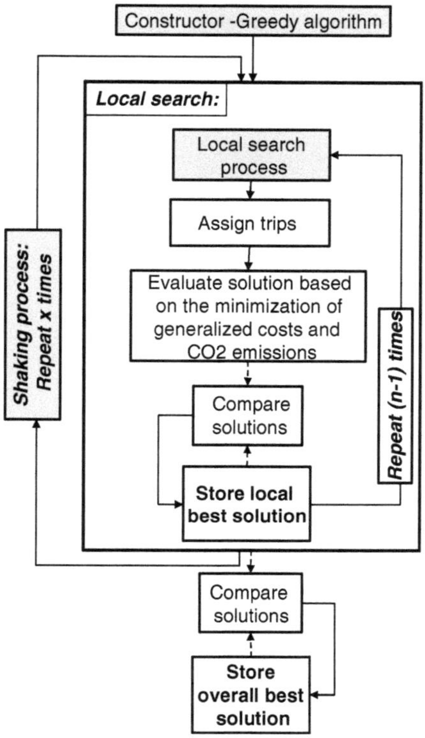

As it can be seen in Fig. 3, the optimization process starts with a constructor, creating a reasonable initial solution, which is done by using a greedy algorithm. Based on that initial solution, the model runs two different cycles: an inner local search process, and an outer shaking process. The local search process tries the optimize the solution by searching for better solutions on the search space vicinity of the initial solution, while the shaking process is used to make the solution "jump" to a different point in the search space, in order to avoid being stuck in a local optimum.

The constructor process is based on a greedy algorithm, which iteratively improves the links with the highest perceived improvement benefit to their maximum possible level, until there is no more available budget for improvements. The formula that is used to measure the perceived improvement benefit of each link, which was freely defined by the authors, is the following:

$$Improvment\_Benefit = \left(1 + \frac{Volume\_of\_Traffic}{Link\_capacity}\right)^2 * Volume\_of\_Traffic \quad (2)$$

As it can be seen in Eq. 2, the improvement benefit for each link is proportional to the volume of traffic that uses the link and to the relative utilization of the link. Also, the relative utilization of each link is also important, as links which are over

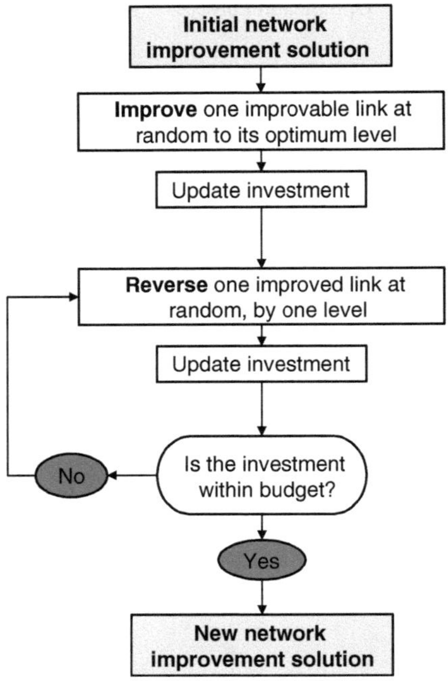

Fig. 4 Local search process

their capacity will benefit the most with a capacity increase, allowing them to be used by more traffic. If an improvable link has no traffic passing throw it, which is the case in links which represent the mere possibility of building a link, the model attributes it an improvement benefit marginally bigger than zero, as an improvement in such a link may be beneficial. The algorithm iteratively improves the links with the higher value of improvement benefit to their maximum level until there is no more budget available to make new improvements, constructing an initial network optimization solution.

The algorithm that is used for the local search process, which is the core of the whole optimization process, is schematized in Fig. 4.

The local process takes an initial network improvement solution and makes a small change in it, by proposing a new solution in the search space vicinity of the initial solution. This is done by improving an improvable link at random to its optimum level, and then iteratively reversing a link at random by one level, until the investment is within budget. The new solution is then tested to see if it is better than the current solution, in which case it becomes the new current solution, and this cycle is repeated by as many times as defined by the user, as it can be seen in Fig. 3.

In order to avoid being stuck in local optimums, the model has a shaking algorithm that makes the solution that comes out of each local search process "jump" to a different place in the search space, from where a new local search process can be performed. This process consists in the reversal of two thirds of the improvement operations that were originally done, followed by the iterative improvement of random links to their optimum level, until the budget is reached or

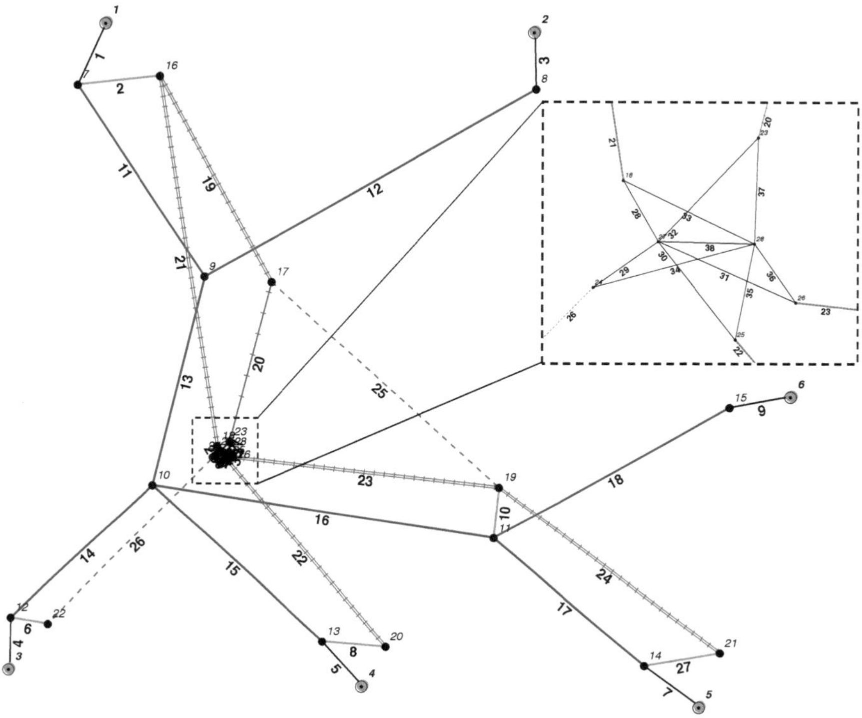

Fig. 5 Network map

exceeded. This process creates a new random solution, which is significantly different from the original solution, and that will likely have left its original search space vicinity, avoiding being stuck in local optimums.

5 Application of the Model

5.1 Description of the Network and Considered Scenarios

In order to test and evaluate its performance, the developed model was applied to a network created by the authors, which is schematized in Fig. 5.

It is a relatively simple network, with six traffic generating poles (centroids), which are represented as large green dots, road and rail links, which are the plain blue and crossed/dashed red lines respectively, and four intermodal terminals, which are links 2, 6, 8 and 27, represented in orange. Links 1, 3, 4, 5, 7 and 9 are connectors which link the centroids to the road network, with the concentration of nodes in the convergence of links 20, 21, 22, 23 and 26 representing a congested rail node, as exemplified in Fig. 2, where the virtual link is link 38. There are

Table 1 Summary of links

Link level	Link type				
	Type 0–connector	Type 1—road link	Type 2—Rail link	Type 3—intermodal terminal	Type 4–rail node virtual Link
0	–	–	Possible link	Possible link	–
1	Centroid to rail	Road link	Non-electrified single line; Max. train length = 450 m	Intermodal terminal—capacity level 1	Rail node—capacity level 1
2	Centroid to road	–	Non-electrified single line; Max. train length = 450 m	Intermodal terminal—capacity level 2	Rail node—capacity level 2
3	Port to rail	–	Non-electrified single line; Max. train length = 750 m	Intermodal terminal—capacity level 3	Rail node—capacity level 3
4	Port to road	–	Non-electrified single line; Max. train length = 450 m	Intermodal terminal—capacity level 4	Rail node—capacity level 4
5	Zero cost connector—for rail nodes	–	Non-electrified single line; Max. train length = 750 m	–	Rail node—capacity level 5
6	–	–	–	–	Rail node—capacity level 6

Table 2 O/D matrices

O/D	General cargo(ton)/Intermodal/Passenger trains (trains)					
	1	2	3	4	5	6
1	0/0/0	0/0/0	40000/0/0	25000/0/7	22500/55000/5	35000/0/0
2	0/0/0	0/0/0	30000/0/0	20000/0/0	27500/0/0	15000/0/0
3	40000/0/0	30000/0/0	0/0/0	0/0/0	0/0/0	0/0/0
4	25000/0/7	20000/0/0	0/0/0	0/0/0	0/0/0	0/0/0
5	22500/55000/5	27500/0/0	0/0/0	0/0/0	0/0/0	0/0/0
6	35000/0/0	15000/0/0	0/0/0	0/0/0	0/0/0	0/0/0

various possible link levels for each link type, as it can be seen in Table 1, and only some of them are used in this network.

By consulting Fig. 5 it is possible to see that links 25 and 26 are represented as dashed lines, which means they are level 0, representing just the possibility of building a link. All the other rail links are level 4, with the exception of link 20, which is level 2. As for the intermodal terminals as well as the virtual link representing the congested rail node, they are all level 1. The O/D matrices with the demand for freight between the six centroids as well as the movement of passenger trains, which are all inputs of the model, can be consulted in Table 2.

The values that were considered for the network improvement costs, attributes of the different links levels and O/D matrices were chosen by the authors for this specific application, being reasonable indicative values. For the sake of simplification, the link attributes for both intermodal cargo and generalized cargo were considered equal, with both types of cargo being measured in tons. Regarding the definition of CO_2 emissions, they were quantified as grams per km in the road and rail links, and as grams per moved ton of cargo in intermodal terminals. As for the links capacity, in the case of intermodal terminals it is measured in tons of moved cargo, while in rail links and congested rail nodes it is measured in number of trains.

5.2 Application Results and Discussion

The network optimization model was applied to the network under the following conditions: regarding the assignment process, the traffic was introduced into the network in 20 interactions; as for the optimization process, each local search process consisted of 50 cycles, and the shaking process considered 50 shaking cycles The optimization program took approximately 16 min to finish in a dual core 2.5 GHz processor, which is a reasonable amount of time for a network of this size with this relatively high number of improvement possibilities. The relative weights that were given to generalized costs and CO_2 emissions minimization were 2 and 1, respectively, and two different scenarios were considered: one with a total available budget of 250 million monetary units, and another with a total budget of 500 million monetary units.

Table 3 Results of the optimization process

	Scenario 1					Scenario 2				
Total investment (million monetary units)	247					484				
Percentage of reduction in total generalized cost	-0.1390%					-3.5632%				
Percentage of change in total CO2 emissions	-1.9433%					-23.8656%				
Combined weighted improvment percentage	-0.7404%					-10.3306%				
	Id	Link type	Original link status	Improved link status	Levels of improvment	Id	Link type	Original link status	Improved link status	Levels of improvment
Improvment solution	1	0	2	2	0	1	0	2	2	0
	2	3	1	1	0	2	3	1	3	2
	3	0	2	2	0	3	0	2	2	0
	4	0	2	2	0	4	0	2	2	0
	5	0	2	2	0	5	0	2	2	0
	6	3	1	1	0	6	3	1	1	0
	7	0	2	2	0	7	0	2	2	0
	8	3	1	1	0	8	3	1	1	0
	9	0	2	2	0	9	0	2	2	0
	10	3	1	1	0	10	3	1	1	0
	11	1	1	1	0	11	1	1	1	0
	12	1	1	1	0	12	1	1	1	0
	13	1	1	1	0	13	1	1	1	0
	14	1	1	1	0	14	1	1	1	0
	15	1	1	1	0	15	1	1	1	0
	16	1	1	1	0	16	1	1	1	0
	17	1	1	1	0	17	1	1	1	0
	18	1	1	1	0	18	1	1	1	0
	19	2	4	5	1	19	2	4	5	1
	20	2	2	5	3	20	2	2	2	0
	21	2	4	4	0	21	2	4	4	0
	22	2	4	4	0	22	2	4	4	0
	23	2	4	4	0	23	2	4	4	0
	24	2	4	4	0	24	2	4	5	1
	25	2	0	0	0	25	2	0	5	5
	26	2	0	0	0	26	2	0	0	0
	27	3	1	1	0	27	3	1	3	2
	28	0	5	5	0	28	0	5	5	0
	29	0	5	5	0	29	0	5	5	0
	30	0	5	5	0	30	0	5	5	0
	31	0	5	5	0	31	0	5	5	0
	32	0	5	5	0	32	0	5	5	0
	33	0	5	5	0	33	0	5	5	0
	34	0	5	5	0	34	0	5	5	0
	35	0	5	5	0	35	0	5	5	0
	36	0	5	5	0	36	0	5	5	0
	37	0	5	5	0	37	0	5	5	0
	38	4	1	2	1	38	4	1	1	0

The results and improvement solutions obtained for each scenario, as well as the type and original level of each link, can be consulted in Table 3. The solutions obtained for each of the two scenarios are considerably different, which reflects the complexity of the network optimization process, as a bigger budget allows for more ambitious network interventions.

In the solution obtained for scenario 1 the virtual link 38, which represents the congested rail node, is improved in order to have a higher capacity, which is justified by the fact that in the original network configuration this node is congested. Also, rail links 19 and 20 are improved to their best possible level, in order to reduce rail transport costs. By contrast, on the solution obtained for scenario 2 the rail node link is not improved. This is justified by the fact that the higher available budget allowed for the construction of link 25, which is a new rail link that diverts rail traffic from the congested node, meaning that it no longer needs a capacity improvement. The construction of this new rail link, combined with the improvement of rail links 19 and 24, makes the rail mode much more competitive for certain routes. This causes a sharp rise in the amount of cargo that uses intermodal terminals 2 and 27, which therefore need to be improved in order to accommodate for this traffic growth.

The obtained results were in line with what was expected, with the model demonstrating the necessary adaptability needed to handle a problem as complex as this optimization problem. This is patent in the very different outcomes that were obtained for the two scenarios, which are justified by the fact that the bigger budget of scenario 2 enabled the adoption of a radically different solution.

6 Conclusions

In this chapter, an innovative network optimization model is presented, being subsequently applied to a network, under two different scenarios. The model is conceived for a strategic level of planning, modeling the major road and rail links, as well as specific congested rail nodes and intermodal terminals. It is an innovative model in the fact that it is not limited, unlike existing freight network optimization models which are usually limited by only allowing for either the construction or the improvement of links, or by having a limited search space. This model allows for both the improvement of existing links as well as the construction of new ones from scratch, and can be applied to very different networks without a limit on the number or variety of network improvement possibilities.

The practical application of the model produced satisfactory results, highlighting the model's adaptability, as it was able to optimize the network investment according to the available budget for each of the two tested scenarios, by delivering considerably different solutions.

Acknowledgments The present study was financed by the Portuguese Science and Technology Foundation.

References

1. Arnold, P., Peeters, D., Thomas, I.: Modelling a rail/road intermodal transportation system. Transp. Res. Part E **40**, 255–270 (2004)
2. Beuthe, M., Jourquin, B., Geerts, J.-F., Há, C.K.à.N.: Freight transportation demand elasticities: a geographic multimodal transportation network analysis. Transp. Res. Part E **37**, 253–266 (2001)
3. Crainic, T.G.: Service network design in freight transportation. Eur. J. Oper. Res. **122**, 272–288 (2000)
4. Crainic, T.G., Florian, M., Léal, J.-E.: A model for the strategic planning of national freight transportation by rail. Transp. Sci. **24**(1), 1–24 (1990)
5. Crainic, T.G., Laporte, G.: Planning models for freight transportation. Eur. J. Oper. Res. **97**, 409–438 (1997)
6. Floyd, R.W.: Algorithm 97: shortest path. Commun. ACM **5**(6), 345 (1962)
7. Guélat, J., Florian, M., Crainic, T.G.: A multimode multiproduct network assignment model for strategic planning of freight flows. Transp. Sci. **24**(1), 25–39 (1990)
8. Jourquin, B.: A multi-flow multi-modal assignment procedure applied to the european freight transportation networks. Stud. Reg. Sci. **35**, 929–945 (2005)
9. Jourquin, B., Beuthe, M.: Transportation policy analysis with a geographic information system: the virtual network of freight transportation in Europe. Transp. Res. Part C **4**(6), 359–371 (1996)
10. Maia, L.C., Couto, A.F.d.: Development of a strategic freight network optimization model. In: XXV ANPET - Congresso de Pesquisa e Ensino em Transportes, pp. 443–453. Belo Horizonte, Brasil (2011)
11. Maia, L.C., Couto, A.F.d.: A new freight traffic assignment model, considering both capacity constraints and a variable perception of costs. In: 15th Meeting of the Euro Working Group on Transportation. Paris, France (2012)
12. Oum, T.H.: Derived demand for freight transport and inter-modal competition in canada. J. Transp. Econ. Policy **13**, 149–168 (1979)
13. Santos, B.F., Antunes, A.P., Miller, E.J.: Interurban road network planning model with accessibility and robustness objectives. Transp. Plan. Technol. **33**(3), 297–313 (2010)
14. Tsamboulas, D., Moraitis, P.: Methodology for estimating freight volume shift in an international intermodal corridor. Transp. Res. Rec. **2008**, 10–18 (2007)
15. Wigan, M., Southworth, F.: Whats wrong with freight models and what should we do about it? In: Transportation Research Board 85th Annual Meeting, Washington DC, USA (2006)
16. Yamada, T., Russ, B.F., Castro, J., Taniguchi, E.: Designing multimodal freight transport networks: a heuristic approach and applications. Transp. Sci. **43**(2), 129–143 (2009)
17. Zhang, G., Lu, J.: Genetic algorithm for continuous network design problem. J. Transp. Syst. Eng. Inf. Technol. **7**(1), 101–105 (2008)

An Integrated Approach for the Design of Demand Responsive Transportation Services

Rui Gomes, Jorge Pinho de Sousa and Teresa Galvão

Abstract Providing quality public transportation can be extremely expensive when demand is low, variable and unpredictable. Demand Responsive Transportation (DRT) systems try to address these issues with routes and frequencies that may vary according to observed demand. The design and operation of DRTs involve multiple criteria and have a combinatorial nature that prevents the use of traditional optimization methods. We have developed an innovative Decision Support System (DSS) integrating simulation and optimization, to help design and operate DRT services, minimizing operating costs and maximizing the service quality. Experiments inspired in real problems have shown the potential of this DSS.

Keywords Public transport · DRT · Multiple-objectives · Simulation · Heuristics

1 Introduction

Transportation systems are a key factor for economic sustainability and social welfare, but providing quality public transportation may be extremely expensive when demand is low, variable and unpredictable, as it is the case of disperse rural areas or some periods of the day in urban areas. Buses circulating with very low occupancy rates mean high costs for the operators, often leading to low frequencies and, as a consequence, social exclusion, low perceived quality and degradation of the image of public transportation. Demand Responsive Transportation

R. Gomes (✉) · J. P. de Sousa · T. Galvão
Faculdade de Engenharia da Universidade do Porto, Porto, Portugal
e-mail: rgomes.5@netc.pt

J. P. de Sousa · T. Galvão
INESC TEC, Porto, Portugal

J. F. de Sousa and R. Rossi (eds.), *Computer-based Modelling and Optimization in Transportation*, Advances in Intelligent Systems and Computing 262,
DOI: 10.1007/978-3-319-04630-3_17,

(DRT) services address this problem with routes and frequencies that may vary according to the actual observed demand. Due to this added flexibility, the service provided by the operators becomes more efficient, with routes planned closer to their start, and vehicles with characteristics that better suit the mobility requirements of potential users [1]. DRT systems can also operate in a complementary way to other transportation systems, feeding traditional regular systems in strategically chosen points. The advantages of DRTs in terms of social cohesion, mobility, traffic, or environment, are fairly obvious. However, in terms of financial sustainability and quality level, the design of this type of services may be rather complicated. Moreover, in terms of operation, DRTs can be viewed as rather dynamic systems, requiring the adaptation of solutions in real-time, in a multiple criteria context.

The problems of designing and operating DRT services are closely related to the Vehicle Routing Problem [2], and in particular to the Dial-A-Ride Problem [3]. In the DARP, one is interested not only in minimizing the operating costs or the distance travelled by the vehicles but also (and this is sometimes more important) in maximizing the quality of the service, based on indicators such as the average passenger waiting time or the on-board (ride) passenger time [4]. Dial-a-Ride services can operate in a static or in a dynamic mode. In the static mode, all requests are known before-hand, whereas in the dynamic mode, requests are gradually revealed along the service operation, with routes and schedules having to be adjusted to meet the demand [5]. In practice, however, "pure" dynamic services are not common since some requests are usually known a priori.

Besides involving multiple conflicting objectives, managing DRT services can also be a strongly dynamic problem [6], requiring the adaptation of solutions in real-time. Given the complexity of these problems [7], optimization methods are highly time-consuming, ruling out their usefulness in practice. Moreover when we consider multiple criteria, the "optimal" solution is in general meaningless because it is impossible to satisfy all (usually conflicting) objectives simultaneously [8]. As service design is critical for the success of DRT systems, decision-makers need to understand well how different ways of operating the service affect its performance.

In this work, a general modeling framework for planning and managing DRT services was developed, starting with a comprehensive analysis of European best practices. Based on this framework, an innovative Decision Support System (DSS) was implemented, to help design services, minimizing operating costs and maximizing the service quality.

2 Problem Description

In the Dynamic Vehicle Routing problem for DRT (DVRDRT) we assume that passengers specify origins and destinations from a set of pre-defined possible route points, a pickup time window, and a desired arrival time. Moreover they are to be served by a fleet of vehicles of equal capacity (number of seats). Each possible route point, with the exception of the depot, can be a pickup point, a delivery point, or both. At a given pickup location, different passengers entering the vehicle can have different destinations (many-to-many). Several users can be simultaneously transported in one vehicle, e.g., a mini-bus. The vehicles start and end their trips at a single depot and transportation requests can be received at any time, from any origin. Since different users have different transportation needs, each point (stop) along the route can have multiple (possibly overlapping) time-windows (both for pickup and for delivery). In association with the real-time arrival of new requests this may require several visits to a stop, at different periods. This is a major difference from all known variants of the VRP and the DARP problems—and quite a fundamental one, thus requiring it to be handled in a new way. Finally, pickup time-windows must be respected and delivery time-windows can be violated at a penalty cost.

We have developed a Decision Support System based on a Dynamic Vehicle Routing model that integrates simulation and optimization. Our approach aimed at finding a good overall service design by running a simulation of several demand-offer scenarios.

3 Service Design

At a strategic level, the DRT service design can be viewed as being related with the "stochastic dimension" of the Dynamic Vehicle Routing problems for DRT (DVRDRT), whereas the "dynamic dimension" is associated to the operational level of the service. Moreover, randomness features of the DVRDRT are, in particular, related to: space (the geographical locations of origins and destinations), time (the arrival process of requests) and travel (expected travel time between two points in the net-work).

3.1 Best Practices in DRT Services

DRT services are flexible in terms of route, schedule, vehicle allocation, operator, booking or payment systems, and passenger category. Hence a wide range of different concepts are possible. In [1], authors conclude that there is a strong link between the flexibility of a particular service and the function provided and [9] show that DRTs can be structured in quite different ways, and serve very different needs depending on the geographical areas considered. Therefore, when studying best practices, DRT services need to be classified according to service type and according to service area.

We have made a European wide DRT survey that shows a strong relationship between the function of a DRT service and both the route flexibility and the market it serves: stand alone or substitution services mostly serve special user groups, whereas interchange DRTs serve a more general market. Moreover DRTs that serve a more general public or both general and special groups tend to be more flexible than those that do not. Another somehow strong relationship was also found between the served geographical area and the type of users—the survey indicates that in large urban areas DRTs are mostly used to serve special user groups whereas in rural/small urban areas DRTs tend to serve both special groups and the general public.

According to the survey, in rural areas, around 50 % of the DRT services operate using predefined stops in a given geographical area or in door-to door scenarios: 38 % of the services in rural areas operated by performing deviations on a scheduled service to predefined routes in a corridor only, or in combination with a more flexible route scenario, and the remaining services operate using predefined stops in a corridor. In small urban areas, 20 % of the services operated using predefined stops in a corridor and 40 % predefined stops in an area/door-to-door. Finally, for medium and large urban areas, around 67 % of the services offered predefined stops in area scenarios or door-to-door.

3.2 Simulation for DRT Services

The flexibility of DRT systems, although being a possible answer to some typical problems with traditional public transport systems, can cause several organizational problems. For example, the number and type of requests can require a very large number of vehicles; very sparse requests can be hard to combine efficiently; the quality of the service in terms of delivery/pickup time and travel duration cannot be guaranteed with the available resources or because unpredictable events occur. In this work we have developed a simulation system to study these effects. The system integrates four models, described below, covering the service area, trip requests, vehicles, and real time events.

Service Area Model. The simulated road network is a graph defined by a set of nodes, representing the available stops, and a set of links, representing the roads connecting the stops. When a vehicle enters a given link, the travel time is randomly selected from a lognormal distribution with mean and standard deviation as functions of the time the vehicle entered the link—as in [10]. In future developments of the system, this information may be fed directly from detailed microscopic network conditions.

Trip Request Model. The objective of the trip request model is to generate trip requests that are consistent with the considered geographical area and with the road network within which the service operates. To the authors' knowledge, the trip requests generation problem for analyzing the performance of a DRT in a realistic environment has not been addressed in a fully satisfactory way, except for some contributions such as [11].

The simulation system proposed in this work generates two types of transportation requests: a priori (or advanced) requests and real-time requests. The common attributes of both types are: the number of seats, desired pick-up time and pick-up location, desired delivery time and delivery location. Real-time requests have an additional attribute: booking time. Let n be the number of real-time requests and h the service horizon. Real-time requests arrivals are modeled as a Poisson process, as this seems to be a general assumption for transportation related works [12], with parameter $\lambda = n/h$. For each problem instance, one can define the degree of dynamism (DOD) as the ratio between the real-time requests over the total number of requests. Different instances can be generated with different DODs: 0, 10, 20,..., 100 %. We have made a survey on European DRT services that showed that the shortest booking time limit was 15 min, so we also use 15 min, and then randomly select booking times with uniform probability between 15 and 60 min. The default time window size is 10 min, for this is the smallest value also found in the aforementioned survey. We do not consider dwell time at each stop. By adding the booking time limit to the requests' arrival time to the system we have the users' desired pickup time. For the definition of the desired delivery time, we use the mean trip time and standard deviation for a given service area as found by mobility studies.

We generate origin and destination locations of the requests following the spatial distribution in the OD matrices of the service area, as given by the available mobility studies for different operation periods.

Vehicle Model. The simulation system deals with two types of vehicles: own fleet vehicles and subcontracted vehicles. It is possible to define a fleet size or let the system calculate the optimal size for the service scenario at hand. Subcontracted vehicles will normally be taxis that the operator can hire, in case it runs out of vehicles, to satisfy extra transportation requests. These subcontracted vehicles may have different (usually higher) costs.

Real-time Events. The real-time events can be, broadly, categorized in customer-related events and vehicle related events. Customer-related events include new real-time requests, cancelations and no-shows, and are assumed to follow a Poisson distribution Vehicle related events are, basically: reaching a stop, breakdowns during service (also assumed to follow a Poisson distribution) and delays.

4 Service Operation

We followed a traditional approach for the dynamic problem that consists in solving static scenarios when a new request arrives [5]. However, the routing algorithm must be fast enough to (re)calculate a solution in case requests arrive in a quick sequence. Route sections already traversed until the arrival of the new request and already accepted requests (already picked or not) are, obviously, unchangeable. Thus the problem is to re-optimize the remaining part of the initial solution after the insertion of the new request(s). With time window constraints, the insertion of a new request in real-time is more complex: sometimes this new request cannot be accepted because it is not possible to include it in any of the existing routes, or there is no other vehicle available to start a new route.

DRT problems involve multiple objectives and can be strongly dynamic, requiring the (re-)design of solutions in real-time. These problems are NP-hard and therefore optimal solutions cannot be reached in useful time. To obtain an approximation of the Pareto front we have designed a parallel heuristic procedure that constructs a feasible route through a reactive greedy random approach, followed by a local improvement phase.

4.1 Heuristic Approach

A Greedy-constructive Algorithm. Each feasible transportation request has an origin, a destination and pickup and delivery times. Having a set of requests, the algorithm tries to find routes respecting a set of constraints. The problem objectives are defined along two perspectives: the vehicle and the passengers. A Node Ranking Function (NRF) has been defined to find, at each iteration, the next "best" node to be inserted into the route under construction, taking into account the two aforementioned perspectives. For the vehicles' perspective, the major factors for determining the next "best" node are the distance to all other nodes from the current position and the number of passengers on those nodes. From the passengers' perspective, the major factors to be considered are the number of passengers in the bus having as destination a given node, and the time windows on the other nodes. For each of these factors, weights are assigned to account for the preferences of the decision maker. Let *W* be the set of all nodes in the problem and *NW* the subset of nodes not yet in the solution. The NRF is defined as:

$$\begin{aligned} &\forall i \in NW, NRF[i] = \\ &\alpha_d \times CRL[i] + \alpha_p \times NRL[i] + \alpha_v \times DRL[i] + \alpha_t \times TRL[i] \end{aligned} \tag{1}$$

Cost Rank List (*CRL*) is an ordered list of the normalized travel costs to each node—*CRL*[*i*] is the cost from the current node to node $i \in NW$. Number of passengers Rank List (*NRL*) is the ordered list of the number of passengers at each node. Delivery Time Rank List (*DRL*) is the ordered list of delivery lower time limit at each node. Time Rank List (*TRL*) is the list of pickup lower time limit at each node. All lists values were normalized according to the observed values. The node with the highest NRF is added to the route under construction at the end of each iteration and the process is repeated.

Improvement phase. In terms of local-search based heuristics, the possibility of having multiple pickups and/or deliveries at each stop with multiple (possibly overlapping) time-windows makes the definition of neighborhood structures non-trivial and their implementation computationally very complex [13]. For the improvement phase, we have therefore set up a combination of three mechanisms:

```
Parameters:GRASP_max_iterations
while (num_iterations<GRASP_max_iterations)
        choose α_k parameter with probability p(α_k) = 1/m ,k = 1,..,m
        initialize S = {}
        Construction phase:
                Calculate S using NRF and α_k for the RCL
        If mod(num_iterations,200)==0 then p(α_k) = q_k/Σ_{j=1}^m q_j , k = 1,...,m
        Calculate the solution cost  F(S) = (Σ_{i=1}^m C(R_i) + Σ_{j=1}^u C(U_j))
        Improvement phase:
                using S do:
                        forward slack time
                        nearby stops analysis
                        simple 2-exchange procedure
                until F(S'') < F(S) or elapsed_time>allowed_running_time
                if F(S'') < F(S)  then S = S'
        update best solution found S*: if F(S) < F(S*) then S* = S
end-while
return best solution S*
```

Fig. 1 Parallel reactive-GRASP type meta-heuristic

the forward slack time, a "nearby-stops" analysis, and a simple 2-exchange procedure. After computing the dead times available through the route (forward slack time), the "nearby-stop" analysis takes each route in the solution set and tries to find in-between stops that appear later in that route and can be served in the meantime. The last improvement is a simple 2-exchange procedure based on the k-interchange procedure by [14].

A GRASP Type Meta-heuristic. The NRF algorithm was embedded in a GRASP type [15] meta-heuristic and was implemented in parallel, mainly because our need of real time solutions. We have also implemented a reactive mechanism that reacts to the produced solutions and tries to adjust the greediness-randomness balance [16]. The construction strategy based on the evaluation of the elements to be inserted in the solution at each iteration according to the NRF function. As new nodes are added to the solution routes, this function takes the already solution into account and, instead of always choosing the "best" node, there is a random choice between the best elements. This initial solution is then used by the local improvements phase in a first-best procedure. The process is repeated a specified number of iterations in parallel.

Figure 1 presents our Parallel Reactive-GRASP for the Dynamic Vehicle Routing problem for Demand Responsive Transportation (DVRDRT), where at every 200th iteration, the probabilities of the parameter that controls "reactiveness" are updated.

4.2 Heuristic Assessment

For the GRASP-type heuristic it is important to know:

- what is its real-time performance;
- what are the major factors affecting the performance;
- what is the competitive ratio of the algorithm;
- what are the effects of assigning different weights to the decision criteria.

Being a "new" problem, to the best of our knowledge there are no "off-the-shelf" benchmark data sets. So, our decision at this stage was to use randomly generated instances for the city of Porto, Portugal, with different number of clients, different special distributions, and different degrees of dynamism with different pickup and delivery time windows. Computational tests were done using an Intel Core Duo running at 1,67 GHz, 2 GB RAM memory, and the adjustment of the α parameter that controls greediness/randomness level at every 100th algorithm iteration. The number of parallel threads running the algorithm was dynamically set to 8. Results on these instances look very promising, both in terms of cost savings and in terms of computational efficiency. These results seem to highlight that, as expected, the major factor affecting the heuristic approach run time is the number of passengers. In terms of performance, each 1000 iterations take less than 1 s, for 20 transportation requests in Porto area. Another observation is the linear increase in running time with the number of iterations, the running time for each iteration being constant—this is in line with literature results for GRASP-based algorithms.

Moreover, one common way to evaluate algorithmic approaches for Dynamic Vehicle Routing problems is to use the competitive analysis framework [17]. Our approach was to increase the degree of dynamism, in order to understand how the over-all solution cost increases when compared to having all information in advance and, as such, provide an empirical estimate of the competitive ratio of the algorithm. For the same problem, the solution cost of a 90 % degree of dynamism scenario, with requests distributed evenly throughout the planning horizon, is around 45 % higher than the static scenario with all information known a priori (0 % degree of dynamism).

5 Decision Support System

5.1 Logic Architecture

The Decision Support System (DSS) presented in this work has a client–server logic architecture based on the Three Tier Distribution pattern [18], in order to better structure the distribution of application functionality among distributed processing contexts. With this architectural pattern we aim at achieving a flexible,

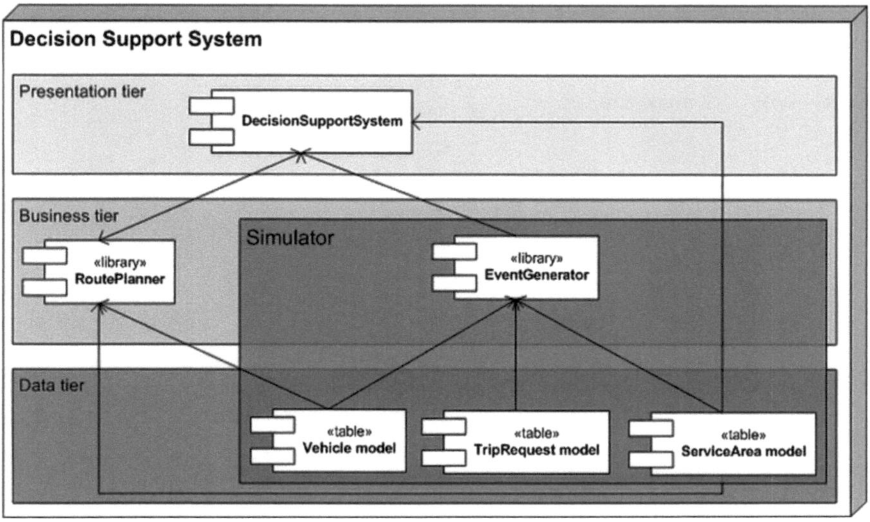

Fig. 2 Integrated high level logic architecture

evolutionary, scalable architecture, supporting technological interoperability. As we could not risk having network performance dependency, we chose to tight the coupling of the server: the three tiers are run on the same platform machine, although in different processes.

Users of the DRT service must specify a request (origin, destination, pickup time, and delivery time) using a request client subsystem. A transportation requests client is a thin-client running on any location, on any platform and implemented on any technology. The Simulator could also run in any location as a stand-alone application, but due to the data requirements, we chose to integrate the Simulator on the DSS: they share the same user interface, the same middle-tier and the same data access tier. The Simulator generates time-ordered travel requests based on the trip request model. These requests are the inputs to the real-time multi-objective heuristic procedure that tries to satisfy each request taking into account: (a) the multiple perspectives of the different stakeholders stated via the DSS; (b) a fleet of vehicles with their associated locations and other attributes (vehicle model); and (c) the expected trip times (service area model). The heuristic approach is also responsible for updating the status of the vehicles and the corresponding set of data: assigned route and schedule, visited stops, network link currently being travelled, current speed, current position, and possible delays.

Figure 2 shows the high-level logic architecture integration of the Simulator in the DSS, highlighting the three-tier architecture.

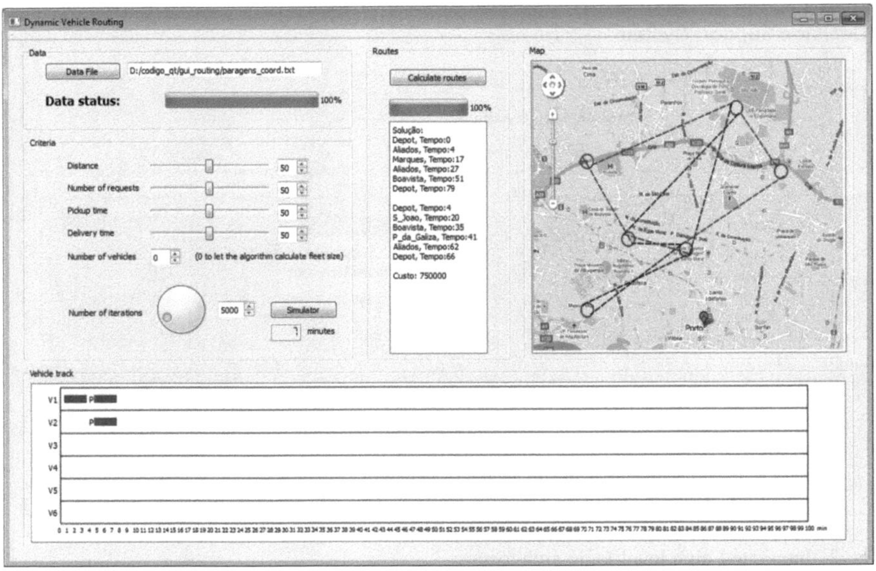

Fig. 3 Decision support system graphical user interface

5.2 Graphical User Interface

The DSS Graphical User Interface (GUI) supports both the visualization of routes and the definition of the weights for the chosen criteria. Vehicle routing can be triggered at any time by the service operator but it can also be started automatically each time a new transportation request arrives in real time.

Vehicle routing requires the service operator to specify his perspectives/preferences, assigning weights to the different criteria: travel distance minimization, maximization of the number of requests served, minimization of passenger waiting time and, finally, minimization of passenger on-board ride time.

The routes in the produced solution are displayed on the area map (Fig. 3). The total solution cost is also displayed. The service operator can check information available on the stops along the routes displayed on the map at a given time, such as the number of passengers waiting at the stop, and the number of passengers who specified that stop as their destination.

6 Experiments

One of the main purposes of simulation is to obtain a better understanding of the behavior of a system in a given set of conditions, under some level of uncertainty. The performance of the system can be determined by observing what happens on

Table 1 Simulation parameters

Parameters	Example values
Decisions	
Fleet size	3
Vehicle capacity	27
External taxi vehicles	2
Time window	10 min
Fare	€1
Distance minimization weight	0.2
Requests maximization weight	0.15
Pickup time weight	0.55
Delivery time weight	0.1
Data	
Vehicle	
Vehicle commercial speed	16 km/h
Vehicle cost per km	€0.32
Taxi speed	30 km/h
Taxi cost per km	€0.60
Service area	
Mean travel time	35 min
Standard deviation travel time	17 min
Source zones (from area's OD matrix)	9, 8, 6, 4, 1
Sink zones (from area's OD matrix)	7, 6, 5, 3, 2, 1
Requests	
Number of a priori requests	100
Degree of dynamism	0.1
New requests birth rate	Poisson ($\lambda = 0.3$)
Cancelations statistical distribution law	Poisson ($\lambda = 0.05$)
No-shows statistical distribution law	Poisson ($\lambda = 0.01$)

the network, during simulation, with different conditions. The results of the simulation runs will also provide guidelines to help operators of public transport to design DRT services.

We have simulated 2 hours of a DRT service that operates during the night time in the city of Porto. The service operates between midnight and 2:00 am, every day. The service booking is open during the day time to receive a priori requests to be served, and is open during service operation time to receive real-time transportation requests. It is assumed that passengers specify origins and destinations from a set of pre-defined possible stops, a pickup time, and a desired arrival time. Moreover, they are to be served by a fleet of 3 vehicles of equal capacity (27 seats each) and 2 sub-contracted external vehicles (taxis). Each possible route point, with the exception of the depot, can be a pickup point, a delivery point, or both. At a given pickup location, different passengers entering the vehicle can have different destinations (many-to-many). Several users can be simultaneously transported in one vehicle. The vehicles start and end their trips at a single depot.

Table 1 presents the parameters and example values used for "base case" simulation runs for our study.

This scenario, with the DRT service and simulated demand structure presented, the operator should consider between 4 and 7 fixed vehicles, subcontract external taxis to cope with the un-satisfied real-time requests (with a capacity between 14 and 20 seats), and set a time-window between 10 and 15 min.

7 Conclusions

Demand Responsive Transportation (DRT) systems aim at delivering quality public transportation in low, variable and unpredictable demand scenarios, for which traditional public transportation services may be extremely expensive. DRT systems contribute to this goal by providing routes and frequencies that may vary according to the actual observed demand. These systems can be strongly dynamic, requiring the adaptation of solutions in real-time, in a multiple criteria context. Service design is critical for the success of DRT services and decision-makers need to understand well how different ways of operating the service affect its performance.

A general modeling framework for planning and managing DRT services was developed, starting with a comprehensive analysis of European best practices. Based on this framework, a Decision Support System (DSS) has been designed. The DSS uses simulation models and heuristics to produce a set of efficient solutions. The proposed simulation system can be used as a tool to study how different ways of operating the service affect its performance, in a given scenario. Moreover the parallel reactive GRASP based constructive heuristic algorithm developed in this work allows the planner to set different weights for the different factors, assuring that the multiple perspectives of the different stakeholders are taken into account, thus improving the decision-making process. In fact our DSS shows a performance level that allows the generation of good solutions in real-time, in order to cope with the dynamism degree of the problem. Its architecture allows for loosely coupled clients, such as the transportations requests client or the simulator, promoting interoperability between different technologies.

Simulation experiments with a small DRT service for the city of Porto have shown the usefulness of the approach for designing and managing DRT services. Decision makers may use the developed Decision Support System to find the best combination of the decision parameters to design DRT services that meet an envisaged cost level and a desired quality of service. For a given demand structure, decision makers should experiment different time window sizes, number of vehicles their capacity, and number of outsourced taxis. These parameters directly influence fixed costs, total distance travelled, number of requests satisfied/unsatisfied and mean pickup delay, just to name a few performance indicators. Also, different weights for the criteria result in different cost structures and service quality and, as such, should be analyzed using the Decision Support System.

Acknowledgments The work of the first author was supported by the Portuguese National Science Foundation (FCT) under grant SFRH/BD/42974/2008.

References

1. Brake, J., Mulley, C., Nelson, J., Wright, S.: Key lessons learned from recent experience with flexible transport services. Transp. Policy **14**(16), 458–466 (2007)
2. Dantzig, G., Ramser, J.: The truck dispatching problem. Manage. Sci. **6**(1), 80–91 (1959)
3. Cordeau, J.F., Laporte, G.: The dial-a-ride problem: models and algorithms. Ann. Oper. Res. **153**(1), 29–46 (2007). (178EF Times Cited:3 Cited References Count:43)
4. Paquette, J., Cordeau, J.F., Laporte, G.: Quality of service in dial-a-ride operations. Computers & Industrial Engineering, In Press, Corrected Proof (2010) doi: 10.1016/j.cie. 2008.07.005
5. Psaraftis, H.: Dynamic vehicle routing: status and prospects. Ann. Oper. Res. **61**(1), 143–164 (1995) Cited By (since 1996): 107 Export Date: 26 May 2010 Source: Scopus
6. Larsen, A.: The dynamic vehicle routing problem. Ph.D. Thesis, Technical University of Denmark (2000)
7. Lenstra, J., Kan, A.: Complexity of vehicle routing and scheduling problems. Networks **11**(2), 221–227 (1981)
8. Branke, J., Deb, K., Miettinen, K., Sowiski, R.: Multiobjective Optimization: Interactive and Evolutionary Approaches. Springer, Heidelberg (2008)
9. Potts, J., Marshall, M., Crockett, E., Washington, J.: TCRP report 140—a guide for planning and operating flexible public transportation services. Technical Report 9780309154802 0309154804. Transportation Research Board, US (2010)
10. Taniguchi, E.: City logistics : network modelling and intelligent transport systems. Pergamon, Amsterdam (2001)
11. Deflorio, F.: Simulation of requests in demand responsive transport systems. IET Intel. Transport Syst. **5**(3), 159–167 (2011)
12. Larson, R., Odoni, A.: Urban Operations Research. Prentice Hall, Englewood Cliffs (1981)
13. Kindervater, G., Savelsbergh, M.: Vehicle Routing: Handling Edge Exchanges. In: Aarts, E.H.L., Lenstra, J.K. (eds.) Local search in combinatorial optimization, pp. 311–336. Wiley (1997)
14. Psaraftis, H.: k-Interchange procedures for local search in a precedence-constrained routing problem. Eur. J. Oper. Res. **13**, 391–402 (1983)
15. Feo, T., Resende, M.: A probabilistic heuristic for a computationally difficult set covering problem. Oper. Res. Lett. **8**(2), 67–71 (1989)
16. Resende, M., Ribeiro, C.: Greedy randomized adaptive search procedures. In: Glover, F., Kochenberger, G.A. (eds.) International Series in Operations Research & Management Science, vol. 57, pp. 219–249. Springer, New York (2003)
17. Larsen, A., Madsen, O., Solomon, M.: Classification of dynamic vehicle routing systems. In: Giaglis, G.M., Minis, I., Tarantilis, C.D. (eds.) Operations Research/Computer Science Interfaces Series, vol. 38, pp. 19–40. Springer, US (2007)
18. Hirschfeld, R.: Three-tier distribution architecture. In: Proceedings of PLoP'96 (1996)

Simulation of Crowd Dynamics in Panic Situations Using a Fuzzy Logic-Based Behavioural Model

Mauro Dell'Orco, Mario Marinelli and Michele Ottomanelli

Abstract Tragic events in overcrowded situations have highlighted the importance of the availability of good models for pedestrian behaviour under emergency conditions. Crowd models are generally macroscopic or microscopic. In the first case, the crowd is considered to be like a fluid, so that its movement can be described through differential equations. In the second case, the collective behaviour of the crowd is the result of interactions among individual elements of the system. In this paper, we propose a microscopic model of crowd evacuation that incorporates the fuzzy perception and anxiety embedded in human reasoning. A Visual C++ application was developed to evaluate the outcomes of the model. The model was tested in scenarios with presence of a fixed obstacle. Simulation results have been analyzed in terms of door capacity and compared with an experimental study.

Keywords Pedestrian behaviour · Fuzzy logic · Micro-simulation modelling · Evacuation simulation

1 Introduction

Understanding competitive egress behaviours can be helpful in avoiding tragic events: effective egress models are useful both in designing large venues and in calculating their working conditions during emergencies. The simulation of

M. Dell'Orco (✉) · M. Marinelli · M. Ottomanelli
D.I.C.A.T.E.Ch., Technical University of Bari, Via E.Orabona, 4, 70126 Bari, Italy
e-mail: dellorco@poliba.it

M. Marinelli
e-mail: m.marinelli@poliba.it

M. Ottomanelli
e-mail: m.ottomanelli@poliba.it

J. F. de Sousa and R. Rossi (eds.), *Computer-based Modelling and Optimization in Transportation*, Advances in Intelligent Systems and Computing 262,
DOI: 10.1007/978-3-319-04630-3_18,

pedestrian motions within an area in presence of obstacles or dangerous events, like fire or explosion, and description of factors that make a pedestrian able to determine autonomously the path to the target destination, are thus crucial problems in evacuation studies.

Pedestrian motion is difficult to describe in terms of simple models, due to phenomena like jamming and clogging, lane formation and oscillations at bottlenecks in counterflow or collective patterns of motion at intersections [14]. However, many interesting models have been developed, usually grouped into two main types: Cellular Automaton models and Social Forces models.

The first approach considers mainly the geometrical aspect of path finding, generally neglecting the dynamic aspects of path planning and seeking. The second approach is based on the idea of attraction and repulsion forces, like magnetic ones; that is, provided that individuals have a positive charge, the method assigns a negative charge to the target destination, and positive charges to the obstacles.

In this chapter we present an innovative approach that allows modelling, through verbal variables and linguistic rules, the imprecise manner of reasoning of humans in decision-making and in facing panic or emergency conditions. The model has been compared with an experimental study by Daamen and Hoogendoorn [3] in terms of door capacity. The results obtained are encouraging and push towards further developments.

2 Existing Evacuation Models

Although pedestrian motion can appear chaotic, it is possible, starting from observed data, to define two different sets of behavioural rules, according to whether the situation considered is 'normal' or 'on panic'.

In *normal conditions*, pedestrians feel a strong dislike for detouring, even if their path is crowded. However, there is some evidence showing that pedestrians normally choose the fastest path, not their shortest one [4]. They consider detours if they can increase their walking comfort or decrease the effort to reach their destination [8]. Pedestrians prefer to proceed at the speed that they can maintain with minimum effort, unless they need to hurry to reach their destination on time [21]. The distribution of desired speeds in crowds is in Gaussian curves [9], in which the average speed depends on the situation [19], as well as on sex, age, time of day, reason for the movement, surroundings, etc. [21]. Pedestrians proceed at a certain distance from each other and from borders like walls and/or obstacles [1]. The density increases around special attraction areas. Individuals knowing each other can form groups that behave like a single pedestrian. The size of these groups follows the Poisson distribution [2]. Individuals are inclined to imitate crowd behaviour. As a consequence, the alternatives are often forgotten or neglected [13, 20]. When density is medium–high, the crowd motion shows some analogies to gas, fluid and granular flow motion. For example, on the boundary between two crowds moving in opposite directions, viscose friction and speed gradients are observable.

In *panic conditions*, individuals' speeds increase above normal, interactions between persons become highly physical and movements are uncoordinated [16]. At exits, clogging and collisions occur, as well as rainbow-like arching structures [10]. Therefore, physical interactions increase so much that they can reach dangerous levels of pressure (even 4500 N/m), able to break barriers down. Resulting evacuation is slow because of persons stumbling, hurting themselves or obstructing each other. Herd behaviour is manifested, with underutilisation of other exits [20].

The pedestrian flow models can be divided into two main groups, macroscopic and microscopic, according to whether the pedestrian movement is traced implicitly or explicitly, and the performances in the considered area are expressed in an aggregate or disaggregate way.

In macroscopic models, the crowd is like a fluid whose behaviour can be described through Navier-Stokes or Boltzmann equations, representing the variations in speed and density over time [9].

Microscopic models, on the other hand, describe the behaviour of the individual in space and time; the collective phenomena come out through interactions among the single elements of the system. However, the analytical complexity of these models could make their use really time consuming.

Although macroscopic models are not particularly complex, they are scarcely adaptable since they cannot represent the interactions among individuals. Moreover, the hydrodynamic analogy is valid only for normal conditions; panic conditions, uncoordinated movements and inappropriate behaviours of individuals invalidate the analogy. Therefore, this chapter focuses straightaway on the microscopic models.

Microscopic models can be divided into Spatial Grids models and Forces-based models. The models belonging to the first group can be subdivided into cost-benefit analysis models and Cellular Automaton model. The second group includes the Magnetic Forces model and the Social Forces model. An additional model outside the previous groups is the Adaptive Control model.

3 The Proposed Model

The basic idea in this work is that pedestrian motion in a given area depends on factors related to human perception and reasoning process.

A pedestrian reacts against fixed or mobile elements around himself/herself as soon as they disturb the space around his/her position. For the sake of simplicity, in this work we assume that the perturbation generated by a fixed element is exactly equal to the volume it takes up, while mobile elements generate a volume that varies over time intervals, according to their speed and path. Therefore, in this case two different perturbations are considered: one due to presence, and another one due to movement.

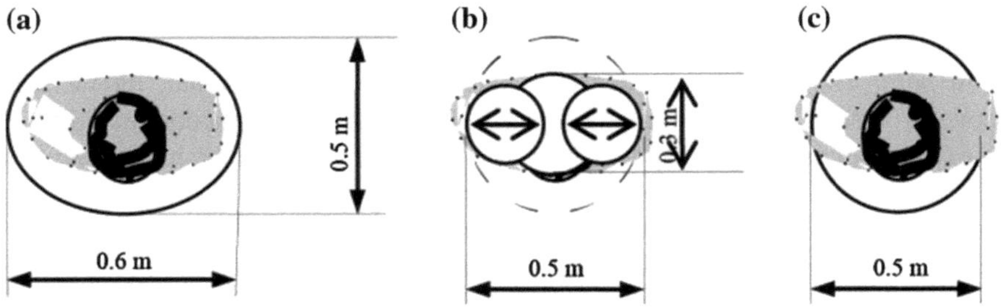

Fig. 1 Pedestrian body models: **a** HCM recommended pedestrian body ellipse; **b** SIMULEX median pedestrian body model; **c** body model used in this work

Consider now a reference frame whose x, y plane is integral with the motion plane, and its z axis directed to the height. Since the proposed model is a 2D model, the perturbation induced by a fixed obstacle is modelled only in the x, y plane as its trace on that plane.

Pedestrians and, more generally, mobile elements, generate:

- A perturbation due to their presence, having the shape of their overall physical dimensions (presence perturbation). The Highway Capacity Manual (HCM) recommends modelling the trace of a standing person as an ellipse of 0.5 m by 0.6 m (Fig. 1a). SIMULEX [12] contains different body types, made of three circles (Fig. 1b) whose diameters are 0.2 for shoulders and 0.3 for torso in the median case. For sake of simplicity, in this chapter we have used, as body model, the circumference circumscribed to the SIMULEX model (Fig. 1c).
- A perturbation due to motion (advancing perturbation). We assumed for this kind of perturbation a semicircular shape, pointing at the pedestrian's advancing direction (Fig. 2).

Actually, this shape shows a spatial isotropy that is not completely effective in case of pedestrian motion. More proper shapes, like triangular or semi-elliptical ones, could be considered, but at the cost of a heavier calculation. Basically, this area represents the space that each person wants to keep free around himself/herself during motion.

In order to do this, pedestrians can change speed and direction of their motion, although they usually try to keep path and speed as steady as possible. It is worth noting that the size of this area is not constant, but a function of the crowd density for computing the pedestrian speed. The crowd density d is a function of the radius $r_v(t)$ of the perturbation area given by the following relation:

$$r_v(t) = \Delta t \cdot v_p(t-1) \tag{1}$$

where Δt is the time step, $v_p(t-1)$ is the pedestrian velocity at t−1. So the pedestrian speed is determined by the crowd density into the perturbation area, and using the results obtained by Weidmann [21] to set up this relationship quantitatively:

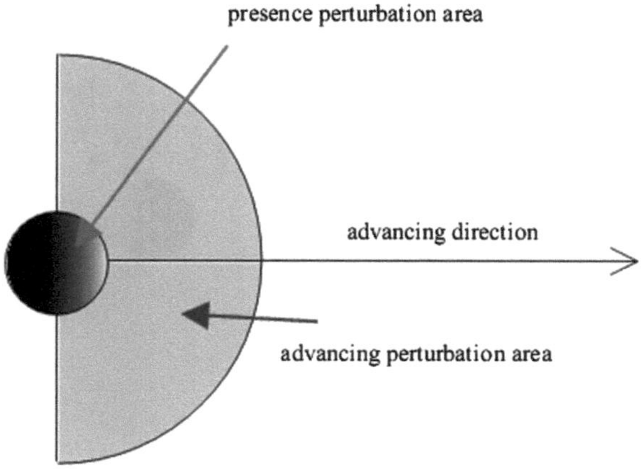

Fig. 2 The pedestrian model

$$v_p(t) = v_m \cdot e^{(-0.74 \cdot d(t))} \tag{2}$$

where v_m is the unimpeded velocity of pedestrians. According to Simulex Technical Reference (2000), we have considered three different ranges of unimpeded velocity:

- 1.4–1.8 m/s for male;
- 1–1.4 m/s for female;
- 0.5–1 m/s for elderly.

To avoid pedestrian overlapping a collision detection method was developed. This method is based on the "point inclusion in polygon test" in order to check the presence of other pedestrians in the perturbation area. It is also used to find the presence of fixed obstacles or fire.

Fixed obstacles are modelled as a series of points obtained sampling the obstacles' edges. So their presence is checked finding one or more samples in the perturbation area. The fire is modelled as a polygon, so its presence is checked verifying intersections between the fire polygon and perturbation area. This presence causes strong deviations from the preferred path in order to make pedestrians move away from fire. The resulting pedestrian motion is obtained through a Fuzzy Inference Engine, which gets as input all these information and gives as output the final deviation angle.

3.1 The Fuzzy Inference Model

The approximate reasoning of the Fuzzy Inference Engine simulates human reasoning. In our case, the Fuzzy Controller will provide, for each pedestrian and for each time step, the deviation from the original path, exactly as the human brain

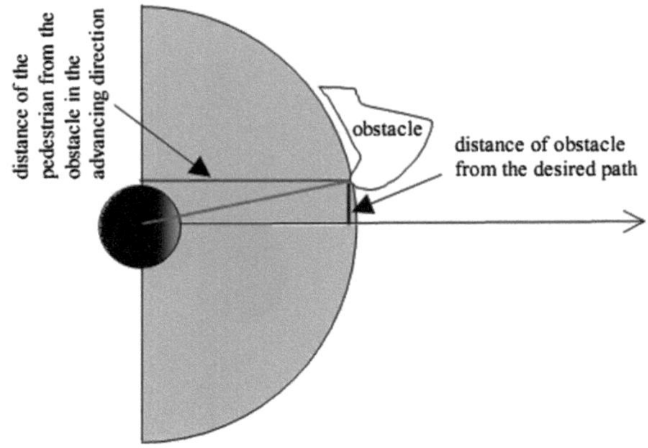

Fig. 3 Perturbation model

would do, as soon as information about an obstacle on the path is received: that is, as soon as an obstacle enters the pedestrian's advancing perturbation area.

The most relevant information in making a decision about the amount of deviation, is the distance in the advancing direction of the pedestrian from an obstacle, located on the left and/or on the right side, and the distance of the obstacle from the desired path (Fig. 3).

Therefore, the crisp input for the fuzzifying interface is made up of:

- the distance from the closest right side obstacle in the advancing direction (right_obst);
- the distance from the closest left side obstacle in the advancing direction (left_obst);
- the distance of the closest right side obstacle from the desired path (right_dev);
- the distance of the closest left side obstacle from the desired path (left_dev);
- the distance from exit (dist_exit);
- the normalized difference between the number of obstacles on the left and right side into the perturbation area (diff_obst);

while the output is the new direction of the motion, in terms of angular deviation from the previous path.

In Fig. 4, the fuzzy values of input attributes are reported.

The Fuzzy Inference System (FIS) provides the direction of motion (direction), in terms of the deviation attributes reported in Fig. 5, through a set of 15 if-then fuzzy rules like the following three:

- *if* ***left_obst*** *is* ***far*** *and* ***right_obst*** *is* ***far*** *then* ***direction*** *is* ***straight ahead***
- *if* ***left_obst*** *is* ***far*** *and* ***right_obst*** *is* ***close*** *and* ***left_dev*** *is* ***close*** *and* ***right_dev*** *is* ***far*** *then* ***direction*** *is* ***right***
- *if* ***left_obst*** *is* ***far*** *and* ***right_obst*** *is* ***close*** *and* ***left_dev*** *is* ***far*** *and* ***right_dev*** *is* ***close*** *then* ***direction*** *is* ***far left***

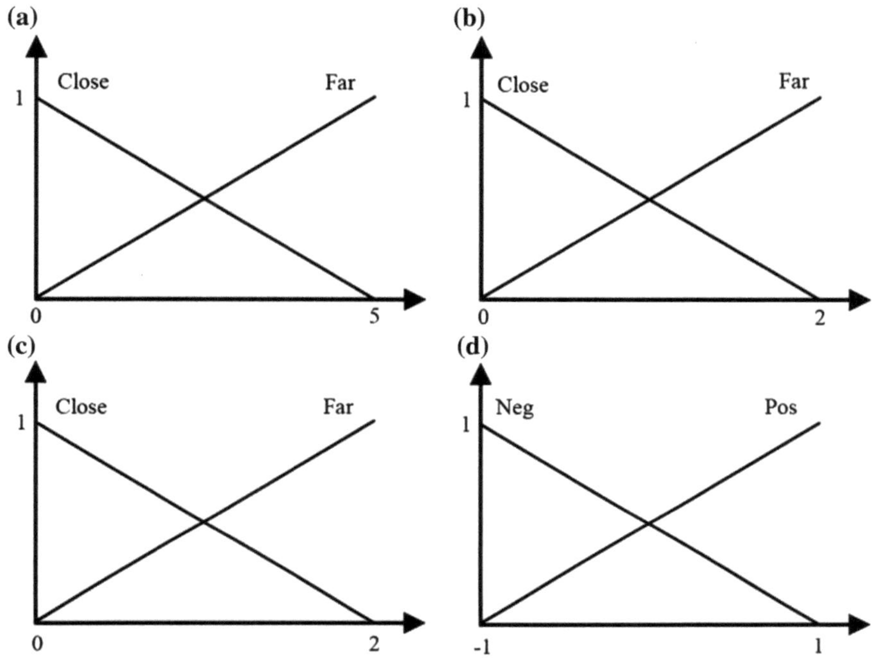

Fig. 4 Attributes of the input to the motion model: **a** pedestrian's distance from the obstacle (m); **b** distance of the obstacle from the path (m); **c** distance from exit (m); **d** normalized difference between the number of obstacles on the *right* and *left side* (m)

The resulting output value is obtained through the "centre of gravity" defuzzification method. Since the proposed model is a time-discrete, microscopic one, we usually refer to a generic pedestrian, unless otherwise specified, in an emergency condition at time t. For the p-th pedestrian, the initial path to exit is the straight line connecting his/her initial position to the centre of the exit (x_e, y_e).

Motion towards the exit and path adjustments to avoid mobile and fixed obstacles can then be modelled through the motion model. When density is high, as persons start pushing each other and clogging the way, the 'faster-is-slower' effect [6], and rainbow-like arching structures [10] take place at exits. Then, jostling and pushing result in lowering the egress rates with respect to regular outflow conditions; in other words, there is a delay $t_{d,p}$ in the egresses of pedestrians that basically depend on their stress level.

As the concept of stress level is clearly fuzzy, we should now set up two new Fuzzy Inference Systems (FIS): one to calculate the stress level, and another one to get $t_{d,p}$ from the stress level. But, provided that this delay depends only on the stress level, and that verbal values of characteristics of both delay and stress level are equal, we can arrange the two FIS into one only, replacing the characteristic 'stress level' in the first FIS with the characteristic 'delay' from the second one.

The characteristics of input for this FIS (Fig. 6) are:

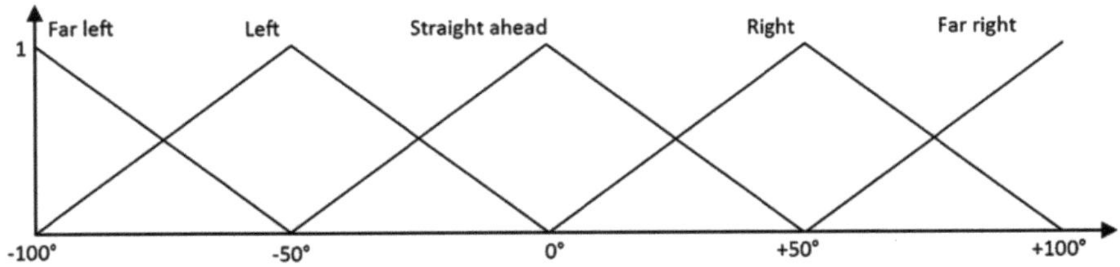

Fig. 5 Attributes of the output of the motion model: direction of motion (*angles measured clockwise*)

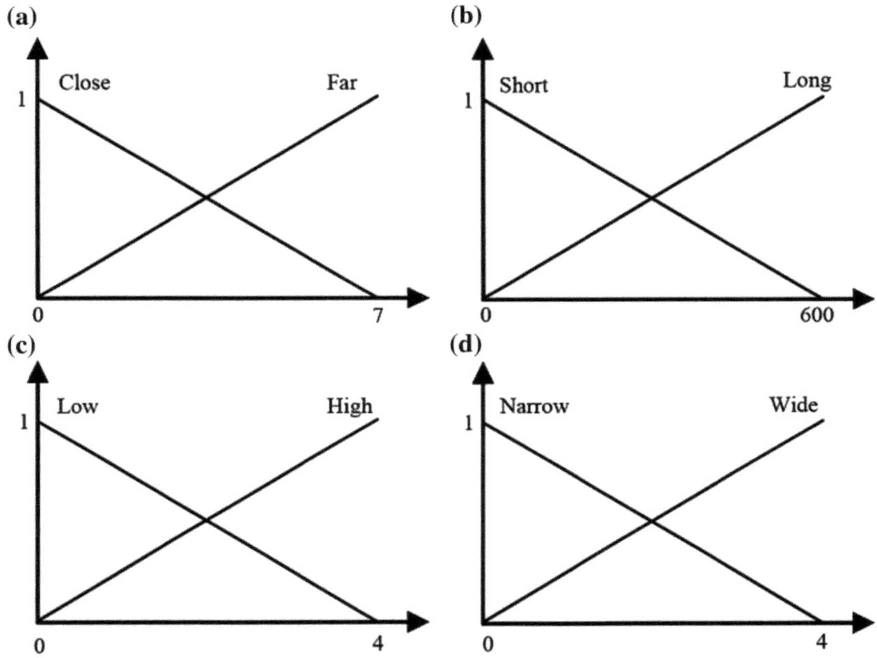

Fig. 6 Attributes of the input to the delay model: **a** distance from exit (m); **b** elapsed time (s); **c** density (pers/m^2); **d** exit width (m)

- distance from the exit (distance);
- time elapsed since the evacuation started (elapsed_time);
- pedestrian density (density);
- exit width (exit_width).

It provides the output 'delay in egress' (delay), in terms of the attributes reported in Fig. 7, through the following set of 12 if…then fuzzy rules:

- *if* ***distance*** *is* ***far*** *and* ***elapsed_time*** *is* ***long*** *and* ***density*** *is* ***high*** *and* ***exit_width*** *is* ***wide*** *then* ***delay*** *is* ***high***

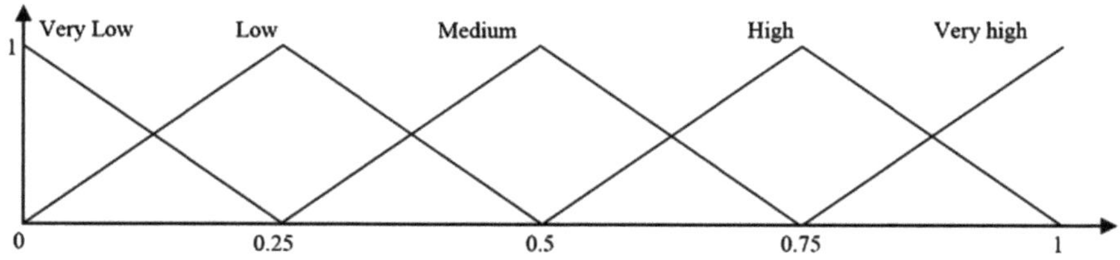

Fig. 7 Output for delay model: delay in egress

- *if* ***distance*** *is* ***far*** *and* ***elapsed_time*** *is* ***long*** *and* ***density*** *is* ***low*** *and* ***exit_width*** *is* ***wide*** *then* ***delay*** *is* ***medium***
- *if* ***distance*** *is* ***close*** *and* ***elapsed_time*** *is* ***long*** *and* ***density*** *is* ***high*** *and* ***exit_width*** *is* ***wide*** *then* ***delay*** *is* ***low***

In a generic multiple-exit case, pedestrians' paths depend also on the choice of exit. The choice model proposed in this work is based on the 'herding behaviour': in a panic situation, the individual is inclined not to behave autonomously, but to imitate and follow the surrounding persons [13, 20].

This behaviour is more evident as panic increases. For a low level of panic, a great number of individuals are still able to choose autonomously the best exit but, as soon as their stress level increases, more and more persons imitate other persons around them, discarding any rational behaviour.

Using the previous single-exit egress model repeatedly, the individual delay and the density, with respect to each exit, at time t can be determined. We can replace, as stated before, the concept of stress level with the concept of delay and then, through another FIS, calculate the preference for each exit. The inputs of this FIS (Fig. 8) are:

- delay (delay)
- distance from the exit (distance)
- pedestrian density (density).

The output 'preference' (preference) (Fig. 9) is provided through the following set of 15 if-then fuzzy rules like the following three:

- *if* ***delay*** *is* ***very low*** *and* ***distance*** *is* ***close*** *and* ***density*** *is* ***low*** *then* ***preference*** *is* ***very high***
- *if* ***delay*** *is* ***very low*** *and* ***distance*** *is* ***close*** *and* ***density*** *is* ***high*** *then* ***preference*** *is* ***medium***
- *if* ***delay*** *is* ***very high*** *and* ***distance*** *is* ***far*** *and* ***density*** *is* ***low*** *then* ***preference*** *is* ***very low***

In this way, we calculated the preference for each exit for all people involved in evacuation. Of course, each pedestrian will select the exit for which he/she has the highest preference, and move towards it. The motion model will then provide the path, and the single-exit egress model can be applied to the exits, to calculate the number of persons exiting at time $t + \Delta t$.

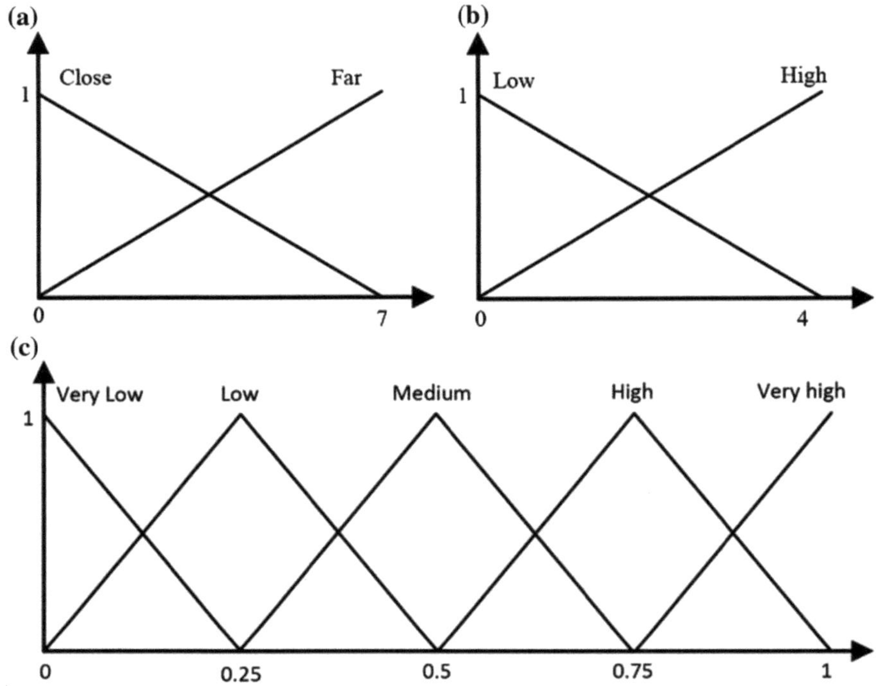

Fig. 8 Attributes of the input to the preference model: **a** distance from exit (m); **b** density (pers/m^2); **c** delay in egress (s)

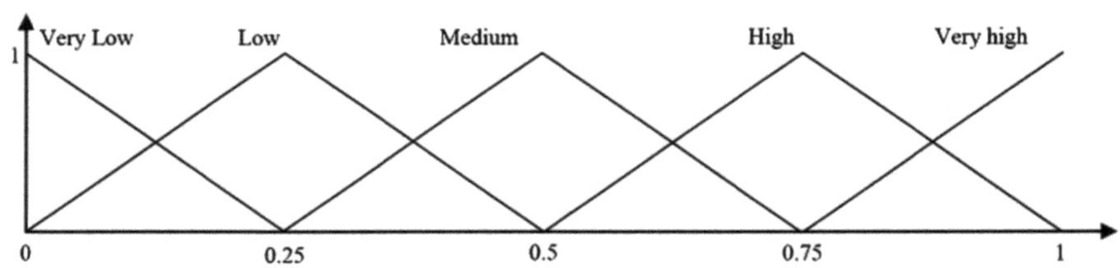

Fig. 9 Output for preference model

4 Simulation Results

We have developed a Visual C++ application to test the proposed simulation model. This application provides functionalities in order to load different scenarios and save the resulting simulation data. This application is based on a multi-threading approach in order to take advantage of the potential of actual multi-core architectures. Each thread computes a single pedestrian's position, so it represents a single pedestrian's mind which works in parallel with the others. At the end of every single time step (iteration), threads are synchronized to acquire and display

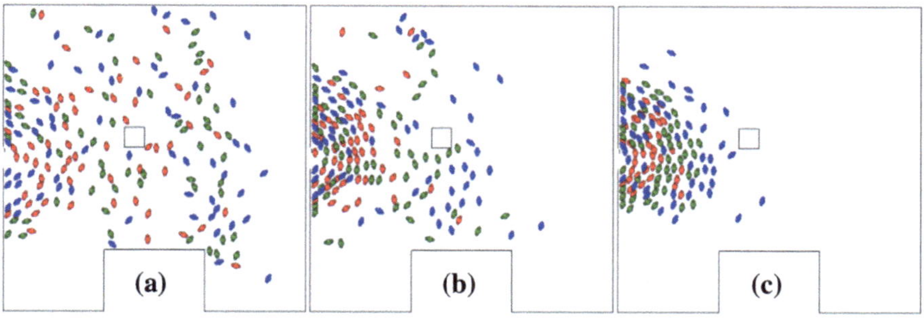

Fig. 10 First scenario. Snapshot of the simulation, at time **a** t = 5 s, **b** t = 12 s, **c** 21 s

position data. This approach turned out to have a better performance than a simple sequential approach, in which pedestrians' positions are computed sequentially in time, using a single processor.

To evaluate outcomes of the proposed model, we have considered two different scenarios. In the first scenario, the evacuation of 200 pedestrians from a square room 15 by 15 m is reproduced. The room has only one exit 1 m in width, placed in the middle of the western wall. A fixed obstacle is in the centre of the room, and another one is leaning against the southern wall. The pedestrians are assumed to be randomly scattered in the room and equally distributed between the three different classes (man, woman, elderly).

Figure 10 shows three snapshots of the simulation at different times. We can observe at the exit the rainbow-like arc making process and bursting exit behaviour. The resulting cumulative curve of pedestrians' flow is presented in Fig. 11a. Moreover, Fig. 11b shows the resulting trajectories formed during the simulation starting from all the positions occupied by pedestrians in the room.

In the second scenario, the room is the same of the first scenario but it has a second exit on the northern wall.

Like in the first scenario, the two exits are 1 m in width, and one obstacle is present in the centre of the area. The positions of pedestrians have been assumed to be randomly scattered and equally distributed between the three different classes. The resulting evacuation time for this scenario is 46 s, about half of the first scenario (81 s) for the same density, indeed here the number of doors is doubled. In Fig. 12, three snapshots of the simulation are reproduced, while Fig. 13a shows trajectories obtained during the simulation.

To complete the evaluation of the simulation model, we have carried out an analysis based on door capacity compared with the experimental study by Daamen and Hoogendoorn [3]. They made laboratory experiments to make a comparison with the Dutch national building code ("Building decree") sets requirements to the width of emergency doors. Our comparison takes into account door capacity values they obtained for a scenario with door opening of 1 m in width and an average population with a stroboscopic stress level. We have run our two scenarios for ten times to obtain different door capacity values as they made for their

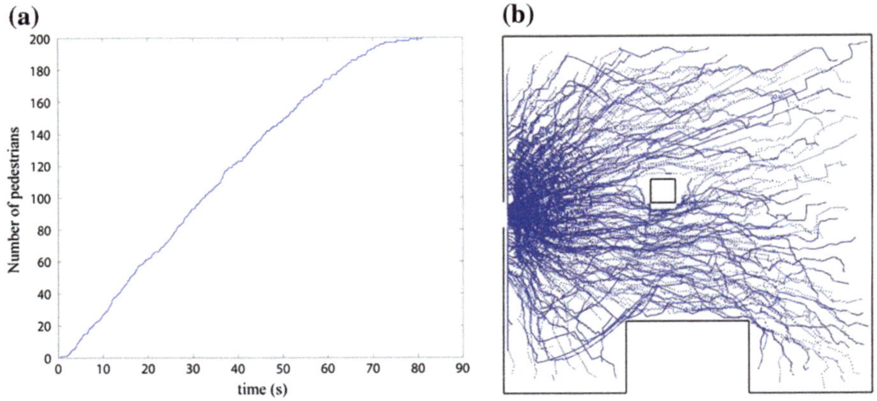

Fig. 11 **a** Cumulative curve related to the simulation of the first scenario; **b** positions occupied by pedestrians during the simulation of the first scenario

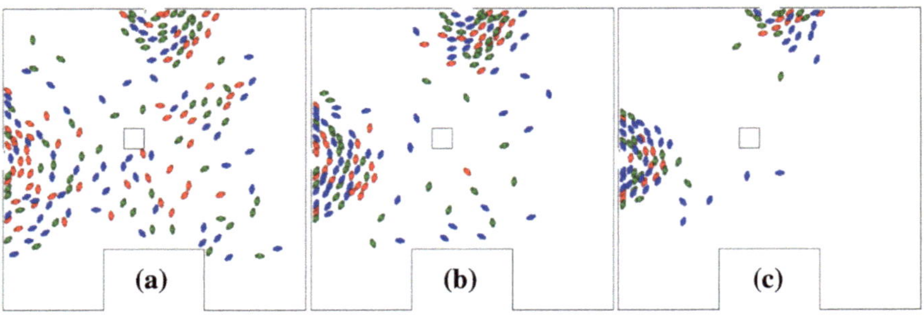

Fig. 12 Second scenario. Snapshot of the simulation, at time **a** t = 5 s, **b** t = 12 s, **c** 21 s

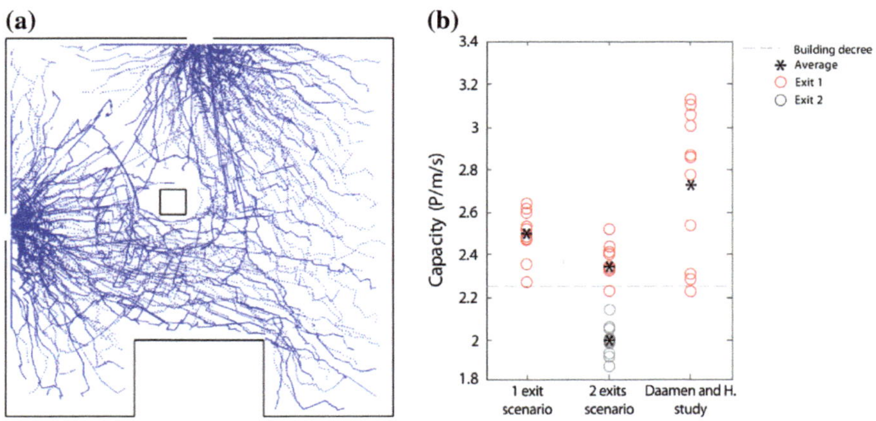

Fig. 13 **a** Positions occupied by pedestrians during the simulation of the second scenario; **b** capacity obtained by simulations compared with Daamen's study

experiments. Figure 13b shows the results of the simulations for each scenario compared with Daamen's study. We can observe that the average capacity value obtained (2.5 P/m/s) in the first scenario is close to Daamen's study (2.75 P/m/s). For the second scenario, we have obtained a different capacity value for each door highlighting that door positioning also affects the resulting capacity.

5 Conclusions

In this paper, a microscopic, time-discrete, behavioural model of crowd dynamics based upon Fuzzy Logic is presented. The model proved to be able to reproduce typical phenomena of the crowd evacuation, as rainbow-like arching structures that are expected to show up in these cases. It is also able, in the case of multiple exits, of handling the choice behaviour taking into account the 'herding behaviour' factor. These facts make the authors confident of the ability of the model to predict the egress dynamics in a quantitative and reliable way. This result is also supported by simulated door capacity values when compared with an experimental study.

Further developments concern the adaptation of the model to large areas, which can take in a big number of pedestrians, preserving the computational performances. We have to point out that the model somehow suffers from lack of real-world data, necessary to calibrate the values of attributes of fuzzy input and output. However, it can be considered a tool to be used with confidence in designing the safety of large infrastructures.

References

1. Brilon, W., Großmann, M., Blanke, H.: Verfahren für die berechnung der leistungsfähigkeit und qualität des verkehrsablaufes auf straßen (Methods for the calculation of the capacity and quality of traffic flow in streets). Straßenbau und Straßenverkehrstechnik Series Number 669, Chap. 13, Ministry of Traffic, Bonn (1993)
2. Coleman, J.S., James, J.: The equilibrium size distribution of freely-forming groups. Sociometry **24**, 36–45 (1961)
3. Daamen, W., Hoogendoorn, S.: Capacity of doors during evacuation conditions. Procedia Eng. **3**, 53–66 (2010)
4. Ganem, J.: A behavioral demonstration of Fermat's principle. Phys. Teach. **36**, 76–78 (1998)
5. Gipps, P.G., Marksjo, B.: A micro-simulation model for pedestrian flows. Math. Comput. Simul. **27**, 95–105 (1985)
6. Helbing, D., Farkas, I., Vicsek, T.: Simulating dynamical features of escape panic. Nature **407**, 487–490 (2000)
7. Helbing, D., Farkas, I.J., Molnar, P., Vicsek, T.: Pedestrian and Evacuation Dynamics. In: Schreckenberg, M., Sharma, S.D. (eds.) Simulation of pedestrian crowds in a normal and evacuation situations. Springer, New York (2002)
8. Helbing, D., Keltsch, J., Molnar, P.: Modelling the evolution of human trail systems. Nature **388**, 47–50 (1997)
9. Henderson, L.F.: The statistics of crowd fluids. Nature **229**, 381–383 (1971)

10. Henein, C.M., White, T.: Multi-Agent and Multi-Agent-Based Simulation: Joint Workshop MABS 2004. In: Davidsson, P., Logan, B., Takadama, K. (eds.) Agent-based modelling of forces in crowds. Springer, Heidelberg (2004)
11. Hoogendoorn, S.: Pedestrian flow by adaptive control. In: Proceedings of TRB 2004 Annual Meeting (2004)
12. IES.: Simulex Technical Reference, Evacuation modeling software. Integrated Environmental Solutions, Inc (2000)
13. Keating, J.: The myth of panic. Fire J. **77**, 57–61, 147 (1982)
14. Kirchner, A., Schadschneider, A.: Simulation of evacuation process using a bionics-inspired cellular automaton model of pedestrian dynamics. Phys. A **312**, 260–276 (2002)
15. Klir, G.J., Folger, T.A.: Fuzzy Sets, Uncertainty and Information. Prentice Hall, New Jersy (1988)
16. Mintz, A.: Non-adaptive group behaviour. J. Abnorm. Soc. Psychol. **46**(2), 150–159 (1951)
17. Nelson, H.E., MacLennan, H.A.: Handbook of Fire Protection Engineering. In: DiNenno, D.J. (ed.) Emergency movement. SFPE, Quincy (1995)
18. Okazaki, S.A.: Study of pedestrian movement in architectural space, Part 1: pedestrian movement by the application of magnetic models. Trans. Architectural Inst. Jpn. **283**, 111–119 (1979)
19. Predtetschenski, W.M., Milinski, A.I.: Personenströme in Gebäuden: Berechnungsmethoden für die Projektierung (Pedestrian Flow in Buildings: Calculation Methods for Design). Müller, Köln–Braunsfeld (1971)
20. Quarantelli, E.: The behavior of panic participants. Sociol. Soc. Res. **41**, 187–194 (1957)
21. Weidmann, U.: Transporttechnik der Fußgänger (Transportation technique for pedestrians). Schriftenreihe des Instituts für Verkehrsplanung, Transporttechnik, Straßen- und Eisenbahnbaunumber 90, ETH Zürich, Switzerland (1993)
22. Zadeh, L.A.: Fuzzy sets, inform. Control **8**, 338–353 (1965)
23. Zimmermann, H.J.: Fuzzy Sets Theory—And Its Applications. Kluwer Academic Publisher, Norwell (1996)

Bilevel O/D Matrix Adjustment Formulation Using High Convergence Assignment Methods

A. Reyes, L. M. Romero and F. G. Benitez

Abstract The Frank-Wolfe algorithm has been for years the most widely used method for solving the traffic assignment problem (TAP). In the last decade there have been new proposals for the resolution of the TAP. It has been shown that these algorithms are feasible for large scale problems with very high convergence, much higher than the achieved by the Frank-Wolfe algorithm. The O/D matrix adjustment problem based upon traffic counts can be formulated as a bilevel optimization problem in which the TAP is the lower level. The convergence of the TAP and the computational cost can be critical because the number of TAPs to be solved during each step of the process is very high. This paper exploits the possibilities offered by new TAP methods in the O/D matrix adjustment problem. Numerical examples on medium-sized networks using the new proposed methods are presented.

Keywords O/D matrix adjustment · Traffic assignment problem · Origin based algorithm

1 Introduction

Currently, transport systems in urban areas are characterized by their great complexity. Every user has the ability to reach its destination through a variety of routes with different transport modes and generally in an affordable time.

A. Reyes (✉) · F. G. Benitez
Transportation Engineering, University of Sevilla, Sevilla, Spain
e-mail: areyes@us.es

F. G. Benitez
e-mail: benitez@esi.us.es

L. M. Romero
Transportation Engineering, AICIA, Sevilla, Spain
e-mail: l_m_romero@esi.us.es

J. F. de Sousa and R. Rossi (eds.), *Computer-based Modelling and Optimization in Transportation*, Advances in Intelligent Systems and Computing 262,
DOI: 10.1007/978-3-319-04630-3_19,

To carry out transportation planning all travel combinations among the different origins and destinations of the network must be considered. In addition, for each origin—destination (O/D) pair the modeler must take into account all transportation modes which connect it. This leads to a combinatorial problem on number of centroids and number of available modes.

The main goal of this research is to design an estimating O/D matrix algorithm based on a previous outdated matrix, the prior matrix. This matrix comes from a survey process, it's a result form either a previous project or the output of a distribution model. This prior matrix is adjusted so that, when it's assigned to the network, modeled link flows resemble the observed volumes measured on the real network.

This chapter is organized as follows. Section 2 introduces the O/D matrix estimation problem. Section 3 presents the bilevel programming methods and Sect. 4 proposes a formulation to solve them. In Sect. 5 the estimation procedure is applied to real data and results are reported and discussed. Section 6 concludes the paper with orientation of future work.

2 The O/D Matrix Estimation Problem

Most of the methods designed to solve the estimation matrix problem using traffic counts present the following generic form:

$$\underset{g,\,v}{Minimize} \quad F(g,v) = \gamma_1 F_1(g,\ \bar{g}) + \gamma_2 F_2(v,\ \bar{v}) \tag{1}$$

$$s.t. \quad v = P(g)g \tag{2}$$

$$g \in \Omega$$

being $\gamma_1,\ \gamma_2$ weighting factors, where $\gamma_1 \geq 0,\ \gamma_2 \geq 0,\ \gamma_1 + \gamma_2 = 1$, reflecting the relative reliability in the information provided by $\bar{g}\ y\ \bar{v}$. Constraints (2) represent the traffic assignment process. Each row of the matrix $P(g)$ represents the proportions of trips corresponding to all O/D pair for a traffic count. Ω is the feasible set of O/D matrices, usually defined by non-negativity constraints of its elements.

Functions $F_1(g,\ \bar{g})$ and $F_2(v,\ \bar{v})$ represent generalized distance measures between estimated O/D matrix g and previous O/D matrix $\bar{g}$, and between estimated link flow v and observed link flow $\bar{v}$, respectively. Functions F_1 and F_2 are assumed to be convex. Generally they are defined as euclidean, entropic or maximum likelihood distances, and designed to take into account the randomness of collected data.

The O/D matrix adjustment problem can be classified into two categories according to the dependence/independence of link trips on the O/D matrix, and the proportional/non-proportional methods respectively. For methods that use non-proportional assignment, [9] formulated the matrix estimation problem as a bilevel programming problem. A matrix estimation problem is solved at the upper level and a user equilibrium assignment problem at the lower level. The solution is achieved using iterative methods that solve alternately each level and sharing parameters between them.

Although each level consists of a convex programming problem with a unique solution, the bilevel program is generally non-convex. The solutions obtained by the existing methods, in the best case scenario, are local optima. Even so, the bilevel programming approach at least ensures consistency between the assignment assumptions made in the estimating process and those made in extrapolating link flow measurements to the unmeasured links ([3]).

The O/D matrix estimation is generally recognized as an indeterminate problem. There are a large number of matrices which, when assigned to the network, produce exactly the traffic count volumes. Thus the matrix estimation problem is reduced to select a particular solution within an indeterminate nonlinear equations system. This provides complexity to the problem and a challenge for researchers. All works in this field focus on finding a formulation that discriminates among this set of feasible O/D matrices which, reproducing the traffic count volumes, most likely represents the situation in the observed transport system.

3 Bilevel Programming Methods

This is the most widespread formulation because of its theoretical and computational properties. From the theoretical point of view this methodology uses non-proportional assignment, being therefore suitable for transport networks with congestion problems. The non-proportional assignment produces a solution that is more consistent with the traffic count volumes and user behavior hypothesis. It can be implemented in any assignment equilibrium method based on the deterministic user hypothesis, stochastic user, system optimal, and other assumptions.

The most numerous contributions in bilevel programming applied to the O/D matrix estimation were developed during the 1990s. References [8, 9] combined the maximum entropy model with user equilibrium constraints, requiring only a subset of traffic counts. For this purpose Fisk formulated the problem with variational inequalities. Reference [12] suggested a heuristic method that, despite having weak theoretical base, provides good practical results. This author proposes a least squares estimation method and approximates the gradient of the objective function with respect to the O/D matrix under the assumption that the proportions matrix $P(g)$ is locally constant. Reference [14] showed that it is possible to integrate the functions of generalized least squares and maximum entropy with a user equilibrium assignment problem following a bilevel optimization program. Reference [8] investigated different gradients; one of their methods, as [12], considers that the objective function is composed of only the distance function F_2 ($\gamma_1 = 0$ according to (1)). To improve Spiess's gradient they use second order information obtained from the hessian of the objective function. Reference [4] proposed to solve the problem by the augmented Lagrangian technique; he provided theoretical justifications to explain the results of Spiess and suggested alternative algorithms for solving the problem. Reference [7] proposed to reduce the distortion of the O/D matrix including trip matrix restrictions within the objective function, solving large scale networks minimizing the amount of information stored. Reference [11], based on the original idea introduced by [8], showed that it is possible to generate the real gradient under certain assumptions, and in case the objective function is differentiable at the current point, then the Jacobian is exact.

3.1 Search Direction Calculation

The property that mainly characterizes bilevel programming methods is the search direction used at the upper level. Due to the complexity of its calculation, different authors have implemented heuristic methods that simplify this process.

The objective function $F(g)$ can be divided into two components, one depends on g and the other depends on $v(g)$:

$$F(g) = \gamma_1 F_1(g) + \gamma_2 F_2(v(g)) \tag{3}$$

The *i-th* component of its gradient $\nabla F(g)$ can be expressed as:

$$\begin{aligned}\nabla_i F(g) &= \frac{\partial F(g)}{\partial g_i} = \gamma_1 \frac{\partial F_1(g)}{\partial g_i} + \gamma_2 \frac{\partial F_2(v(g))}{\partial g_i} = \gamma_1 \frac{\partial F_1(g)}{\partial g_i} + \gamma_2 \sum_{a \in \bar{A}} \frac{\partial F_2(v(g))}{\partial v_a} \cdot \frac{\partial v_a(g)}{\partial g_i} \\ &= \gamma_1 \frac{\partial F_1(g)}{\partial g_i} + \gamma_2 \sum_{a \in \bar{A}} \frac{\partial F_2(v(g))}{\partial v_a} \cdot J_{a,i}\end{aligned} \tag{4}$$

where i is a O/D pair that belongs to the set of O/D pairs, P; and a is an element of the set $\bar{A}$ of measured or observed links, subset of the set A of all the network links, $\bar{A} \subset A$.

The partial derivatives $\partial F_1(g)/\partial g_i$ and $\partial F_2(g)/\partial v_a$ are easily calculated when objective functions and link cost functions are continuous and regular. The difficulty lies in the computation of the components of the jacobian matrix $J_{a,i}$. The element *(a,i)* of the Jacobian matrix represents the variation of flow in link a with respect to the number of trips of O/D pair *i*.

The heuristic method proposed in [12] uses the gradient method for the O/D matrix estimation problem. Assuming that the paths probabilities are locally constant, an expression of $J_{a,i}$ is obtained, given by:

$$J_{a,\,i} = \frac{\partial v_a}{\partial g_i} = \sum_{k \in K_i} \delta_{ak} p_k = \frac{\sum_{k \in K_i} \delta_{ak} h_k}{\sum_{k \in K_i} h_k} \quad \forall a \in A,\ i \in P \tag{5}$$

where K_i represents the set of paths of O/D pair i, δ_{ak} is the Kronecker index which identifies whether link a is part of route k and p_k is the route choice proportions computed as the fraction of trips of O/D pair i which use route k.

References [5, 15] solve the Jacobian calculation using the sensitivity analysis of [13]. Such sensitivity analysis provides an analytical expression of the Jacobian that can be used to calculate the gradient. However, the formulation has been criticized because the expression involves the inversion of non-invertible matrices, and therefore, in general, has no analytical solution.

References [8, 11] carry out the implementation of a gradient wherein the Spiess approach is the starting point. These authors replace the Spiess's linear approximation by a quadratic approximation; in case the objective function is differentiable at the current point, this approach provides accurate gradients.

4 Proposed Formulation

Focusing on the approach performed by [12, 14] in computing the descent direction, it can be stated that the methodology is not accurate for transport problems under congestion. These approaches do not take into account the relationships between O/D pairs through saturated links.

According to Eq. (5), when no path $k \in K_i$ of O/D pair i traverses link a, the value of $J_{a,i}$ will be zero. So although there was a discrepancy in link a between the flow assigned and the observed volume, the estimation problem would not modify the value of g_i to reduce this discrepancy along the whole process. However, the congestion level of any route $k \in K_i$ can affect indirectly the volume of link a. For example in the small network depicted in Fig. 1, considering links 1 and 5 as observed links, according to Eq. (5) the flow from origin B to destination D does not affect to potential discrepancies between their observed volumes and modeled values v_1 and v_5. But an increase in the B–D flow clearly alters the equilibrium among the users of pair A–C. Travel time in link 4, t_4, has been increased, and route composed of links 2–4–5 becomes more costly than route formed by link 1, and therefore flow is transferred from route 2–4–5 to route 1. This behavior of the transportation systems, absolutely common in congested networks, is not explained by expression (5), which tends to keep the flow constant between a great number of O/D pairs, forcing to alter O/D pairs with a direct effect in the discrepancies.

Based on the previous arguments, a new formulation is proposed. This approach overcomes the shortcomings of Spiess and Yang et al. methods, maintaining its efficiency for large scale networks. The core of the proposed approach is based on a very well-known numerical methodology, the finite difference method. By means of this technique, the Jacobian matrix is computed in a column-based schedule. The algorithm applies an infinitesimal increment to each O/D pair i and subsequently a new user equilibrium situation is computed. With this new flow pattern, the flow increments of all links of the network are directly obtained. This flow increment vector corresponds to the *i-th* column of Jacobian matrix J. This action

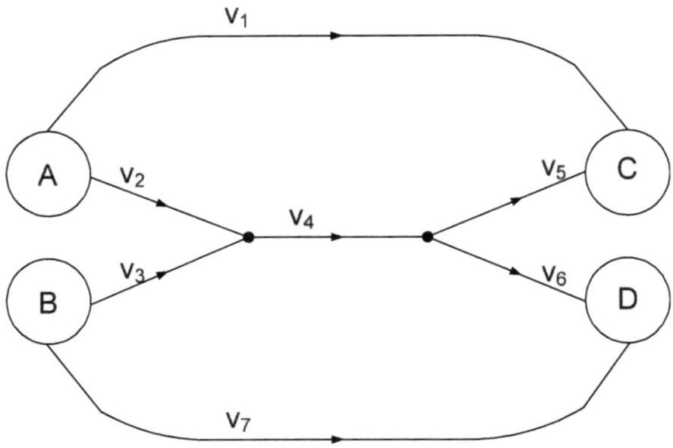

Fig. 1 Small network

must be repeated for each O/D pair. The computation of descent direction is reduced to solve new steady states after O/D pairs are infinitesimally increased. This new user equilibrium state is substantially simpler to obtain than launching a new assignment process because the previous solution is known, and the changes in the O/D matrix are very small. In this case, it can be assumed that none of the O/D pair change its paths and, at the same time, these trip matrix infinitesimal increments generate minimal deviation in the steady state so the number of O/D pairs affected is limited. In this situation, the algorithms can take advantage of *warm start* strategies, reducing considerably the computation times and making feasible the application of this methodology even to large networks.

In bilevel programming, the lower level is defined by the traffic assignment algorithm used. Several methods have been tested in this level: [1, 6, 10]. After solving this stage, internal variables are stored for later use in computing the Jacobian matrix. When using the algorithm proposed by Dial, the stored information is the acyclic bush and links flow; Florian's method can keep the set of paths and Bar-Gera's algorithm saves the set of pairs of alternative segments generated.

In the methodology proposed simplified versions of the lower level traffic assignment algorithm are implemented for the Jacobian computing once the variables in the lower level are stored. These new traffic assignment methods, unlike the Frank and Wolfe's method, show high convergence solutions and high computational efficiency.

5 Tested Cases

Four cases have been tested, ranging from small to large network scale. The first two cases are synthetic simulated ones, the third other corresponds to an existing real case which have been synthetically modified, the fourth case is an empirical real case.

Table 1 Prior O/D matrix

O/D pair	Trips
A–C	400
A–D	200
B–C	0
B–D	300

5.1 Small Scale Network

The results derived by the proposed formulation when applied to a small-scale network are presented. A comparison between the proposed formulation and Spiess's method is made. In this case a small network is simulated by a graph consisting of seven directed links and four centroids (A, B, C and D).

The links of the network connect four O/D pairs: A–C, A–D, B–C and B–D. Table 1 reproduces the O/D trip matrix.

The network has just one traffic count location on link v_5 which has captured a total of 300 vehicles. Assigning the prior O/D matrix a discrepancy of 240 vehicles on this link appears, four times higher than currently modeled. Table 2 shows the results obtained with the adjustment methods tested.

Five different adjustment algorithms were tested. Method 1 implements the algorithm proposed by Spiess, a modification of method 1 is implemented in methods 2 and 3 where null cells of the prior matrix are preloaded with 1 and 10 trips, respectively, so these methods take into account null elements in the O/D matrix adjustment. Methods 4 and 5 use the methodology proposed in this paper to calculate the Jacobian by means of the finite difference method. The traffic assignment problem employed in the lower level is based on Florian's method. The objective function implemented by all tested methods just depends on link flows ($\gamma_1 = 0$ according to (1)). The F_2 distance function between measured and modeled flows used in Methods 1–4 is the classical euclidean distance; method 5 utilizes an entropy maximization function.

Methods 1 and 2 give the same solution in which pair A-C trips have been modified. Spiess's algorithm only adjusts the O/D pairs that have been intercepted by some traffic count, it does not consider the influences an O/D pair has on a traffic count not contained in the paths. The variation of trips in an O/D pair affects the paths flow of others O/D pairs as the user equilibrium is modified. Therefore, an O/D pair can affect the flow of a link even when the link does not belong to the paths connecting the said O/D pair.

If null elements of O/D matrix are not initially preloaded, Spiess's algorithm does not take them into account. Moreover, according to the matrices obtained by methods 2 and 3, the preload of the O/D matrix must be considerable in order to the Spiess's algorithm affect the null O/D pairs. Preloading zero elements of the O/D matrix is not recommended as the number of null elements is usually high and this action adds trips to the network thereby increasing the level of congestion.

Table 2 Adjusted O/D matrices with different methods

O/D pair	Prior matrix	Estimated matrices				
		Method 1	Method 2	Method 3	Method 4	Method 5
A–C	400	767	767	749	544	545
A–D	200	200	200	200	172	171
B–C	0	0	0	13	168	168
B–D	300	300	300	300	295	295
Total trips	900	1267	1267	1262	1179	1179

In the first 3 methods a total of 10 iterations are needed to reach convergence. Methods 4 and 5 are solved with a single iteration since the problem is very simple and the approximation of the Jacobian provides real values for the displacement of the O/D matrix during the adjustment iteration.

Table 2 shows that methods 4 and 5 generate coincident adjusted O/D matrices. Also it is found that, after adjusting using traffic counts, all O/D pairs have been modified. This comes because the influences of all O/D pairs on the traffic counts are taken into account by the approaches implemented.

The assignment of the prior O/D matrix identifies a deviation of 240 vehicles between observed and modeled flow in link v_5. In the matrix estimation problem, methods 1, 2 and 3 add 367 trips to the O/D matrix while the proposed methods increase in 279 the total number of trips; therefore the proposed approaches solve the adjustment problem with a less divergence between adjusted and prior matrices.

5.2 Medium Scale Network

In this section the models are applied to a mid-size real network. The selected network belongs to Tiergarten, a central area of Berlin. The topological information for network design is taken from [2]. The original network has been altered by increasing the number of trips between areas in order to aggravate congestion problems.

The Tiergarten network consists of 747 links, 344 nodes and 26 centroids (650 OD pairs); 10 % of the links are supposed to be monitored with traffic counts (75 links), therefore 75 traffic count data will be used to estimate the value of 650 matrix elements.

In this case, three different methods are evaluated. The algorithms used here correspond to methods 1, 4 and 5 described in the previous section. A total of 1000 iterations are carried out and results are presented. Figure 2 shows the objective function evolution and Table 3 compares the most important outcomes of the methods implemented.

The objective function and R^2 values are significantly better in methods 4 and 5. Following the evolution of the objective function, method 1 requires a greater

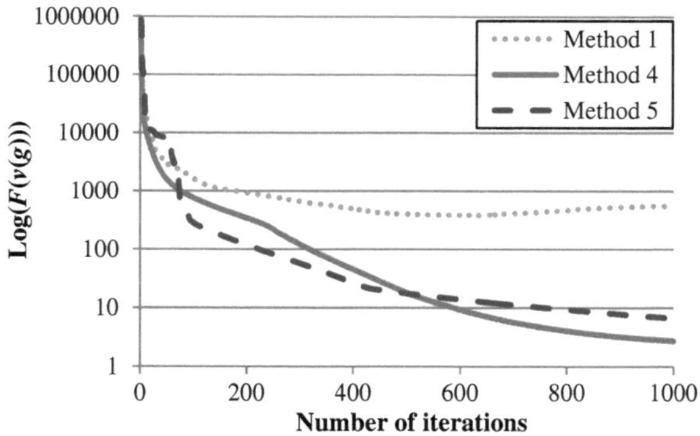

Fig. 2 Evolution of the objective function

Table 3 Results

	Method 1	Method 4	Method 5
Objective function	552.7	2.7	6.6
R^2	0.9739	0.9923	0.9903
Time (seconds)	2421.0	573.9	1454.7
Total modified trips	4922.56	3823.01	3852.84
Mean of elements variation	29.3 %	21.2 %	21.4 %
Element with higher variation	321 %	112 %	159 %

number of iterations to reach the values obtained with methods 4 and 5. Under the same number of iterations, the runtime of methods 1 and 5 are clearly superior to the time spent by method 4.

The last three rows presented in Table 3 give an idea about the displacement applied to the O/D matrix during the adjustment. There is a larger divergence of the O/D matrix in method 1.

It can be concluded that methods 4 and 5 improve method 1 based on the fact that minimizing the objective function results in a smaller divergence between the matrices involved.

5.3 Large Scale Network

The proposed algorithm has been tested in Madrid network. Table 4 shows the main properties of the network used.

In this section the algorithm's efficiency is the only feature tested. The traffic assignment method used in Sects. 5.1 and 5.2 is not efficient when the O/D matrix adjustment process is applied to large scale networks. In this case, computational cost increases excessively due to paths storage. To avoid paths enumeration [1, 6]

Table 4 Large scale networks

	Chicago sketch	Madrid
Links	2.950	17.544
Traffic counts	500	491
Nodes	933	7.966
Centroids	387	1.180
O/D pairs	149.382	1.391.220

based assignment algorithm are implemented in the lower level. After several test, Bar-Gera based algorithm provide better results than Dial's. Bar-Gera's method achieves solutions with less execution time.

Madrid network is tested with the same upper level than used in methods 1–4 and in lower level the traffic assignment problem is based on Bar-Gera's methodology. To minimize the objective function of Madrid network the estimation algorithm takes 35.6 h to complete 30 iterations. This execution time includes thirty traffic assignment problem resolutions and thirty Jacobian computing processes. Usually, to compute a lower level iteration it is necessary to spent 0.9 h approximately when the problem is executed with high convergence; therefore, 70–80 % of the O/D matrix estimation problem is due to the traffic assignment problem. It can be concluded that the bilevel estimation problem is an expensive process mainly because of the traffic assignment problem.

6 Conclusions

The matrix adjustment problem has been analyzed for congested networks with user equilibrium assignment hypothesis by formulating a bilevel optimization problem. The proposed formulation has been solved by implementing an algorithm based on the gradient method. This method is characterized by a new numerical method for calculating the Jacobian.

The proposed estimation model has been tested for networks of small, medium and large scale with satisfactory results. A comparison with Spiess's method has been carried out and the advantages and disadvantages between the two methodologies have been described.

The bilevel estimation problem is suitable for transport networks with congestion problems and produces a solution that is more consistent with the traffic count volumes and user behavior hypothesis. However, it can be an expensive process due to the implementation of the traffic assignment problem. The applicability of the proposed technique to large networks depends on the algorithm used to solve the lower level. The method based on [1] has proved to be the most convenient for these large networks.

It is proposed, as future work, to further analyze the goodness of matrix adjustment with different objective functions formulated in the technical literature and consider the convenience of selecting different sets of traffic counts.

Acknowledgments This research was funded in part by both the Spanish Ministry of Science (National Plan Programs TRA2012-36930, FPI BES-2009-025110) and the Spanish Ministry of Public Works (P63/08). One of the authors, L.M. Romero, acknowledges the support of the grant PTQ-11-04952 from the Spanish Ministry of Science through the INNCORPORA-PTQ Programme.

References

1. Bar-Gera, H.: Traffic assignment by paired alternative segments. Transp. Res. **44B**, 1022–1046 (2010)
2. Bar-Gera, H.: Transportation Networks Test Problems. http://www.bgu.ac.il/~bargera/tntp/
3. Bell, M.G.H., Iida, Y.: Transportation Network Analysis. Wiley, Chichester (1997)
4. Chen, Y.: Bilevel programming problems: analysis, algorithms and applications. PhD Thesis, Publication 984. Centre de Recherche sur les Transports, Université de Montréal, Montréal, Canada (1994)
5. Denault, L.: Étude de deux méthods d'adjustement de matrices origine-destination à partir des flots des véhicules observés (in French). Report CRT-991, Mémoire D'étudiant. Centre de recherche sur les transports (CRT), Université de Montréal, Montréal, Québec, Canada (1994)
6. Dial, R.: A path-based user-equilibrium traffic assignment algorithm that obviates path storage and enumeration. Transp. Res. **40B**, 917–936 (2006)
7. Doblas, J., Benitez, F.G.: An approach to estimating and updating origin–destination matrices based upon traffic counts preserving the prior structure of a survey matrix. Transp. Res. **39B**, 565–591 (2005)
8. Drissi-Kaitouni, O., Lundgren, J.: Bilevel origin–destination matrix estimation using a descent approach. Technical Report lith-matr92-49. Department of Mathematics, Institute of Technology, Linkoping, Sweden (1992)
9. Fisk, C.S.: On combining maximum entropy trip matrix estimation with user optimal assignment. Transp. Res. **22B**, 66–79 (1988)
10. Florian, M., Constantin, I., Florian, D.: A new look at projected gradient method for equilibrium assignment. Transp. Res. Rec. **2090**, 10–16 (2009)
11. Lundgren, J.T., Peterson, A.: A heuristic for the bilevel origin-destination matrix estimation problem. Transp. Res. **42B**, 339–354 (2008)
12. Spiess, H.: A descent based approach for the OD matrix adjustment problem. *Publication* 693, Centre de recherche sur les transports, Université de Montréal, Montréal, Canada (1990)
13. Tobin, R.L., Friesz, T.L.: Sensitivity analysis for equilibrium network flows. Transp. Sci. **22**, 242–250 (1988)
14. Yang, H., Sasaki, T., Iida, Y., Asakura, Y.: Estimation of origin-destination matrices from link traffic counts on congested networks. Transp. Res. **26B**, 417–434 (1992)
15. Yang, H.: Heuristic algorithms for the bi-level origin-destination matrix estimation problem. Transp. Res. **29B**, 1–12 (1995)

Part V
Traffic Modelling, Control and Network Traffic Management

The aim of *modelling traffic* is to eliminate congestion by controlling traffic-light cycle structure and duration. The problem of *control* must be solved by both analytical methods and simulations, with continuous models. *Network traffic management* consists of techniques to attain optimal performance for diverse classes of users to satisfy their needs and those of the public. Network reliability can be enhanced effectively by improving key links with limited resources. *Wakabayashi and Fang* demonstrate the advantage of their improved importance index, and suggest a combination strategy of that index and cost-benefit analysis, in order to make improvements to comparative analyses. *Rossi et al.* describe the development of a gap-acceptance model based on fuzzy system theory, specifically applicable to traffic entering a roundabout. The work of *Petrik et al.* analyses the impact of volume-delay function inputs and parameters on the uncertainty of a four-step model traffic forecast by link, and identifies the main contributors of errors. *Vasconcelos et al.* present a procedure to calibrate the Gipps car-following model based on macroscopic data, extending previous approaches to explain the effect of driver variability in speed-flow relationships. *Cantarella et al.* apply genetic algorithms to solve signal setting design at a single junction. The same authors also discuss the effectiveness of multicriteria versus monocriteria optimization, and report the advantages of using Non-Dominated Sorting Genetic Algorithms.

Comparison of Importance Indices for Highway Network Reliability Improvement Combined with Cost–Benefit Analysis

Hiroshi Wakabayashi and Shuming Fang

Abstract The 2011 Japan Earthquake and the 2008 Sichuan Earthquake disrupted many sections in the highway networks and consequently blocked emergency and rescue activities in the wide area. Thus, it is very important to construct a highly reliable highway network in advance for not only abnormal but also normal period. Network reliability can be improved effectively by improving key links under limited resources. Once such key links are identified, network reliability can be efficiently improved and maintained. For identifying the key links, several importance indices such as Birnbaum's importance (*RI*) and criticality importance (*CI*) have been proposed. However, use of these indices is found to lead to unreasonable results in the small networks in this chapter. Thus use of *RI* and *CI* is inappropriate for an actual network. Therefore, this chapter demonstrates the advantage of the improved importance index (*CIW*) proposed by authors. Then this chapter proposes a combination strategy of *CIW* and cost–benefit analysis (B/C), and provides comparative analyses.

Keywords Network reliability improvement · Reliability importance · Criticality importance · Cost–benefit analysis · Boolean absorption

1 Introduction

A highly reliable traffic network is very important for both abnormal and normal periods. This fact is recognized in many earthquakes such as the 1991 Roma Prieta in San Francisco, the 1994 Northridge in Los Angeles, the 1995 Kobe in West

H. Wakabayashi (✉)
Faculty of Urban Science, Meijo University, Nijigaoka 4-3-3, Kani,
Gifu 509-0261, Japan
e-mail: wakabaya@urban.meijo-u.ac.jp

S. Fang
School of Information and Electrical Engineering, Shandong Jiaotong University,
Jinan Shandong 250357, China

J. F. de Sousa and R. Rossi (eds.), *Computer-based Modelling and Optimization in Transportation*, Advances in Intelligent Systems and Computing 262,
DOI: 10.1007/978-3-319-04630-3_20,

Japan, the 2008 Sichuan in China and the 2011 off the Pacific coast of Tohoku (the 2011 Japan) Earthquakes. Network reliability can be improved and maintained effectively by improving the most important key links in the network. Once such important links are identified, network reliability can be efficiently improved and maintained. However, the methodology is not developed for either what link has the highest priority in quantity or what is the best order for retrofitting links in the network. Thus this chapter addresses the importance indices for finding the key links.

Several importance indices such as Birnbaum's importance (*RI*) and criticality importance (*CI*) have been proposed for finding the key links. However, these indices have their own shortcomings described in Sect. 2. Therefore, this chapter introduces a newly improved importance index, *CIW*. In addition, the budget constraint for retrofitting network is also important. Thus, a cost–benefit analysis, namely, cost–link reliability improvement analysis, combined with these importance indices is developed. On the basis of this cost–benefit analysis, the most effective importance index can be identified. In addition, it is very difficult to calculate the terminal reliability of the traffic network and the values of importance indices of all links when the network becomes huge. Thus, we develop calculation algorithm combined with Boolean algebra and path sets for computing the terminal reliability and the values of importance indices for large and complex networks.

This chapter firstly addresses the importance indices of *RI*, *CI*, and *CIW* for improving network reliability. Secondly, an efficient calculation algorithm with a partial differential is proposed for calculating these importance indices on the basis of the calculation algorithm for Boolean absorption (CABA). It permits an automatic calculation for the terminal reliability and the values of *RI*, *CI*, and *CIW* of all links, even for complex networks. Using the CABA, the processes of network improvement with *RI*, *CI*, and *CIW* are compared for some types of networks. Thirdly, a method of cost–link reliability improvement analysis combined with the previously proposed indices will be presented in order to compare the efficiency of these importance indices. Series network, parallel network, a simple bridge network and a field-shape network will be discussed. Depending on the cost–link reliability function, the behavior of the network improvement process will differ, and the proposed strategy combining *CIW* index and B/C is the best when the network is large. Finally, concluding remarks of our method for effective and efficient network improvement are discussed.

2 Current Importance Indices for Network Reliability Analysis

The concept of importance has long been proposed in the field of systems engineering, but has appeared in only a few papers [1]. Importance is defined as the degree of magnitude that improvement in the reliability of a link contributes for system reliability. The importance indices proposed in this chapter are on the basis of the connectivity reliability.

2.1 Terminal Reliability

The connectivity reliability of a highway network is defined as the probability that two given nodes over the network are connected with a certain service level of traffic for a given time period. Similarly, link reliability in the network is defined as the probability that the traffic reaches a certain service level for a given time period. Terminal reliability, R, is given by an expression using minimal-path sets, as follows:

$$R(\mathbf{r}) = E[\phi(\mathbf{X})] = E[1 - \prod_{s=1}^{p} (1 - \prod_{a \in P_s} X_a)], \tag{1}$$

where P_s is the s-th minimal-path set, p is the total number of minimal-path sets, X_a is the binary indicator variable for link a (Wakabayashi and Iida, [5, 6, 9, 10], and $\phi(\mathbf{X})$, $\mathbf{X}$ and $\mathbf{r}$ are a structural function, vector representations for X_a and r_a:

$$X_a = \begin{cases} 1, & \text{if link } a \text{ provides a certain service level,} \\ 0, & \text{otherwise.} \end{cases} \tag{2}$$

Link reliability, r_a, is defined as

$$r_a = E[X_a]. \tag{3}$$

The connectivity reliability of a traffic network depends on the network structure and the link reliabilities. Therefore, two basic approaches have been taken to improve network reliability: to improve the network structure or to improve the reliability of the links. The focus here is on identifying what links should be improved to maximize the improvement in network reliability.

2.2 Reliability Importance

2.2.1 Definition of Reliability Importance

In order to find out the key link for improving the terminal reliability most efficiently, the Reliability Importance (*RI*) (Birnbaum [2]) was proposed as

$$RI_a = \frac{\partial R(\mathbf{r})}{\partial r_a}, \quad 0 \le RI_a \le 1. \tag{4}$$

RI indicates the impact of an improvement in link reliability, i.e. the increase or decrease in the reliability of link *a* affects the increase or decrease in the terminal reliability. *RI* is also known as the Birnbaum's structural importance.

2.2.2 Merits and Demerits of Reliability Importance

Although *RI* has the potential to improve network reliability, it has a shortcoming that we will discuss in this section.

For the case of two links in a series network, the terminal reliability R_{AB} is given by Eq. (5);

$$R_{AB} = r_1 r_2 \tag{5}$$

where r_1 and r_2 are the values of reliability for links 1 and 2, respectively.

Reliability importance for a series network, RI_1 and RI_2, are obtained from Eqs. (4) and (5) as

$$RI_1 = r_2 \text{ and } RI_2 = r_1. \tag{6}$$

It follows that

$$RI_1 > RI_2, \text{ if } r_1 < r_2. \tag{7}$$

Equation (7) indicates that in the case of a series-type network, improving the less reliable link is most effective for improving terminal reliability. This fact is easily expanded for large series-type networks. This result for improving, managing and reconstructing a network is the expected result.

For the case of two links in a parallel network, however, the terminal reliability R_{AB} is given by Eq. (8);

$$R_{AB} = 1 - (1 - r_1)(1 - r_2). \tag{8}$$

RI_1 and RI_2, for these two links in a parallel network, are obtained from Eqs. (4) and (8) as

$$RI_1 = 1 - r_2 \text{ and } RI_2 = 1 - r_1. \tag{9}$$

It follows that

$$RI_1 < RI_2, \text{ if } r_1 < r_2. \tag{10}$$

The result from Eq. (10) indicates that in the case of a parallel-type network, improving the more reliable link will be more effective for improving terminal reliability. Usually, however, it is difficult to improve a more reliable link, whereas it is rather easy to improve a less reliable link. This result for improving, managing, and reconstructing network is counter to what one would expect.

2.3 Criticality Importance

2.3.1 Definition of Criticality Importance

Because of the shortcoming of reliability importance, the Criticality Importance index (*CI*) was proposed as the ratio of the proportional improvement in the network reliability to the proportional improvement in the link reliability [4]:

$$CI_a = \lim_{\Delta r_a \to 0} \left\{ \frac{\Delta R_{AB}(\mathbf{r})/R_{AB}(\mathbf{r})}{\Delta r_a / r_a} \right\} = \frac{\partial R_{AB}(\mathbf{r})}{\partial r_a} \times \frac{r_a}{R_{AB}(\mathbf{r})} = RI_a \frac{r_a}{R_{AB}(\mathbf{r})}. \quad (11)$$

2.3.2 Merits and Demerits of Criticality Importance

CI also has shortcomings that we will discuss in this section.

For the case of two links in a series network, *CI* is given from Eqs. (4)–(6) and (11) that

$$CI_1 = \frac{r_1 r_2}{R} = CI_2. \quad (12)$$

This result suggests that the criticality importance index is the same for both links in a series network. However, in a series network, it would be reasonable to strengthen a less reliable link, thus this is a shortcoming of the criticality importance index. In addition, it provides no information to distinguish the link between the two links for improving network reliability.

For the case of two links in a parallel network, *CI* is given from Eqs. (4), (8), (9) and (11) that

$$CI_1 = \frac{r_1 - r_1 r_2}{R}, \quad (13)$$

and

$$CI_2 = \frac{r_2 - r_1 r_2}{R}. \quad (14)$$

It follows that

$$CI_1 < CI_2, \ \text{if} \ r_1 < r_2. \quad (15)$$

Therefore, *CI* index also indicates that in the case of a parallel-type network, improving a more reliable link gives a greater increase in the terminal reliability of the network. The results for a parallel network provided by both the *RI* and the *CI* suggest that a less reliable link should be ignored in a parallel system. In other words, the people who live along a less reliable link would be neglected both before and after a disaster. This is not a reasonable planning for disaster prevention and reduction. Thus, this result is not as one would expect.

2.4 *Advanced Criticality Importance Proposed by Wakabayashi*

2.4.1 Definition of Advanced Criticality Importance Proposed by Wakabayashi

The Reliability Importance (*RI*) cannot reflect the fact that it is more difficult to improve a more reliable link than a less reliable link. In addition, *RI* and *CI* ignore the improvement of the less reliable link in paralleled network. Thus it is convenient to define the importance as the proportion of the marginal change in terminal reliability against the marginal change in link unreliability. Changing the definition of equation in the reliability engineering, the advanced criticality importance *CIW* proposed by Wakabayashi [11] is introduced as Eq. (16);

$$CIW_a = \lim_{\Delta q_a \to 0} \left\{ -\frac{\Delta R(\mathbf{r})/R(\mathbf{r})}{\Delta q_a/q_a} \right\} = -\frac{\partial R(\mathbf{r})}{\partial q_a} \times \frac{q_a}{R(\mathbf{r})} = RI_a \times \frac{(1-r_a)}{R(\mathbf{r})}, \tag{16}$$

$$q_a = 1 - r_a, \tag{17}$$

where $q_a (= 1 - r_a)$ is the unreliability of link a.

2.4.2 Merits and Demerits of Advanced Criticality Importance Proposed by Wakabayashi

For the case of two links in a series network, *CIW* is given from Eqs. (4)–(6) and (16) that

$$CIW_1 = \frac{1 - r_1}{r_1}, \tag{18}$$

and

$$CIW_2 = \frac{1 - r_2}{r_2}. \tag{19}$$

It follows that

$$CIW_1 > CIW_2, \text{ if } r_1 < r_2. \tag{20}$$

Thus, in a series-type network, the *CIW* proposed by Wakabayashi [11] has the same property as *RI*, and this property from Eq. (20) is exactly as one would expect.

For the case of two links in parallel, *CIW* is given from Eqs. (4), (8), (9) and (16) that

$$CIW_1 = \frac{(1 - r_1)(1 - r_2)}{r_1 + r_2 - r_1 r_2} = CIW_2. \tag{21}$$

From Eq. (21), although the criticality importance proposed by Wakabayashi made more progress than the index proposed by Henley and Kumamoto [3], this index is the same for both links in a parallel network. Thus it provides no information to distinguish the potential link to be improved for improving network reliability.

The importance indices, *RI*, *CI*, and *CIW* discussed above, because of their own shortcomings, cannot be directly used to select the most important key link of a traffic network. Therefore, a good solution cannot be obtained by these indices for evaluating the improvement of network reliability. In addition, although the cost–benefit ratio is also important [7], these indices cannot predict the increase in cost for improving link reliability when link reliability increases. Although Wakabayashi [11] proposed *CIW* index, he did not state explicitly how to use *CIW*. In this chapter, we propose the strategy combining *CIW* and cost–link reliability improvement analysis as stated in Sect. 3 for improving network reliability.

3 A Method for Cost–Benefit Analysis for Improvement of Traffic Network Reliability

According to the criticality importance proposed by Wakabayashi described in Sect. 2, the less reliable link in a series network should be improved in accordance with Eq. (20). However, the result from Eq. (21) provides no distinguishable information as to which link should be improved firstly in a parallel network. Thus, a method to determine the cost of the reliability improvement of the traffic network will be proposed in this section.

We will assume two cases of invest strategies to improve the link reliability (cost–reliability function). Differential of cost is assumed as linear and quadratic for general discussion as follows:

- *Case 1*: The cost for improving a link of higher reliability is more than to improve a link of lower reliability, and the investment to increase the same degree of the link reliability varies according to the link reliability. *Case 1* of the investment strategy is shown as Eq. (22), where the initial value of $C_1 = 50{,}000$ and $C_{10} = 2500$.

$$\frac{\partial Cost_a}{\partial r_a} = C_1 * r_a + C_{10}, \tag{22}$$

- *Case 2*: The investment to increase the same degree of the link reliability is cumulative and varies according to a quadratic function of the link reliability. *Case 2* of the investment strategy is shown as Eq. (23), where the initial value of $C_2 = 250{,}000$, $C_3 = 50{,}000$, and $C_{20} = 5000/3$.

$$\frac{\partial Cost_a}{\partial r_a} = C_2 * r_a^2 + C_3 * r_a + C_{20}, \tag{23}$$

The effect of an improvement in network reliability, which requires a cost increase, may not be obvious in the short term. Thus, a simple cost–benefit function that shows the improvement of the network reliability against the cost increase for a long time is defined as follows:

$$Eff(Y,F) = \sum_{n=1}^{Y} \frac{1}{(1+d)^n} * \frac{(R_{AB} - R_{AB0})}{Cost_{AB}} * F, \tag{24}$$

where,

Y	Number of years to invest;
d	Cost discount, which is assumed to be a constant value;
F_n	The conversion cost benefit of the increased traffic volume obtained by the reliability improvement in the n-th unit time (In this chapter, the unit time is one year);
R_{AB0}	Original network reliability;
$Cost_{AB}$	Cost increase to improve the network reliability from R_{AB0} to R_{AB}.
$Eff(Y)$	The efficiency of cost benefit obtained by the reliability improvement of traffic systems in Y years.

To simplify the calculation, d is assumed to be 0, and $F_1 = F_2 = \cdots = F_Y = F$ is assumed. Thus, Eq. (24) evolves into Eq. (25). There, the value of F is only a virtual data to discuss the efficiency of cost benefit obtained by the reliability improvement and it should be tested and verified in the actual application.

$$Eff(Y,F) = \frac{(R_{AB} - R_{AB0})}{Cost_{AB}} * Y * F. \tag{25}$$

4 A Calculation Algorithm for Boolean Absorption (CABA) for Terminal Reliability, *RI*, *CI* and *CIW*

The main idea of this algorithm is to directly expand Eq. (1) automatically [5, 9]. Only 1 bit of memory is used to store each random variable of every link of the network. And a minimal-path set can be stored as a decimal number in a memory unit with 32-bits. This idea prevents the intermediate expansion of both terms and memory in the process of expanding Eq. (1). In addition, *RI*, *CI* and *CIW* are obtained by Eqs. (4), (11) and (16), respectively. Thus, the calculation for the importance indices of link can also be achieved by this algorithm. The algorithm is as follows:

- *Step 1*: Let p be the number of minimal-path sets to be used in this calculation. Store these minimal-path sets. Here, every minimal-path set that is composed of links expressed as binary numbers is stored as a decimal number. For example, minimal path set $\alpha = X_1X_2X_5X_{10}$, that is, {1, 2, 5, 10}, is expressed as the binary

Path Sets		Memory Variable		Bit Variable																
				16	15	14	13	12	11	10	9	8	7	6	5	4	3	2	1	
$X_1X_2X_5X_{10}$	→	$2^{10-1}+2^{5-1}+2^{2-1}+2^{1-1} = 531$	→	0	0	0	0	0	0	1	0	0	0	0	1	0	0	1	1	
$X_1X_4X_9X_{12}$	→	$2^{12-1}+2^{9-1}+2^{4-1}+2^{1-1} = 2313$	→	0	0	0	0	1	0	0	1	0	0	0	0	1	0	0	1	
$X_3X_8X_{11}X_{12}$	→	$2^{12-1}+2^{11-1}+2^{8-1}+2^{3-1} = 3204$	→	0	0	0	0	1	1	0	0	1	0	0	0	0	1	0	0	(+
		$2^{12-1}+2^{11-1}+2^{10-1}+2^{9-1}+2^{8-1}+2^{5-1}+2^{4-1}+2^{3-1}+2^{2-1}+2^{1-1} = 3999$	←	0	0	0	0	1	1	1	1	1	0	0	1	1	1	1	1	
		"3999" means $X_1X_2X_3X_4X_5X_8X_9X_{10}X_{11}X_{12}$																		

Fig. 1 Example of Boolean absorption and storage processes for terminal reliability: $X_1X_2X_5X_{10}*X_1X_4X_9X_{12}*X_3X_8X_{11}X_{12} = X_1X_2X_3X_4X_5X_8X_9X_{10}X_{11}X_{12}$

number 0000001000010011 (read this figure from the right). At this step, the number is translated into a decimal number then memorized; the binary number 0000001000010011 is stored as the decimal number 531 $(= 2^{1-1}+2^{2-1}+2^{5-1}+2^{10-1})$. This procedure permits reduction in the size of the memory region used in the computer.

- *Step 2*: Let $m = 1$.
- *Step 3*: Any product composed of m minimal-path sets (obtained in the expansion of Eq. (1) into $2^p - 1$ terms) is expressed as $(-1)^m \cdot \alpha_{s_1} \cdot \alpha_{s_2} \cdot \cdots \cdot \alpha_{s_n}$.

Arrange this product by Boolean absorption in terms of links. For example, the product of the minimal-path sets {1, 2, 5, 10}, {1, 4, 9, 12}, and {3, 8, 11, 12} is translated into the memory variable 3,999, which indicates $X_1X_2X_3X_4X_5X_8X_9X_{10}X_{11}X_{12}$. This procedure is demonstrated in Fig. 1.

On the basis of the memory variable of the product for terminal reliability, the memory variable for the *RI* of all links can be calculated and stored in other locations. If the corresponding bit of X_a does not exist in the memory variable of the product for terminal reliability, the memory variable for the *RI* of link a translates into 0, otherwise, the corresponding bit of X_a in the memory variable of the product for terminal reliability is translated into 0, and the new memory variable is stored in other locations as the memory variable for the *RI* of link a. For example, the product of RI_1 is $X_2X_3X_4X_5X_8X_9X_{10}X_{11}X_{12}$ according to the memory variable 3,999 of terminal reliability, thus the memory variable of RI_1 is 3,998. This process demonstrates partial differential in Eq. (4). However, the memory variable of RI_6 is 0 because link 6 does not exist in the memory variable 3,999. This procedure is demonstrated in Fig. 2.

- *Step 4*: Combine link terms. The products generated in *step 3* are checked as to whether the same product has been generated in the preceding process. For the above examples, the numbers 3,999 and 3,998 are checked as to whether the same number exists in the same locations. When the same product exists, the coefficient of the product is updated; when not, it is newly stored.
- *Step 5*: Iterate *steps 3* and *4* for all combinations of $(-1)^m \cdot \alpha_{s_1} \cdot \alpha_{s_2} \ldots \alpha_{s_n}$. The number of iterations is $\begin{pmatrix} p \\ m \end{pmatrix}$.
- *Step 6*: Iterate *steps 3* through *5* for $m = 2, 3, \cdots, p$.
- *Step 7*: Each number in the storage region corresponds to each term in the polynomial expression of X_a, for which Boolean absorption has already been

Sequence of Terminal Reliability is $X_1X_2X_3X_4X_5X_8X_9X_{10}X_{11}X_{12}$, Memory Variable of Terminal Reliability is 3999																		
		Bit Variable																
Example of Reliability Importance for Link 1		16	15	14	13	12	11	10	9	8	7	6	5	4	3	2	1	
The binary number of "3999"	→	0	0	0	0	1	1	1	1	1	0	0	1	1	1	1	1	
To determine whether the corresponding bit of Link 1 exist or not in "3999":																		
This bit is translated into "0" when it exists	→	0	0	0	0	1	1	1	1	1	1	1	1	1	1	1	0	(×
The memory variable for *RI* of Link 1 is "3998"	←	0	0	0	0	1	1	1	1	1	0	0	1	1	1	1	0	
"3998" means $RI_1 = X_2X_3X_4X_5X_8X_9X_{10}X_{11}X_{12}$																		
Example of Reliability Importance for Link 6		16	15	14	13	12	11	10	9	8	7	6	5	4	3	2	1	
The binary number of "3999"	→	0	0	0	0	1	1	1	1	1	0	0	1	1	1	1	1	
To determine whether the corresponding bit of Link 6 exist or not in "3999":																		
All bits are translated into "0" when it does not exist	→	0	0	0	0	0	0	0	0	0	0	0	0	0	0	0	0	(×
The memory variable for *RI* of Link 6 is "0"	←	0	0	0	0	0	0	0	0	0	0	0	0	0	0	0	0	
"0" means $RI_6 = 0$																		

Fig. 2 Examples of the partial differential process of CABA for reliability importance (*RI*): *Upper* example indicates $RI_1 = \partial(r_1r_2r_3r_4r_5r_8r_9r_{10}r_{11}r_{12})/\partial r_1$, and *lower* example indicates $RI_6 = \partial(r_1r_2r_3r_4r_5r_8r_9r_{10}r_{11}r_{12})/\partial r_6$

carried out. If the number 3,999 remains in the storage region for terminal reliability, the corresponding term, $X_1X_2X_3X_4X_5X_8X_9X_{10}X_{11}X_{12}$, exists in the polynomial expression for terminal reliability. Similarly, if the number 3,998 remains in the storage region for reliability importance, the corresponding term, $X_2X_3X_4X_5X_8X_9X_{10}X_{11}X_{12}$, exists in the polynomial expression for reliability importance. Therefore, the value of the terminal reliability and reliability importance are obtained by substituting the value for the link reliability into the corresponding terms. On the basis of terminal reliability of the object network and reliability importance of every link in the object network, *RI*, *CI* and *CIW* are calculated and stored into the corresponding storage regions.

5 Numerical Examples

A method for the cost–benefit analysis for the improvement of traffic network reliability was proposed in Sect. 3, and three investment strategies for improving link reliability are assumed to find the key link for improving the network reliability. In this section, a simple bridge network and a field-shape network will be selected to carry out the cost–benefit analysis for the improvement of network reliability.

5.1 Comparison of Importance Indices Combined with Cost–Benefit Analysis for a Simple Bridge Network

A simple bridge network that has four nodes and five links will be simulated as shown in Fig. 3. The minimal-path sets of this network are $P_1 = \{1, 2\}$, $P_2 = \{3, 4\}$, $P_3 = \{1, 5, 4\}$, and $P_4 = \{3, 5, 2\}$. The independent minimal-path set is a series network system [8], thus, the reliability of the terminal reliability between node A and B is given by Eq. (1) as Eq. (26):

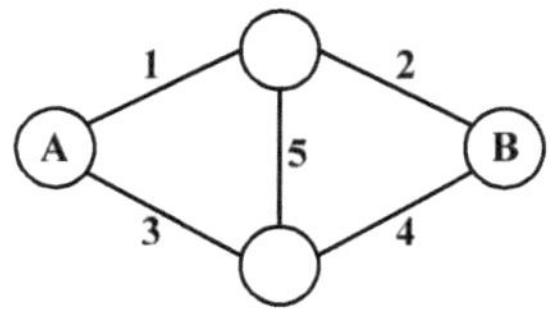

Fig. 3 Simple bridge network

$$R(\mathbf{r}) = E[1 - (1 - X_1X_2)(1 - X_3X_4)(1 - X_1X_5X_4)(1 - X_3X_5X_2)], \quad (26)$$

The exact value of the network reliability for the bridge network is obtained as Eq. (27) by using CABA.

$$\begin{aligned} R(\mathbf{r}) &= r_1r_2 + r_3r_4 + r_1r_5r_4 + r_3r_5r_2 - r_1r_2r_3r_4 - r_1r_2r_4r_5 \\ &\quad - r_1r_3r_4r_5 - r_1r_2r_3r_5 - r_2r_3r_4r_5 + 2r_1r_2r_3r_4r_5 \end{aligned} \quad (27)$$

Three strategies for selecting the most important key link for maximizing the improvement of network reliability on the basis of the same investment strategy are presented:

- Some link should be selected as the most important key link according to *RI*;
- Some link should be selected as the most important key link according to *CI*;
- Some link should be selected as the most important key link according to *CIW*.

Although the combinations of the original link reliability set for five links are infinite for finding the most important key link, only the following three cases will be discussed due to page limits:

Case BA: The difference of the original reliability between link 1 and link 2 is great and the terminal reliability of the primary minimal-path {1, 2} is the same as the terminal reliability of the primary minimal-path {3, 4}. In this case, $r_{10} = 0.3$, $r_{20} = 0.8$, $r_{30} = 0.6$, $r_{40} = 0.4$ and $r_{50} = 0.5$ are assigned.

Case BB: The difference of the original reliability between link 1 and link 2 is small and the difference of the original reliability between link 3 and link 4 is also small, at the same time, the terminal reliability of the primary minimal-path {1, 2} is almost the same as the terminal reliability of the primary minimal-path {3, 4}. In this case, $r_{10} = 0.5$, $r_{20} = 0.3$, $r_{30} = 0.4$, $r_{40} = 0.4$ and $r_{50} = 0.9$ are assigned.

Case BC: The terminal reliability of the primary minimal-path {1, 2} is different greatly from the terminal reliability of the primary minimal-path {3, 4}. And $r_{10} = 0.7$, $r_{20} = 0.6$, $r_{30} = 0.1$, $r_{40} = 0.3$ and $r_{50} = 0.2$ are assigned.

In order to simplify the calculation of the cost–benefit analysis, the investment strategies of *Case 1* and *Case 2* will be used for this simple bridge network. The parameters of Eq. (25) are assigned as $Y = 50$ (months), $F = 100{,}000{,}000$ in order to carry out the cost–benefit analysis of the improvement of network reliability.

The results of cost–benefit analysis of three Cases of *Case BA*, *Case BB* and *Case BC* according to these importance indices are shown in Table 1. From Table 1, the results are obtained as follows:

Table 1 The results of cost–benefit analysis for network reliability improvement of the simple bridge network

Cases	Items	Results			
Case BA	Terminal reliability of network	Importance indices	*RI*	*CI*	*CIW*
		R_{AB}	0.7730	0.7510	0.7680
	Total cost of Case 1	Efficiency	7.9030	5.1860	8.0000
		Ranking	2	3	1
	Total cost of Case 2	Efficiency	2.0590	1.0670	2.1820
		Ranking	2	3	1
Case BB	Terminal reliability of network	Importance indices	*RI*	*CI*	*CIW*
		R_{AB}	0.6840	0.6790	0.6620
	Total cost of Case 1	Efficiency	7.0640	6.3330	7.6660
		Ranking	2	3	1
	Total cost of Case 2	Efficiency	1.7450	1.4250	2.1810
		Ranking	2	3	1
Case BC	Terminal reliability of network	Importance indices	*RI*	*CI*	*CIW*
		R_{AB}	0.7390	0.7250	0.7390
	Total cost of Case 1	Efficiency	8.8810	7.9460	8.8810
		Ranking	1	3	1
	Total cost of Case 2	Efficiency	1.9600	1.6470	1.9600
		Ranking	1	3	1

(1) The network reliability improvement according to *RI* is the highest when the improving iteration times are same and the improved degree of link reliability is the same every time; therefore, *RI* is the best if the attention of improvement in this simple bridge traffic system is paid only to the network reliability improvement.
(2) The cost for improving network reliability according to *CI* is the most costly.
(3) The Efficiency of network reliability improvement according to *CIW* is the best in *Case BA*, *Case BB*.
(4) The Efficiency of network reliability improvement according to *CIW* and *RI* is the same in *Case BC*.
(5) The Efficiency of network reliability improvement according to *CI* is almost the worst in all cases.

5.2 Comparison of Importance Indices Combined with Cost–Benefit Analysis for a Simple Field-Shape Network

A simple field-shape network includes 9 nodes and 12 links shown in Fig. 4.

The independent path set is a set of links in a series system. The reliability of one path set is a combination of the link reliability. The network can be considered as a parallel system composed by all the independent path sets. Once the key path set is found, the most important key link belonging to the key path set can be found according to the importance indices. However, it is complicate to lay out the

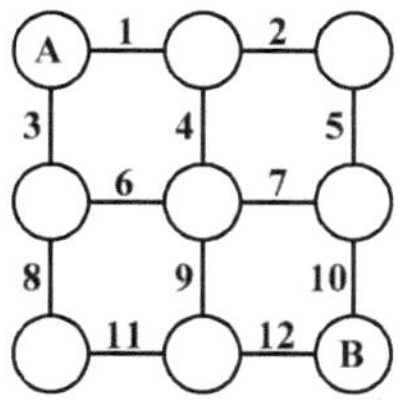

Fig. 4 Field-shape network

expression of terminal reliability of this field-shape network. It is also difficult to calculate the exact value of terminal reliability of this field-shape network.

In this section, three strategies according to the importance indices of *RI*, *CI* and *CIW* are presented for improving network reliability on the basis of the same investment strategy. The investment strategies of *Case 1* and *2* will be used for the field-shape network and the parameters of Eq. (25) are assigned as $Y = 50$ (months), $F = 100{,}000{,}000$.

Although the combinations of the original reliability of 12 links are infinite to find the most important key link, only the following three cases will be discussed due to page limits:

Case FA: The original reliability of all links is the same and the original terminal reliability of the primary minimal-path sets is the same. In this case, the link reliability of all links is assigned as 0.5.

Case FB: The original terminal reliability of two primary minimal-path sets is relatively greater than the original terminal reliability of other primary minimal-path sets. In addition, the difference of the original terminal reliability between these two primary minimal-path sets is little. Therefore, in this case, $r_{10} = 0.6$, $r_{20} = 0.6$, $r_{30} = 0.4$, $r_{40} = 0.7$, $r_{50} = 0.5$, $r_{60} = 0.7$, $r_{70} = 0.4$, $r_{80} = 0.4$, $r_{90} = 0.3$, $r_{100} = 0.5$, $r_{110} = 0.5$, and $r_{120} = 0.4$ are assigned.

Case FC: The original terminal reliability of some primary minimal-path sets is very greater than the original terminal reliability of other primary minimal-path sets. In addition, the original link reliability of links located in this primary minimal-path set is different. In this case, $r_{10} = 0.4$, $r_{20} = 0.6$, $r_{30} = 0.2$, $r_{40} = 0.1$, $r_{50} = 0.8$, $r_{60} = 0.1$, $r_{70} = 0.1$, $r_{80} = 0.3$, $r_{90} = 0.1$, $r_{100} = 0.5$, $r_{110} = 0.3$, and $r_{120} = 0.2$ are assigned.

Before selecting the key link from this field-shape network, the values of importance indices of every link should be calculated by using CABA algorithm. And then, the efficiency of cost–benefit analysis according to the importance indices of *RI*, *CI* and *CIW* is compared. The results of cost–benefit analysis of three *Cases* of *Case FA*, *FB* and *FC* according to these importance indices are shown in Table 2. From Table 2, the results are obtained as follows:

(1) The improvement of network reliability according to *RI* is the highest when the improving iteration times are same and the improved degree of link reliability is the same every time; therefore, *RI* is the best if the attention of improvement in the field-shape traffic system is paid only to an improvement of network reliability and the cost of improvement may not be considered.

Table 2 The results of cost–benefit analysis for network reliability improvement of the field-shape network

Cases	Items	Results			
Case FA	Terminal reliability of network	Importance indices	*RI*	*CI*	*CIW*
		R_{AB}	0.4070	0.3921	0.3966
	Total cost of Case 1	Efficiency	3.8215	2.8748	3.8549
		Ranking	2	3	1
	Total cost of Case 2	Efficiency	0.9696	0.6216	1.0670
		Ranking	2	3	1
Case FB	Terminal reliability of network	Importance indices	*RI*	*CI*	*CIW*
		R_{AB}	0.4980	0.4191	0.4083
	Total cost of Case 1	Efficiency	4.3643	3.9124	4.6987
		Ranking	2	3	1
	Total cost of Case 2	Efficiency	1.0220	0.8459	1.2890
		Ranking	2	3	1
Case FC	Terminal reliability of network	Importance indices	*RI*	*CI*	*CIW*
		R_{AB}	0.2430	0.1998	0.2430
	Total cost of Case 1	Efficiency	4.5314	2.4334	4.5314
		Ranking	1	3	1
	Total cost of Case 2	Efficiency	1.2431	0.5261	1.2431
		Ranking	1	3	1

(2) The cost for improving network reliability according to *CI* is the most costly.
(3) From *Case FC*, the results of selecting the most important key links according to *RI* and *CIW* are the same when the original terminal reliability of particular primary minimal-path set is much greater than the original terminal reliability of other primary minimal-path sets.
(4) The Efficiency of network reliability improvement according to *CIW* is the best in *Case FA*, *FB* and *FC* by using the investment strategy of *Case 1* and *2*. It means that the advanced criticality importance is the best by using cost–benefit analysis for improving network reliability. This result also can be certificated by any example of the original reliability of all links in this field-shape network.
(5) The Efficiency of network reliability improvement according to *CI* is almost the worst in all cases.

From Tables 1 and 2, *RI* is the best of importance index for finding the most important key link when the cost for improving the whole network reliability may not be considered. However, if the cost for improving the whole network reliability is limited, the importance index of *CIW* should be used to find the most important key link on the basis of overall consideration of the cost and the network reliability improvement.

6 Conclusion

In this chapter, firstly, in order to discuss the improvement of the reliability of a traffic network, the current indices of reliability, including *RI*, *CI* and *CIW*, were introduced and the merits and demerits of these indices were pointed out. Secondly, a method for a cost–benefit analysis, on the basis of the cost–reliability function, was proposed for improvement of the reliability of the traffic network. Thirdly, the calculation algorithm for Boolean absorption (CABA) for terminal reliability and reliability importance were developed. Finally, three numerical examples for the simple parallel network, the simple bridge network and the field-shape network were simulated on the basis of the calculation algorithm for Boolean absorption and the cost–reliability function. From these simulations, general conclusions can be obtained as follows:

(1) In a very simple network, the most important key link should be selected according to the investment strategy for the traffic network reliability improvement.
(2) In a general network, *RI* is the best of importance index for finding the most important key link when the cost for improving the whole network reliability may not be considered. However, if the cost for improving the whole network reliability is limited, the importance index of *CIW* should be used to find the most important key link on the basis of overall consideration of the cost and the network reliability improvement.

These conclusions are, however, on the basis of only limited types of networks. As a future study, more types of traffic network should be studied for finding the most important key link. In addition, an enormous amount of CPU time and memory size is still needed even when using the CABA for calculating the terminal reliability and the importance indices for a large-scale network. Therefore, an approximate method for calculating these importance indices should be studied on the basis of the use of partial minimal path sets. This further work is now carried out.

References

1. Barlow, R.E., Proschan, F.: Statistical Theory of Reliability and Life Testing: Probability Models. Holt, Rinehart and Winston, New York (1975)
2. Birnbaum, Z.W.: On the Importance of Different Components in a Multi-Component System. Multivariate Analysis II. Academic Press, New York (1969)
3. Henley, E.J., Kumamoto, H.: Reliability Engineering and Risk Assessment. Prentice-Hall Inc, Englewood Cliffs (1981)
4. Henley, E.J., Kumamoto, H.: Probabilistic Risk Assessment: Reliability Engineering. Design and Analysis. Institute of Electrical and Electronics Engineer, New York (1992)

5. Iida, Y., Wakabayashi, H.: An efficient calculation method to obtain upper and lower bounds of terminal reliability of road networks using Boolean algebra. Proceeding of JSCE, No. 395/IV-9, pp. 75–84 (1988) (in Japanese)
6. Iida, Y., Wakabayashi, H.: An approximation method of terminal reliability of road network using partial minimal path and cut sets. Proceeding of the 5th WCTR, Yokohama, vol.4, pp. 367–380 (1989)
7. Nicholson, A.: Optimizing network terminal reliability. Proceedings of 3rd International Symposium on Transport Network Reliability. Delft University, Netherlands (2007)
8. Wakabayashi, H., Iida, Y.: An efficient evaluation method for road network reliability in disaster. International Symposium on Natural Disaster Reduction and Civil Engineering, JSCE, pp. 397–405 (1991)
9. Wakabayashi, H., Iida, Y.: Upper and lower bounds of terminal reliability of road networks: an efficient method with Boolean algebra. J. Nat. Disaster Sci. **14**(1), 29–44 (1992)
10. Wakabayashi, H., Iida, Y.: Improvement of road network reliability with traffic management. In: Liu, B., Blosseville, J.M. (eds.) Transportation Systems: Theory and Applications of Advanced Technology, pp. 603–608. Pergamum Elsevier Science, UK (1994)
11. Wakabayashi, H.: Network reliability improvement: reliability importance. Proceedings of 2nd International Symposium on Transport Network Reliability. University of Canterbury, Christchurch, New Zealand (2004)

Fuzzy Logic Models of Gap-Acceptance Behavior at Roundabouts

Riccardo Rossi, Massimiliano Gastaldi, Gregorio Gecchele and Claudio Meneguzzer

Abstract Gap-acceptance behavior at intersections has been extensively studied in the field of traffic theory and engineering using various methods. An interesting application of gap-acceptance theory regards roundabouts, which differ from ordinary unsignalized intersections in terms of geometry and driving behavior. Several studies on gap-acceptance at roundabouts can be found in the literature, but, to our knowledge, the fuzzy logic approach has never been used to analyze this type of problem. This chapter describes the development of a gap-acceptance model based on fuzzy system theory and specifically applicable to traffic entering a roundabout. As an alternative to probabilistic discrete choice models, fuzzy system based models can be considered to be appropriate for describing gap-acceptance behavior at roundabouts, because they allow to represent the uncertainty and vagueness that characterizes various aspects of the choice situation under study. Possible applications of fuzzy logic models of gap-acceptance behavior include roundabout entry capacity estimation and use in the context of traffic micro-simulation software. The study is based on data derived from on site observations carried out at a roundabout near Venice, Italy. The performance of the model, evaluated using the Receiver Operating Characteristic (ROC) curve analysis, indicates that fuzzy models can be considered an alternative to the use of random utility models.

Keywords Roundabout · Gap-acceptance · Fuzzy system · Fuzzy logic · ROC curve analysis

---

R. Rossi (✉) · M. Gastaldi · G. Gecchele · C. Meneguzzer
Department of Civil, Environmental and Architectural Engineering,
University of Padova, Via Marzolo 9 35131 Padova, Italy
e-mail: riccardo.rossi@unipd.it

J. F. de Sousa and R. Rossi (eds.), *Computer-based Modelling and Optimization in Transportation*, Advances in Intelligent Systems and Computing 262,
DOI: 10.1007/978-3-319-04630-3_21,

1 Introduction

In studies of vehicular gap-acceptance behavior, the choice to accept or reject a gap of a certain size is generally considered the result of a driver decision process which includes, as inputs, subjective estimates of a set of explanatory variables, given specific objective factors.

These subjective evaluations are usually affected by a high degree of uncertainty, which can be properly treated both by classical probabilistic models, e.g. Logit [12, 26, 27] and by fuzzy system theory [22].

The main objective of the chapter is to develop a model of gap-acceptance based on fuzzy system theory specifically applicable to traffic entering a roundabout. The correct modeling of the gap-acceptance behavior has a strong impact on the accuracy of capacity estimates obtained by micro-simulation models or, more generally, in intersection operational analysis. The Receiver Operating Characteristic (ROC) curve analysis [4, 10, 13] is adopted to evaluate the performance of the model, which is identified using video survey data derived from on-site observations at a roundabout near Venice, Italy.

This work is an extension of previous studies, conducted by the authors, focused on modeling gap-acceptance behavior at "T" intersections using both random utility and fuzzy logic approaches [20–24].

The chapter is organized as follows. Section 2 briefly summarizes past studies concerning gap-acceptance behavior. Section 3 describes the experimental data used in this study. Section 4 describes the main characteristics of the fuzzy logic model for representing gap-acceptance behavior at a roundabout. Section 5 deals with the ROC curve analysis of the gap-acceptance model. Concluding remarks are presented in Sect. 6.

2 Related Work

Several studies on gap-acceptance behavior can be found in the literature and a number of them focused on gap-acceptance models at roundabouts, with specific attention to critical gap estimation [18, 32].

The gap-acceptance problem considered in this chapter refers to the situation in which a driver, starting from the approach of a roundabout, wants to perform an entry maneuver. Essentially, this requires the choice between two mutually exclusive actions: to accept or reject a gap (or lag) of a given size in the circulating traffic stream. In this chapter the term gap refers to the time interval between two successive vehicles passing a section of the circulating roadway (measured from the rear bumper to the front bumper), while the term lag refers to the residual part of the first gap that faces the driver starting from the approach of a roundabout.

Evidently, such a choice is the result of a decision process affected by objective and subjective variables, including gap/lag size, driver characteristics (e.g. driving experience, gender and age [26, 27, 29]), geometric characteristics of the roundabout and its approaches, and specific aspects of the choice situation (e.g. waiting time of vehicles on the entry leg, position of vehicles on the circulatory roadway, and destination exit of the entering vehicle [1, 16, 19].

In previous studies, gap-acceptance behavior at intersections and roundabouts has been described using probabilistic discrete choice models, such as Logit [2, 12, 15, 17, 24, 26–28] or Probit [3, 11], or Neural Networks [14].

Using a different approach [20–23], Fuzzy Logic can deal with the ambiguity which affects the gap-acceptance decision process. The subjective and imprecise evaluations made by the driver are described by fuzzy systems theory using verbal expressions. A set of "if–then" rules is built from the fuzzy knowledge base, properly describing the cause–effect mechanism of the decision process. The rules are generally easy to interpret, because they are expressed in verbal terms. Moreover the use of Fuzzy Logic is attractive since other variables, that cannot easily be incorporated in the utility function of a probabilistic model, can be included in a fuzzy logic model (e.g. drivers' characteristics that are vague by nature such as driving style or state of anxiety).

3 Data Collection and Analysis

The experimental data used in the analysis are gap-acceptance observations (driver decisions) at a roundabout located in a sub-urban area near Venice (Fig. 1), which have been video-recorded during a 2-h peak period. Using an application software the images have been processed identifying the arrival and departure at the stop line (SL in Fig. 1) of each vehicle entering from the roundabout approach, the arrival at the conflict point (C9-2) with the trajectory of circulating vehicles, together with the vehicle category. The data have been organized in a database and the following information associated to each driver decision has been extracted:

- IT = type of time interval (lag or gap);
- IS = time interval size (in seconds);
- V = category of entering vehicle (car, light goods vehicle, heavy goods vehicle);
- waiting time of entering vehicles (queuing delay plus stop-line delay) on the roundabout approach;
- category of circulating traffic stream vehicle closing the interval;
- driver decision (interval acceptance or rejection).

Table 1 summarizes the gap-acceptance data recorded during the survey and used for the analysis.

The identification of the Fuzzy model has been carried out using the stratified holdout approach [30]. The full dataset has been divided in a calibration dataset (70 % of data) and a validation dataset (30 % of data), to calibrate the model and

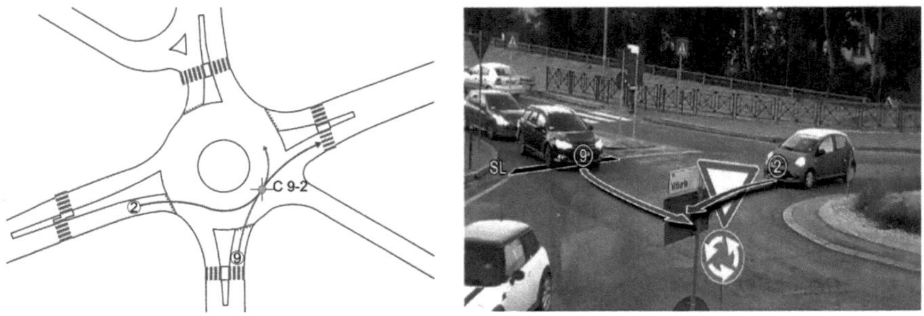

Fig. 1 Layout and picture of the analyzed roundabout

Table 1 Characteristics of the dataset

Type of interval	Acceptances	Rejections	Total	Average number of decisions per approaching driver
Gap	381	337	718	
Lag	442	381	823	1.87
Total	823	718	1,541	

evaluate its performance, respectively. This procedure allows to measure correctly the predictive capabilities of the model, because it is well known that using the same data for estimation and validation could lead to an optimistic evaluation of model performance. Data were randomly sampled from the full dataset, maintaining approximately the same proportion of output classes (acceptance and rejection), type of intervals (gap and lag) and category of entering vehicles (car, light goods vehicle, heavy goods vehicle).

4 Fuzzy Model Identification

The size of the time interval between vehicles on the circulatory roadway is the most important factor affecting gap-acceptance behavior, as widely reported in literature, and drivers evaluate this variable in subjective terms. For these reasons in this work we consider time interval as a fuzzy variable. The fuzzy gap-acceptance model (GA_F) has been developed from experimental data using FisPro, an open-source software available for free on the Internet [7]. The membership functions of the premise and consequence fuzzy sets are identified with the Hierarchical Fuzzy Partitioning algorithm [6] and the rules of inference with the so-called FPA (Fast Prototype Algorithm [5]). In Fig. 2 the fuzzy sets of the premises and of the consequence are shown. The fuzzy knowledge base is represented by three triangular and two trapezoidal membership functions in the domain of the time interval size, by three "singletons" in the domain of the crisp variable representing the category

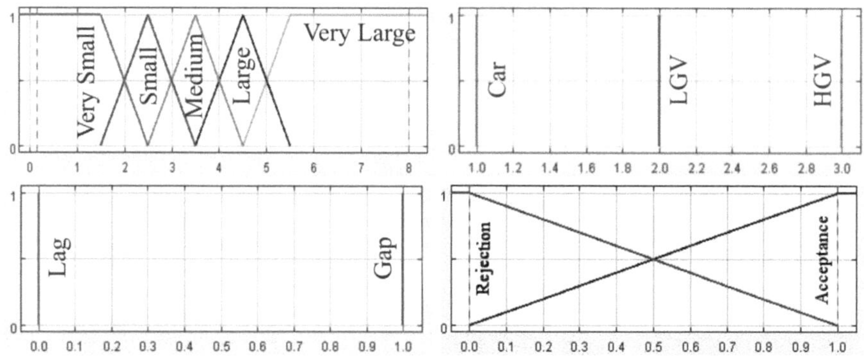

Fig. 2 GA_F model. Premise and consequence fuzzy variables

Table 2 Synthetic description of the decision rules

Kind of rule	Model rules			
	Interval size	Vehicle type	Interval type	Decision
Non compensatory	If IS is VS	–	–	then Rejection
Non compensatory	If IS is S	–	–	then Rejection
Non compensatory	If IS is VL	–	–	then Acceptance
Compensatory	If IS is M	and VT is CAR	–	then Acceptance
Compensatory	If IS is M	and VT is LGV	and IT is L	then Acceptance
Compensatory	If IS is M	and VT is LGV	and IT is G	then Rejection
Compensatory	If IS is M	and VT is HGV	–	then Rejection
Compensatory	If IS is L	and VT is CAR	–	then Acceptance
Compensatory	If IS is L	and VT is LGV	–	then Acceptance
Compensatory	If IS is L	and VT is HGV	and IT is L	then Acceptance
Compensatory	If IS is L	and VT is HGV	and IT is G	then Rejection

IS, Interval size: *VS*, Very small; *S*, Small; *M*, Medium; *L*, Large; *VL*, Very large
VT, Vehicle type: *CAR*, Car; *LGV*, Light goods vehicle; *HGV*, Heavy goods vehicle
IT, Interval type: *G*, Gap; *L*, Lag
"–" means all cases of the variable

of the entering vehicle, and by two "singletons" in the domain of the crisp variable representing the type of interval. The output is described by two triangular membership functions representing acceptance and rejection, respectively. Eleven rules have been identified using Mamdani's product-sum inference; they are shown in compact form in Table 2. A satisfactory value of goodness-of-fit has been obtained ($R^2 = 0.76$).

The fuzzy output variable "acceptance" is defuzzified using the centroid method [9] in order to obtain an "acceptance index" of a certain gap/lag. Using the "acceptance index", it is possible to build "acceptance curves" that allow to use the model as predictive tool (and to validate it over the validation sample). When a gap/lag of a certain size has an acceptance index greater than or equal to the 0.5 threshold, it is considered "acceptable", otherwise it is considered "unacceptable".

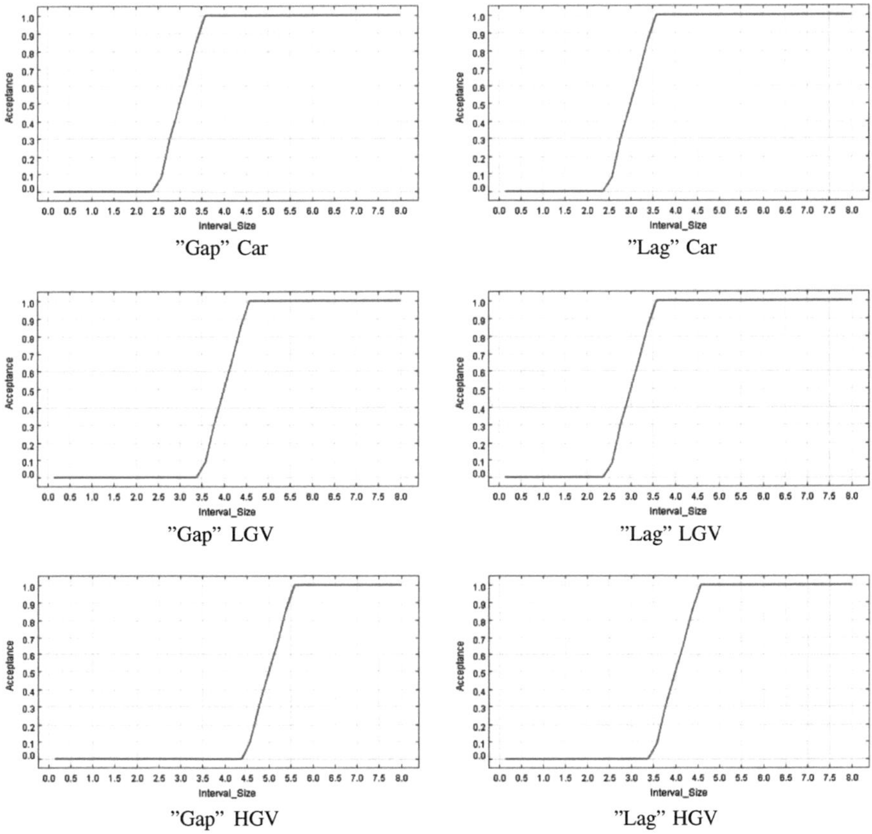

Fig. 3 GA_F model. Acceptance curves as a function of "Time Interval Size" for different vehicle types and interval types

From the acceptance curves shown in Fig. 3, some trends regarding the relationships among the premise variables and gap-acceptance behavior are visible. The interval size IS, which is included in all inference rules (compensatory and non-compensatory), is the most important attribute for the driver decision. "Small" or "very small" intervals are always rejected and "very large" intervals are always accepted by drivers, for any interval type and vehicle type. Similarly, the type of vehicle entering the roundabout affects the gap-acceptance choice in accordance with the differences existing among the performance of vehicles. For cars there is no difference in the acceptance of gap/lag intervals, while LGVs and HGVs accept lag-type intervals of smaller size than gap-type intervals. Furthermore, the size of intervals accepted by cars is smaller than that of intervals accepted by LGVs and HGVs, which need more time to complete the entering maneuver.

5 Analysis of the Model Results

The predictive ability of the model has been tested by means of the ROC curve analysis [4], a method used in various research fields for evaluating and comparing the discriminatory power of models having binary outputs [10, 13], including Logit and Fuzzy models [25]. Few examples are found in transportation applications [21, 23, 31].

The basic idea of ROC curve analysis may be explained by considering an experiment with only two possible outcomes, 1 and 0, that are denoted as positive and negative outcomes. In the GAF models the two outcomes are the acceptance (positive) and the rejection (negative) of a certain gap/lag, therefore four cases are possible:

- True Positive (TP): the model predicts an acceptance and the driver accepted a gap/lag of a certain size;
- False Positive (FP): the model predicts an acceptance and the driver rejected a gap/lag of a certain size;
- True Negative (TN): the model predicts a rejection and the driver rejected a gap/lag of a certain size;
- False Negative (FN): the model predicts a rejection and the driver accepted a gap/lag of a certain size.

The probability of correctly identifying positive outcomes is the True Positive Rate (TPR), and the probability of correctly identifying negative outcomes is the True Negative Rate (TNR). They are calculated by:

- True Positive Rate (TPR) = number of TP/(number of TP + number of FN);
- True Negative Rate (TNR) = number of TN/(number of TN + number of FP).

Another metric commonly used is the False Positive Rate (FPR), which is calculated by:

- False Positive Rate (FPR) = 1-TNR = number of FP/(number of TN + number of FP).

The discriminatory power of the model increases as both TPR and TNR increase. The ROC curve describes the relationship between TPR, also called "sensitivity", and (1-TNR), also called "1-specificity", for all possible classification thresholds. Since the "1-specificity" is the FPR, the ROC curve describes the relationship between the "percentage of hits" and the "percentage of false alarms" obtained with the model. The analysis of ROC curve can be useful to determine the best cut-off value of the variable of interest. In the gap-acceptance case the best cut-off value is expected to be in correspondence of a value of the acceptance index equal to 0.5 for the GA_F model. The results confirm these assumptions, as shown in Fig. 4.

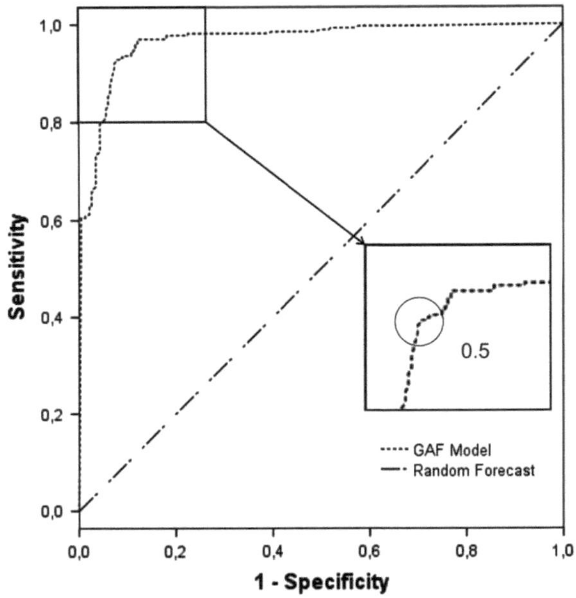

Fig. 4 ROC Curve for GA_F model. Detail for the 0.5 threshold

Table 3 Performance of GA_F model

AUC	TPR	TNR	Precision	Percent right	F-Measure	Youden index
0.966	0.947	0.883	0.903	91.7 %	0.925	0.830

It is known that the area under the ROC curve (AUC) is related to the accuracy of the model predictions, and increases with it; in particular, when this area is equal to one the model produces perfect forecasts, and when it is equal to 0.5 the model produces random forecasts (no discriminatory power). The AUC is equivalent to the Gini coefficient = 2*AUC-1, and also to the Mann–Whitney–Wilcoxon two-independent sample non-parametric test statistic [8].

Additional performance metrics adopted are the precision metric, that represents the percentage of correct acceptance predictions, the F-measure, that is the harmonic average of Precision and TPR, the percent right (or accuracy), that is the percentage of correct predictions globally made. The Youden Index is a powerful metric in cases where both specificity and sensitivity are equally important in the model analyzed. It takes a maximum value of 1 when the model has specificity and sensitivity equal to 1, i.e. it is a perfect model. A good model should have high values of metrics AUC, TPR, TNR, Precision, F-measure, Percent right and Youden Index, and low values of FPR and FNR. The metrics computed for the GA_F model (Table 3) show a good capability of the estimated model to represent the observed decisions.

6 Concluding Remarks

This chapter deals with the development of a model of gap-acceptance based on fuzzy system theory and specifically applicable to traffic entering a roundabout. The main objective of the chapter is to evaluate the accuracy of Fuzzy model predictions with reference to a specific case study. Experimental data have been collected at a real roundabout, focusing on an entry maneuver. The results have been evaluated using the ROC curve analysis. Some remarkable findings are that:

- in absolute terms the fuzzy model shows good capability of representing real driver gap-acceptance behavior;
- the fuzzy model appears very simple and easy to generalize to other gap-acceptance situations (changing inference rules or shape and domain of the membership functions): the description of the decision process appears reasonable, since drivers elaborate in a short period of time a limited number of information about gap-acceptance characteristics;
- with reference to variables commonly used in gap-acceptance analysis, the descriptive capability of the model appears substantially coherent with previous results reported in the literature (obtained using probabilistic models, e.g. logit model).

Nevertheless, there are some directions in which this work could be extended:

- comparative analysis of fuzzy models and probabilistic models (e.g. Logit or Probit) with reference to entry maneuver from a roundabout approach;
- applications to intersection capacity analysis and micro-simulation models, including tests of computation efficiency of fuzzy and probabilistic models;
- analysis of other factors that could affect gap-acceptance behavior (roundabout geometry, speed and type of approaching vehicles in the circulating traffic stream, driver's personal characteristics, past involvement in car accidents, fatigue, etc.);
- sensitivity analysis of model results (acceptance index for fuzzy models) with respect to model parameters.

Acknowledgments The authors acknowledge the technical support of Engineer Alberto Sarto, Mr. Stefano Borgato and Mr. Alessandro Giachi.

References

1. Adebisi, O., Sama, G.N.: Influence of stopped delay on driver gap acceptance behavior. J. Transp. Eng.-ASCE **115**(3), 305–315 (1989)
2. Cassidy, M., Madanat, S.M., Wang, M.H., Yang, F.: Unsignalized intersection capacity and level of service: revisiting critical gap. Transp. Res. Rec. **1484**, 16–22 (1995)
3. Daganzo, C.F.: Estimation of gap acceptance parameters within and across the population from direct roadside observation. Transport. Res. B-Meth. **15B**, 1–15 (1981)

4. Fawcett, T.: An introduction to ROC analysis. Pattern Recognit. Lett. **27**(8), 861–874 (2006)
5. Glorennec, P.-Y.: Algorithmes d'apprentissage pour systems d'inférence floue. Editions Hermès, Paris (1999)
6. Guillaume, S., Charnomordic, B.: Generating an interpretable family of fuzzy partitions. IEEE T. Fuzzy Syst. **12**, 324–335 (2004)
7. Guillaume, S., Charnomordic, B.: Learning interpretable Fuzzy Inference Systems with FisPro. Int. J. Inf. Sci. **181**(20), 4409–4427 (2011). doi:10.1016/j.ins.2011.03.025
8. Hanley, B., McNeil, J.: The meaning and use of the area under a receiver operating characteristics curve. Radiology **143**, 29–36 (1982)
9. Klir, G., Yuan, B.: Fuzzy sets and fuzzy logic. Theory and applications. Prentice Hall PTR, Upper Saddle River (1995)
10. Lloyd, C.: The use of smoothed ROC curves to summarize and compare diagnostic systems. J. Am. Stat. Assoc. **93**, 1356–1364 (1998)
11. Mahmassani, H., Sheffi, Y.: Using gap sequences to estimate gap acceptance functions. Transport. Res. B-Meth. **15B**, 143–148 (1981)
12. Maze, T.: A probabilistic model of gap acceptance behavior. Transp. Res. Rec. **795**, 8–13 (1981)
13. Obuchowski, N.: Receiver operating characteristic curves and their use in radiology. Radiology **229**, 3–8 (2003)
14. Pant, P.D., Balakrishnan, P.: Neural network for gap acceptance at stop-controlled intersections. J. Transp. Eng.-ASCE **120**, 432–446 (1994)
15. Pollatschek, M.A., Polus, A., Livneh, M.: A decision model for gap acceptance and capacity at intersections. Transport. Res. B-Meth. **36**, 649–663 (2002)
16. Polus, A., Kraus, J., Reshetnik, I.: Non-stationary gap acceptance assuming drivers learning and impatience. Traffic Eng. Control **37**, 395–402 (1996)
17. Polus, A., Shiftan, Y., Shmueli-Lazar, S.: Evaluation of the waiting-time effect on critical gaps at roundabouts by a Logit model. Eur. J. Transp. Infr. Res. **5**(1), 1–12 (2005)
18. Polus, A., Shmueli, S.: Analysis and evaluation of the capacity of roundabouts. Transport. Res. Rec. **1572**, 99–104 (1997)
19. Polus, A., Shmueli-Lazar, S., Livneh, M.: Critical gap as a function of waiting time in determining roundabout capacity. J. Transp. Eng.-ASCE **129**(5), 504–509 (2003)
20. Rossi, R., Gastaldi, M., Gecchele, G.: Development of gap acceptance fuzzy models using data from driving simulator experiments. In: Ayyub, B.M. (ed.) Vulnerability, Uncertainty, and Risk: Analysis, Modeling, and Management. Proceedings of the ICVRAM 2011 and ISUMA2011 Conferences, Hyattsville, Maryland, 11–13 Apr 2011, pp. 138–146 (2011). ASCE, doi:10.1061/41170(400)17
21. Rossi, R., Gastaldi, M., Gecchele, G., Meneguzzer, C.: Comparative analysis of random utility models and fuzzy logic models for representing gap-acceptance behavior using data from driving simulator experiments. Procedia Soc. Behav. Sci. **54**, 834–844 (2012). doi:10.1016/j.sbspro.2012.09.799
22. Rossi, R., Gastaldi, M., Gecchele, G., Meneguzzer, C.: Logit model versus fuzzy logic model for representing gap-acceptance behavior. WSC16 2011 Online Conference on Soft Computing in Industrial Applications (2011)
23. Rossi, R., Gastaldi, M., Gecchele, G., Meneguzzer, C.: Transferability of fuzzy models of gap-acceptance behavior. In: Gaspar-Cunha, A., Takahashi, R., Schaefer, G., Costa, L. (eds.) Soft Computing in Industrial Applications, Advances in Intelligent and Soft Computing, vol. 96, pp. 379–390. Springer, Berlin (2011)
24. Rossi, R., Meneguzzer, C., Gastaldi, M.: Transfer and updating of Logit models of gap-acceptance and their operational implications. Transport. Res. C-Emer. **28**, 142–154 (2013). doi:10.1016/j.trc.2011.05.019
25. Tang, T.-C., Chi, L.-C.: Predicting multilateral trade credit risks: comparisons of Logit and Fuzzy Logic models using ROC curve analysis. Expert Syst. Appl. **28**, 547–556 (2005)
26. Teply, S., Abou-Henaidy, M.I., Hunt, J.D.: Gap acceptance behaviour—aggregate and Logit perspectives: Part 1. Traffic Eng. Control **9**, 474–482 (1997)

27. Teply, S., Abou-Henaidy, M.I., Hunt, J.D.: Gap acceptance behaviour—aggregate and Logit perspectives: Part 2. Traffic Eng. Control **10**, 540–544 (1997)
28. Toledo, T.: Driving behavior: models and challenges. Transport Rev. **27**, 65–84 (2007)
29. Wennel, J., Cooper, D.F.: Vehicle and driver effects on junction gap acceptance. Traffic Eng. Control **22**(12), 628–632 (1981)
30. Witten, I.H., Frank, E.: Data Mining: Practical Machine Learning Tools and Techniques, 2nd edn. Morgan Kaufmann, San Francisco (2005)
31. Yau, J.-T., Chou, E., Lin, J.-D., Yu, J.: Pavement overlay effectiveness and optimal timing determination using receiver operating characteristic curve and data envelopment analysis methods. In: TRB 87th Annual Meeting Compendium of Papers DVD (2008)
32. Yin, D., Qiu, T.Z.: Comparison of macroscopic and microscopic simulation models in modern roundabout analysis. Transport. Res. Rec. **2265**, 244–252 (2011)

The Influence of the Volume–Delay Function on Uncertainty Assessment for a Four-Step Model

Olga Petrik, Filipe Moura and João de Abreu e Silva

Abstract This work analyzes the impact of volume–delay function inputs and parameters on the uncertainty of a four-step model traffic forecasts by link and identifies the relevant major error contributors. For that different specifications of the volume–delay function (including its choice and parameters probability distribution) and road capacity as an uncertain element of the input are considered. The uncertainty is expressed in form of variance of the link flows forecast provided by the model for different links types. To illustrate the analyses, a case study data from Aveiro, a medium sized city in Portugal, is used. The results suggest that the capacity variation has higher impact on the final uncertainty (up to 6 % of the coefficient of variation in average for all types of links) than the volume–delay function parameters (up to 3 % of the coefficient of variation) and the links with lower speed limits are affected most.

Keywords Uncertainty analysis · Four-step model · Volume–delay function

1 Introduction

Transport models imply a variety of uncertainties originated from the variability of natural processes and imperfection of knowledge about the studied phenomena. The resulting inaccuracy of the forecasts produced based on these models affects the viability of transportation projects. This means that the risk and uncertainty assessment are of significant importance at the project planning stage. The results of the assessment allow making decisions regarding acceptance or non-acceptance of uncertainty for investors, policymakers and project planners and possible ways

---

O. Petrik (✉) · F. Moura · J. de Abreu e Silva
CESUR/DECivil, Instituto Superior Técnico, Technical University of Lisbon,
Av. Rovisco Pais 1049-001 Lisbon, Portugal
e-mail: olga.petrik@ist.utl.pt

J. F. de Sousa and R. Rossi (eds.), *Computer-based Modelling and Optimization in Transportation*, Advances in Intelligent Systems and Computing 262,
DOI: 10.1007/978-3-319-04630-3_22,

of its mitigation. The latter usually can be performed by gaining supplementary information (e.g. conducting a survey, collecting expert judgments, etc.), which imposes an additional financial burden to the project cost. And thus, the decision maker should thoroughly scope for which information the improvement efforts need to be concentrated on, which implies that the sources of largest errors should be identified first.

This work analyzes the impact of volume–delay function (VDF) specifications and parameters on the uncertainty of a four-step model and identifies the relevant major error contributors. To illustrate our analyses, we use as a case study data from the city of Aveiro which is a medium sized city in Portugal with a multi-modal transportation system available.

1.1 Background

The amount of studies on uncertainty analysis in transportation is fairly limited [1, 2] and most practitioners continue to operate with point estimates despite the importance of risk and uncertainty assessment, especially in case of large-scale transportation projects. In transport demand models the uncertainty originated from each of many model inputs, parameters, and model structure choice propagates through several modeling steps compounding or attenuating itself. This propagation depends on the uncertainty nature: as non-deterministic inputs and parameters are combined in the model, aleatory[1] uncertainty (due to inherent variability) associated with them partially cancel itself out. At the same time the epistemic uncertainty (due to lack of knowledge) can contain systematic error and, thus, not necessarily attenuates itself [3].

Most studies on uncertainty analysis in the four-step model (or some of its steps) are based on probabilistic approaches with application of a simple Monte Carlo simulation [4, 5], Latin Hypercube sampling method [1] or equal-probability bins [6, 7]. Cheung and Polak [8] and Sevcikova et al. [9] implemented methods based on Bayesian probability theory.

Some authors in their studies on uncertainty analysis in the four-step model used simplistic algorithms and did not take into account correlations among the inputs [10, 11], did not postulate probability distributions [12] or applied scenario analysis with no probabilities associated with them [13]. Zhao and Kockelman [4] in a more systematic study quantified the output uncertainty by propagating the variability of model inputs and parameters through the four step model components and showed that the uncertainty grew through the first three modeling steps (trip generation, trip distribution and modal split) and was partially reduced on the forth, traffic assignment, step. Prodhan and Kockelman [6], Krishnamurty and

[1] Some researchers argue that the uncertainty associated to the inherent variability should be called "random" or "stochastic" rather than aleatory. However, the discussion of a more appropriate name is out of the scope of this research.

Kockelman [5] and Clay and Johnston [7] developed similar studies taking into account the land-use component and introducing some feedbacks between the modeling steps. Nielsen and Knudsen [14] in their analysis of uncertainty propagation through the four-step model included discussion on how capacity restrictions and feedback cycle applied on the assignment stage influence the final results' uncertainty. Sevcikova et al. [9] applied Bayesian melding approach to a transport demand model integrated with the land-use component for analyzing uncertainty of travel times in case of tearing down the Alaskan Way Viaduct in Seattle. All the authors who systematically analyzed the uncertainty propagation through the four-step model concluded that the outcome uncertainty is of a high order of magnitude and might significantly affect the demand modeling forecast results. The same has been confirmed by Flyvbjerg [15] in his ex-post analysis comparing the modeling outcomes with the actual observations after the projects have been implemented.

There is often a possibility for the demand modeler or the decision-maker to reduce the input and calibration uncertainty by gaining additional information, every piece of which has its cost. In this case the uncertainty analysis, by identifying the major contributors to the outcome model uncertainty, helps to determine the data elements on which the decision-maker should concentrate most. The amount of systematic studies aiming at determining the most important sources of uncertainty is very limited and the authors admit that to generalize their results more research should be done (e.g., [4–6]).

This study is a part of uncertainty analysis research which aims to identify possible sources of uncertainty in a four-step travel demand model and quantify the outcome uncertainty associated with each of these sources and with all of them together. This research presents an analysis of the impact of the VDF, its parameters and specifications on the uncertainty of a four-step model and identifies the relevant major error contributors within the function inputs and parameters. For that, we consider different specifications of the VDF (including parameters' variation) as well as link capacity uncertainty. As a result of the parameters and specification variations related to the VDF we obtain the output variations for link flows depending on their type (motorways, urban arterials or urban streets). The uncertainty is expressed in the form of variance of the forecasts provided by the model and in terms of differences from the base scenario in link flows and travel times. We also compare the uncertainty due to the VDF inputs and parameters to the results of our previous work [16] where we analyzed the similar impact of the impedance function on the forecasted link flows.

1.2 *Four-Step Model and Volume–Delay Function as a Part of It in a Context of Uncertainty Assessment*

The classical four-step model allows the evaluation of the travel demand and respective assignment to a transport network and is widely used, especially by practitioners [17] in travel demand modeling due to its simplicity and low data intensity, and despite criticisms from academic researchers regarding its capacity to adequately represent mobility patterns. The four-step model also "provides a point of reference to contrast alternative methods" [18]. The complexity of the model leads to numerous sources of uncertainty associated with each of the parameters, inputs, and with the model structure. In Petrik et al. [16] we present a detailed inventory of possible sources of input uncertainty involved in each stage of the four-step modeling process.

The VDF (also called link cost function, amid other terminologies) quantifies the variation of travel time according to volume-to-capacity ratio at the link level in the traffic assignment step. In this step, the demand is assigned to the traffic network so that the demand and supply are equilibrated in a way that travelers cannot find better alternative routes to reach desired destinations from departing origins [19]. The network and the modal characteristics at link and path levels are inputs from the supply side and the Origin–Destination matrix derived at the previous model steps expresses the demand input. The representation of costs, the specification and parameters of the VDF and the chosen traffic assignment algorithm are the sources of model uncertainty at this step. Other sources of uncertainty at the assignment step are node-link representation of the road network; consideration of cost perception on the aggregate level; assumption that the travelers possess perfect information about costs and the road network; temporal variations in demand and traffic [18].

The VDF can reflect the dependence between the costs on a link (usually referred to as generalized costs) and all the link volumes in the network or just between the costs on a link and its traffic flow. The former functional relationship is more suitable for urban case but involves more difficulties in its estimation and incorporation into the traffic assignment step [18].

Many different functions have been proposed by researchers and practitioners with no common agreement on the "best" type as countries with distinctive highway environment, demographic, economic and other characteristics can have different VDFs [20]. The calibration of VDF can be based on the conventional statistical methods using as an input the observed travel time and corresponding volume data for the links (e.g., [21]), or based on bilevel programming using only observed link volume counts [20]. There are mathematical and behavioral properties which VDF has to possess in order to be successfully incorporated into the traffic assignment including realism, non-negative monotonic increase, continuity, differentiability among the others (see, e.g., [18]). Branston [21] presents a review of different VDF specifications developed by researchers and practitioners. Among the most widely used today are the exponential function proposed by Smock [22] for the Detroit Area, the BPR VDF devised by the Bureau of Public Roads [23] and its modifications, the conical VDF [24]. All of these VDF are good trade-offs between the specification parsimony and its ability to adequately represent the actual cost-volume relationships. There are also many functions of more sophisticated form which are claimed to give better fit to the observations, especially in urban conditions, such as a function presented by Akçelik [25] function (allows to account for delay junctions) or by Jastrzebski [26] (reduces over-assignment in case of large trip matrices).

Usually VDF is specified as a product of free-flow travel time (t0) and a congestion function normalized with respect to the link capacities [24]:

$$t(v) = t_0 \cdot \left(\frac{v}{c}\right),$$

where v is the link flow and c is the link capacity.

The concept of link capacity and its units has been slightly changing as the traffic assignment research and practice is developing (for a detailed review of these changes, see Ref. [20]). Today link capacity is usually expressed as the maximum number of vehicles per unit time or as a set of maximum capacities for different levels of service, link types and driving conditions. According to the Highway Capacity Manual (HCM) of 2000 [27], the definition of capacity is: "The maximum sustainable flow rate at which vehicles or persons reasonably can be expected to traverse a point or uniform segment of a lane or roadway during a specified time period under given roadway, geometric, traffic, environmental, and control conditions; usually expressed as vehicles per hour, passenger cars per hour, or persons per hour." The HCM approach implies considering a base capacity per lane type that is then degraded based on the expected effects of a series of characteristics of the road.

As link capacity is difficult to measure precisely and there are many factors affecting it, it is a potential source of the input uncertainty in the VDF, together with its parameters and specification; and it brings a logical question to which extent the uncertainty related to the link capacity will affect the outcome of the model. For example, the British Highway Agency in its manual for traffic capacity evaluation for urban roads [28] states that because of variety of factors affecting the capacity its flows may differ up to ±10 % than the values presented in the manual. On the other hand, Dheenadayalu et al. [29] showed in his analysis of different approaches to urban road capacity measurement that the coefficient of variation of the root mean square error can be from 20 up to 91 % for different methods.

2 Uncertainty and Sensitivity Analysis

2.1 Building the Model

The case study for this research is the urban road network of Aveiro, which is a medium sized city in Portugal. We assign light and heavy vehicles trips between 163 zones to the network consisting of 3,660 links. The four-step model is built skipping the classical generation step, which is conditioned by the available data, namely the mobility survey, from which the information on trip generation is obtained directly. Based on the survey we obtain the OD-matrix and correct it by means of t-flow fuzzy algorithm [30] to match the traffic counts available on some of the links (mainly, trunk roads). The resulting OD-matrix is used to estimate a new flow at the distribution step taking into account generalized cost as an argument of an impedance function and using a doubly constrained gravity model. Since we do not consider public transport in this analysis, the OD-matrix obtained after the distribution step is plugged directly into the traffic assignment step bypassing the modal split step. We apply the occupancy rate so the trips in OD-matrix are presented in vehicles. For the assignment, we use the Frank–Wolfe equilibrium algorithm [31] being the one that provides faster convergence towards an optimum. This aspect is crucial when many simulation runs are to be performed. We do not model a feedback from the assignment step to the trip distribution to avoid increase of the model complexity.

In order to analyze the variability of the outcome associated with different specifications of the VDF we use two popular versions of impedance function: BPR and the conical VDF. BPR function is the most widely used today and it is defined as:

$$t(v) = t_0 \cdot \left(1 + \alpha\left(\frac{v}{c}\right)^{\beta}\right),$$

where α and β are the calibration coefficients, t_0 is a free-flow travel time and v/c is the ratio of current link volume (v) to the maximum link capacity (c). The BPR function is a good trade-off between the function parsimony and representation of the cost–flow relationships, however, it has some disadvantages as slowing down the convergence, causing numerical problems such as a loss of precision, non-unique solutions for the highly congested links [24]. The conical function developed by [24] solves most of these issues as it guaranties uniqueness of the link volumes, and the steepness of the congestion curve represented by this VDF is limited, which allows avoiding problems with highly congested links. The conical function is defined as:

$$t(v) = t_0 \cdot \left(2 - \beta + \alpha\left(1 - \frac{v}{c}\right) + \sqrt{\alpha^2\left(1 - \frac{v}{c}\right)^2 + \beta^2}\right),$$

where α and β are the calibration parameters, α is larger than 1 and $\beta = (2\alpha - 1)/(2\alpha - 2)$.

2.2 Uncertainty Analysis

We assess the uncertainties originated from the use of the VDF in the model associated with its inputs and parameters and changes in uncertainty in case of different VDF specifications. The uncertainty is measured in the link flows obtained after assigning OD trips depending on the category of the link. The three link types are:

- Urban streets, for which free flow speed is equal or below 50 km/h;
- Urban arterials, for which free flow speed is below 80 km/h (that includes also interurban highways);
- Motorways, for which free flow speed is below 120 km/h.

2.2.1 Model Uncertainty

The model uncertainty is associated with the model specification and parameters [32]. The base scenario is the BPR VDF with the parameters defined for Portugal based on the experts judgment, namely with $\alpha = 0.25$ and $\beta = 4$. We measure how the link flows' change in case when BPR is defined differently for each type of the link using the parameters calibrated for Portugal by Viegas et al. [33], as presented in Table 1.

We calculate the differences in link flows and travel times for these two methods (BPR with constant parameters and with different ones depending on the link type). Then we compute the mean values and the standard deviations for the absolute values of these differences for each of the three groups of the links. The mean values of the link flows are normalized with respect to the link capacities and of the link travel times are normalized to the corresponding free-flow times. The results are presented in Table 2.

As one can see from Tables 1 and 2, the motorways, for which the BPR parameters have changed less than for the other groups of links (from $\alpha = 0.25$ to $\alpha = 0.65$ and from $\beta = 4$ to $\beta = 4.8$), are affected least, both in terms of link flow and link travel time, while the urban streets have changed more, especially in its link travel times.

Similar comparison has been performed for the BPR VDF with constant parameters ($\alpha = 0.25$, $\beta = 4$) and the Conical VDF with constant $\alpha = 4$. As one can see from Table 3, the differences are relatively small if to compare with the link capacities and free-flow link travel times; they are smaller than in the case of different BPR functions. That means that the change of the VDF specification did not change the result of the traffic assignment substantially. The biggest differences are present in the link travel times and flows for the urban streets.

The final test has been performed to estimate the variations in the final links flows due to uncertainty in the VDF parameters: α and β in case of BPR and α for the conical function. For the BPR VDF we increased incrementally α from 0.10 to 2 and β from 1 to 10, and for the conical VDF α has been increased from 2 to 12, with 50 steps for each of the three tests. The results are summarized in Table 4.

The results suggest that the urban streets are affected most by changes in the parameters of BPR VDF and the conical VDF, up to 7 % of coefficient of variation for the latter case. Changes in β seem to have less influence on the link flows outcome for the motorways and urban streets. Although not addressed here,

Table 1 The values of BPR volume–delay function calibrated for different types of links [33]

Link type	α in BPR function	β in BPR function
Motorways	0.65625	4.8
Urban arterials	1	1.5
Urban streets	1.28571	1

Table 2 The normalized mean values and the standard deviations for the absolute values of the differences in link flows and travel times calculated for BPR VDF with constant parameters ($\alpha = 0.25$ and $\beta = 4$) and with different parameters depending on the link type

Link type	Link flow differences		Link travel time differences	
	Mean	Standard deviation	Mean	Standard deviation
Motorways	0.001	0.004	0.144	0.732
Urban arterials	0.002	0.005	0.154	0.208
Urban streets	0.008	0.022	0.361	3.007

Table 3 The normalized mean values and the standard deviations for the absolute values of the differences in link flows and travel times (hours) calculated for BPR VDF with constant parameters ($\alpha = 0.25$ and $\beta = 4$) and the conical VDF with constant $\alpha = 4$

Link type	Link flow differences		Link travel time differences	
	Mean	Standard deviation	Mean	Standard deviation
Motorways	0.001	0.002	0.217	0.661
Urban arterials	0.002	0.004	0.157	0.594
Urban streets	0.003	0.011	0.296	2.394

Table 4 Coefficients of variation (the standard deviation as percentage of the mean) for link flows as a result of traffic assignment with variation of the parameters of BPR and conical VDF

Link type	Coeff. of variation, % BPR, β from 1 to 10	Coeff. of variation, % BPR, α from 0.10 to 2	Coeff. of variation, % Conical, α from 1 to 12
All links	1.279	2.904	3.421
Motorways	0.136	0.119	0.227
Urban arterials	0.582	0.426	0.335
Urban streets	2.492	6.070	7.131

volume assignment of smaller urban streets is chiefly influenced by the choice of connectors between zone centroids and urban streets. Also important is the number of connectors to each centroid, how far these reach urban streets are and the distribution of trips across these connectors. It is common to find highly penalized links because all trips are assigned to one single urban street when in reality it is not so. This will be analyzed in future research.

Table 5 Coefficients of variation (the standard deviation as percentage of the mean) for link flows as a result of traffic assignment with variation of the link capacities and parameters of BPR

Link type	Coeff. of variation, % variation of link capacities	Coeff. of variation, % variation of link capacities and parameters of VDF
All links	6.436	5.275
Motorways	0.790	0.797
Urban arterials	1.473	0.911
Urban streets	12.794	10.448

2.2.2 Input Uncertainty

Links capacity is the source of the input uncertainty for VDF, as it was discussed above. In order to simulate the variation of the capacities, we postulate a uniform probability distribution assuming that the capacity of each link might change within ±30 % range of the value used in the base scenario. The magnitude of the capacity variation was chosen based on the results of studies discussed in Sect. 1.2. We apply Monte Carlo simulation with 150 runs. Table 5 contains the changes in the link flows for two tests: variation of link capacity only, and variation of link capacity and parameters together.

For all the link groups the uncertainty is slightly attenuated in case of variation of both link capacities and the BPR function parameters if to compare with the case when only the link capacities are varied. The variation of the link flows for the urban streets (12.8 %) is also larger than in tests with only parameters variation.

2.3 Sensitivity Analysis

The sensitivity analysis for the case of the BPR function variations in inputs and parameters has been performed in order to define the major contributors to the final uncertainty. For this we calculate and compare the linear regression coefficients (Pearson) and rank correlations (Spearman's rho) between each input or parameter and the corresponding outcome. The higher these values are for a certain input variable or parameter, the bigger the impact of its variation has on the overall model uncertainty.

In Fig. 1 the results of the sensitivity analysis are presented in a tornado chart, where values on the horizontal axis show the amount of change in the output due to one unit variation the standard deviation of each input. Figure 1 illustrates the mean impact of input and parameters' variation on the flow of each link, considered individually (i.e., average of all individual impacts). Strong linear relationships are not present (the highest value of R equals to −0.6, which corresponds to R-squared of 0.36, in the case of link capacity variation) while the Spearman's rank correlation gives stable (statistically significant for the confidence level of

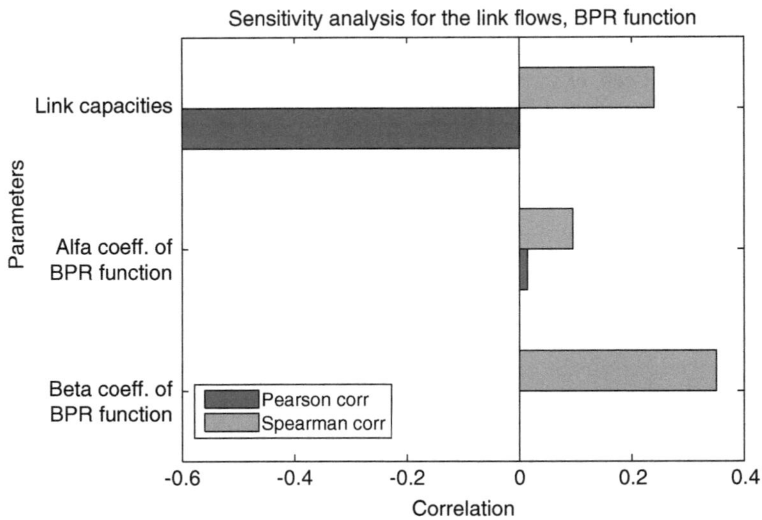

Fig. 1 Sensitivity of the output (link flow value for the non-empty network links) to changes in input and parameters of BPR function

95 %) results for the beta parameter of the VDF and for the link capacities variation. The order of magnitude of the Spearman's rank correlation coefficient does not suggest strong non-linear relationships among the corresponding variables. For the case of alpha coefficient of the BPR function, the sensitivity statistics are not stable, which means that a similar simulation with the same model could generate a different result.

3 Major Findings, Discussion and Conclusions

This study presents an uncertainty and sensitivity analysis related to different volume–delay function specifications and corresponding parameters' and input variation. More specifically, two different specifications (BPR function and conical VDFs) and their parameters, and link capacities as an uncertain input into the VDF are considered. The major findings of this study are:

- In case of substitution of the uniform BPR VDF with three different BPR functions depending on the link types or with the conical VDF the mean values of the result differences in link travel times normalized to the related free flow times are up to 0.4 with the standard deviation up to 3 (for the urban streets). However, this does not affect the link flows in a substantial way as the observed resulting differences in link flows are extremely small when compared with the corresponding link capacities (the normalized mean values are less than 0.01 with the standard deviation less than 0.03 for all three groups of the links).

- The observed variability of the outcome link flows as the BPR and the conical function parameters change incrementally for the motorways and urban arterials are also relatively small represented by the coefficient of variation of less than 0.6 %, while in case of the urban streets it is within 7 %. The variation of the parameter of the conical function has the highest impact on the link flow variations except for the urban arterials.
- Higher output uncertainty for the urban streets, which are links with lower speed-limits. This could be expected intuitively, since the links with lower-speed limits have usually smaller capacity and thus the travel time might be more sensitive to the changes along these links and, consequently, the link flow established as a result of equilibrium in traffic time.
- Relatively low order of magnitude of the output uncertainty in the link flows due to the parameters change compared with the uncertainty due to the variation of link capacities, especially for urban streets. In the case of simultaneous variation of the BPR function parameters and the link capacities, the uncertainty has slightly attenuated itself when compared with the case of only link capacity variation.
- Results of sensitivity analysis confirmed mostly an insignificant influence of the parameters variation on the links flows but moderate linear correlation in case of changes in the link capacities.

Thus, our results suggest that the uncertainty related to the link capacity input may lead up to 12 % variation of the final link loads' assignment, ceteris paribus, while the VDF parameters variation has much less impact on the outcome uncertainty. This in general is in compliance with previous research that demonstrated that lower speed links are the mostly affected by the input and parameters variations [16]. There is a distinction with the results of the analysis of the uncertainty related to the impedance function parameters and inputs in the previous study, as it showed that the inputs variation had much smaller impact on the outcome uncertainty than the parameters one, and that was found out not to be the case for VDF. For the decision-maker or modeller aiming to reduce the outcome uncertainty, the findings from this study suggest that attention should be more focused on the precise estimation of the link capacities rather than on the VDF parameters calibration.

The order of magnitude of the measured uncertainty might be case specific (i.e., both for each network analysed and corresponding loads, and for the ranges of variation assumed). Also the reason why the urban streets are affected most for the considered network is the shape of the volume–delay function curve whereby in case of increasing flow and approaching the congestion level the rate of change of the function (travel time) increases very fast. And as for a city of Aveiro size the urban streets are the ones which are usually congested, they are also the ones most sensitive to any perturbations in VDF inputs and parameters while for other types of cities and networks motorways or urban arterials might be affected most. Another reason why the urban streets are subject to greatest variations can be the impact of the choice of connectors. The effect observed for the case study whereby the type of the volume–delay function does not affect much the traffic assignment results might be also case specific and thus a theoretical study should be conducted in order to generalize this conclusion.

There are also limitations of this study due to the lack of knowledge about the inputs probability distributions, parameters and relationships among them. We can expect some criticism for applying an "older generation" four-step model instead of, for instance, an activity based model, not addressing the feedback from the traffic assignment step to the previous steps and a simplistic approach to the link capacity concept. The choice of the four-step model can be justified by the fact that it is still commonly applied by practitioners due to its simplicity and lower data intensity. In terms of uncertainty, the choice for the simpler model may lead to increased impact of the model structure and specification uncertainty while in the case of choosing a more complex model the result may be dominated by the parameter uncertainty [1]. The feedback from the assignment step has been omitted purely for simplicity reasons; however, it can be included in further research. Moreover, independently from the model used, this research suggests a methodology and an example of its application for measuring uncertainty and defining the main contributors to it in a travel demand model. For practical usefulness similar tests should be done for every element of the model.

As the main goals of the uncertainty and risk analysis are quantifying the output uncertainty and determination of its major sources, for further research we would suggest applying similar methodology for analyzing and comparing the impact of other elements of different transportation models (four-step, activity-based) and their variations and uncertainty propagation through the model steps. The methodology should also be applied on more complicated multimodal urban and interurban networks.

Acknowledgments This research was developed in the framework of the EXPRESS Research Project (MIT/SET/0023/2009) sponsored by the Portuguese national research funds through FCT/MCTES (PIDDAC) and co-financed by the European Regional Development Fund (ERDF) under the Operational Agenda for Competitiveness Factors—COMPETE. We acknowledge the collaboration of the Municipality of Aveiro that allowed the use of the mobility survey data. This survey was performed by the transportation consultants Way2Go for the Aveiro's Urban Mobility Plan. We are grateful to an anonymous EWGT referee for comments that helped to improve this chapter.

References

1. De Jong, G., Daly, A., Pieters, M., Miller, S., Plasmeijer, R., Hofman, F.: Uncertainty in traffic forecasts: literature review and new results for The Netherlands. Transportation **34**, 375–395 (2007)
2. Rasouli, S., Timmermans, H.: Uncertainty in travel demand forecasting models: literature review and research agenda. Transp. Lett. Int. J. Transp. Res. **4**(1), 55–73 (2012)
3. Ferson, S., Ginzburg, L.R.: Different methods are needed to propagate ignorance and variability. Reliab. Eng. Syst. Saf. **54**, 133–144 (1996)

4. Zhao, Y., Kockelman, K.M.: The propagation of uncertainty through travel demand models: an exploratory analysis. Ann. Reg. Sci. **36**, 909–921 (2001)
5. Krisnamurthi, S., Kockelman K.M.: Propagation of uncertainty in transportation-land use models: an investigation of DRAM-EMPAL and UTPP predictions in Austin, Texas. Proceedings of the 81st Annual Meeting of the TRB, Washington, DC (2006)
6. Pradhan, A., Kockelman, K.M.: Uncertainty propagation in an integrated land-transportation modeling framework: output variation via UrbanSim. Transp. Res. Rec. J. Transp. Res. Board **1805**, 128–135 (2002)
7. Clay, M.J., Johnston, R.A.: Multivariate uncertainty analysis of an integrated land use and transportation model: MEPLAN. Transp. Res. Part D **11**(3), 191–203 (2006)
8. Cheung, K., Polak, J.: A Bayesian approach to modelling uncertainty in transport infrastructure project forecasts. Presented at the European Transport Conference, Noordwijk, The Netherlands (2009)
9. Sevcikova, H., Raftery, A., Waddell, P.: Uncertain benefits: application of Bayesian melding to the Alaskan Way Viaduct in Seattle. Transp. Res. Part A **45**, 540 (2011)
10. Ashley, D.J.: Uncertainty in the context of highway appraisal. Transportation **9**, 249–267 (1980)
11. Matas, A., Raymond, J., Ruiz, A.: Traffic forecasts under uncertainty and capacity constraints. Transportation **39**, 1–17 (2012)
12. Bonsall, P.W., Champerowne, A.F., Mason, A.C., Wilson, A.G.: Transport modeling: sensitivity analysis and policy testing. Prog. Plann. **7**, 153–237 (1977)
13. Boyce, A.M., Bright, M.J.: Reducing or managing the forecasting risk in privately-financed projects. Presented at the European Transport Conference, Strasbourg, France (2003)
14. Nielsen, O. A., Knudsen, M.A.: Uncertainty in traffic models, Presented at the European Transport Conference, Strasbourg, France (2006)
15. Flyvbjerg, B., Skamris Holm, M.K., Buhl, S.L.: Inaccuracy in traffic forecasts. Transp. Rev. **26**(1), 1–24 (2006)
16. Petrik, O., Moura, F.M.M.V., Abreu e Silva, J.: The influence of the impedance function on uncertainty propagation through a four-step model. Presented at the European Transport Conference, Glasgow (2012)
17. Wardman, M.: Meta-analysis of European values of time. Presented at IATBR, Toronto (2012)
18. Ortúzar, J.D., Willumsen, L.G.: Modeling Transport, 4th edn. Wiley, Chichester (2011)
19. Wardrop, J.G.: Some theoretical aspects of road traffic research. Proceedings of the Institution of Civil Engineers, Part II, 1, pp. 325–362 (1952)
20. Suh, S., Park, C., Kim, T.J.: A highway capacity function in Korea: measurement and calibration. Transp. Res. **24A**, 177–186 (1990)
21. Branston, D.: Link capacity functions: a review. Transp. Res. **10**, 223–236 (1976)
22. Smock, R.J.: An iterative assignment approach to capacity restraint on arterial networks. Highw. Res. Board Bull. **156**, 1–13 (1962)
23. Bureau of Public Roads: Traffic Assignment Manual. Urban Planning Division, US Department of Commerce, Washington, DC (1964)
24. Spiess, H.: Conical volume delay functions. Transp. Sci. **24**(2), 153–158 (1990)
25. Akcelik, R.: Travel time functions for transport planning purposes: Davidson's function, its time-dependent form and an alternative travel time function. Aust. Road Res. **21**, 49–59 (1991)
26. Jastrzebski, W.P.: Volume delay functions. Presented at 15th International EMME/2 Users' Group Conference, Vancouver, Canada (2000)
27. Bureau of Public Roads: Highway Capacity Manual: Practical Applications of Research. US Department of Commerce, Washington, DC (2000)
28. The Highway Agency: Design Manual for Roads and Bridges (DMRB), vol. 5, Sect. 1, Part 3 TA 79/99 Amendment No 1 2/1 (1999)
29. Dheenadayalu, Y., Wolshon, B., Wilmot, C.: Analysis of link capacity estimation methods for urban planning models. J. Transp. Eng. **130**(5), 568–575 (2004)

30. PTV (Planung Transport Verkehr AG): VISUM User Manual 11.0 Fundamentals
31. Frank, M.; Wolfe, P. (1956). An algorithm for quadratic programming. *Naval Research Logistics Quarterly*, *3(1-2),* 95–110 (2009)
32. Walker, W.E., Harremoës, P., Rotmans, J., van der Sluijs, J.P., van Asselt, M.B.A., Janssen, P., Krayer von Krauss, M.P.: Defining uncertainty: a conceptual basis for uncertainty management in model-based decision support. Integr. Assess. **4**(1), 5–17 (2003)
33. Viegas, J.M., Macário, R., Martins, P.: PETS deliverable 12—Pricing European transport system—case study: Lisbon, the crossing of River Tagus. Research Project for the Commission of the European Communities, Transport—DGVII RTD Programme (1999)

Hybrid Calibration of Microscopic Simulation Models

Luís Vasconcelos, Álvaro Seco and Ana Bastos Silva

Abstract This chapter presents a procedure to calibrate the Gipps car-following model based on macroscopic data. The proposed method extends previous approaches in order to account for the effect of driver variability in the speed–flow relationships. The procedure was applied in a real calibration problem for the city of Coimbra, Portugal, as part of a broader calibration framework that also includes a conventional optimization based on a genetic algorithm. The results show that the new methodology is promising in terms of practical applicability.

Keywords Car-following · Gipps · Speed · Flow · Density · Calibration · Microscopic · Macroscopic

1 Introduction

Microscopic simulation models are practical traffic analysis tools used today in a wide range of areas and applications. These models usually include components related to the infrastructure, the demand, and associated behavioral models. These components and models have complex data requirements and numerous model parameters. Some of these parameters have a clear physical meaning and can be unambiguously identified. However, in most cases, parameter estimation is a real

L. Vasconcelos (✉)
Department of Civil Engineering, Polytechnic Institute of Viseu, Viseu, Portugal
e-mail: vasconcelos@estv.ipv.pt

Á. Seco · A. B. Silva
Department of Civil Engineering, University of Coimbra, Coimbra, Portugal
e-mail: aseco@dec.uc.pt

A. B. Silva
e-mail: abastos@dec.uc.pt

J. F. de Sousa and R. Rossi (eds.), *Computer-based Modelling and Optimization in Transportation*, Advances in Intelligent Systems and Computing 262,
DOI: 10.1007/978-3-319-04630-3_23,

challenge, for the following reasons: (*a*) the estimation of some parameters requires observational techniques that are not available to most model users (for example, the detailed calibration of the car-following model requires detailed position and speed time series of individual drivers, which are particularly difficult to collect); (*b*) model users always have to make compromises between development cost and expected model detail and precision; this requires, for example, making assumptions about the distribution law of a parameter and obtaining representative values for that distribution from a sample, instead of using directly measurable values; (*c*) all traffic simulation models have specification errors and sometimes it is difficult to decide whether the best parameters are: the ones that can be directly estimated from field data, or the ones that make the model perform at its best, and offset the model specification errors.

Given the above, model calibration based on the direct measurement of individual vehicle/driver characteristics is a very specific approach followed mostly by the model developers, while end-users rely on the use of easily measurable traffic data, such as counts and speeds at detectors. Traditionally, model parameters are iteratively adjusted within known plausible limits until a satisfactory correspondence between model and field data is achieved. When done manually this is tedious and time consuming, even when some engineering judgment is used to reduce the number of attempts. Approaches that are more systematic are based on automatic calibration. These approaches regard model calibration as an optimization problem in which a combination of parameter values that best satisfies an objective function is searched. The objective function is formulated as a black-box model and the solution is searched using heuristics. The computational complexity is exponential [1] and therefore the optimization procedure requires a large number of simulation runs, which are generally very costly, depending on the size of the network and the traffic conditions simulated. Thus it is usual to reduce the optimization complexity by selecting a sub-set of parameters either by engineering judgment or by more systematic techniques, such as sensitivity analyses or analyses of variance [2], and by further limiting the range of possible values for each parameter.

Within microscopic simulation models, a problem of special interest is the calibration of the car-following sub-model. Some authors have presented alternative calibration procedures in recent years, based on macroscopic variables. The idea is to derive the traffic stream models that correspond to microscopic flow models and then fit those models to traffic data (speeds and counts) provided by loop detectors. This is a very promising approach because it is fast and simple, as it does not require simulation runs, and it focuses on a few parameters related to steady-state operations. Despite the potential of this calibration procedure, it has only seldom been used in practical applications. The two likely main reasons for this are: first, the methodology is still not known or understood by most practitioners; second, it is unable to reproduce some important features of traffic flow, such as the progressive reduction of average speed with the density, in both congested and uncongested regimes. In this chapter we are presenting an improved calibration procedure. First, we show that the variability in the drivers' desired speed must be accounted for if the uncongested branch of the field speed-flow data is to be accurately reproduced; second, we derive a look-up table that enables that effect to be accounted for in macroscopic relationships. Finally, we demonstrate how this procedure can be used in a real-world calibration problem.

2 Macroscopic Calibration of Gipps' Car-Following Model

2.1 The Gipps Model

Gipps' car-following model is the most commonly used model from the collision avoidance class of models. Models of this class aim to specify a safe following distance behind the leader vehicle. Gipps' model is mostly known for being the building block of the Aimsun microscopic simulator [3]. It consists of two components: acceleration and deceleration sub-models, corresponding to the empirical formulations illustrated by Eqs. (1) and (2), which give the speed of each vehicle at time t in terms of its speed at the earlier time.

$$v_n^{acc}(t+\tau) = v_n(t) + 2.5\,a_n\tau\left(1 - \frac{v_n(t)}{v_n^{\max}}\right)\sqrt{0.025 + \frac{v_n(t)}{v_n^{\max}}} \tag{1}$$

$$v_n^{dec}(t+\tau) = -b_n\left(\frac{\tau}{2}+\theta\right) + \sqrt{b_n^2\left(\frac{\tau}{2}+\theta\right)^2 + b_n\left\{2[x_{n-1}(t) - x_n(t) - S_{n-1}] - \tau v_n(t) + \frac{v_{n-1}(t)^2}{b'_{n-1}}\right\}} \tag{2}$$

where τ is the reaction time, θ is a safety margin parameter, $v_n(t)$ and $v_{n-1}(t)$ are, respectively, the speeds of vehicles n (follower) and $n-1$ (leader) at time t, $v_n^{\max}$ and a_n are respectively the follower's desired speed and maximum acceleration, b_n and b'_{n-1} are respectively the most severe braking that the follower wishes to undertake and his estimate of the leader's most severe braking capability ($b_n > 0$ and $b'_{n-1} > 0$), $x_{n-1}(t)$ and $x_n(t)$ are respectively the leader's and the follower's longitudinal positions at time t, and S_{n-1} is the "leader's effective length", that is, the leader's real length L_{n-1} added to the follower's desired inter-vehicle spacing at stop s_{n-1} (between front and rear bumpers). SI units are used unless otherwise stated.

The speed of vehicle n at time $t+\tau$ is given by the minimum of $v_n^{acc}(t+\tau)$ and $v_n^{dec}(t+\tau)$. If vehicle n has a large headway the minimum speed is given by Eq. (1) and the vehicle accelerates freely according to a law derived from empirical traffic data, tending asymptotically to the desired speed. In other cases the minimum speed is given by Eq. (2). This speed allows the follower to come to a stop, using its maximum desired deceleration b_n, without encroaching on the safety distance. In this derivation it is assumed that the leader brakes according to

b'_{n-1} and that the follower cannot commence braking until a reaction time τ has elapsed. It is also assumed that drivers use a delay θ to avoid always braking at the maximum deceleration rate. Gipps set this parameter to $\theta = \tau/2$ at an early stage of is derivation and it is usually implicit in the above equations.

The vehicles' positions can then be easily updated with a trapezoidal integration scheme by setting the time step to the reaction time τ.

2.2 Calibration Based on Gipps' Steady-State Equations

A particular solution of the Gipps car-following formulation can be obtained for uniform flow. In this case it is assumed that, for a given section, traffic does not vary with time. This happens when all vehicles have the same characteristics and the simulation period is long enough to allow the stabilization of speeds and headways. Wilson [4] derived the following expression for the space headway h_s (distance between front bumpers of two consecutive vehicles) in steady-state:

$$h_s = S + v(\tau + \theta) + \frac{v^2}{2}\left(\frac{1}{b} - \frac{1}{b'}\right), \quad \forall\, v < v^{\max} \tag{3}$$

This function is strictly increasing in the domain $[0, v^{\max})$ and is multi-valued when $v = v^{\max}$. From this expression the macroscopic variables of flow, speed and density (q, v, k) can be easily obtained. In particular, noting that the density is the inverse of the space headway, $k = 1/h_s$, and keeping in mind the fundamental equation of traffic flow $q = kv$ we obtain the speed–flow relationship:

$$q = \frac{v}{S + v(\tau + \theta) + \frac{v^2}{2}\left(\frac{1}{b} - \frac{1}{b'}\right)}, \quad \forall\, v < v^{\max} \tag{4}$$

Wilson demonstrated that when $b > b'$ the car-following model may become unphysical and produce multiple solutions for the same set of parameters. Consequently, b should be set to b' or less [5]. Taking the usual substitution $\theta = \tau / 2$, the q–v relationship—Eq. (4) takes five parameters (v^{max}, τ, S, b and b'). If it is further assumed $b = b'$ then it takes only three parameters). Theoretically, the calibration of these three parameters should be straightforward: first, v^{max} would be set to the observed mean speed of vehicles during very low volume conditions; second, S would be set to the inverse of jam density (measured, for example, from aerial photos) and, finally, the reaction time τ would be manually adjusted to make the curvature of the theoretical curve fit the observations. However, as illustrated in Fig. 1 (case $b = b'$), applying this procedure to real detector data usually results in a good fit in the congested regime and a sub-optimal fit in the uncongested regime. This is because the steady-state solution of Gipps' model predicts constant speed in the uncongested branch and, for $b = b'$, capacity at the free-flow speed. However, numerous field studies show that speed is sensitive to flow and speed-at-capacity is

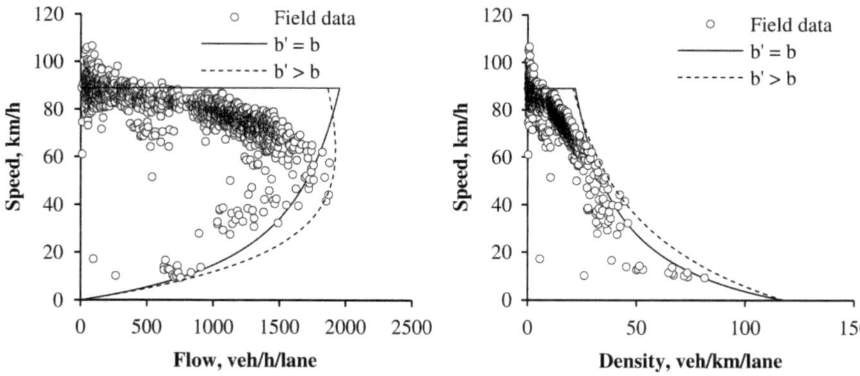

Fig. 1 Adjustment of the q–v and q–k relationships to field data (A44 freeway, Portugal): $v^{\max} = 89$ km/h, $S = 8.5$ m, case 1: $b = b' = 3$ m/s^2, $\tau = 1.0$ s; case 2: $b = 3$ m/s^2, $b' = 3.6$ m/s^2, $\tau = 0.6$ s; $\theta = \tau/2$

lower than free-flow speed. Punzo and Tripodi [6] noted that the fit can be improved by adopting a value $b' > b$. In fact, this relation increases the curvature of the congested branch and results in a speed-at-capacity lower than free-flow speed (Fig. 1, case $b' > b$). Punzo and Tripodi (op. cit.) also suggested that, to be suitable for real applications, steady-state relationships must be generalized to multi-class traffic flows and they proposed an analytical solution for two vehicle classes. Despite being a promising approach, the resulting formulation was also unable to predict speed variation in the uncongested branch of the q–v relationship.

Surprisingly, the solution to this problem had been indicated earlier by Gipps [7]. Remembering that the deduction of the steady-state relationships was based on the assumption of uniform flow, Gipps found that the distribution of desired speeds affects the shape of the uncongested branch. That is, the final steady-state behavior should be seen as the interaction of the average steady-state parameters with the effect of variability. Following this lead, Farzaneh and Rakha [8] investigated the effect of desired speed variability on the INTEGRATION micro-simulator, based on the Van Aerde steady-state relationship [9], and concluded that model users can control the curvature of the uncongested steady-state behavior to a limited extend by adjusting the coefficient of variation of the desired speed distribution. According to Lipshtat [10], this is because of two opposing effects: on one hand, higher variability in the speeds means more occasions when faster vehicles are delayed by slower ones; on the other hand, more variability also leads to greater distances between successive vehicles, making lane changing easier.

To investigate the effect of the desired speed variability on the macroscopic relationships, we performed a sensitivity analysis in which was assumed that the desired speed of each vehicle released into the network follows a normal distribution with fixed coefficient of variation (CV = 10 %) for three different average values ($v^{\max}$= 50, 70 and 90 km/h). The remaining values were set as follows: $\tau = 0.75$ s, $\theta = \tau/2$ (hard-coded in Aimsun), $S = 5$ m, $b = 4$ m/s^2, $b' = 4$ m/s^2

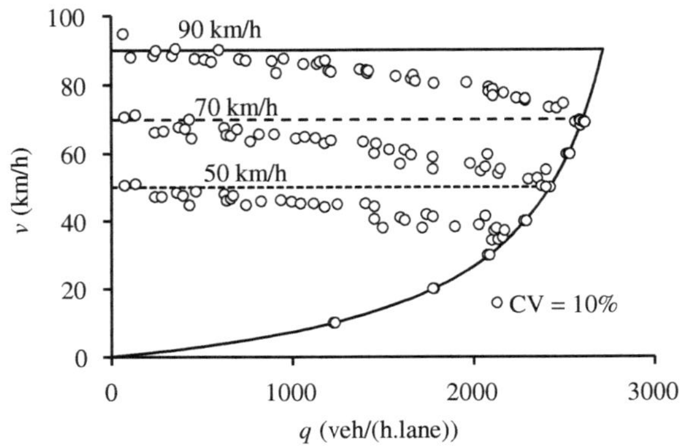

Fig. 2 Sensitivity analysis of the Gipps steady-state parameters

(see Fig. 2). It becomes clear that the adoption of a lower desired speed leads to a lower capacity and that variability in the desired speeds leads to a *q*–*v* diagram in which speeds decrease with flow in the uncongested branch, thus leading to a capacity reduction and speed-at-capacity lower than free-flow speed, as observed in the real world. Therefore, it seems reasonable to express the expected simulation macroscopic relationships as functions of the parameters $v^{\max}$, τ, S and also CV (coefficient of variation of the desired speed).

The introduction of the CV allows the simplification $b = b'$ and the consequent elimination of these two parameters from the calibration process. The problem of this approach is that the effect of variability in the desired speeds is the result of random interactions between vehicles which may depend on several factors such as the road geometry, driver and vehicle characteristics. In order to make this calibration approach simulation free, the following section describes the construction of a look-up table that enables the shape of the *q*–*v* and *k*–*v* curves for a given CV to be analytically identified.

2.3 Derivation of a Look-Up Table to Describe the Uncongested Regime

The first step towards the construction of the look-up table was to decide on the shape of the *q*–*v* curve in the uncongested regime. This question was previously addressed by Wu [11], who proposed a macroscopic model in which the equilibrium speed-flow-density relationships are described as a superposition of homogeneous states. Specifically, in the uncongested part of the diagram (fluid traffic) Wu's model considers two possible states: vehicles moving freely at their desired speed (free state) or bunched vehicles traveling in succession (convoy state, or fluid platoon). Assuming pure stationary conditions and equal distribution of traffic in all lanes, the speed in the uncongested regime can be expressed as a

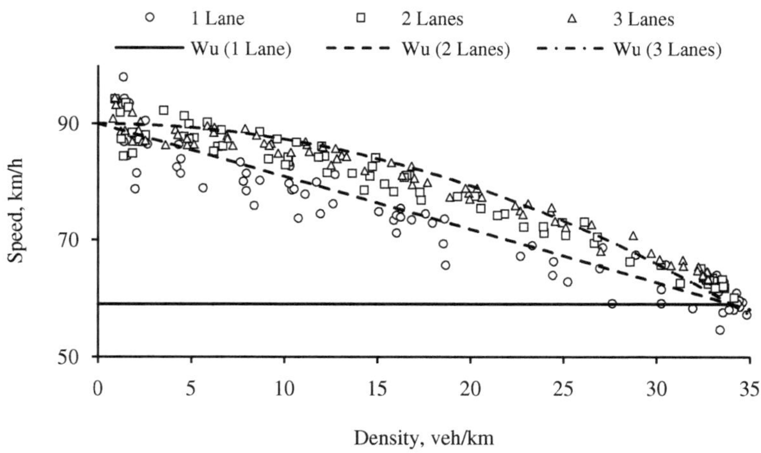

Fig. 3 Density–Speed relationships in the uncongested regime: Wu's model (*lines*) versus simulation results (*markers*)

function of the density k, the free-flow speed v^{max}, the number of lanes N, the speed v_{ko} and density k_{ko} within the platoons:

$$v(k) = v_{\max} - (v_{\max} - v_{ko})\left(\frac{k}{k_{ko}}\right)^{N-1} \quad \text{for} \quad k \leq k_{ko} \tag{5}$$

This indicates that the k–v relationship is linear in a two-lane road (one-way), quadratic in a three-lane road and so on (Fig. 3). The case $N = 1$ is especially interesting: according to this formulation, the speed on the road is v_{ko} regardless of the density, that is, all vehicles are in platoons. Naturally, this only happens under pure stationary conditions. On real world roads, fast vehicles travel in platoons only for a limited stretch of the road.

In order to better understand how simulation results agree with Wu's formulation, k–v measurements were obtained at the middle section of a 1000 m link, for different demand levels and numbers of lanes ($v^{max} = 90$ km/h, CV $= 0.15$, $N = \{1, 2, 3\}$). The resulting capacity parameters were $k_C = 34$ veh/(km·lane) and $v_C = 59$ km/h. When these parameters are used in (5) we conclude that the one-lane model must be rejected, as it underestimates speeds at all densities below k_C, the two-lane model (linear) provides a good-fit for one-lane and two-lane roads, while the three-lane model (quadratic), is a good fit with the simulation results on two-lane and three-lane roads.

Taking Wu's linear model for the k–v relationship (which is reasonable for up to three lanes), we can obtain the capacity parameters as the intersection of the uncongested and congested branches.

For the uncongested branch—linear model with intercept v^{max} and slope b: $v = v^{\max} - k\,b$. At the congested branch—resulting from Gipps' spacing–speed equation: $v = (1 - kS)/(1.5\,\tau k)$. The intersection of the two curves occurs for:

$$k \equiv k_C = \frac{3v^{\max}\tau + 2S - \sqrt{(3v^{\max}\tau + 2S)^2 - 24b\tau}}{6b\tau} \quad (6)$$

Finally, the slope of the k–v relationship was obtained from a full factorial experiment involving the following simulation parameters: Number of lanes: 2; Section lengths: 500, 1000 and 1500 m; Measurement locations: every 50 m; Effective vehicle lengths: 5, 7.5 and 10 m; Reaction times τ: 0.50, 0.75, 1.00 and 1.25 s; Mean desired speeds v^{max}: 50, 70, 90 and 120 km/h; Coefficients of variation: 5, 10, 15, 20, 25 and 30 %. For each combination of parameters a very high demand was set at the input centroid, thus assuring the observation of capacity conditions; the simulation was allowed to run for 30 min. Speed and flow measurements were obtained at each detector at the end of each 5 min period. Capacity parameters were calculated as the average of the last five measurements, but the first period was excluded, in order to assure stable flow. The analysis of the results made it possible to conclude that the section length has less effect on the slope than the other parameters and that the measurement locations could thus be clustered in three regions: initial (first third), middle (second third) and final (last third). The experimental conditions dictate that the resulting look-up table (Table 1) is roughly discretized, requiring crossed interpolation to find the slope b for specific cases outside the discrete domain.

3 Application

Two sites in Coimbra, Portugal, were selected to illustrate the calibration procedure (see Fig. 4). The first is a 4.2 km section of the EN-111A road, between the Geria and Choupal roundabouts. This road is one of the main entries to the city center, used mostly by passenger vehicles. It only has one lane in each direction and gets congested during the morning rush hour. The second site, included for verification purposes, is an off-ramp of the Rainha Santa Bridge, in the W-E direction (350 m), which connects to a single-lane roundabout with intense opposing traffic.

Given the objectives of this study, every effort was made to prevent the introduction of measurement errors, particularly ones related to traffic generation. A video camera was installed near the Geria Rbt to record the type and passage times of approaching vehicles under free-flow conditions. A second camera was placed at the Choupal Rbt to record the passage times of entry and opposing vehicles and, finally, a microwave Doppler detector was installed between the Parque and Choupal roundabouts to collect macroscopic data on approaching vehicles (1-min average speeds and flows). Travel times were measured at 5-min intervals by recording the times at which selected vehicles passed the Geria and Choupal roundabouts. There were some minor differences in the counts for each of these periods, but these were corrected by assuming a corresponding number

Table 1 Look-up table for the slope of the k–v relationship (partial view)

Slope of the upper k–v line, b ($m^2 \cdot veh^{-1} \cdot seg^{-1}$)			Detector location CV of the desired speed distribution														
			Initial					Intermediate					Final				
S (m)	v^{max} (km/h)	τ (s)	5 %	10 %	15 %	20 %	25 %	5 %	10 %	15 %	20 %	25 %	5 %	10 %	15 %	20 %	25 %
5.0	50	0.75	32	56	77	91	96	27	51	72	92	104	18	36	56	73	93
		1	45	73	91	108	111	35	62	84	110	117	22	41	61	86	101
		1.25	51	82	113	125	125	39	72	104	127	132	23	46	74	95	107
	70	0.75	65	105	137	158	161	51	91	125	161	175	34	66	98	136	167
		1	81	135	169	189	190	63	115	151	195	208	39	78	113	162	194
		1.25	105	151	201	210	212	79	131	178	215	239	45	86	127	162	223
	90	0.75	110	169	213	236	239	85	144	193	237	269	57	106	154	204	270
		1	132	215	268	280	280	103	182	242	284	319	65	128	186	232	320
		1.25	158	255	319	315	317	123	216	285	344	367	74	143	208	278	353
7.5	50	0.75	32	63	85	112	118	29	58	81	114	124	18	39	60	88	105
		1	38	72	102	127	134	34	66	96	126	141	21	43	69	92	116
		1.25	55	93	121	144	150	40	81	113	144	158	22	51	78	102	125
	70	0.75	60	108	153	186	202	51	98	143	185	223	34	70	110	149	210
		1	89	135	182	213	226	67	116	164	211	259	41	77	121	163	241
		1.25	102	166	216	245	254	76	139	192	242	274	43	89	136	177	231
	90	0.75	114	169	232	270	287	88	148	212	268	315	58	109	167	223	304
		1	138	223	281	322	332	106	187	249	324	367	66	128	188	264	342
		1.25	165	266	342	373	375	125	222	304	378	424	74	145	219	297	385

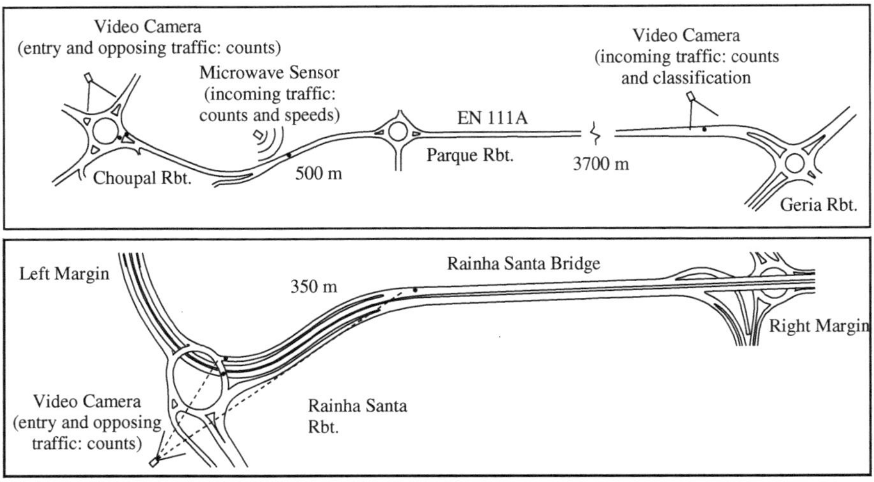

Fig. 4 Layout of application sites

of U-turns at the Parque Rbt, according to an exponential distribution. A video camera was installed at a high point at the Rainha Santa Bridge to simultaneously record the passage times of the entry and opposing traffic. This data was used to generate vehicles with a 1-s precision in the simulator, overriding the default generation algorithm.

Two sets of parameters were considered for the calibration process: the first set is directly related to the car-following behavior (the reaction time τ, the desired inter-vehicle spacing at stop s and the average and standard deviation values of the speed acceptance factor (μ_{sa}, σ_{sa}), which relates the free flow speed v_{max} with the posted speed); the second set is mostly related to interrupted flow and driver behavior at intersections (maximum acceleration a, normal deceleration b, reaction time at stop *RTS* and maximum give way time *GWT*). *RTS* is the time it takes for a stopped vehicle to react to the acceleration of the vehicle in front, *GWT* is used to determine when a driver starts to get impatient if he/she cannot find a gap. With the exception of the abovementioned speed acceptance factor, the vehicles were assumed to be identical.

The first part in the calibration process was to fit the macroscopic relations to the speed-flow data collected with the microwave sensor (Fig. 5). After a few manual attempts, the parameters $S = 6.0$ m ($s = 2.0$ m, $L = 4.0$ m) and $\tau = 1.20$ s were identified as those providing the best adjustment of the lower branch of the q–v curve (4) to the field data (specifically, to the lower side of the point cloud, to minimize the effect of driver heterogeneity). In order to plot the upper branch, the free flow speed (average—54.2 km/h, standard deviation—9.2 km/h, coefficient of variation—17 %) was directly measured from field data during periods of very low traffic ($q < 120$ veh/h). The relation between the free-flow speed and posted speed ($\mu_{sa} = 54.2/50 = 1.08$) was assumed constant and

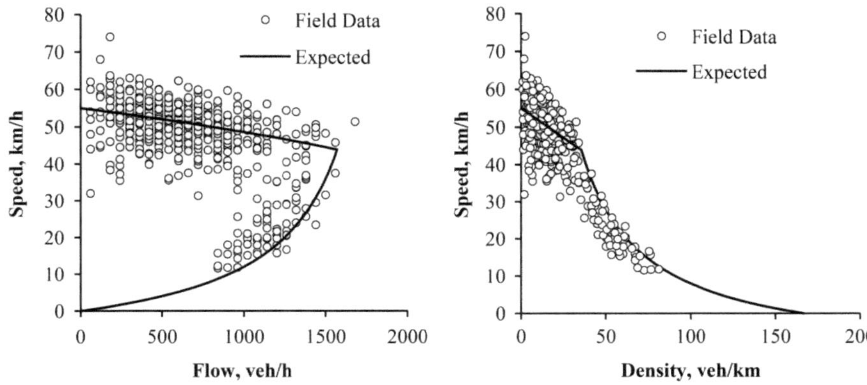

Fig. 5 Field observations and expected Speed–Flow and Density–Flow relationships at EN111A, between Parque and Choupal roundabouts

used to estimate the free flow speed at the upstream section, where the posted speed is 90 km/h. Finally, an approximation for the *v–k* slope was obtained from the look-up table (slope $b \approx 85$). The resulting curve (Fig. 5) for the uncongested branch follows the field data satisfactorily, thus validating the choice of parameters. The intersection with the lower branch—Eq. (6) yields the capacity parameters: $v_c = 43.1$ km/h, $k_c = 36.3$ veh/km, $q_c = 1565$ veh/h. The second part addressed the parameters *a*, *b*, *RTS* and *GWT*. As none of these parameters is involved in the steady-state relationships or can be easily measured from field observations, the calibration takes the form of an optimization problem. In this application a procedure based on a genetic algorithm (GA) was followed. This is a widely used technique that utilizes ideas from natural evolution to effectively find good solutions for combinational parametric optimization problems. The parameter calibration problem was formulated in the following optimization framework [1]: minimize $f|_{a,b,RTS,GWT}(\mathbf{M}^{obs}, \mathbf{M}^{sim})$subject to the constraints $2 \le a \le 6\,\text{m/s}^2$, $2 \le b \le 6\,\text{m/s}^2$, $0.5 \le RTS \le 3.0\,\text{s}$ with $RTS \ge \tau$ and $0 \le GWT \le 30$ s,where *f* is the goodness of fit function that measures the distance between the observed and simulated measurements, $\mathbf{M}^{obs}$ and $\mathbf{M}^{sim}$.

We were particularly interested in understanding how Aimsun was able to predict the variation of queue lengths during the simulation period and so the vector **M** was defined as the vehicle density between the Geria and Choupal roundabouts at each 1-min observation/simulation period *i*. To measure the goodness of fit, the GEH statistic was preferred to other more commonly used measures such as the root mean square error due to its self-scaling property, which allows the use of a single acceptance threshold when comparing a wide range of traffic volumes. Its definition is:

$$MGEH = \frac{1}{N}\sum_{i=1}^{N}\sqrt{\frac{2(M_i^{obs} - M_i^{sim})}{M_i^{obs} + M_i^{sim}}} \tag{7}$$

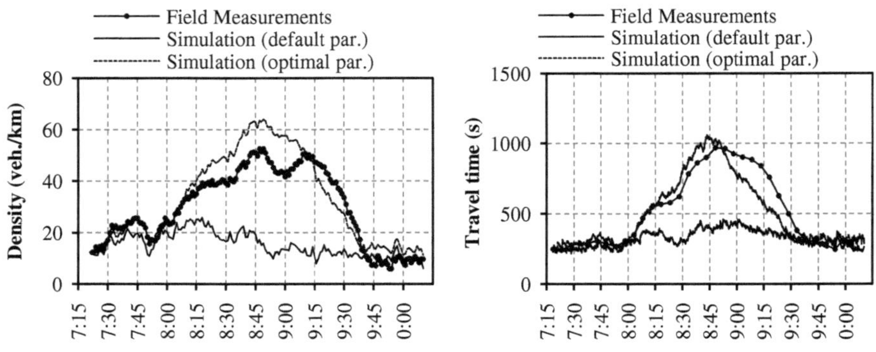

Fig. 6 Time-series of Density and Travel Time at the first site (EN-111A): field observations versus simulation outputs

The optimization framework was implemented in Matlab using the built-in genetic algorithm tool. The algorithm starts by generating an initial population (100 individuals, each of which corresponds to a set of 4 parameters). For each individual, a *Python* script modifies the corresponding parameters in Aimsun, simulates the model in console mode and compares the observation and simulation outputs to compute the fitness value. When all individuals are evaluated, the GA generates a new population: besides elite children, who correspond to the individuals in the current generation with the best fitness values, the algorithm creates crossover children by selecting vector entries from a pair of individuals, and mutation children, by applying random changes to a single individual [12]. After 74 generations and approximately 3.5 h of computing time, the algorithm reached a convergence condition and returned the optimal solution: $a = 5.95$ m/s^2, $b = 3.78$ m/s^2, $RTS = 1.29$ s and $GWT = 1.35$ s. The field and simulation measurements of the density and travel times are presented in Fig. 6, left and right panels, respectively. Despite a slight overestimation of the maximum values, the simulation with optimized parameters generates a satisfactory fit to the field data, particularly in comparison with the simulation using the default parameters. It is also relevant to note the robustness of the estimation: similar results were obtained for both variables, even though only density was explicitly evaluated by the fitness function.

Finally, we looked at how the results of this calibration can be used in a different location. The traffic at the second site—Rainha Santa Brige—has a vehicle mix similar to EN-111A, so the corresponding full set of optimized parameters was used for the first attempt (see Fig. 7). The model overestimated the density in the peak period, indicating lack of capacity at the downstream roundabout. A visual inspection of the simulation revealed excessively cautious maneuvers at the roundabout entry. This was not a key issue at the first site because the opposing traffic was very light, making the model relatively insensitive to the parameters affecting the entry maneuvers. A few manual adjustments

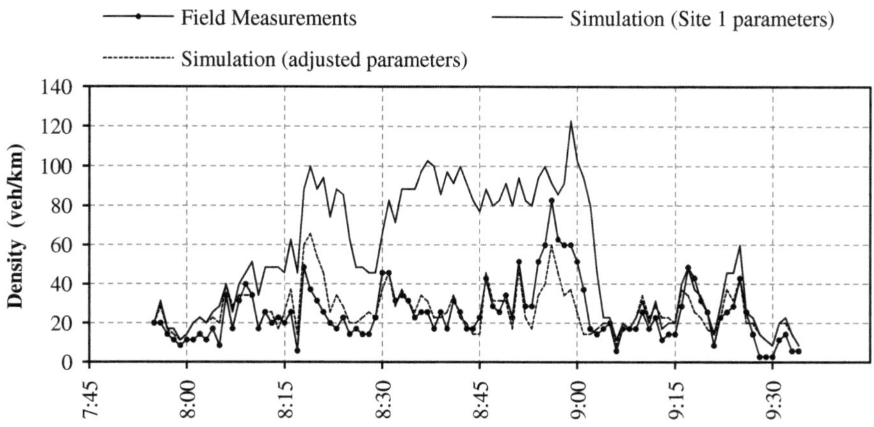

Fig. 7 Time-series of Density at the second site (Rainha Santa Bridge): field observations versus simulation outputs

were thus made to improve the gap-acceptance behavior: τ was reduced to 0.95 s and *RTS* was updated by assuming the same relation between these parameters; the maximum give way time *GWT* was set to a very low value, thus reducing the waiting time at the yield line. The simulation with the adjusted parameters now closely follows the field density time-series. These results suggest that the macroscopic estimation procedure yields relatively robust parameters, that is, they can be applied to different locations with satisfactory results, provided that the traffic has a similar composition at both locations. The traditional methods, however, being based on "blind" optimization, disregard the parameters' physical meaning and can lead to biased estimates that reduce the explanatory power of the model.

4 Conclusions

Current macroscopic calibration procedures for the Gipps car-following model assume constant speeds in the uncongested regime, whereas numerous field studies indicate that the speed-flow diagram is curved in both the congested and uncongested regimes. In this chapter we have shown that it is essential to account for the variability in the drivers' desired speed distribution to accurately reproduce the uncongested part of the field speed–flow data. We used a vast set of simulations to construct a look-up table that relates the slope of the density–flow relationship with the vehicles' desired speed variability and with other variables that also have a significant effect in the macroscopic relationships, such as the detector location and the vehicles' effective length and reaction time. This procedure was used in a real-world calibration problem, as part of a broader calibration framework that also includes a conventional optimization based on a genetic algorithm. The car-following parameters resulting from the macroscopic calibration proved to be robust, which enabled their use at different locations with only minor adjustments, and showed that the new methodology is promising in terms of practical applicability.

Acknowledgments This work was supported by FCT (Portugal) under the R&D project PTDC/SEN-TRA/122114/2010 (AROUND—Improving Capacity and Emission Models of Roundabouts).

References

1. Ciuffo, B., Punzo, V., Torrieri, V.: Comparison of simulation-based and model-based calibrations of traffic-flow microsimulation models. Transp. Res. Rec. J. Transp. Res. Board, **2088**(-1), 36–44 (2008)
2. Punzo, V., Ciuffo, B.: How parameters of microscopic traffic flow models relate to traffic dynamics in simulation. Transp. Res. Rec. J. Transp. Res. Board, **2124**(-1), 249–256 (2009)
3. Casas, J., et al.: Traffic simulation with Aimsun. In: Barceló, J. (ed.) Fundamentals of Traffic Simulation, pp. 173–232. Springer, New York (2010)
4. Wilson, E.: An analyses of Gipps' car-following model of highway traffic. IMA J. Appl. Math. **66**(5), 509–537 (2001)
5. Rakha, H., Wang, W.: Procedure for calibrating the gipps car-following model. In: TRB 2009 Annual Meeting. Transportation Research Board, Washington DC (2009)
6. Punzo, V., Tripodi, A.: Steady-state solutions and multiclass calibration of Gipps microscopic traffic flow model. Transp. Res. Rec. J. Transp. Res. Board, **1999**(-1), 104–114 (2007)
7. Gipps, P.G.: A behavioural car-following model for computer simulation. Transp. Res. B, **15**(B), 105–111 (1981)
8. Farzaneh, M., Rakha, H.: Impact of differences in driver-desired speed on steady-state traffic stream behavior. Transp. Res. Rec. J. Transp. Res. Board, **1965**(-1), 142–151 (2006)
9. Rakha, H., Crowther, B.: Comparison of Greenshields, pipes, and Van Aerde car-following and traffic stream models. Transp. Res. Rec. J. Transp. Res. Board **1802**(1), 248–262 (2002)
10. Lipshtat, A.: Effect of desired speed variability on highway traffic flow. Phys. Rev. E **79**(6), 066110 (2009)
11. Wu, N.: A new approach for modeling of fundamental diagrams. Transp. Res. Part A Policy Pract. **36**(10), 867–884 (2002)
12. The MathWorks, Inc.: Genetic Algorithm and Direct Search Toolbox for Use with MATLAB®: User's Guide. MathWorks (2005)

Signal Setting Design at a Single Junction Through the Application of Genetic Algorithms

Giulio Erberto Cantarella, Stefano de Luca, Roberta Di Pace and Silvio Memoli

Abstract The purpose of this chapter is the application of Genetic Algorithms to solve the Signal Setting Design at a single junction. Two methods are compared: the monocriteria and the multicriteria optimisations. In the former case, three different objectives functions were considered: the capacity factor maximisation, the total delay minimisation and the total number of stops minimisation; in the latter case, two combinations of criteria were investigated: the total delay minimisation and the capacity factor maximisation, the total delay minimisation and the total number of stops minimisation. Furthermore, two multicriteria genetic algorithms were compared: the Goldberg's Pareto Ranking (GPR) and the Non Dominated Sorting Genetic Algorithms (NSGA-II). Conclusions discuss the effectiveness of multicriteria optimisation with respect to monocriteria optimisation, and the effectiveness of NSGA-II with respect to the GPR.

Keywords Signal setting · Optimisation modelling · Genetic algorithms · Metaheuristics · Single junction

---

G. E. Cantarella · S. de Luca (✉) · R. Di Pace · S. Memoli
Department of Civil Engineering, University of Salerno, Via Giovanni Paolo II 132, 84084 Fisciano, SA, Italy
e-mail: sdeluca@unisa.it

G. E. Cantarella
e-mail: g.cantarella@unisa.it

R. Di Pace
e-mail: rdipace@unisa.it

S. Memoli
e-mail: smemoli@unisa.it

J. F. de Sousa and R. Rossi (eds.), *Computer-based Modelling and Optimization in Transportation*, Advances in Intelligent Systems and Computing 262,
DOI: 10.1007/978-3-319-04630-3_24,

1 Problem Statement

The Signal Setting Design (SSD) is usually addressed through optimisation models where the decision variables are the signal timings, while the network layout is usually assumed given.

Existing contributions on the SSD, address the following problems: single junction optimisation or network optimisation. Furthermore, SSD can be generally addressed considering the junctions as isolated or interacting within a network. In the former case, the green timings, their scheduling and the cycle length are calculated without considering the influence of upstream on downstream junctions, in the latter case, the interaction between successive junctions must be considered [8].

Optimisation models for SSD can be solved through feasible direction algorithms, still these algorithms may fail to find the optimal solution when the objective function is not convex, and/or has several local optimal points. Moreover, multicriteria optimisation, which is receiving an increasing attention, can hardly be solved through traditional feasible direction algorithms.

On the basis of these points of weakness, in the last years several metaheuristics derived from biological metaphors, such as evolutionary algorithms and swarm intelligence, have been developed first to deal with discrete optimisation, and then extended to cope with continuous optimisation possibly with several optimal points. These methods include genetic algorithms (GAs), differential evolution (DE), ant colony (ACO), bee colony (BCO), bacteria foraging algorithms (BFO), etc.

According to the literature particular attention has been given to the GAs which have been applied to monocriteria network SSD in several cases (see [13, 21, 16]) while few contributions can be find with respect to the application of GAs to multicriteria SSD [5, 24]. Furthermore, in some cases, the multicriteria network SSD has been carried out by using the weighting coefficients in order to combine more objective functions in a unique objective function (see [11]).

On the basis of previous considerations more investigations need to be made on multicriteria optimisation with respect to both the single junction and the network SSD. In particular, this chapter aims at investigating the application of genetic algorithms to multicriteria SSD at a single junction (see [9]).

The chapter is organised as follows: in Sect. 2, the SSD Background is summarised; in Sect. 3, the Problem is described; in Sect. 4, the Monocriteria and Multicriteria GAs are briefly discussed; in Sect. 5, the Numerical Results are shown and finally in Sect. 6, the Conclusions and research perspectives are summarised.

2 Background

In case of single junction SSD in undersaturation conditions, two main problems with different sets of variables may be defined: the green timing and the green timing and scheduling. In the former case, the optimisation variables are the green timings and (probably) the cycle length while the stage matrix is fixed; in the latter case, the stage sequences are also considered as optimization variables.

In accordance with the literature, the green timing problem has been solved by the application of: (i) the Equisaturation principle [27]; (ii) the total delay minimisation, formulated as a convex programming model ([2] and solved by the program SIGSET described in Ref. [4]); (iii) the capacity factor maximisation, formulated as a linear programming (LP) model ([3]; solved by the program SIGCAP described in Ref. [6]).

The green timing and scheduling, in undersaturation conditions, can be addressed through the capacity factor maximisation or the total delay minimisation, and several methods based on discrete linear or continuous non linear optimisation techniques have been proposed [6, 7, 14, 19]. The oversaturation conditions for a single junction can be addressed through total delay minimisation during the entire oversaturated period and not over the cycle length, by looking for the best phase switching strategies, such as the semi-graphical methods [15], where the Pontryagin maximum principle is applied to derive analytical solutions of the optimal trajectories. Reference [20] proposed the bang bang control method which attempted to find an optimal switch over point during.

Finally, most papers address the monocriteria SSD [23, 26], while the multicriteria SSD method has not been discussed in depth [22, 25].

Summing up the aim of this chapter is threefold:

(i) preliminarily investigate the effect of some algorithms parameters on the algorithms effectiveness;
(ii) compare monocriteria optimisation with multicriteria optimisation;
(iii) compare two multicriteria optimisation algorithms;

In case (ii), three objectives functions were considered: the capacity factor maximisation, the total delay minimisation and the number of stops minimisation; in case (iii), two combinations of criteria were compared: the total delay minimisation and the capacity factor maximisation, the total delay minimisation and the total number of stops minimisation.

3 Problem Description

This chapter, as stated in the introduction, aims at solving the monocriteria and the multicriteria single junction SSD with given stage matrix through the application of genetic algorithms. The general framework is described below.

3.1 Variables

The main definitions and notations are introduced below. Let:

c be the cycle length, assumed known;
t_j be the duration of stage j, a decision variable;
t_{ar}, be the so-called all red period at the end of each stage to allow the safe clearance of the junction, assumed known;
Δ be the approach-stage incidence matrix (or stage matrix for short), with entries $\delta_{kj} = 1$ if approach k receives green during stage j and 0 otherwise, assumed known;
l_k be the lost time for approach k, assumed known;
$g_k = \sum_j \delta_{kj} t_j - t_{ar} - l_k$ be the effective green for approach k;
$r_k = c - g_k$ be the effective red for approach k;
q_k be the arrival flow for approach k, assumed known;
s_k be the saturation flow for approach k, assumed known;
$(s_k \times g_k)/(c \times q_k)$ be the capacity factor for approach k.

So far, each solution is described by a vector/chromosome with entries/genes given by the stage lengths.

3.2 Constraints

Some constraints are introduced in order to guarantee:
stage durations being non-negative

$$t_j \geq 0 \qquad \forall j$$

effective green being non-negative

$$g_k \geq 0 \qquad \forall k$$

This constraint is usually guaranteed by the stage duration non-negative; it must be observed that in case of variable cycle length, for too small durations of it, say $c \leq \sum_j MAX_k(\delta_{kj} l_k + t_{ar})$, with regard to the values of all-red period length and lost times.

Consistency among the stage durations and the cycle length

$$\mathrm{S}_j t_j = \mathrm{c}$$

the minimum value of the effective green timing

$$g_k \geq g_{min} \qquad \forall k$$

A further constraint may be included in order to guarantee
the capacity factor must being greater than 1

$$0 \leq (s_k \times g_k)/(c \times q_k) - 1 \qquad \forall k$$

Such a constraint may be added only after having checked that the maximum junction capacity factor is greater than one, otherwise a solution may not exist whichever is the objective function/s.

3.3 Objective Functions

In the monocriteria SSD three objective functions were considered: (i) the Capacity Factor (CF) to be maximised; (ii) the Total Delay (TD) to be minimised; (iii) the Number of stops (NS) to be minimised. In the multicriteria SSD, the optimisation can be carried out with respect to any combination of the above introduced objective functions. In the following results are described for some combinations of two objective functions only: (v) the Total Delay, to be minimised and the Total Number of stops, to be minimised; (vi) the Total Delay, to be minimised and the Capacity Factor, to be maximized.

All defined multicriteria problems were solved by applying the Goldberg's Pareto ranking method (see [17]) and the NSGA-II (see [12]).

The effectiveness of each optimisation method was evaluated on the base of following performance index: (i) the Capacity Factor (CF); (ii) the Total Delay (TD); (iii) the Number of stops (NS), this indicator was introduced given its effect on the air pollution.

The objective functions in the optimisation problems were computed as follows junction capacity factor as

$$CF = MIN_k(s_k \times g_k) \, / \, (c \times q_k) \tag{1}$$

total delay applying the two terms Webster formula (see [27]) as

$$\begin{aligned} TD = & S_k q_k \times (0.45 \times c \times (1 - g_k/c)^2/(1 - q_k/\, s_k) \\ & + q_k \times 0.45/\,(s_k \times g_k/\,c) \times ((g_k/\,c) \times (s_k/\,q_k) - 1))) \end{aligned} \tag{2}$$

total number of stops applying the Akçelik formula [1] as:

$$NS = MAX_k 0.9 \times q_k \times (1 - g_k/c) \, / \, (1 - q_k/s_k) \tag{3}$$

It is worth remembering that the delay from the two-terms Webster formula and the number of stops from the Akçelik formula are convex with respect to variables g_k/c, thus ratios t_j/c are the decision variables actually used.

4 Monocriteria and Multicriteria GAs

GAs search the optimal solution by simulating the evolution of a "population" (of individuals), mimicking the basic principle of bacteria evolution. Each solution is described by a vector of decision variables called a chromosome made up by genes. Usually the most adopted approach for genes generation is based on decodification of binary code for each one, however the degree of accuracy of decodification depends on the landscape of objective functions: in case of stable landscape decodification should be avoided and the direct generation of stages duration should be applied. The optimisation is carried out through an iterative process of random reproduction of the individuals (solutions) in the population based on the fitness function [18].

After reproduction each chromosome may be modified by genetic operators, *the crossover* and *the mutation*. This iterative procedure is repeated until some conditions (e.g. number of iterations or of improvement of the best solution) are satisfied. Main parameters to fully specify a GAs are the population size, the crossover probability (or rate), and the mutation probability (or rate).

In case of multicriteria optimisation some major considerations need to be made with respect to procedure described above and regarding the selection criteria based on the fitness function evaluation. In fact, during the iterative procedure, the fitness function can be computed with respect to one or more criteria.

Multicriteria GAs are often called Multiobjective Evolutionary Algorithms (MOEAs), and they can be implemented following one of the below methods: (i) *The Aggregating Functions method,* which defines how all the objectives are combined into a single one by the application of weighting coefficients; (ii) *The Pareto-based methods,* that incorporate the estimation of the Pareto front in the selection mechanism; these methods include the Goldberg's Pareto Ranking, the Non dominated Sorting Genetic Algorithm (NSGA-II), the Niched Pareto Genetic Algorithm etc.

In this chapter two multicriteria GAs were implemented: the Goldberg's Pareto Ranking and the NSGA-II. The Goldberg's Pareto Ranking introduces the *rank based* fitness assignment; in particular the fitness function (*f*) for a given chromosome *j* is calculated by the *linear ranking* [4] approach as follows:

$$f(j) = 2 - SP + 2\,(SP - 1)\,(rank(j) - 1)/N$$

where

SP is the selective pressure fixed to 1.5;
N is the population size;
rank(j) is computed from the chromosome dominance hierarchy.

This algorithm was compared with the NSGA-II in which an additional criterion is introduced for selecting among solutions with the same ranks. Each solution is attached to a value of *crowding distance* which is computed by the Eulerian distance between the vector of the fitness functions of the given solution and the vector of the best fitness functions (i.e. the best value among all solutions); if two or more solutions have the same rank value, selection at successive steps, is based on the best value of the crowding distance.

$c = (1.5\,L + 5) / (1 - Y)$

where

$L = \Sigma_j\, l_{k(j)} \geq 0;$

$y_j = q_{k(j)} / s_{k(j)} \geq 0;$

$Y = \Sigma_j\, y_j$, $k(j)$ is the reference approach of

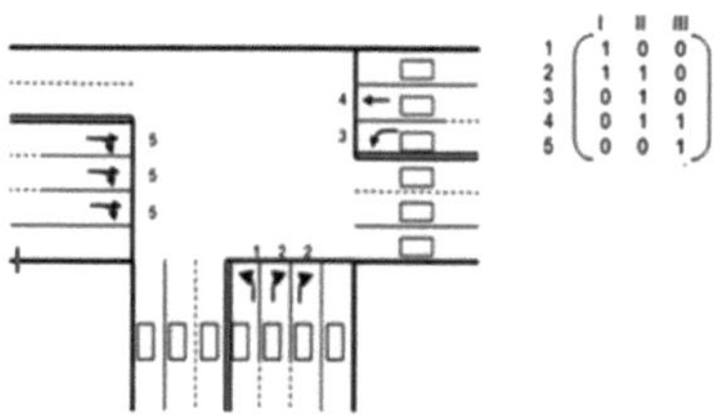

approach	q_k(pcu/h)	s_k(pcu/h)	c(s)
1	400	1518	
2	350	2122	
3	270	1511	64
4	125	1145	
5	550	4392	

Fig. 1 Junction layout; stage matrix; characteristics of the junction

5 Numerical Results

This section discusses results obtained for SSD of the T junction with layout described in Fig. 1. In Fig. 1 the stage matrix and the main input data are showed. The cycle length was computed following the Webster indication.

First monocriteria SSD problem was addressed. The population size, the crossover rate and the mutation rate were assessed by observing the speed of convergence towards the (sub-)optimal solution. The optimal value of objective function versus population size (6; 20; 40; 60; 80; 100) were evaluated with regard to each crossover rate (PC) and/or mutation rate (PM) pair {(0.7;0.01), (0.6;0.1), (0.8,0.001), (0.8,0.0001), (0.9,0.001), (0.9,0.0001)} (see below Fig. 2).

Numerical results are shown in Table 1 in which each row shows stage durations obtained with one of the objective functions described above, together with values of the other objective functions; best values are in bold. Population size values, the crossover probability/rate (PC), and the mutation probability/rate (PM) are given in the table.

At first it can be observed that values of stage durations are affected by the criterion considered as objective functions; by comparing obtained results the best value of each indicator is consistent with the optimisation criterion (i.e. in case of CF optimisation, $CF = 1.35$ and this is the best value of CF among all).

Furthermore, about other indicators, in case of TD minimisation lower value of TD is obtained ($TD = 9.14$ veh); in case of CF maximisation, higher value of capacity factor is reached ($CF = 1.35$); finally, in case of NS minimisation lower value of NS is obtained ($NS = 397.88$ stops/h). Furthermore, the proposed approach was also validated by comparing GAs results with SIGCAP and SIGSET results.

Starting from monocriteria SSD results, multicriteria SSD problem was investigated preliminary assessing the speed of convergence with respect to parameter values (i.e. population size and PC, PM pairs). The GPR and the NSGA-II methods were performed and results are shown in Table 2.

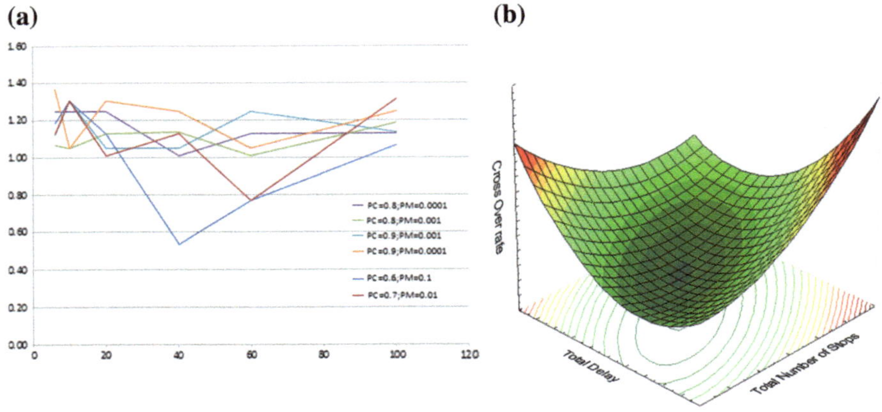

Fig. 2 Sensitivity analysis of optimisation criteria with respect to parameters (population size; PC; PM). **a** Landscapes of objective function (for given pair of PM & PC) versus population size; **b** Tridimensional surface of TD & NS with respect to different values of PC (for given values of PM and population size)

Table 1 Results of monocriteria SSD—Population size = 40

Criterion	PC/PM	TD (veh-h/h)	CF	NS (stops/h)	t_1(s)	t_2(s)	t_3(s)
TD	0.8/0.001	9.14	1.32	459.76	28	20	16
CF	0.9/0.0001	9.15	1.35	459.76	28	20	16
N_{vq}	0.9/0.0001	12.01	1.16	397.87	23	18	23

Table 2 Results of multicriteria SSD (GPR; NSGA-II); Population size = 40

Opt. Meth.	C^*_1	C^*_2	PC/PM	TD (veh-h/h)	CF	NS (stops/h)	t_1(s)	t_2(s)	t_3(s)
GPR	TD	CF	0.9/0.0001	9.45	1.26	450.93	26	21	17
		NS	0.8/0.001	10.35	1.17	442.08	29	17	18
NSGA-II	TD	CF	0.9/0.0001	9.23	1.33	450.92	27	20	17
		NS	0.8/0.001	10.18	1.23	468.60	31	18	15

*C** Criterion considered in the objective function

In particular, three kinds of evaluations can be made: (i) with respect to the effectiveness of each multicriteria optimisation, results obtained by combining TD and CF are better than results obtained by combining TD and NS; (ii) with respect to the comparison between NSGA-II and GPR, more satisfactory results can be obtained by the application of the first one in particular with regard to TD and CF; (iii) finally, with respect to the comparison between monocriteria (based on CF) and multicriteria methods (based on TD & CF), no significant differences can be appreciated between the first one ($TD_{mono} = 9.15$ veh; $CF_{mono} = 1.35$; $NS_{mono} = 459.76$ stops/h) and the second one (TD & CF: $TD_{multi} = 9.23$ veh;

$CF_{multi} = 1.33$; $NS_{multi} = 450.92$ stops/h); however CF and TD were better in case of monocriteria optimisation than multicriteria optimisation, while a lower (and better) value of total number of stops can be appreciated in case of NSGA- II than monocriteria optimisation.

6 Conclusions and Research Perspectives

This chapter is addressed to preliminary investigate the GAs application to the monocriteria or multicriteria SSD for a single junction.

In case of monocriteria optimisation three criteria were considered based on total delay minimisation, capacity factor maximisation and total number of stops minimisation. Monocriteria GAs based on CF maximisation and TD minimisation were compared to SIGCAP and SIGSET methods in order to consolidate their effectiveness of the method. In case of multicriteria optimisation the combination of total delay minimisation and capacity factor maximisation and the combination of total delay minimisation and total number of stops minimisation were considered.

Two algorithms were compared: the GPR and the NSGA-II. On the base of obtained results it can be observed that NSGA-II outperforms GPR. Furthermore, monocriteria GAs were compared to multicriteria GAs (in particular with NSGA-II method). Results show that no significant differences can be appreciated with respect to the values of indicators ($TD_{mono} = 9.14$ veh; $CF_{mono} = 1.35$ *vs.* $TD_{multi} = 9.23$ veh; $CF_{multi} = 1.33$) while better results of total number of stops are carried by the application of multicriteria optimisation ($NS_{multi} = 450.92$ - stops/h) than monocriteria optimisation ($NS_{mono} = 459.76$ stops/h).

In future papers, procedures with cycles length optimisation will be addressed. Furthermore obtained results will be compared to other metaheuristics (such as the Simulated Annealing). In particular, the interest in further investigations about application of other metaheuristics to SSD, is related to the complexity of GAs implementation; in fact in this case an high number of parameters (such as crossover rate, mutation rate, population size) must be set preliminarily.

Finally, the integration of multicriteria SSD within the traffic assignment problem with variable demand (see [10]) will be investigated.

References

1. Akcelik, R.: Traffic Signals: Capacity and Timing Analysis. Research Report ARR No. 123. ARRB Transport Research Ltd., Vermont South, Australia (1981)
2. Allsop, R.E.: Delay minimising settings for fixed time traffic signals at a single junction. J. Inst. Math. Appl., **8**, 164–185 (1971)
3. Allsop, R.E.: Estimating the traffic capacity of a signalized road junction. Transp. Res., **6**, 245–255 (1972)

4. Baker, J.E.: Adaptive selection methods for genetic algorithms. Proceedings of the 3rd International Conference on Genetic Algorithms and Applications. In: Grefenstette, J.J. (ed.), New Jersey. Lawrence Erlbaum: Hillsdale, pp. 100–111 (1985)
5. Benekohal, R.F., Waller, S.T.: Multiobjective traffic signal timing optimization using non-dominated sorting genetic algorithm. In: Intelligent Vehicle Symposium, Proceedings IEEE 9, pp. 198–203 (2003)
6. Cantarella, G.E., Improta, G.: A nonlinear model for control system design at an individual signalized junction. Proceedings of the Conference of the Operation Research Italian Society , pp. 709–722 (1983)
7. Cantarella, G.E., Improta, G.: Capacity factor or cycle time optimization for signalized junctions: a graph theory approach. Transp. Res. B, **22B**, 1–23 (1988)
8. Cantarella, G.E., Di Pace, R., Memoli, S., de Luca, S.: The network signal setting problem: the coordination approach vs. the synchronisation approach. Computer Modelling and Simulation (UKSim), 2013 UKSim 15th International Conference, pp. 575–579 (2013a). doi:10.1109/UKSim.2013.99
9. Cantarella, G.E., Di Pace, R., Memoli, S., de Luca, S.: The application of multicriteria genetic algorithms for signal setting design at a single junction. 8th EUROSIM Congress on Modelling and Simulation, pp. 472–477 (2013b). doi:10.1109/Eurosim.2013.85
10. Cantarella, G.E., de Luca, S., Di Gangi, M., Di Pace, R.: Stochastic equilibrium assignment with variable demand: literature review, comparisons and research needs. WIT Trans. Built Environ. **130**, 349–364 (2013)
11. Ceylan, H., Bell, M.G.H.: Genetic algorithm solution for the stochastic equilibrium transportation networks under congestion. Transp. Res. Part B **39**, 169–185 (2005)
12. Deb, K., Pratap, A., Agarwal, S., Meyarivan, T.: A fast and elitist multiobjective genetic algorithm: NSGA-II. IEEE Trans. Evol. Comput. **6**(2), 182–197 (2002)
13. Foy, M.D., Benekohal, R.F., Goldberg, D.E.: Signal timing determination using genetic algorithms. Transp. Res. Rec. **1365**, 108–115 (1992)
14. Gallivan, S., Heydecker, B.G.: Optimising the control performance of traffic signals at a single junction. Transportation Research B, **8**, 357-370 (1988)
15. Gazis, D.C.: Optimal control of a system of oversaturated intersections. Oper. Res. **12**(6), 815–831 (1964)
16. Girianna, M., Benekohal, R.F.: Using genetic algorithms to design signal coordination for oversaturated networks. Intell. Transp. Syst. **8**, 117–129 (2004)
17. Goldberg, D.E.: Genetic Algorithms in Search, Optimization, and Machine Learning. Addison-Wesley, Reading (1989)
18. Holland, J.H.: Adaptation in Natural and Artificial System. The University of Michigan Press, Ann Arbor (1975)
19. Improta, G., Cantarella, G.E.: Control system design for an individual signalized junction, Transp. Res. B, **18**, 147–168 (1984)
20. Michalopoulos, P., Stephanopolos, G.: Oversaturated signal system with queue length constraints. Transp. Res. **11**, 413–421 (1977)
21. Park, B., Messer, C.J., Urbanik II, T.: Traffic signal optimization program for oversaturated conditions: genetic algorithms approach. Transp. Res. Rec. **1683**, 133–142 (1999)
22. Putha, R., Quadrifoglio, L., Zechman, E.: Comparing ant colony optimization and genetic algorithm approaches for solving traffic signal coordination under oversaturation conditions. Comput Aided Civil Infrastruct. Eng. **27**, 14–28 (2012)
23. Renfrew, D., Xiao-Hua, Yu.: Traffic Signal Optimization Using Ant Colony Algorithm, pp. 1–7. IEEE, Brisbane (2012)
24. Sun, D., Benekohal, R.F., Waller, S.T.: Multiobjective traffic signal timing optimization using non-dominated sorting genetic algorithm in Intelligent Vehicle Symposium. Proc. IEEE **9**, 198–203 (2003)

25. Sun, D., Benekohal, R.F., Waller, S.T.: Bi-level programming formulation and heuristic solution approach for dynamic traffic signal optimization. Comput. Aided Civil Infrastr. Eng. **21**, 321–333 (2003)
26. Teklu, F., Sumalee, A., Watling, D.P.: A genetic algorithm approach for optimising traffic control signals considering routing. J. Comput. Aided Civil Infrastr. Eng. **22**, 31–43 (2007)
27. Webster, F.V.: Traffic signal settings. Road Research Technical Paper, 39, HMSO, London

Part VI
Transportation Planning

Transportation planning traditionally involves the evaluation, assessment, design and positioning of transport facilities. However, planners are increasingly expected to adopt multi-disciplinary approaches, moving from mere technical analysis to promoting sustainability through integrated transport policies. *Carotenuto and Paradisi* argue that demand-responsive transport systems (DRTS) seem to be the solution for the trade-off between flexibility and efficiency. *Mesa et al.* examine a transit line in which a fleet of trains circulates and stops at stations according to a pre-set timetable known by users arriving at these stations to board the trains. The authors determine service rescheduling, forced by any kind of occurrence, in order to minimize the loss of passengers who require transfers between different lines at interchange stations. *Kalic et al.* present results from both fuzzy logic and sensitivity analysis to model trip generation and trip distribution processes in the field of air transport. The chapter of *Castaldi et al.* deals with random utility models, used to represent the choice of route alternatives when changes occur in a network. They extend results available in the literature on the computation of transition probabilities and conditional welfare measures in cases of imperfect before/after correlation of random terms and changing choice sets.

Testing a Heuristic for a Flexible Transport System

Pasquale Carotenuto and Leonardo Paradisi

Abstract The concept of innovation in transport systems requires the satisfaction of two main objectives: the service flexibility and the costs minimization. The demand responsive transport systems (DRTS) seem to be the solution for the trade-off between flexibility and efficiency. They require the planning of travel paths (routing) and customers pick-up and drop-off times (scheduling) according to received requests, respecting the limited capacity of the fleet and time constraints (hard time windows) for each network's node, and the service time of the system. Even considering invariable conditions of the network a DRTS may operate according to a static or to a dynamic mode. In the static setting, all customers' requests are known beforehand and the DRTS returns routing and scheduling solutions by solving a Dial-a-Ride Problem (DaRP) instance which derives from the Pick-up and Delivery Problem with Time Windows (PDPTW). In reality, the static setting may be representative of a phase of reservation occurred the day before the execution of the service. In the dynamic mode, customers' requests arrive when the service is already running and, consequently, the solution may change whilst the vehicle is already travelling. In this mode it is necessary that the schedule is updated when each new request arrives and that this is done in a short time to ensure that the potential customer will not leave the system before a possible answer. In this work, we use an algorithm able to solve a dynamic multi-vehicle DaRP by managing incoming transport demand as fast as possible. The heuristics is a greedy method that tries to assign the requests to one of the fleet's vehicles finding each time the local optimum. The feature of this work is that, in addition to finding a plan schedule, it can be used for sizing the number of vehicles required to satisfy a percentage of demand that may be established before. Vehicles will be employed

P. Carotenuto (✉)
Consiglio Nazionale delle Ricerche—Istituto per le Applicazioni del Calcolo M. Picone, via dei Taurini 19, 00185 Roma, Italy
e-mail: carotenuto@iac.cnr.it

L. Paradisi
University of Rome "Tor Vergata", via del Politecnico 1, 00133 Roma, Italy

J. F. de Sousa and R. Rossi (eds.), *Computer-based Modelling and Optimization in Transportation*, Advances in Intelligent Systems and Computing 262,
DOI: 10.1007/978-3-319-04630-3_25,

only when strictly necessary, in this way the costs will be minimized. The work is enriched by a series of tests with different values of the fleet's vehicles and their capacity, of time windows and of incoming requests' number. Finally, a set of performance indicators evaluate the solution planned by the heuristics.

Keywords Public transport · Demand responsive transport systems · Dial-a-ride problem · Heuristics

1 Introduction

Within the city logistics problems, there are many solutions aimed to the realization of a flexible transport system. Car sharing and Responsive Demand Transport service are examples of services capable to adapt themselves to the requirements of a specific user and satisfy the non-systematic part of mobility demand for which, the conventional transport services cannot be both of a good quality and economical in comparison to the private transport [1, 11].

Demand Responsive Transport provides a fleet composed of small buses. Travelers book their trip through an operating center indicating the origin and the destination of the trip, the desired departure and arrival time and any other special requirements. The operating center evaluates the travelers requests and if accepted, it defines the service schedule and sends it to the bus drivers [12, 18].

The problem of working out optimal service paths and times is called a Dial-a-Ride Problem (DaRP), which derives from the well-known Vehicle Routing Problem (VRP) [2, 6, 17, 21], with the addition of precedence constraints between pick-up and drop-off locations and time windows [10, 16].

Their computational complexity makes both DaRP and VRP as NP-hard problems, so attempts to develop optimal solutions have been limited to simple and small-size problems.

The distinguishing feature of heuristic methods is to find a solution in a polynomial time, which makes them suitable for dynamic applications. The heuristic method finds a solution that is not necessarily the best but may be acceptable according to the parameters set by those who are looking for a good plan. Who implements algorithms of this kind aims to find a good solution in a short time rather than find an optimal in a very long time. This is the case of a Demand Responsive Transport System (DRTS) modeled with a DaRP.

Clearly, the DaRP has to take into account specific constraints as we are considering people instead of goods. As a consequence, each customer specifies a possible pickup and delivery time, as well as an upper bound on the riding time and narrow time windows [8, 9]. Furthermore, in this context, customers often formulate two requests per day, specifying an outbound request from the pick-up point to a destination and an inbound request for the return trip.

A further and fundamental analysis regards the static or dynamic nature of the problem. A DRTS, in fact, may operate according to a static or to a dynamic mode.

In the static setting all the customer requests are known beforehand, and the DRTS produces, by solving a DaRP instance, the tour each vehicle has to make, respecting the pick up and delivery time windows while minimizing the solution cost [5, 13, 15, 22]. In the dynamic mode, the customer requests arrive over time to a control station and, consequently, the solution may also change over time [3, 4, 7, 14].

In practice, the dynamic approach can be tackled in two ways. The first one relates to all those procedures which face a dynamic problem by dividing the overall time frame into smaller intervals for which an optimization procedure is adopted in order to solve several static problems [19]. Every time, the algorithm runs with the information relating to the specific static problem. The second one refers to the insertion procedures: when a new request arrives, the algorithm tries to assign it to one of the operating vehicle without excluding travelers requests already in the schedule or violating the related constraints and trying to make the inclusion in the best possible position.

2 The Solution Approach

Our designed algorithm (see also [7] for the general algorithmic structure) tries to solve a dynamic DaRP. The optimization procedure (see Fig. 1) starts when arrives a new request, but without utilizing any information on future transport requests. This is commonly referred to as an on-line algorithm. Each time a request is accepted, the system updates the current solution by changing the schedule of the selected vehicle. To accept a new request, the algorithm must find an admissible schedule for it, does not refuse any transport requests previously accepted and does not violate any operational constraint. At any moment must be known the position and the remaining capacity of the vehicles.

Regarding the operational mode of the service, our algorithm can operate in two ways in "booking" mode or in "running" mode:

- Booking mode: the algorithm processes the requests on-line, considering the requirements of each customer without knowing or considering request that will be subsequently received, the pickup and delivery service is not yet operational.
- Running mode: the algorithm works as before but it must consider the position of the vehicle to evaluate the incoming requests because the service has already started.

2.1 The Booking Mode

During the booking phase, customers who will use the DRTS have the opportunity to reserve their trip in the previous day, or also they have the possibility to book their travel on a regular basis (subscription) for several day. The system immediately process the customer demand and gives them a fast response, refusing or accepting

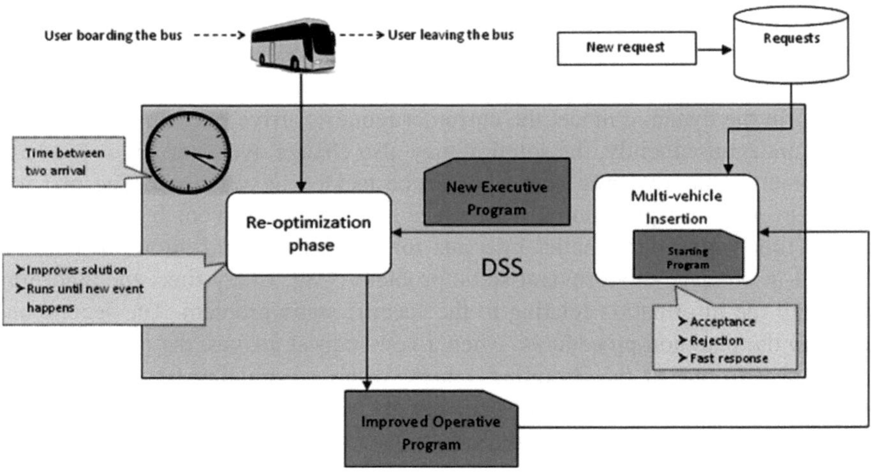

Fig. 1 Algorithm architecture

the request, evaluating the possibility of carrying out the service. During this phase, the requests arrive dynamically, but the vehicles are still in the depot. The algorithm in assessing the transportation requests and preparing the relative schedule, should not take into account the position of the vehicles. The algorithm work in this way until the end of the booking phase. By this time, the set of requests is completely known and all the vehicles are assigned. A re-optimization problem can run, solving a static DARP. Usually there is a large amount of time between the end of booking phase and the start of the running phase. Therefore it is possible to run an algorithm to solve a static DARP (also for the whole night), in order to evaluate the best schedule based on the requests arrived to the control center.

We follow now the procedure of processing a request to introduce another feature of our algorithm in the booking phase. As shown in the Fig. 1, when a customer makes a new booking to demand a trip for the next day, the request will be immediately evaluated by the algorithm and the results sent to the customer. Therefore the customer in very short time will know if his request has been accepted or refused. If accepted, the request is inserted in the last schedule that takes in account all the previous requests. If the request is the first, it will be inserted in an empty schedule.

We have realized until now a sort of pre-processing task that takes into account the arrived requests and the vehicle availability for the next day. Supposing now that we have enough time between the arrival of a request and the following one and to have a fast algorithm to evaluate again the schedule and try to improve the solution as in our case. Following the procedure showed in Fig. 1 we can see that the schedule evaluated in the first phase is the starting point for the re-optimized phase. In the re-optimized phase we use again the same algorithm as the first phase but using a different arrival order of the accepted requests. In this case the algorithm takes a much longer time to improve the solution, but we have time, because we have already notified the response for the last request and we have to wait a new arrival. We are able, in this way to improve the solution found. For this reason, different strategies have been tested and compared, as can be seen from the tests discussed in the following.

2.2 The Running Mode

In the running phase the requests arrive to the control center inside the execution time of the service, in this way the system has to face a dynamic pickup and delivery problem with time windows considering also the vehicles position. In fact, the transport requests come at a certain time of the service and depends on the particular situation in which are located the vehicles at that precise time if they will be accepted and by which vehicle of the fleet they will be served. Also in this case, the system is in a situation of incomplete information. When the vehicles leave the store to reach and satisfy customers request who have already booked their trip, they do not know anything about the demands that will come in the future. The evaluation of the customer's request will be done in ON-LINE mode.

When the operative control center receive a new request, it must determine whether there is the possibility to insert the request in the existing tour and then notify the acceptance, rejection, or the negotiation of the proposal. To do so, the insertion procedure in the operative schedule is called to assign the request to a vehicle in the best way possible. Now both dynamic factors come into play, and the answer to customer must be given in short time. Every time a new request is assigned, the last operative schedule found has to be updated. When the travel allocation procedure is called, it must take account the request arrival time, the fleet location and the sequence of stations that they will visit. It also needs information on the road network and the people who are on the vehicles. In fact, the optimization should take into account that some customers cannot be served by another bus, as they are already on board. In the "running phase", at least one vehicle should change its route to allow the insertion of the new request and allow the trip. It is essential, therefore, a fast and efficient form of communication between the vehicle and the operations center. Modern technologies are providing more sophisticated devices that calculate different parameters of state in real time, overcoming, in performance, the old radio communication systems.

In particular, in the "running phase" the insertion algorithm must take into account the instant in which the requests arrive to the system and also the position of the vehicles at that instant. That because, operatively, we have to assume that the vehicles while they are moving between two stops cannot change their path before reaching the destination node toward which they are directed. The algorithm must ensure that the request made by the user is inserted in the new operative schedule not only after its arrival time but also to the end of the running path of the vehicle on which the request will be inserted.

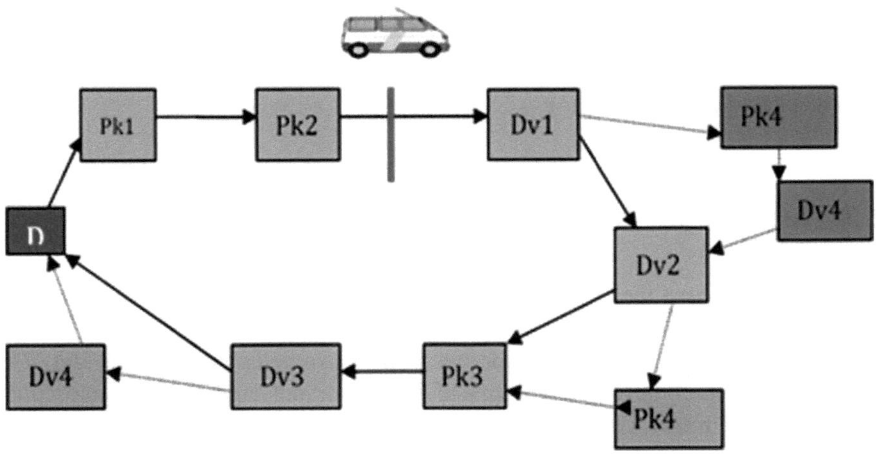

Fig. 2 Time management: insertion point in the running mode

As we can see in Fig. 2 the request can be inserted (both the phase of pickup and delivery) only in the positions subsequent to delivery1, after the bus has led the customer's request 1 to the destination. Figure 2 hypothesize that there are only two possible position to pick up and drop off the customer related to the request 4 (green and blue blocks), the algorithm will choose the one most convenient between them.

3 Test

To validate the algorithm, a series of preliminary tests have been run using random instances and a test network with horizontal and vertical arcs with different travel times: 2 min for horizontal arcs and 1 min for vertical arcs (see [20]. Every test has been done on 20 random instances considering the average value of the fitness function. To build the set of request, a procedure randomly generates a pair of pickup and delivery nodes and the pick up time requested by the user, then the delivery time is established by adding to the pickup time the minimum time necessary to travel between the two nodes.

3.1 Comparison Between the Strategies

In this section we compare three different strategies which aim to increase the target function by reducing the user waiting time at the bus stop. Those strategies are:

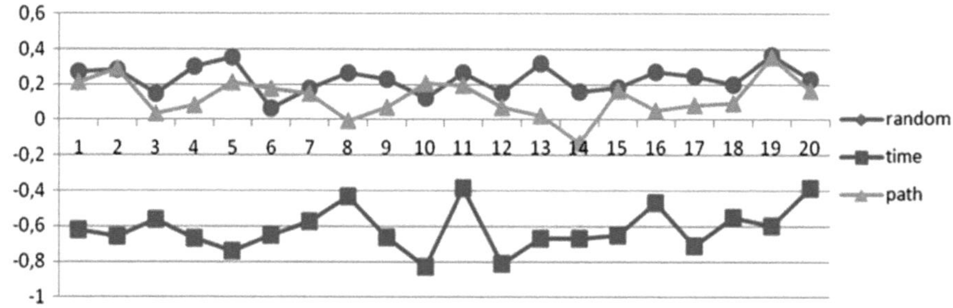

Fig. 3 Fleet of 3 vehicles

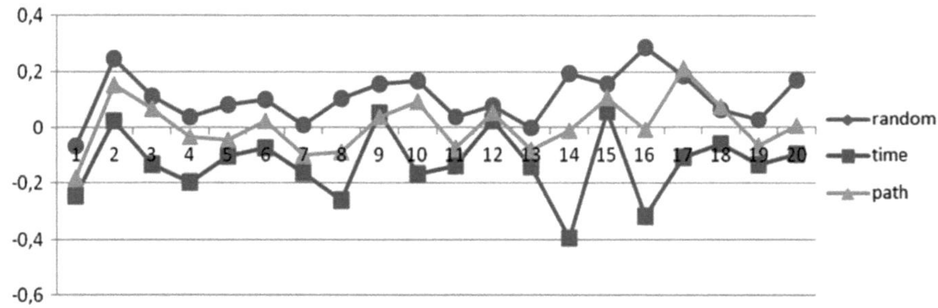

Fig. 4 Fleet of 4 vehicles

- Rearranging the requests randomly (strategy 1)
- Rearranging the requests according to the increasing pickup time (strategy 2)
- Rearranging the requests according to growing path length (strategy 3)

In these tests we can see which strategy increases the objective function value and which finds the best schedule during the re-optimization phase. All the results (related the 3 strategies) are expressed as percentage related to the values realized with the "natural" arrival order of the requests (x axe). The diagrams refer to a system with time windows = 10 min, a service time = 1080 min (the whole day), a number of requests = 250 and a variable number of vehicles for the fleet assigned to the service.

The first diagram (see Fig. 3) refers to a fleet with 3 vehicles, the second (see Fig. 4) with 4, the third (see Fig. 5) with 5 and the last one (see Fig. 6) refers to a fleet with 10 vehicles.

The DRTS procedure finds a first solution through the requests arrival by natural order. Then it runs the increasing time pick up strategy which will find most times a better solution respect the first one. If the new request isn't arrived, the DRTS will execute the increasing lenght path strategy. If, even in this case, while the DRTS is processing a new plan, a request hasn't arrived, the random strategy is launched. This one is launched a fixed number of time. This because, unlike the other strategies, every times it's launched it finds a different solution,

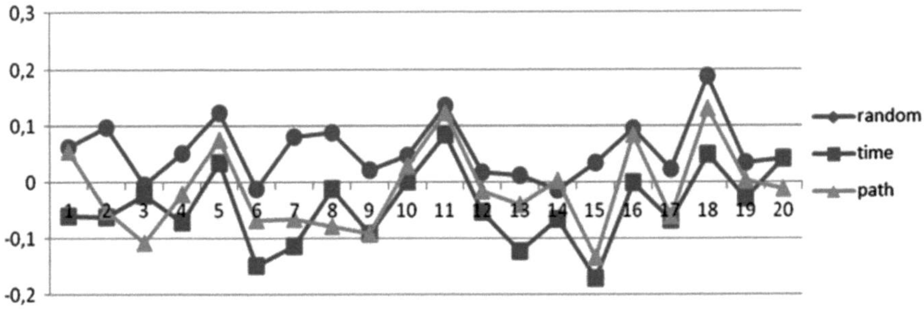

Fig. 5 Fleet of 5 vehicles

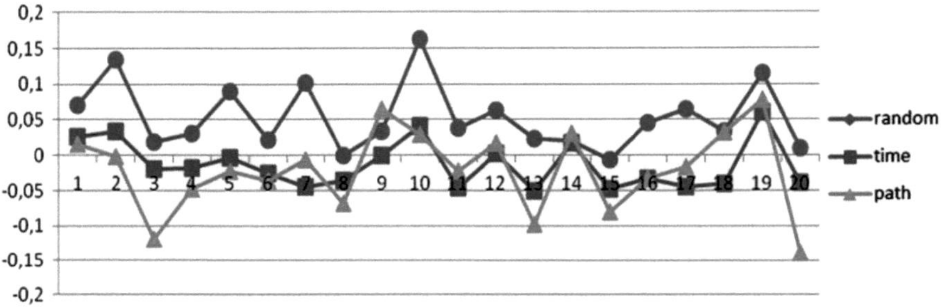

Fig. 6 Fleet of 10 vehicles

which could be better than the last one. The reshuffling number is determinated by the arrival of the new request.

In our cases to evaluate improvement of the reshuffling strategy we have performed it a fixed number of times. It means that in the diagrams it's shown the best solution found in the n instances.

As shown in the diagrams, the random strategy is the best in almost all the cases. In some events it can improve the solution of 40 % and it never worsens in relation with the natural order of the request. This is an important result since in the re-optimization phase we can decide to privilege one strategy rather than another to save precious seconds.

3.2 Speed Test

In the diagram of Fig. 7, we can see how the processing time for each request increases with the number of vehicles used. The reason is because the algorithm has to run tests for each vehicle and then compare the objective function found between all the vehicles scheduled. So it will take more time before it can find the best objective function from all the schedules. We can also observe how the processing time grows if we increase the number of request from 100 to 250.

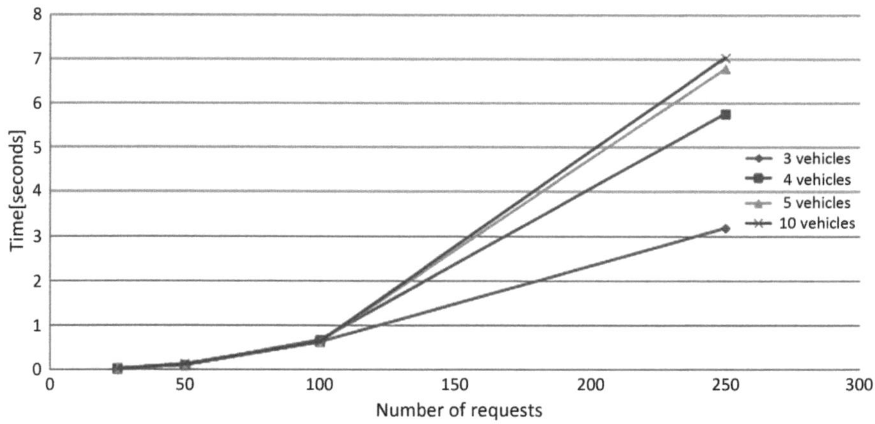

Fig. 7 Speed test

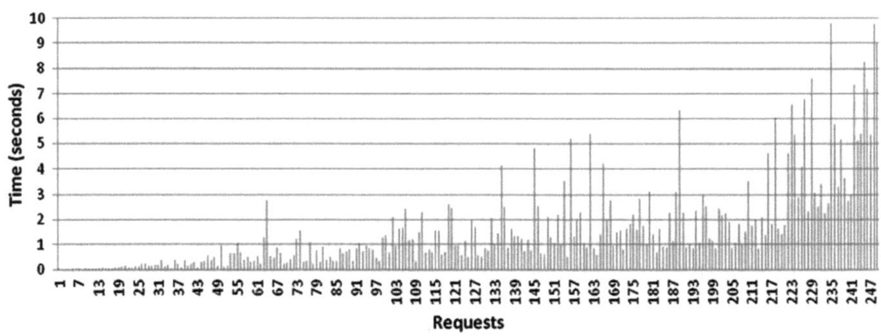

Fig. 8 Insertion speed test

Other tests on the evaluation speed are being made. In Fig. 8 we can see a graph showing the request insertion time in the schedule when there are n−1 requests already accepted. With the increase of accepted requests in the schedule, the computation time increases but average values remain acceptable.

The example refers to a system with time windows = 10 min, a service time = 1080 min (the whole day), a number of requests = 250 and a fleet with 3 vehicles.

3.3 Test on Transportation Request Refusal

Table 1 shows the results of the test on the rejected percentage varying the number of vehicles and the number of requests.

Table 1 Rejected percentage according to the number of requests end vehicles

Rejected %	3 vehicles (%)	4 vehicles (%)	5 vehicles (%)	10 vehicles (%)
25 requests	0, 0	0, 0	0, 0	0, 0
50 requests	0, 0	0, 0	0, 0	0, 0
100 requests	0, 0	0, 0	0, 0	0, 0
250 requests	22, 5	5, 4	0, 5	0, 0

The number of rejected requests is relevant in just one case, with 3 vehicles and 250 requests. In all other instances the number of rejection is really low. The DRTS responds very well in a system with an average number of requests, even if we aren't using a big fleet of vehicles. This feature makes this DRTS a good tool whenever we would like to implement it in a real system, indeed it doesn't need a big number of vehicles.

4 Conclusion

In this chapter we propose a re-optimization phase for the DaRP and described its implementation. The DaRP is part of a DRTS whose goal is to support a public Service Provider in the management of transport activities. In particular, for the DRTS we approached the problem using a solution architecture based on a two-stage algorithm. After setting the parameters of the algorithm, (population size, crossover rate and mutation rate), we have implemented some computational experiments and obtained some interesting preliminary results.

After the tests we have observed the system robustness in a network of this type. If equipped with a greater number of vehicles, the number of requests satisfied is 100 %. Even in the case of 250 requests, with a fleet of 10 vehicles, there aren't users rejected. The curve relating to the use of 3 vehicles grows only after 100 requests, when the system has reached a saturation state of the available resources. Moreover, up to 100 requests, the processing time is less than a second, so the software responds in real-time to the needs of the users.

With a number of requests equal to 250, the calculation time increases, in particular with a number of vehicles greater than 3. This slowness is due to a network computational structure with 25 nodes, horizontal arc cost = 2 and vertical arc cost = 1. A spatial arrangement which is quite difficult to have in reality but that turns out to be a good challenge to analyze how the algorithm interacts in such critical situations.

References

1. Ambrosino, G., Nelson, J.D., Romanazzo, M.: Demand Responsive Transport Services: Towards the Flexible Mobility Agency. ENEA, Rome (2004)
2. Barrie, M.B., Ayechew, M.A.: A genetic algorithm for the vehicle routing problem. Comput. Oper. Res. **30**, 787–800 (2003)
3. Beaudry, A., Laporte, G., Melo, T., Nickel, S.: Dynamic transportation of patients to hospitals. OR Spectr. **32**, 77–107 (2010)
4. Berbeglia, G., Cordeau, J.-F., Laporte, G.: Dynamic pickup and delivery problems. Eur. J. Oper. Res. **202**, 8–15 (2010)
5. Bergvinsdottir, K.B., Larsen, J., Jørgensen, R.M.: Solving the Dial-a-Ride Problem using Genetic algorithms, IMM-Technical report-2004-20. Informatics and Mathematical Modelling, Technical University of Denmark (2004)
6. Bodin, Lawrence, Golden, Bruce, Assad, Arjang, Ball, Michael: Routing and scheduling of vehicles and crews: the state of the art. Comput. Oper. Res. **10**(2), 63–210 (1983)
7. Carotenuto, P., Cis, C., Storchi, G.: Hybrid genetic algorithm to approach the DaRP in a demand responsive passenger service, information control problems in manufacturing 2006. Proceedings of the 12th IFAC International Symposium, Elsevier Science, vol. 3, pp. 349-354. ISBN: 978-0-08-044654-7 (2006)
8. Cis, C.: Logistica urbana: algoritmi genetici ibridi per la realizzazione di servizi innovativi per il trasporto di persone. MBA Thesis, University of Rome Tor Vergata (2004)
9. Coslovich, L., Pesenti, R., Ukovich, W.: A two-phase insertion technique of unexpected customers for a dynamic dial-a-ride problem. EJOR **175**, 1605–1615 (2006)
10. Cordeau, J.-F., Laporte, G.: The dial-a-ride problem: models and algorithms. Ann. Oper. Res. **153**(1), 29–46 (2007)
11. Cubillos, C., Guidi-Polanco, F., Demartini, C.: Multi-agent infrastructure for distributed planning of demand-responsive passenger transportation service. In: Proceeding IEEE International Conference on Systems, Men and Cybernetics, 2013–2017 (2004)
12. Horn, M.E.T.: Fleet scheduling and dispatching for demand-responsive passenger services. Transp. Res. Part C **10**, 35–63 (2002)
13. Jaw, J.J., Odoni, A.R., Psaraftis, H.N., Wilson, N.H.M.: A heuristic algorithm for the multi-vehicle advance request dial-a-ride problem with time windows. Transp. Res. Part B **20**, 243–257 (1986)
14. Jih, W.J., Hsu, Y.: Dynamic vehicle routing using hybrid genetic algorithms. In: Proceedings of the 1999 IEEE Conference on Robotics and Automation, pp. 453–458 (1999)
15. Jorgensen, R.M., Larsen, J., Bergvinsdottir, K.B.: Solving the dial-a-ride problem using genetic algorithms. J. Oper. Res. Soc. **58**, 1321–31 (2007)
16. Nanry, W., Barnes, J.W.: Solving the pickup and delivery problem with time windows using reactive tabu search. Transp. Res. Part B **34**, 107–121 (2000)
17. Prins, C.: A simple and effective evolutionary algorithm for the vehicle routing problem. Comput. Oper. Res. **31**, 1985–2002 (2004)
18. Quadrifoglio, L., Dessouki, M.M., Palmer, K.: An insertion heuristic for scheduling mobility allowance shuttle transit (MAST) services. J. Sched. **10**, 25–40 (2007)
19. Savelsbergh, M.W.P., Sol, M.: The general pickup and delivery problem. Transp. Res. **29**, 17–29 (1995)
20. Taniguchi, E., Thompson, R.G., Yamada, T., van Duin, R.: City Logistics: network modelling and intelligent transport system. Pergamon, Amsterdam (2001)
21. Toth, P., Vigo, D.: The Vehicle Routing Problem. SIAM Monographs on Discrete Mathematics and Applications, Philadelphia (2002)
22. Uchimura, K., Saitoh, T., Takahashi, H.: The dial-a-ride problem in a public transit system. Electron. Commun. Jpn. Part III **82**(7), 30–38 (1999)

Rescheduling Railway Timetables in Presence of Passenger Transfers Between Lines Within a Transportation Network

Juan A. Mesa, Francisco A. Ortega, Miguel A. Pozo and Justo Puerto

Abstract The problem of coordinating transfers consists of determining timetables which ensure the transfer of passengers between trains from different line runs at interchange stations. Two strategies can be considered: (1) Forcing the line runs to be synchronized; that is, a solution can be accepted only if there exists a connection between them, while the goal is minimizing travel times for passengers by using the minimum number of vehicles needed. (2) Minimizing an objective function that penalizes the lack of synchronization between line runs. The problem of transfer coordination turns out to be NP-hard even in the simple case of periodic timetables. Therefore, the problem is usually treated sequentially in two stages: first, determine the frequency of service according to the rate of demand, and then solve the problem of coordination by means of heuristics. This chapter considers a transit line where a train fleet circulates and stops at the stations according to a predetermined timetable which is known by the users. At any instant, passengers arrive at different stations in order to board these vehicles according to an assumed deterministic model of arrivals. In this scenario, a service rescheduling forced by an incidence is determined in order to minimize the loss of passengers who require transfers between different lines at the interchange stations. A case study consisting of a railway line with several equi-spaced stations, where it is possible a connection to

J. A. Mesa
Higher Technical School of Engineering, University of Seville, Camino de los Descubrimientos s/n, 41092 Seville, Spain
e-mail: jmesa@us.es

F. A. Ortega (✉)
Higher Technical School of Architecture, University of Seville, Av. Reina Mercedes 2 41012 Seville, Spain
e-mail: riejos@us.es

M. A. Pozo · J. Puerto
Faculty of Mathematics, University of Seville, c/Tarfia s/n, 41092 Seville, Spain
e-mail: miguelpozo@us.es

J. Puerto
e-mail: puerto@us.es

J. F. de Sousa and R. Rossi (eds.), *Computer-based Modelling and Optimization in Transportation*, Advances in Intelligent Systems and Computing 262,
DOI: 10.1007/978-3-319-04630-3_26,

other lines at intermediate stations is analyzed for different scenarios where the loss of transfers is penalized.

Keywords Transit network · Transport scheduling · Disturbance management · Schedule synchronization

1 Introduction

Timetable design is a central problem in railway planning with many interfaces with other classical problems: line planning, vehicle scheduling, and delay management. The single-line Train Timetabling Problem (TTP) is devoted to obtaining and optimizing timetables of periodic and non-periodic heterogeneous trains that share a railway line with single and multiple track sections.

Given a railway infrastructure provided with different sections along a single transit line, the Train Timetabling Problem (TTP) consists of computing timetables that satisfy existing constraints and that optimize a single/multicriteria objective function for trains of both, passengers and cargo. The railway line may be occupied by other trains whose priority is higher than that of the new ones, and the new trains to be added may belong to different train operators. The requirement for periodicity of the timetables leads to the classification of TTP into Periodic (or cyclic) Train Timetabling and, on the other hand, Non-Periodic Train Timetabling.

In Periodic Timetabling, the timetable is easy to remember for the passengers although its solutions can become inefficient when planning resources such as crews and rolling stock. The mathematical model called Periodic Event Scheduling Problem (PESP) by Serafini and Ukovich [16] is the most widely used in the literature. In PESP, the events are scheduled for one cycle in such a way that the cycle can be repeated according to periodic time windows constraints. The PESP model has been used by authors in [10, 14, 15]. Interesting contributions on efficient railway operation management, oriented to European real contexts, can be found in the ARRIVAL project (http://arrival.cti.gr/, 2009).

Non-Periodic Train Timetabling is especially relevant on heavy-traffic, long-distance corridors where the capacity of the infrastructure is limited due to great traffic densities, as well as in presence of disturbances that can affect to the operativeness of train transit. The non-periodic train timetabling problem has been considered by most authors: [1, 2, 4, 5, 7–9, 11, 13, 17, 18].

Planners usually use running maps as graphic tools to help them in the planning process. A running map is a time–space diagram where possible crossings of trains can be observed. Figure 1 shows a time–space diagram that synthesizes the train expeditions of a piece of the C4 line that belongs to the Madrid commuter railway network (see Fig. 2). The names of the stations (cantons where passengers can board or alight on/from trains) are presented on the left side, and the vertical line represents train speeds when it passes through the sequence of tracks between consecutive stations.

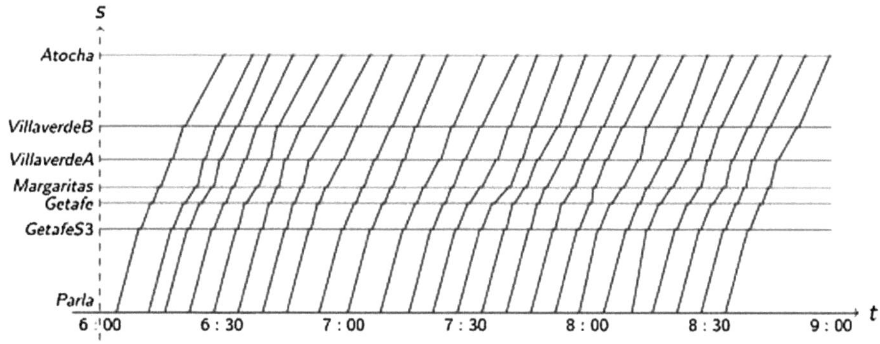

Fig. 1 Twenty-five instances of train schedules along the transit corridor

Fig. 2 Line C4 (Parla-Atocha)

The slope of the polygonal line associated with a train corresponds with the commercial speed of that train (with a stop between stations), and the horizontal segments can be viewed as the stopping time at stations. A transit corridor of high

traffic density will generate in a labyrinthine tangle of polygonal lines, each of which will correspond to the hours of operation of a train, making infeasible a non-automated assessment of the possible alternatives.

A conflict of crossing between trains could take place along the transit corridor. Graphically, a conflict can be represented by means of a pair of two rectilinear segments that intersect in the same canton. If such canton had multiple pathways, it is possible to solve the conflict by imposing a waiting time to some of trains.

In order to be feasible, a timetable has to fulfill a set of constraints that can be classified into three main groups, depending on whether they are concerned with:

- User Requirements (parameters of trains to be scheduled): time windows for departure and arrival times, maximum delay.
- Traffic constraints: running time, crossing, commercial stop, overtaking on the track section, delay for unexpected stop, reception time, expedition time, simultaneous departure.
- Infrastructure constraints: network topology, finite capacity of stations, closing time, headway time.

Many references consider Mixed Integer Problem formulations in which the arrival and departures times are represented by continuous variables and there are binary variables expressing the order of the train departures from each station. The variables chosen to formulate the model must be able to formally express these restrictions, so that only feasible timetables can be considered as possible solutions. There are two main criteria to assess the quality of the solutions: Minimize operating costs (point of view of the operator) and Minimize riding and transfer times (perspective of passengers). Moreover, other complementary objectives can be used, for example: minimize the passenger waiting time in the case of changeovers, balance the delay of trains in both directions, minimize the average delay of new trains with respect to their optimum, etc. Accidents, strike days and other sources of train delays or cancellations force to modify the scheduled timetable when trains in some sections cannot run according to the initial planning. Rescheduling is the process of updating an existing production plan in response to disruptions or other changes [19]. Rescheduling timetables is especially important in heavily used areas because individual events (delays) can easily impact many other trains causing secondary delays to ripple through the network. In order to manage this domino effect when a train is late and reduce the impact on the other trains, controllers must manually adjust the routing of trains. The effectiveness of the rescheduling and train control system at reducing total delay is highly dependent on the specific circumstances (timetable, train routes, topology of the station and occupation of those tracks located before in the bottleneck area). The modification to the timetable should be performed without introducing inconsistencies. In that case, assessing the feasibility of any modification of the existing timetables will necessarily require a computer-aided procedure. In terms of railway operation production plans, the main decisions that must be addressed in the rescheduling process are:

- Provide new reference times for all trains located at specific points in the network (downtime of trains at stations and reference speed on open track).
- Re-routing trains.
- Re-allocate available resources (staff, rolling stock).

The results presented in this chapter are focused mainly on retiming trains that remain operative (the above first point). For that purpose, it will be necessary to change the departure or arrival times at stations and other reference points. Decisions can also include rerouting trains in affected areas by means of cancelling trains, adding supplementary stops or short turns.

Research on rescheduling algorithms has been underway for many years. See, for example, the survey paper by Cordeau et al. [6] and the recent contribution of Canca et al. [3]. In order to successfully use rescheduling algorithms in dense railway networks, it is necessary to analyze the whole production process to determine how new schedules can be most efficiently implemented. The time it takes to complete a rescheduling process (from the point of time when a given threshold is exceeded until the new production plan is applied) for a large network leads to three important questions regarding implementation of the process. Namely, rescheduling: should it be periodic; could it be interrupt-able; should it be implemented in spite of being an infeasible plan?

Each of these questions must be clarified before to completing the rescheduling process. In this chapter, we present an approach to generate acyclic timetables for single line track. This approach is based on geometrical properties associated to topology of a transit corridor.

The chapter is organized as follows. In Sect. 2, a model of graphical representation for train schedules is introduced and a procedure for estimating the number of users associated to feasible schedules is developed. The formulation of the decision model is introduced in Sect. 3, in addition to the subsequent extended scenarios, where transfers between lines that cross the transit corridor are taken into account. Finally, some conclusions are summarized in Sect. 4.

2 Discretizing Time Horizon and Weighting the Feasible Schedules

2.1 *A Graphical Schedule Representation Based on Timetable-Points*

According to Mesa et al. [12], we assume a canonical time unit $h > 0$ (time taken to travel without stopping between two consecutive sections) for generating a uniform mesh of squares of length h in the first quadrant, that can be used to represent the activity map of trains at section (or station) k. Inside this activity map of the k-th stretch/station, each point will indicate the arriving time (x-coordinate) and the leaving time (y-coordinate) of a train. Arrival time of trains can be seen in

Fig. 3 Three trains passing through the station k

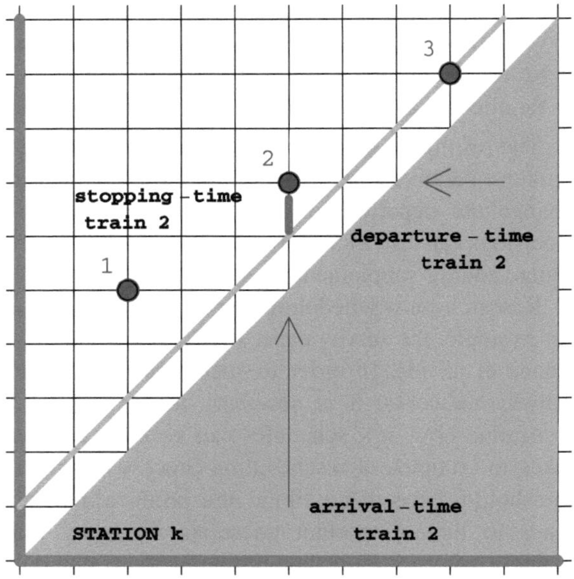

the horizontal axis, while the projection on the vertical axis represents its departure time. In a generic k-station, each train has assigned a unique point (timetable-point, in the following) whose coordinates must be necessarily located in the upper triangle of the first quadrant above the straight line $y = x + h$ (the upper sub-diagonal outlined in green). For instance, Fig. 3 shows data corresponding to three different trains. The first one spends a time equivalent to $2h$ in boarding and alighting passengers. Train 2 uses the minimum time required for that operation, i.e. h. Finally, train 3 does not stop at station k, hence its position is located on the displaced diagonal in the first quadrant.

The sequence of sections (with stops or not for passengers) along the railway line will correspond to a succession of temporary diagrams, where active time-table-points will indicate real arrival-departure timetables. Each timetable-point in the k-th diagram of activity will match to some other feasible timetable-point of the vertical segment that starts from its projection on the diagonal in the $(k+1)$-th activity-map. Moreover, since activity maps have the same homogeneous structure for all stations, a line run can be graphically viewed as a non-decreasing polygonal line that crosses through the sequence of temporary diagrams corresponding to the corridor stations, visiting an only timetable-point per diagram.

2.2 Assuming a Pattern for the Demand Behavior

Assume that arrival/departure times of trains at stations were previously set and are known by users. In that case, we can ensure that these temporal marks mobilize a population of potential travelers towards the station platform, converging in time

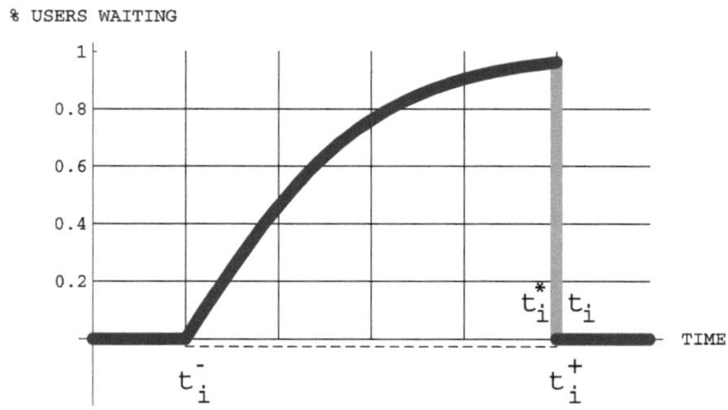

Fig. 4 Usual demand behavior in terms of percentage of user's presence at platform

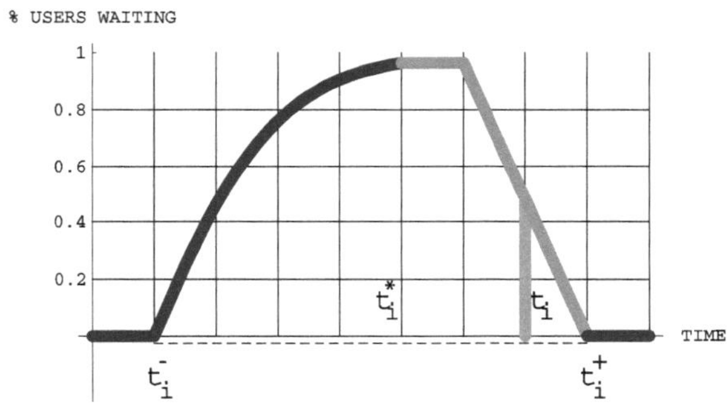

Fig. 5 Demand behavior when train is delayed

with the timely arrival of such train. Figure 4 explains in percentage terms the traveler accumulation on the platform of the station k, due to the imminent arrival of the scheduled train at time t_i. Time interval associated with the arrival of travelers and their consequent accumulation in the platform is denoted by $[t_i^-, t_i^+]$.

Before arriving train t_i at station k, the number of users that reaches the platform with the purpose of boarding on the train is increasing until shortly before the estimated time of train arrival. If the train arrived on time, the whole population placed on platform could be transported as shows Fig. 4. Nevertheless, if train t_i were to be delayed, the reaction of users when they know the existence of such delay would consist of initially waiting along a short certain period of time. Subsequently, the curve that models the percentage of population waiting would appear stabilized. After this period, the traveler population gradually decreases until disappearing. If the train arrived late, only a portion of the population that normally waits could be transported (Fig. 5).

Finally, if the train arrived and departed in advance, only users who were already placed on the platform could take the train. The other passengers will be coming in the usual way, because they were unaware of this schedule change (Fig. 6). The option to wait a certain period of time leads to the possibility of taking the next train.

As illustration, a study case composed of a railway line with several equi-spaced stations, separated from each other by a distance (travel time) equal to 2 min, is considered. It has been assumed an operational time corresponding to the

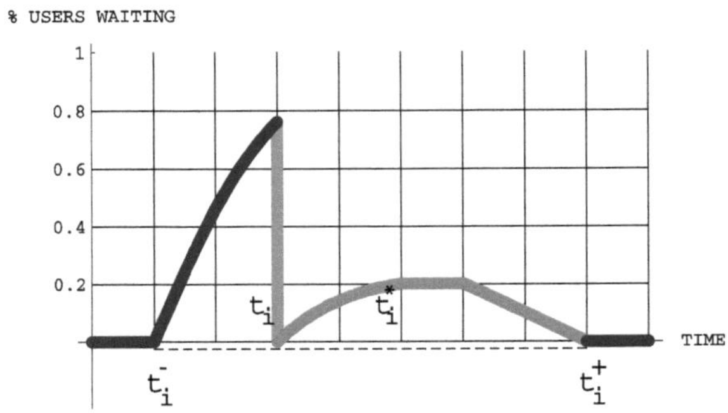

Fig. 6 Demand behavior if train departed in advance

Table 1 Arrival/departure times associated to trains at stations

Station	1	2	3	4	5	6	7
Train 1	8:26/8:28	8:30/8:32	8:34/8:36	8:38/8:42	8:44/8:48	8:50/8:52	8:54/8:56
Train 2	8:38/8:40	8:42/8:44	8:46/8:48	8:50/8:54	8:56/9:00	9:02/9:04	9:06/9:08
Train 3	8:50/8:52	8:54/8:56	8:58/9:00	9:02/9:06	9:08/9:12	9:14/9:16	9:18/9:20

Table 2 Passengers boarding to train at stations

Station number	1	2	3	4	5	6	7
Train 1: Passengers	1,417	1153	664	281	77	39	0
Train 2: Passengers	1,143	756	359	113	23	10	0
Train 3: Passengers	2,131	1,204	488	117	18	7	0

morning interval (8:20–9:30) with partitions of size h (1 min). Initially, there are three vehicles to run along the line and the arrival/departure time at stations are known by users (Table 1).

In the trip distribution along the corridor, it has been assumed that the first stations are mainly trip generators, while the ending station is an attractive destination. In real instances, this setting is commonly associated to a transit line which connects far residential areas with the city center. Attractiveness levels between the first nodes and the final station have been assumed to be decreasing with respect to the distance between them. According to the above consideration, a time-dependent origin-destination matrix has been randomly built for a population of 10,000 users. From it, the number of users accessing each train station is shown in Table 2.

Assume that, as consequence of an incident, the system operator must reduce the fleet size by one unit. Rescheduling train timetables must minimize the loss of users, by introducing advances or delays in the original schedules of the two vehicles which will remain operative. According to the model proposed in the article, the following distribution of passengers that access to stations is shown in Fig. 7 and, if train were not punctual, population waiting for boarding can be deterministically estimated (Fig. 8). Since it is assumed that the user loss for

TIMING	1	2	3	4	5	6	7	8	9	10	11	12	13	14	15	16	17	18
ST 7	0	0	0	0	0	0	0	0	0	0	0	0	0	0	0	0	0	0
ST 6	0	0	0	0	0	0	0	0	0	0	0	0	15	25	32	36	39	0
ST 5	0	0	0	0	0	0	0	0	0	0	30	50	63	72	77	0	21	15
ST 4	0	0	0	0	0	0	0	110	182	229	261	281	0	44	73	92	105	113
ST 3	0	0	0	0	250	430	542	616	664	0	140	233	293	333	359	0	191	316
ST 2	0	0	451	747	942	1069	1153	0	296	490	617	701	756	0	471	780	983	1116
ST 1	554	918	1157	1314	1417	0	447	740	933	1060	1143	0	833	833	1381	1740	2131	0

TIMING	19	20	21	22	23	24	25	26	27	28	29	30	31	32	33	34	35
ST 7	0	0	0	0	0	0	0	0	0	0	0	0	0	0	0	0	0
ST 6	9	6	8	9	10	0	6	6	6	6	7	0	0	0	0	0	0
ST 5	19	21	23	0	7	12	15	17	18	0	0	0	0	0	0	0	0
ST 4	0	46	76	96	108	117	0	0	0	0	0	0	0	0	0	0	0
ST 3	398	453	488	0	0	0	0	0	0	0	0	0	0	0	0	0	0
ST 2	1204	0	0	0	0	0	0	0	0	0	0	0	0	0	0	0	0
ST 1	0	0	0	0	0	0	0	0	0	0	0	0	0	0	0	0	0

Fig. 7 User's presence at platform waiting a punctual arrival

TIMING	1	2	3	4	5	6	7	8	9	10	11	12	13	14	15	16	17	18
ST 7	0	0	0	0	0	0	0	0	0	0	0	0	0	0	0	0	0	0
ST 6	0	0	0	0	0	0	0	0	0	0	0	0	15	25	32	36	39	36
ST 5	0	0	0	0	0	0	0	0	0	0	30	50	63	72	77	72	72	65
ST 4	0	0	0	0	0	0	0	110	182	229	261	281	251	274	255	202	105	113
ST 3	0	0	0	0	250	430	542	616	664	616	683	653	553	333	359	333	484	549
ST 2	0	0	451	747	942	1069	1153	1069	1237	1237	1068	701	756	701	1088	1269	1278	1116
ST 1	554	918	1157	1314	1417	1314	1603	1658	1486	1059	1143	1059	1766	2120	2186	1978	2131	1976

TIMING	19	20	21	22	23	24	25	26	27	28	29	30	31	32	33	34	35
ST 7	0	0	0	0	0	0	0	0	0	0	0	0	0	0	0	0	0
ST 6	36	32	23	9	10	9	10	10	9	5	7	5	5	4	2	0	0
ST 5	49	22	23	22	26	26	23	16	18	16	14	11	7	0	0	0	0
ST 4	105	138	149	140	108	117	108	95	76	46	0	0	0	0	0	0	0
ST 3	538	452	488	452	396	316	191	0	0	0	0	0	0	0	0	0	0
ST 2	1204	1116	983	780	471	0	0	0	0	0	0	0	0	0	0	0	0
ST 1	1740	1380	833	0	0	0	0	0	0	0	0	0	0	0	0	0	0

Fig. 8 User's presence at platform waiting a delayed arrival

railway system is only caused by decisions of putting advanced or delayed schedules, the sequence of blue cells indicates optimal reprogramming of the two feasible schedules.

If the solution obtained by applying this methodology (8,127 user served) is compared with the result obtained when the train that serves to the least number of users is cancelled (7,596), passenger loss is reduced by six.

3 Model Formulation

Assuming the previous pattern, a new timing for train arrivals at stations along the transit line can be determined in order to take advantage of overlapping demand curves generated from neighboring timetable-points. As a result of a disruption on the network, the subsequent rescheduling of train timetables can be based in this fact with the aim of minimizing the loss of passengers. Two scenarios can be considered, depending on that passengers can (or not) require transfers toward/from other network lines at particular times.

3.1 Scenario 1: Without Transfers

3.1.1 Indices and Sets

$i \in I$ index identifying trains of set I which run along the transit corridor
$k \in K$ index identifying cantons (or stations) of set K
$u, v \in T$ indices identifying the discrete time horizon T
$(u, v) \in M_k$ coordinates corresponding to temporary map M at station k.

3.1.2 Parameters

a_v^{ik} user population available to boarding train i at station k and at time v.

3.1.3 Decision Variables

x_{uv}^{ik} binary variable equals to 1 if train i is located at point (u, v) at station k 0, otherwise.

3.1.4 Formulation of the Model Without Transfers

$$\max \sum_{i\in I}\sum_{k\in K}\sum_{(u,v)\in M_k} a_v^{ik}\, x_{uv}^{ik} \tag{1}$$

s.t.

$$\sum_{i\in I}\sum_{(u,v)\in M_k} x_{uv}^{ik} = |I|; \quad k\in K \tag{2}$$

$$\sum_{k\in K}\sum_{(u,v)\in M_k} x_{uv}^{ik} = |K|; \quad i\in I \tag{3}$$

$$\sum_{i\in I}\sum_{u'<v} x_{u'v}^{ik} \le 1, \quad \sum_{i\in I}\sum_{v'>u} x_{uv'}^{ik} \le 1; \quad (u,v)\in M_k, \quad k\in K \tag{4}$$

$$x_{uv}^{ik} \le \sum_{v'>v} x_{(v+1)v'}^{i(k+1)}; \quad (u,v)\in M_k, \quad k\in K\ (k\ne |K|) \tag{5}$$

$$\sum_{i\in I}\sum_{u'<v,\, v'>u} x_{u'v'}^{ik} \le n_k - x_{uv}^{ik}; \quad (u,v)\in M_k, \quad k\in K \tag{6}$$

$$x_{uv}^{ik}\in\{0,\ 1\}; \quad (u,v)\in T, \quad i\in I, \quad k\in K \tag{7}$$

The objective function maximizes the number of users that can be transported along the rail corridor, picking them up at their respective stations during the time interval that they are waiting on platforms. In this sense, the objective function (1) maximizes the mobility of customers that use the railway system. Constraints (2) establish that the number of train schedules to be located must be exactly $|I|$. Restrictions (3) force passage through each station (with or without stopping) for all trains to be determined. Constraints (4) indicate that there can be no train arriving/departing from the k-th station if there was just another train operating. Restrictions (5) establish that if there is a timetable-point located at position

(u, v) of the temporary map for the k-th station, then there must be another timetable point, at the $(k + 1)$-th station, on the v-th column. Limitation of the number of trains that can operate, according to the existing number of tracks, is indicated by means of constraints (6). Finally, restrictions (7) state the binary nature of the decision variables of integer linear programming model.

From the temporary map corresponding to the first station to that of the last station (respectively labeled 0 and $|K|$), a directed acyclic graph $G = (V, E)$ can be built connecting feasible sequence of timetable-points corresponding to adjacent activity maps. Nodes Q_k (u_k, v_k) and Q_{k+1} (u_{k+1}, v_{k+1}) are connected by an arc of set E if both timetable-points satisfy the model constraints (2)–(6). By means of this geometric meaning, the determination of an optimal train timetable will be equivalent to find an optimal solution to the problem of locating a sequence of $|K|$ timetable-points (one for each temporary map) which maximizes the user mobility. Subsequently, the application of a longest path algorithm on graph $G = (V, E)$ will generate an effective schedule for the line. This decision model uses $|J| \cdot |K| \cdot |M|$ variables, where $|M|$ represents the maximum number of feasible timetable-points (u, v) in the upper triangles of temporary maps associated to stations of corridor.

3.2 Scenarios Preserving/Rewarding Transfers Between Transit Lines

If transit corridor intersects with others transit lines at specific stations, as is shown in Fig. 9, the determination of new timetables should ensure the transfer of passengers between trains from different line runs at such interchange stations.

Two strategies can be considered:

- Imposing synchronization between the timetables of these lines; that is, a solution can be accepted only if the connection between them is feasible (Scenario 2.1).
- Rewarding the possibility of providing transfers for passengers of external lines towards concurrent expeditions of the internal line by means of a weighting factor $\gamma > 1$ (Scenario 2.2).

The geometrical model above developed can be slightly adapted in order to catch the consideration of transfers. For this purpose, we define different new indices and parameters. Let $j \in J$ be the index that identifies trains of other transit lines concurrent with lines runs of set I. Let $s \in S \subset K$ be the index that enumerates the subset of stations that allow transfers to the travelers. Let $F_s(u, v) \subset M_s$ be the subset of timetable-points in the temporary map M of station s where transfers between two transit lines can be carried out. For instance, Fig. 10 shows the timetable-point (filled in red) of another line (line A) when arrives/departs at/from station 4 at times $u = 4$ and $v = 8$, respectively. If the synchronization between the timetables of these lines were imposed, the feasible subset of

Fig. 9 Corridor intersected with external lines

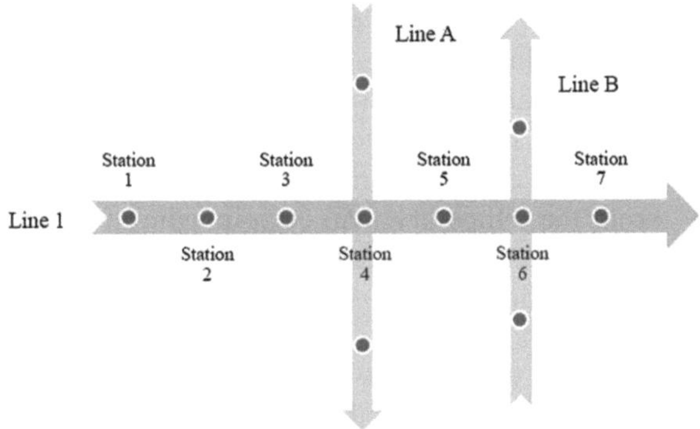

Fig. 10 Candidate locations for timetable-points which ensure transfers

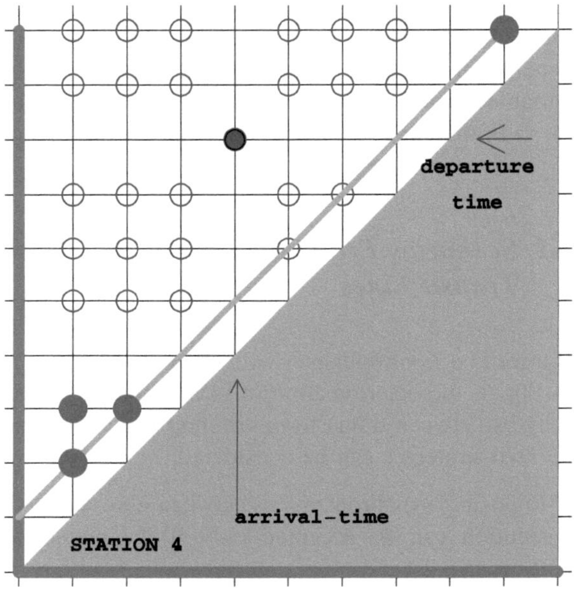

timetable-points (i.e., F_s (4, 8)), where transfer is preserved, would coincide with the set of unfilled points in magenta color. Consistently with the notation used for decision variables in the model, let y_{uv}^{js} be a binary input data which is equal to 1 if train j (of an external line whose arrival/depart timetables are given) is located at timetable-point (u, v) at station; otherwise, its value would be 0.

Scenario 2.1: Constraints (8) establish that if there is an active (i.e., $y_{uv}^{js} = 1$) timetable-point located at position (u, v) of the temporary map for the s-th station of an outside line, then there must be at least another active timetable-point at the same station for synchronizing transfers from/toward line runs i of the inner transit line I.

$$y_{uv}^{js} \leq \sum_{i \in I} \sum_{(u',v') \in F_s(u,v)} x_{u'v'}^{is}; \quad j \in J,\ (u, v) \in M_s,\ s \in S \tag{8}$$

Therefore, objective (1) and constraints (2)–(8) constitute a procedure for maximizing the number of passengers who enter in the system after rescheduling, once transfers towards other external lines is preserved.

Scenario 2.2: For this context, it is necessary to distinguish between users who enter in the system from outside and passengers who previously entered into the system with the certainty of being able to make a transfer to another line already. Objective to maximize must take into account this division of populations and asymmetrically favor one over the other population by using a weighting factor $\gamma > 1$. Let b_{uv}^{jk} be a real input data which represents the population available to transferring from train j at station k and at timetable-point (u, v). The objective (1) after being modifying is

$$\max \sum_{i \in I} \sum_{k \in K} \sum_{(u,v) \in M_k} (a_v^{ik} + \mu \sum_{j \in J} \sum_{(u',v') \in F_k(u,v)} b_{u'v'}^{jk})\, x_{uv}^{ik} \tag{9}$$

We must remark that if k is not an interchange station, then $F_k(u, v) \equiv \varnothing$ and the second additive term is cancelled. Therefore, objective (1) and constraints (2)–(8) constitute a procedure for maximizing mobility of travelers who enter in the system after rescheduling, by ensuring the option of transferring from/towards other external lines at interchange stations.

4 Conclusions

A geometric approach to determine the redistribution of service along a rail corridor has been introduced. Motivation for rescheduling railway timetables is caused by the forced reduction of fleet size due to accidents, strikes and other sources of train delays and cancellations. Two scenarios have been presented: a context without considering transfers from/towards other transit lines, and a setting where the existence of transfers between lines must be preserved although the service would have been rescheduled. In the second case, two scenarios have been formulated taking or not into consideration an equitable evaluation between the different collective of users. A common approach for these scenarios has been developed by using a geometrical representation of train timetables at stations. The associated formulations are Integer Linear Programming models, where the number of decision variables can be reduced according to different constraints imposed by the structural and fleet capacities. The theoretical development has been illustrated with a non-sophisticated example in order to clarify the concepts used through the chapter.

Acknowledgments This work was partially supported by Ministerio de Economía y Competitividad (Spain)/FEDER under grants MTM2010- 19576-C02-01 and MTM2012-37048, and by Junta de Andalucía (Spain)/FEDER under excellence projects P09-TEP-5022, FQM-5849 and P10-FQM-5849.

References

1. Barber, F., Ingolotti, L., Lova, A., Tormos, P., Salido, M.A.: Meta-heuristic and constraint-based approaches for single-line railway timetabling. Lect. Notes Comput. Sci. **5868**, 145–181 (2009)
2. Cai, X., Goh, C.J.: A fast heuristic for the train scheduling problem. Comput. Oper. Res. **21**, 499–510 (1994)
3. Canca, D., Barrena, E., Zarzo, A., Ortega, F.A., Algaba, E.: Optimal train reallocation strategies under service disruptions. Proc. Soc. Behav. Sci. **54**, 402–413 (2012)
4. Caprara, A., Monaci, M., Toth, P., Guida, P.: A lagrangian heuristic algorithm for a real-world train timetabling problem. Discr. Appl. Math. **154**, 738–753 (2006)
5. Carey, M., Lockwood, D.: A model, algorithms and strategy for train pathing. J. Oper. Res. Soc. **46**, 988–1005 (1995)
6. Cordeau, J., Toth, P., Vigo, D.: A survey of optimization models for train routing and scheduling. Trans. Sci. **32**, 380–404 (1998)
7. Higgins, A., Kozan, E., Ferreira, L.: Heuristic techniques for single line train scheduling. J. Heuristics **3**, 43–62 (1997)
8. Ingolotti, L., Lova, A., Barber, F., Tormos, P., Salido, M.A., Abril, M.: New heuristics to solve the CSOP railway timetabling problem. Lect. Notes Comput. Sci. **4031**, 400–409 (2006)
9. Jovanovic, D., Harker, P.T.: Tactical scheduling of rail operations: the SCAN-I system. Trans. Sci. **25**, 46–64 (1991)
10. Kroon, L., Peeters, L.: A variable time model for cycling railway timetabling. Trans. Sci. **37**, 198–212 (2003)
11. Mesa, J.A., Ortega, F.A., Pozo, M.A.: Effective allocation of fleet frequencies by reducing intermediate stops and short turning in transit systems. Lect. Notes Comput. Sci. **5868**, 293–309 (2009)
12. Mesa, J.A., Ortega, F.A., Pozo, M.A.: A geometric model for an effective rescheduling after reducing service in public transportation systems. Comput. Oper. Res. **40**, 737–746 (2013)
13. Michaelis, M., Schöbel, A.: Integrating line planning, timetabling, and vehicle scheduling: a customer-oriented heuristic. J. Public Transp. **1**, 211–232 (2009)
14. Nachtigall, K., Voget, S.: A genetic algorithm approach to periodic railway synchronization. Comput. Oper. Res. **23**, 453–463 (1996)
15. Odijk, M.: A constraint generation algorithm for the construction of periodic railway timetables. Trans. Res. B **30**, 455–464 (1996)
16. Serafini, P., Ukovich, W.: A mathematical for periodic scheduling problems. SIAM J. Discr. Math. **2**, 550–581 (1989)
17. Silva de Oliveira, E.: Solving single-track railway scheduling problem using constraint programming. Ph.D thesis, The University of Leeds, School of Computing (2001)
18. Szpigel, B.: Optimal train scheduling on a single track railway. In: Roos, M. (ed.) Proceedings of IFORS Conference on Operational Research 1972, pp. 343–352 (1973)
19. Vieira, G.E., Herrmann, J.W., Lin, E.: Rescheduling manufacturing systems: a framework of strategies, policies, and methods. J. Sched. **6**, 39–62 (2003)

The Fuzzy System Sensitivity Analysis: An Example of Air Travel Demand Models

Milica Kalić, Slavica Dožić and Jovana Kuljanin

Abstract This chapter presents the results obtained by employing both fuzzy logic and sensitivity analysis to model trip generation and trip distribution processes in the domain of air transportation. Qualitative and imprecise information taken from experts represent an invaluable source when objective knowledge on certain process is not available or even does not exist. Thus, fuzzy logic is seen as a convenient mathematical tool that efficiently treats uncertainty in-built in the socio-economic parameters that are selected to describe trip generation and trip distribution problem. The chapter analyzes the sensitivity of fuzzy system solutions obtained by two models in respect to different factors such as domain discretization of input and output variables, various forms of membership function and different approximate reasoning techniques hereby enabling possible improvements to the models.

Keywords Sensitivity analysis · Fuzzy logic · Air travel demand · Trip generation · Trip distribution

1 Introduction

A variety of traffic and transportation problems are successfully solved by employing both stochastic and deterministic models. All these models require mathematical equation in order to be efficiently solved. Fuzzy logic presents one of the mathematical tools for approximate reasoning which is often used in

M. Kalić (✉) · S. Dožić · J. Kuljanin
Faculty of Transport and Traffic Engineering, University of Belgrade,
Vojvode Stepe 305, 11000 Belgrade, Republic of Serbia
e-mail: m.kalic@sf.bg.ac.rs

S. Dožić
e-mail: s.dozic@sf.bg.ac.rs

J. Kuljanin
e-mail: j.kuljanin@sf.bg.ac.rs

J. F. de Sousa and R. Rossi (eds.), *Computer-based Modelling and Optimization in Transportation*, Advances in Intelligent Systems and Computing 262,
DOI: 10.1007/978-3-319-04630-3_27,

engineering when solving problems surrounded with uncertainties and ambiguity. Likewise, many parameters and occurrences in transportation industry are remarkably tied to subjectivity. The results presented are achieved by employing both fuzzy logic and sensitivity analysis to model trip generation and trip distribution processes. The parameters encountered in these two models do not have precisely defined bounds and are characterized by subjectivity judgement. Therefore, the use of fuzzy logic is deemed as essential. Since the entire concept of the fuzzy set theory relies on the idea of describing qualitative and imprecise information taken from experts, fuzzy logic can be seen as a promising tool in modelling the above mentioned processes where expert opinion and experience about key variables plays a significant role. In addition to a mathematical framework that can efficiently deal with the uncertainties naturally inbuilt in the cognitive processes (thinking, reasoning, knowledge acquisition) provided by this theory, fuzzy logic also tends to show possibilities in estimating the observed function from numerical examples. Thus, the chapter underlies the importance of fuzzy logic as an universal approximator mapping an input space to an output space in the process of modelling trip generation and trip distribution.

Trip generation model primarily emphasizes GDP as the economic factor that has the largest impact on overall economic activity. Therefore, GDP is employed as explanatory variable in order to create a robust model that describes air travel demand on an aggregate level (country level). Qualification of GDP as a variable associated with expert knowledge was carried out—the authors have consistently expressed a rule base so as to make the fuzzy system. Trip distribution model evolved in this chapter was also solved by using fuzzy logic. In order to properly describe flows between two countries, it is of vital importance to consider mutual activities that run between two countries such as level of trade, migration of labor flows, historical relationship and level of tourist attraction. The same procedure of taking into account appropriate explanatory variables was performed in order to obtain fuzzy rules. The non-fuzzy method (linear regression) was employed in order to justify the use of the fuzzy logic approach.

The aim of the chapter was to analyze the sensitivity of the fuzzy system solutions obtained by two models in respect to different factors such as domain discretization of input and output variables, various shapes of membership function and different approximate reasoning techniques. By changing one of these factors (while keeping others constant), performance of the fuzzy system is monitored hereby enabling possible improvements to the models. In addition to these experiments, analysis of importance weight assigned to different factors in the model was also carried out.

The rest of the chapter is organized as follows. After introduction, in Sect. 2 literature review is presented. Section 3 gives theoretical background. Section 4 is dedicated to description of trip generation and trip distribution models. The results obtained by sensitivity analysis are given in Sect. 5. Section 6 presents the concluding remarks.

2 Literature Review

It is well-known that ambiguity, uncertainty and vagueness are present in great many of the problems in transportation planning. Teodorović [13] gives an overview of classification and analysis of the results achieved using fuzzy logic to model complex traffic and transportation processes showing the possibilities regarding the further application of fuzzy logic in this field. From the traditional point of view, the transportation planning process consists of four stages: trip generation, trip distribution, modal split and route choice. A variety of different techniques has recently been developed for air passenger demand modeling process. Trip generation model is solved using fuzzy logic by Kalić and Teodorović [9]. Fuzzy rule base in this chapter is generated by learning from numerical example, and the available data set was divided into two subsets: the first was used as the fuzzy rule base and the second served as a testing data subset. The test of the model proved the efficient use of fuzzy logic approach to obtain the closest estimate of the actual number of trips generated in a given area. Kalić and Teodorović [8] also solve trip distribution problem by employing fuzzy logic. Similar to trip generation problem, authors generate a fuzzy rule base by learning from numerical example. Wang and Mendel [16] procedure is used in order to generate a fuzzy rule base. Estimation of the number of air passengers traveling between major industrial cities and given regions has been achieved by testing the previously obtained fuzzy system. In the following research, Kalić and Teodorović [10] improve the previously mentioned model by combining genetic algorithm with fuzzy logic. Genetic operators enable generation of several rule bases thereby allowing the selection of the final fuzzy rule base which gives the best estimate of the actual number of trips. Teodorović and Kalić [14] also consider fuzzy logic to deal with the mode choice problem. These authors generate fuzzy rule base by using numerical data on differences between travel times and travel costs of competitive modes. With regard to route choice, Teodorović and Kikuchi [15] were the first to introduce the fuzzy logic concept by applying this technique in a binary route choice situation where only travel time was considered as a decision-making factor. Additionally, Arslan and Khisty [1] propose a psychometric approach enabling a more proper description of route choice behavior in transportation systems. The model developed by these authors is based on Weber's psycho-physical law of 1834 and a set of fuzzy 'if–then' rules is used to represent a typical driver's psychology for capturing essential preferences, pair-wise, among the alternatives that a driver may consider. Many authors have also considered different methods to improve performance of fuzzy logic systems. Gürocak and de Sam Lazaro [4] propose a method for fine tuning the rule base of a fuzzy system that treats the fuzzy system's rule base as a multivariate function and performs parameter optimization. Furthermore, Kalić [5] analyzed the sensitivity of the fuzzy system solution in respect to domain discretization of input and output variable, various forms of membership functions, different reasoning techniques and a number of defuzzification methods. Kaymak and van Nauta Lemke [12] consider a sensitivity analysis approach to introduce weight factors into decision functions in fuzzy multicriteria decision making.

Many papers have shown that the fuzzy logic can be successfully applied in sequential procedures of passenger demand forecasting. Those papers considered each phase in the transportation planning process (trip generation, trip distribution, modal split and route choice) separately. Kalić and Tošić [11] developed a trip distribution model in air transportation under irregular conditions, when business and recreational trips were almost totally absent (Belgrade Airport case study between 1991 and 2000). The basic assumption in this chapter was that the main impact on trip distribution was the number of people emigrating out of the country. More than one decade later, Kalić et al. [6, 7] developed air travel demand models on different aggregation level by using different approach. This chapter is based on the above-mentioned research in which trip generation has been considered on the country level and trip distribution between origin country and destination countries (country-pair level).

3 Theoretical Background

Accurate prediction of air travel demand has attached considerable attention by many researches and experts employed in aviation industry during the last two decades. Most of the empirical research in this area has focused on assessing the effects of certain economic factors combined with geographical characteristics of the area around an airport and the routes involved in demand for air travel. Economic factors largely reflect economic activity of a specific country compromising variables such as Gross Domestic Product (GDP), level of trade growth, foreign direct investments (FDI), employment and unemployment rate, number of full-time employees, employment composition. On the other hand, many authors emphasize the importance of service related factors that are primarily characteristics of the air transport system and are, in contrast to geo-economic factors, under control of airlines (Grosche et al. [3]). Of the general factors on the side of service (i.e. supply), actual price of air fares, frequency of service, speed of air travel and convenience of air travel are perhaps the most important ones.

4 Models Developed

Two models are developed for passenger volume estimation. The first refers to trip generation and the second tries to describe passenger flows between country pairs. In particular, trip generation model emphasizes the effect of certain economic variables on the aggregate level. For this purpose, GDP was identified as a major factor that significantly affects a country's overall economic activity. Thus, the model developed can be considered as macro model for estimating the air passenger demand. According to Gobrial and Fleming [2], microanalysis of air transportation usually deals with only aggregate measures of socioeconomic variables, economic activities and of the air transportation service and is not normally concerned with specific airport or with origin–destination flows. Therefore, GDP is selected to be an explanatory variable in order to create a robust model which tends to describe air travel demand on the aggregate level. The model developed is highly applicable to countries of South East Europe where small variation in income (GDP) will induce large change in demand.

4.1 Trip Generation Model

Fuzzy logic system is used in order to make a short term forecast of total passenger flow from Serbia taking into account the previously mentioned factor. The experiments regarding the number and the shape of the input and output fuzzy sets were carried out and it was decided to use triangle and trapezoid sets as follows. The membership functions of fuzzy sets Very Very Low, Very Low, Low, Low Medium and Medium are related to GDP, while the membership functions of fuzzy sets Approximately 1, Approximately 2, Approximately 2.5 and Approximately 3 are related to number of passengers (NP).

The output variable of the fuzzy system is the total flow from Serbia. The initial results are obtained by applying MAX–MIN fuzzy reasoning and defuzzification by centre of gravity, Kalić et al. [7]. Also, the results obtained by linear regression are derived from the Eq. (1):

$$NP = -948796 + 3095.454 * GDP \tag{1}$$

Regression statistics are following: $R^2 = 0.84$, $F = 49.22$, t Stat $= -2.05$, p-value $= 0.069$ for intercept coefficient, t Stat $= 7.01$, p-value $= 6.21 * 10 - 5$ for GDP coefficient.

Kalić et al. [7] point out that the model based on fuzzy logic produced very satisfying results (average relative error is 6 %) in comparison to results obtained by regression (average relative error is 8 %).

4.2 Trip Distribution Model

The trip distribution model is also based on fuzzy logic. In order to determine passenger flows between Serbia and 14 selected countries, three input variables are introduced: number of emigrants (EMG), imports by countries in millions of RSD (IMP) and attraction of the destination (ATT). As it is emphasized by Kalić et al. [7] number of emigrants reflects a strong connection between two countries and also a strong passenger flow, while imports indicate economic and business relation between the observed countries. As the third input, a quantitative indicator of attraction, which describes route taking into account tourist attraction of the destination country, historical relationship between two countries, and existence of hub airports in the destination country, is introduced. This indicator is calculated as follows [2]:

$$\text{ATTRACTION OF DESTINATION} = \text{Tourism} + \text{Historical (other) links} + \text{Hub} \quad (2)$$

Values for Tourism factor range from 0 to 6, while Historical (and other) links vary from 0 to 4. Hub variable can take discrete value from the set of $\{0, 2, 4, 6\}$. Detailed explanations regarding this indicator are given by Kalić et al. [7].

Fuzzy logic system is used in order to specify passenger flows between Serbia and European countries. Kalić et al. [7] define the membership functions of input and output variables as follows: the membership functions of fuzzy sets Very small, Small, Medium, Large and Very large is related to EMG, the membership functions of fuzzy sets Low, Medium and High is related to IMP. ATT is described using fuzzy sets Low, Medium and Huge attraction, while the membership functions of fuzzy sets Small, Medium and Large are related to passenger flow between two countries (Fuzzy Logic Flow–FL Flow). The fuzzy rule base is complete and consists of 45 rules. Some of them are presented below:

Rule 1: If EMG is Very small and IMP is Low and ATT is Low, then FL Flow is Small, else

Rule 2: If EMG is Very small and IMP is Low and ATT is Medium, then FL Flow is Small, else …

Rule 21: If EMG is Medium and IMP is Low and ATT is Huge, then FL Flow is Medium, else …

Rule 44: If EMG is Very large and IMP is High and ATT is Medium, then FL Flow is Large, else

Rule 45: If EMG is Very large and IMP is High and ATT is Huge, then FL Flow is Large.

As it is mentioned in previous research (Kalić et al. [7]), the results obtained by applying MAX–MIN fuzzy reasoning and defuzzification by centre of gravity point out a very close correspondence if the real and estimated value of passenger flows are compared.

5 Sensitivity Analysis

Fuzzy logic systems are systems connecting fuzzy concept (fuzzy sets, linguistic variables) and fuzzy logic. Fuzzy logic system, or fuzzy system, consists of fuzzification, fuzzy rule base, fuzzy inference method and defuzzification. This chapter employs a singleton fuzzification assuming crisp inputs and their mapping into fuzzy sets. The first step of the fuzzy system design is the domain discretization of input and output variables. By definition, the domain discretization of input–output variables is the domain division into fuzzy regions within which membership functions domain interval will be chosen. Membership functions of fuzzy sets can assume different shapes, but for the sake of simplicity, triangular shapes are the most often used. In this chapter, three kinds of shapes are used: triangular/trapezoid and Gaussian curves.

In the second step of the fuzzy system design, the fuzzy rule base is generated. Fuzzy rule base design in order to create simpler and more robust approximate reasoning as well as, not so large data set of input and output variables are reasons to generate fuzzy rule base without using the well known methods.

The third step represents the implementation of fuzzy logic, i.e. inference method (reasoning techniques). This chapter uses two fuzzy reasoning techniques: MAX–MIN (min inference) and MAX-DOT (product inference). The output of fuzzy logic is a fuzzy set. Defuzzification is then applied as the process of computing a crisp numerical value from the resulting fuzzy set. This chapter applies three defuzzification methods: the center of gravity, mean of maximum (MOM) and smallest of maximum methods (SOM).

The aim of this chapter is to analyze the sensitivity of the fuzzy system solutions in respect to:

- Domain discretization of input and output variables,
- Various forms of membership functions,
- Different approximate reasoning techniques,
- Number of defuzzification methods, and
- Different rule weights.

The fuzzy system sensitivity analysis is exemplified with trip generation and trip distribution in air transport.

5.1 Sensitivity Analysis: Trip Generation Model

The initial analysis refers to simultaneous changing of the domain discretization of the input GDP and output variables number of passengers (NP). Simultaneous changing of the domain discretization in this case means the same number of fuzzy sets of the input and output variables. For example, the first experiment is conducted with three input fuzzy sets (Very Low, Low and Medium GDP) and three output fuzzy sets (Approximately 1, Approximately 1.65, and Approximately 2.5). Other experiments are conducted in the same manner. The number of experiments is limited to 7 input and 7 output fuzzy sets, because using more than 7 fuzzy sets is not reasonable due to both the nature of data and the problem being considered. For each of these combinations (from: 3 input fuzzy-sets 3 output fuzzy sets to: 7 input fuzzy sets-7 output fuzzy sets), experiments with different membership

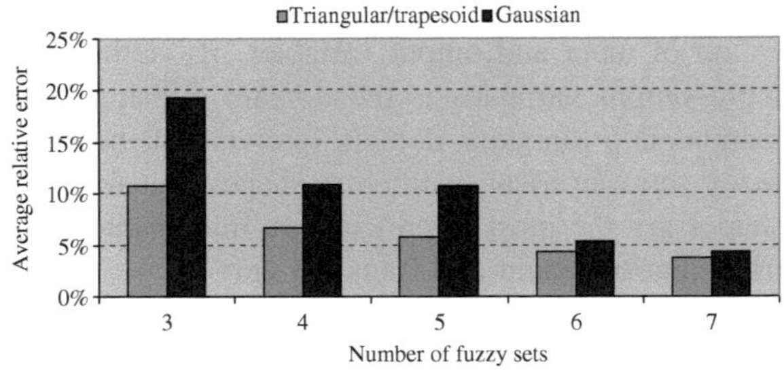

Fig. 1 Solution sensitivity based on simultaneous changing of the domain discretization of the input GDP and output variables NP and for various forms of membership functions

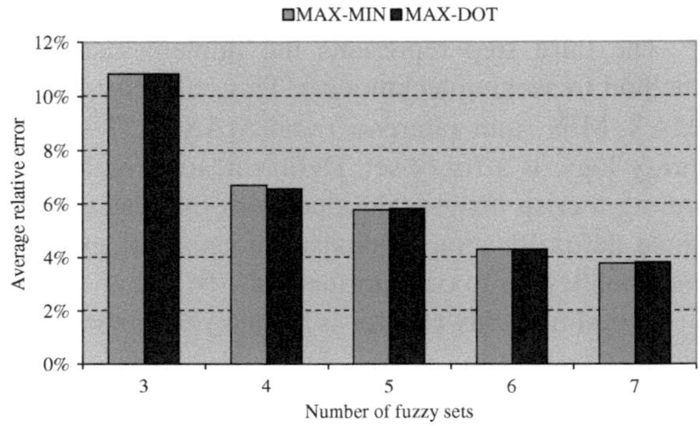

Fig. 2 Solution sensitivity based on changing approximate reasoning techniques for different domain input and output discretization

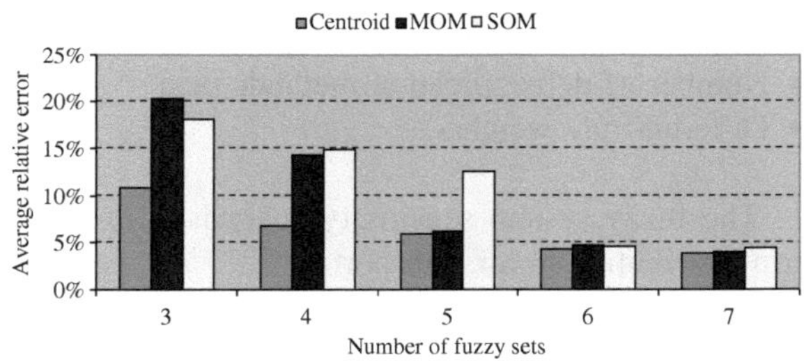

Fig. 3 Solution sensitivity based on changing defuzzification methods for different domain input and output discretization

functions, different defuzzification methods and different approximate reasoning techniques are done.

The results obtained by sensitivity analysis are expected. As seen from the Fig. 1, average relative error decreases by increasing the number of fuzzy sets. These experiments provide a better solution when using triangular/trapezoid forms of membership functions in comparison to Gaussian curve. This chapter uses two fuzzy reasoning techniques, MAX–MIN and MAX-DOT, but results are very similar for different domain input and output discretization (Fig. 2). Figure 3 shows solution sensitivity based on the changing defuzzification methods for different domain input and output discretization. It can be seen that the center of gravity method generates better solution for smaller number of input and output fuzzy sets. Otherwise, solution sensitivity is insignificant (small).

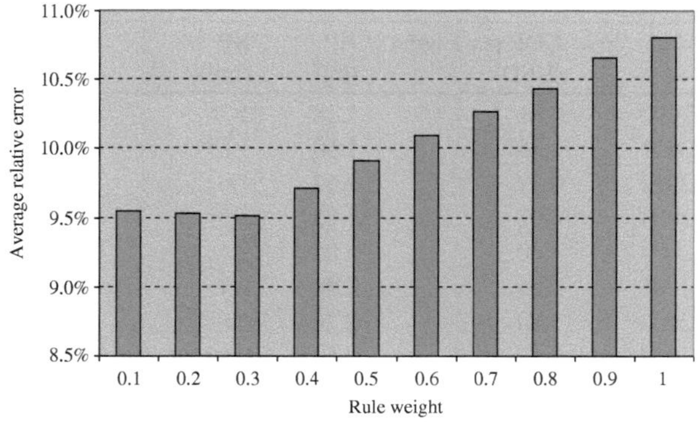

Fig. 4 Solution sensitivity based on changing one rule weight

One of the possible ways to tune fuzzy system is to use rule weights. The weighting of rules is sometimes presented as the measure of importance, influence or reliability. Changing rule weight by step 0.1, in the case of 3 input and output, it is shown (Fig. 4) that the solution sensitivity is small.

Results of the two experiments (5/5—5 input and output fuzzy sets, and 7/7—7 input and output fuzzy sets, MAX–MIN approximate reasoning, and defuzzification by center of gravity) are shown in Table 1. The average relative errors in these examples are 5.8 and 3.8 %, respectively.

5.2 Sensitivity Analysis: Trip Distribution Model

In this model, changing of the domain discretization of the output variable is done for 3, 4 and 5 output fuzzy sets, in the case of triangular/trapezoid fuzzy sets, MAX–MIN reasoning technique and defuzzification with center of gravity (centroid). As can be seen from the Fig. 5, average relative, absolute and squared errors decrease by increasing the number of output fuzzy sets (FL Flow) in this experiment.

Extremely better results are obtained by employing triangular/trapezoid forms comparing triangular/trapezoid forms of membership functions to Gaussian curve. The same applied for different techniques of defuzzification—extremely bad solutions are obtained by using MOM, SOM and LOM (largest of maximum) if compared to centroid.

Two fuzzy reasoning techniques, MAX–MIN and MAX-DOT, are used, and the results obtained by MAX-DOT reasoning are unfavorable if compared to MAX–MIN reasoning technique (Fig. 6).

Considerable improvements are not achieved by weighting of rules, and therefore, the solution sensitivity to this change is insignificant. Table 2 gives input data for number of emigrants (EMG), imports by countries in millions of RSD (IMP) and attraction of destination (ATT), flows by countries (in thousands), flows obtained by fuzzy logic if number of output fuzzy sets is 5 (FL Flow 5) and

Table 1 Input data and two fuzzy logic outputs—trip generation

Year	GDP per Capita (USD)	NP (mil)	NP 5/5 (mil)	Relative error	NP 7/7 (mil)	Relative error
2000	0.81	1.28	1.26	0.02	1.26	0.02
2001	0.85	1.50	1.63	0.09	1.54	0.03
2002	0.89	1.62	1.63	0.01	1.63	0.01
2003	0.92	1.85	1.9	0.03	1.9	0.03
2004	1.00	2.05	2.1	0.02	2.1	0.02
2005	1.06	2.06	2.1	0.02	2.1	0.02
2006	1.10	2.26	2.65	0.17	2.1	0.07
2007	1.17	2.54	2.69	0.06	2.43	0.04
2008	1.22	2.67	2.69	0.01	2.85	0.07
2009	1.18	2.40	2.69	0.12	2.43	0.01
2010	1.20	2.72	2.69	0.01	2.85	0.05
2011	1.19	3.15	2.69	0.15	2.85	0.10

Fig. 5 Solution sensitivity based on changing of the domain discretization of the output variable (FL Flow)

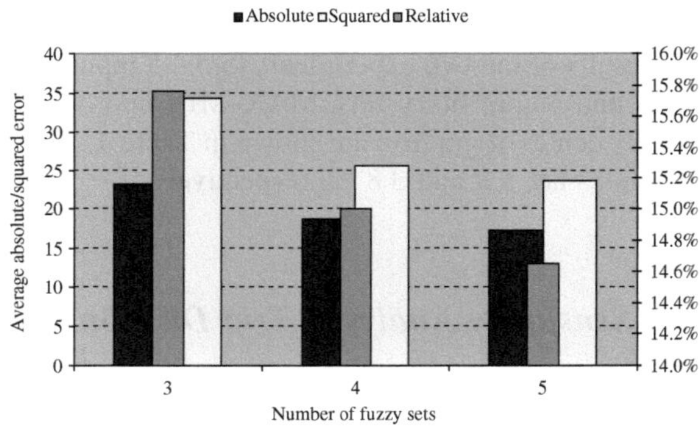

Fig. 6 Solution sensitivity based on changing approximate reasoning techniques for different domain output discretization

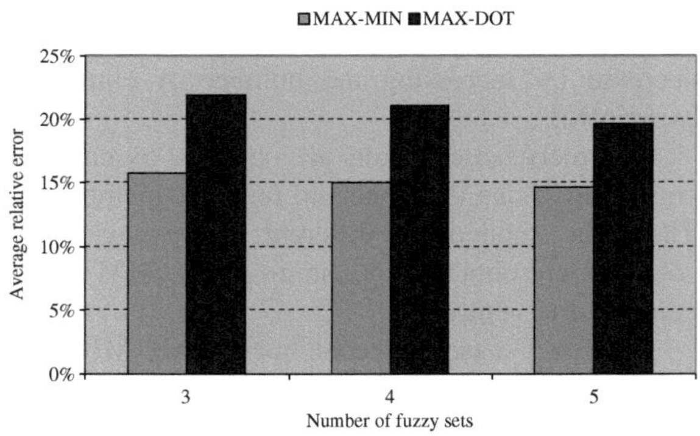

flows obtained if the first rule in the fuzzy rule base is changed. Changing the first rule in the fuzzy rule base enables better results by countries and also reduction of average relative error from 15 to 10 %, for these two experiments respectively.

Table 2 Input data and two fuzzy logic outputs—trip distribution

	EMG (000)	IMP	ATT	Flow (000)	FL Flow 5 (000)	FL Flow 5-rule 1 changed (000)
Austria	300	512	$2 + 1.5 + 2 = 5.5$	203	188	188
Denmark	7	91	$0 + 0 + 0 = 0$	35	23.2	34.1
France	120	482	$4 + 1 + 2 = 7$	129	132	132
Greece	10	228	$6 + 0 + 0 = 6$	77	98.9	98.9
Holland	20	256	$0 + 0 + 2 = 2$	47	42.2	42.2
Italy	102	1,432	$3 + 1 + 0 = 4$	138	150	150
Germany	845	1,768	$1 + 2 + 6 = 9$	538	585	585
United Kingdom	17	197	$1 + 1 + 4 = 6$	136	113	113
FRY Macedonia	1	272	$0 + 0.5 + 0 = 0.5$	47	29.1	43.1
Russian Federation	25	2,157	$0.5 + 2 + 0 = 2.5$	133	150	150
Switzerland	120	202	$0 + 0 + 6 = 6$	177	119	119
Bosnia and Herzegovina	1	556	$0 + 0.5 + 0 = 0.5$	24	23.4	23.4
Turkey	13	325	$6 + 2 + 0 = 8$	107	107	105
Montenegro	1	165	$6 + 4 + 0 = 10$	391	381	381

From planners' perspective, estimation of approximate number of passengers and passenger flow is much more valuable than having precise information about these values. Thus, results presented in Tables 1 and 2 are valuable for further planning activities.

6 Conclusion

Air traffic in Serbia has passed through a dramatic and turbulent period in the last two decades. Political circumstances that appeared in early 90s severely affected the entire economy and led to a sharp drop in the number of passengers at Belgrade Airport. After the breakup of Yugoslavia, with previous population of roughly 23 million inhabitants, six separate countries have been constituted. Therefore, the number of passengers per year of approximately 5 million that was reached in the previously existing country was almost impossible to achieve in such emerging conditions. Thus, traffic volume has reached its minimum of only 339 thousand passengers in 1993, due to the isolation, severe hyperinflation and ban on traffic. Normal air traffic was resumed in 2001 when the national government and international attitudes towards Serbia changed. Although stabilization of the entire air traffic is evident from 2001, tremendous fluctuations in economic and political environment that occurred before that period have disenabled making of sufficiently large time series observations.

Trip generation model tends to roughly describe the entire activity (expressed as a number of passengers) of the population towards air travel behavior. Thus,

GDP is seen as an appropriate variable which allows robustness of the model. In addition, trip generation model takes into account only the dense flow between Serbia and selected countries. Still, some routes are exclusively characterized as tourist and appear periodically as offers on the market. Therefore, these routes have been omitted from the study (such as flow between Serbia and Egypt, Tunis etc.).

According to the fuzzy system sensitivity analysis presented, it can be said that domain discretization of input and output variables, various forms of membership functions, different approximate reasoning techniques and a number of deffuzification methods may have significant influence to fuzzy system solution.

In order to sustain the concept of fuzzy logic system itself, a reasonable number of fuzzy sets for input and output of the models have been examined. Consequently, the number of experiments carried out was not large, with a tendency to obtain a meaningful system that efficiently describes the phenomenon being considered. Sensitivity analysis indicates that both models are highly susceptible to change of different shapes of membership functions, different methods of defuzzification and different reasoning techniques. On the other hand, models do not express sensitivity by employing different rule weights.

However, design of fuzzy systems should compromise simplicity and robustness of the model on the one hand and fuzzy logic system fine tuning as a result of sensitivity analysis on the other hand.

Acknowledgments This research has been supported by the Ministry of Science and Technological Development, Republic of Serbia, as part of the project TR36033 (2011–2014).

References

1. Arslan, T., Khisty, C.J.: A rational reasoning method from fuzzy perceptions in route choice. Fuzzy Sets Syst. **150**(3), 419–435 (2005)
2. Fleming, K., Ghobrial, A.: An analysis of the determinants of regional air travel demand. Transp. Planning Technol. **18**, 37–44 (1994)
3. Grosche, T., Rothlauf, F., Heinzl, A.: Gravity models for airline passenger volume estimation. J. Air Transp. Manag. **13**(4), 175–183 (2007)
4. Gürocak, H.B., de Sam Lazaro, A.: Fine tuning method for fuzzy logic rule base. Fuzzy Sets Syst. 67(2), 147–161 (1994)
5. Kalić, M.: Fuzzy system sensitivity analysis: An example of trip generation in air transportation. In: Proceedings of the 11th Mini-EURO Conference on Artificial Intelligence in Transportation Systems and Science, and 7th EURO-Working Group Meeting on Transportation, Helsinki, Finland (1999)
6. Kalić, M., Dožić, S., Babić, D.: Predicting Air Travel Demand Using Soft Computing: Belgrade Airport Case Study. 15th Euro Working Group on Transportation, Paris (2012a)
7. Kalić, M., Kuljanin, J., Dožić, S.: Air travel demand fuzzy modelling: Trip generation and trip distribution. WSC17 2012 online conference on soft computing in industrial applications anywhere on earth, 10–21 December 2012 (2012b)

8. Kalić, M., Teodorović, D.: Solving the trip distribution problem by fuzzy rules generated by learning from examples. In: Proceedings of the XXIII Yugoslav Symposium on Operations Research, pp. 777–780. Zlatibor, Yugoslavia, (in Serbian) (1996)
9. Kalić, M., Teodorović, D.: A soft computing approach to trip generation modeling. Paper presented at the 9th Mini EURO Conference Fuzzy sets in traffic and transport systems, Budva, Yugoslavia (1997)
10. Kalić, M., Teodorović, D.: Transportation route choice model using fuzzy inference technique. Transp. Plann. Technol. **26**(3), 213–238 (2003)
11. Kalić, M., Tošić, V.: Soft demand analysis: Belgrade Case Study. In: Proceedings of the 8th Meeting of the Euro Working Group Transportation EWGT and Workshop IFPR on Management of Industrial Logistic Systems "Rome Jubilee 2000 Conference", pp. 271–275 Rome, Italy (2000)
12. Kaymak, U., van Nauta, L.: A sensitivity analysis approach to introducing weight factors into decision functions in fuzzy multicriteria decision making. Fuzzy Sets Syst. 97(2), 169–182 (1998)
13. Teodorović, D.: Fuzzy logic systems for transportation engineering: the state of the art. Transp. Res. Part A **33**, 337–364 (1999)
14. Teodorović, D., Kalić, M.: Solving the modal split problem by fuzzy rules generated by learning from examples. In: Proceedings of Information Technologies, pp. 48–54. Žabljak, Yugoslavia, (in Serbian) (1996)
15. Teodorović, D., Kikuchi, S.: Transportation route choice model using fuzzy inference technique. In: Ayyub, B.M. (ed.) Proceedings of ISUMA '90. The First International Symposium on Uncertainty Modeling and Analysis, pp. 140–145. IEEE Computer Press, College Park, Maryland (1990)
16. Wang, L., Mendel, J.: Generating fuzzy rules by learning from examples. IEEE Trans. syst. Man Cybern. **22**, 1414–1427 (1992)

Stochastic User Equilibrium and Analysis of Users' Benefits

Claudia Castaldi, Paolo Delle Site, Francesco Filippi and Marco Valerio Salucci

Abstract When random utility models are used to represent the choice of the route alternatives, the benefits accruing to the users as a consequence of an intervention on the network can be estimated rigorously on the basis of the expectation of the compensating variation. A rigorous disaggregate analysis which considers shares of shifters and non-shifters and attributes benefits to them can be carried out based on transition probabilities and associated conditional expectations of the compensating variation. The chapter extends the results available in the literature on the computation of transition probabilities and conditional welfare measures to cases of imperfect before-after correlation of the random terms and changing choice set. The theoretical results are illustrated with applications to the town bypass case and the Dupuit-Nguyen network.

Keywords Random utility · Route choice · Stochastic user equilibrium · Transition probability · Compensating variation

C. Castaldi · P. Delle Site (✉) · F. Filippi · M. V. Salucci
DICEA Department of Civil, Architectural and Environmental Engineering,
University of Rome La Sapienza, Via Eudossiana 18, 00184 Rome, Italy
e-mail: paolo.dellesite@uniroma1.it

C. Castaldi
e-mail: castaldi@ctl.uniroma1.it

F. Filippi
e-mail: francesco.filippi@uniroma1.it

M. V. Salucci
e-mail: salucci@ctl.uniroma1.it

C. Castaldi · P. Delle Site · F. Filippi · M. V. Salucci
CTL Research Centre for Transport and Logistics, University of Rome La Sapienza, Via Eudossiana 18, 00184 Rome, Italy

J. F. de Sousa and R. Rossi (eds.), *Computer-based Modelling and Optimization in Transportation*, Advances in Intelligent Systems and Computing 262,
DOI: 10.1007/978-3-319-04630-3_28,

1 Introduction

A key problem in transportation planning is the appraisal of interventions on road networks that modify attributes of the route alternatives, such as travel time and money expenditure, or add new routes. The conventional approach to appraisal considers the network equilibrium before and after the intervention and measures the users' benefits associated with the change.

Within a stochastic network equilibrium framework, users are assumed to make route choices according to a random utility model (RUM) which accounts for inter-personal heterogeneity in the perception of the attributes of the route alternatives. The flow of each origin–destination (OD) pair is assigned to the available routes of the OD pair. Stochastic user equilibrium (SUE) defines the state of a congested network where route travel times are consistent with flows (introduced by Daganzo and Sheffi [2]; a review is in Cascetta [1]).

With SUE, users' benefits can be measured at the level of each OD pair using different metrics: the average welfare change, measured by the expectation of the compensating variation, which takes into account changes in both monetary and non-monetary attributes of the route utilities, and the monetary value of the total travel time savings. The expectation of the compensating variation can be computed in simulation, i.e. by drawing from the distribution of the random terms (McFadden [11]), or analytically. McFadden [11] has provided the formula in the case without income effects (i.e. when income does not affect choice), which is the so-called logsum in the case of multinomial logit. Dagsvik and Karlström [3] have provided the formula in the case with income effects.

A disaggregate analysis of users' benefits is used to be carried out with the so-called rule-of-a-half (De Jong et al. [4]; Jara-Díaz [8]) which attributes benefits to route alternatives, to non-shifting users and new users of each route alternative, and to components of the generalised cost of travel. However, the attribution of benefits with the rule-of-a-half is only conventional, while the measurement at the level of the OD pair based on the expectation of the compensating variation is rigorously consistent with the micro-economic foundation of the route choice model.

Only recently, authors have developed the theory that is needed for a rigorous disaggregate analysis of users' benefits with RUMs (De Palma and Kilani [6]). This theory considers the transition choice probabilities and the expectation of the compensating variation conditional on the transitions. These quantities are of practical interest because, in the usual interpretation of RUMs where the random terms are seen as individual specific, they provide, respectively, the shares of shifters and non-shifters and the associated average welfare measures. In addition, it is possible to attribute rigorously to shifters and non-shifters the value of travel time savings. No application of the theory in the transportation or other fields is found to date in the literature.

The theory has been formulated under the restrictive assumptions that the random terms do not change with the intervention, i.e. perfect before-after correlation, and the choice set also does not change. It is justified to consider other before-after correlation patterns to take into account changes in unobserved attributes or intra-personal taste variation. Extension to cases of imperfect before-after correlation of the random terms, with application to modal choice, has been provided by Delle Site and Salucci [5]. Changing choice sets are of interest in the case of new links and have not been tackled yet.

The chapter offers the following original contributions:

- it provides the theoretical framework for the computation of the transition choice probabilities and the associated average welfare measures with extension to cases of changing choice set;
- it applies the framework to the analysis of users' benefits in a network under SUE;
- it illustrates the theory with numerical examples for a two-link network and for the Nguyen-Dupuit network, using a multinomial logit route choice model calibrated with stated preference data, and considering cases of perfect before-after correlation and of independent random terms.

2 Network Equilibrium

Let $G = (N, A)$ be a strongly connected road transportation network, with N and A being the sets of nodes and links, respectively. Let a be the link index. Origins (O) and destinations (D) constitute a subset of N. Let R be the set of OD pairs and r the OD pair index. Let k^r be the set of simple paths of OD pair r, and k the path index.

For each path $k \in k^r$, F_k^rdenotes the corresponding path flow. We denote by z_a the flow on link $a \in A$. The link flows are obtained from the path flows by:

$$z_a = \sum_{r \in R} \sum_{k \in K^r} \delta_a^{K,r} \cdot F_k^r \quad a \in A \tag{1}$$

where $\delta_a^{k,r}$ is the element of the link-path incidence matrix whose value is 1 if path k includes link a, is 0 otherwise.

The demand flow of the OD pair r is denoted by q^r. We have the demand constraints:

$$q^r = \sum_{k \in K^r} F_k^r \qquad r \in R \tag{2}$$

The feasible path flows are all the non-negative F_k^r satisfying the demand constraints (2). Therefore, the set of feasible path flows is non empty, compact and convex.

Let T_k^r denote the travel time on path k of OD pair r. Let t_a denote the travel time on link a. The link travel times are continuous functions of the link flows:

$t_a = t_a(z_a, a \in A)$. The path travel times are obtained from the link travel times by the standard link-additive model:

$$T_k^r = \sum_{a \in A} \delta_a^{k,r} \cdot t_a(z_a, a \in A) \quad k \in K^r, r \in R \tag{3}$$

Route choice is modelled with RUMs. The users of an OD pair perceive a utility on each path. This path utility is a random variable given by the sum of a systematic, i.e. deterministic, component and a random term. The random terms summarise factors that are unobserved by the modeller. The random terms are interpreted as individual specific thus accounting for both inter-individual and intra-individual variability of tastes. The individual-specific random terms may change across repeated choices.

The path systematic utility depends on two attributes: residual income (i.e. the difference between income and the monetary cost paid to use the path) and travel time.

Thus, we have:

$$\begin{array}{l} u_k^r = v_k^r + \varepsilon_k^r \\ v_k^r = \alpha \cdot g(y - c_k^r) + \beta \cdot T_k^r \end{array} \quad k \in K^r, \ r \in R \tag{4}$$

where:

u_k^r is the path perceived utility
v_k^r is the path systematic utility
ε_k^r is the random term
α is the coefficient of the term in residual income
g is an increasing function of the argument
y is the income
c_k^r is the monetary cost of the path
β is the time coefficient.

In the case without income effects the systematic utility reduces to:

$$v_k^r = \alpha \cdot (y - c_k^r) + \beta \cdot T_k^r = -\alpha \cdot c_k^r + \beta \cdot T_k^r \tag{5}$$

since income cancels out of the econometric specification. In this case α is the marginal utility of income.

Users of class j of OD pair r who choose path k are those who perceive this path to maximise their utility. The choice probabilities are defined as:

$$P_k^r = \Pr\left(u_k^r \geq u_j^r \quad \forall j \neq k \in K^r\right) \quad k \in K^r, r \in R \tag{6}$$

We assume that the random terms ε_k^r have a non-degenerate joint probability density function that is continuous, strictly positive, and independent of the path systematic utility. We assume that the choice probabilities are single-valued and continuous in the path systematic utilities:

$$P_k^r = P_k^r\left(v_k^r, k \in K^r\right) \quad k \in K^r,\ r \in R \tag{7}$$

The hypotheses are sufficiently general to admit a range of behavioural assumptions through the form of the joint distribution for the stochastic terms, thus encompassing various additive models, including, but not restricting to, multinomial logit.

A SUE is a solution of the fixed-point problem in the path flows F_k^r:

$$F_k^r = \dot{q^r} \cdot P_k^r\left(F_k^r,\ k \in K^r,\ r \in R\right) k \in K^r,\ r \in R \tag{8}$$

which is obtained by chaining the expressions (4) of the systematic utilities in the path travel times, the expressions (3) of the path travel times in the link travel times, the link travel times in the link flows, and the expressions (1) of the link flows in the path flows. The non-negativity constraints on the path flows are redundant because the probabilities are non-negative.

In the light of the Brouwer's fixed point theorem, a solution to SUE exists since the feasible set is non empty, compact and convex (having taken into account the demand and the non-negativity constraints) and all the functions composed to form the fixed-point formulation are continuous. It is possible to prove that, under an additional assumption on the monotonicity of the link time-flow functions, the solution is unique (see, among the others, Cascetta [1]). SUE can be computed by the method of successive averages (MSA) algorithm.

3 Transition Choice Probabilities and Welfare

3.1 *Transition Choice Probabilities*

For a straightforward derivation of the quantities of interest, we use classical probability theory (introduced by Kolmogorov [9]) which associates events to sets of the random terms.

Consider a n-variate random variable $\mathbf{X} \in \mathbf{R}^n$ in the n-dimensional Euclidean space $\mathbf{R}^n$ with probability density function $f(\mathbf{X})$. Consider an event represented by the set $S \subset \mathbf{R}^n$. The indicator function I is defined as follows:

$$I(\mathbf{X} \in S) = \begin{cases} 1 & \mathbf{X} \in S \\ 0 & otherwise \end{cases} \tag{9}$$

Thus, the probability of the event S is:

$$\Pr(S) = \int \cdots \int_{\mathrm{R}^n} I(\mathbf{X} \in S) \cdot f(X_1, \ldots, X_n) \cdot dX_n \ldots dX_1 \tag{10}$$

We apply these results to the RUM-based representation of path choice. Hereafter, we make reference to a specific OD pair and omit the corresponding index r. We denote by $|K|$ the cardinality of the path set and by $f(\boldsymbol{\varepsilon})$ the probability density function of the random terms $\varepsilon = \left[\varepsilon_1, \ldots, \varepsilon_{|K|}\right]^T$.

Consider the event S_k in the $|K|$-dimensional Euclidean space $\mathbf{R}^{|K|}$of the random terms $\boldsymbol{\varepsilon}$ that path k is chosen:

$$S_k = \left\{u_k \geq u_j \quad \forall j \neq k \in K\right\} = \left\{\varepsilon_j \leq \varepsilon_k + v_k - v_j \quad \forall j \neq k \in K\right\} \tag{11}$$

The probability of choosing path k is:

$$P_k = \Pr(S_k) = \int \ldots \int_{\mathrm{R}^{|K|}} I(\boldsymbol{\varepsilon} \in S_k) \cdot f\left(\varepsilon_1, \ldots, \varepsilon_{|K|}\right) d\varepsilon_{|K|} \ldots d\varepsilon_1 \tag{12}$$

Consider two states of the network, the state before the intervention and the state after. We denote by the prime symbol "/" quantities in the before state, by the double prime symbol "//" quantities in the after state. Notice that the intervention may modify the monetary cost of the paths, the link-flow relationships and the path choice set. As a result, a new SUE is set.

The transition probability from path k to path m is the probability that path k is chosen in the before state and path m is chosen in the after state. Computation of the transition probabilities requires making assumptions on how the random terms are correlated over the choices in the two states. Thus, we need to consider the joint before-after distribution of the random terms. We denote by $h\left(\varepsilon'_1, \ldots, \varepsilon'_{|K'|}, \ldots, \varepsilon''_{|K''|}\right)$ the joint before-after probability density function of the random terms.

Consider the event that path k is chosen in the before state:

$$S'_k = \left\{u'_k \geq u'_j \quad \forall j \neq k \in K'\right\} \tag{13}$$

the event that path m is chosen in the after state:

$$S''_m = \left\{u''_m \geq u''_j \quad \forall j \neq m \in K''\right\} \tag{14}$$

and the event that path k is chosen before and path m after: $S_{k \to m} = S'_k \bigcap S''_m$.

The transition probability $P_{k \to m}$ from path k to path m is:

$$P_{k \to m} = \Pr(S_{k \to m}) = \int \ldots \int_{\mathrm{R}^{|K'|+|K''|}} I\left([\varepsilon', \varepsilon'']^T \in S_{k \to m}\right) \cdot h\left(\varepsilon'_1, \ldots \varepsilon'_{|K'|}, \varepsilon''_1, \ldots \varepsilon''_{|K''|}\right) d\varepsilon''_{|K''|} \ldots d\varepsilon''_1 d\varepsilon'_{|K'|} \ldots d\varepsilon'_1 \tag{15}$$

To compute the transition probability by simulation, i.e. by drawing from the distribution of the random terms, we use the frequency estimator (Lerman and Manski [10]):

$$\hat{P}_{k\to m} = \frac{1}{T} \cdot \sum_{t=1}^{T} I\left(\left[\boldsymbol{\varepsilon}'_t, \boldsymbol{\varepsilon}''_t \right]^T \in S_{k\to m} \right) \tag{16}$$

where t denotes the draw and T the number of draws. The frequency estimator is minimum variance unbiased and strongly consistent. The frequency estimator is referred to by Train [14] as accept-reject simulator because it equals the proportion of draws that are "accept" with respect to the transition regions of the $|K'| + |K''|$-dimensional Euclidean space $\mathbf{R}^{|K'|+|K''|}$ of the before and after random terms.

The transition probabilities are particularly easy to compute when the random terms are independent across choices. In this case the two events $S_k{}'$ and $S_m{}''$ are independent and, therefore, we have:

$$P_{k\to m} = \Pr(S_{k\to m}) = \Pr\left(S'_k\right) \cdot \Pr\left(S''_m\right) \tag{17}$$

The transition probability from k to m equals the probability of choosing k before times the probability of choosing m after.

3.2 Welfare

The interest is in measuring the welfare change associated with the intervention. The income y is assumed to be unchanged. The compensating variation is considered. The compensating variation is defined as the income that needs to be taken from the individual in the state after the change in order to bring her to the condition of utility of the state before the change.

The random compensating variation cv conditional on the vector of before random terms and the vector of after random terms satisfies:

$$\max_{k\in K'} \left[v'_k + \varepsilon'_k \right] = \max_{k\in K''} \left[v''_k (y - cv) + \varepsilon''_k \right] \tag{18}$$

where $v''_k(y - cv)$ denotes that the systematic utility $v_k{}''$ is a function of the compensated income $y - cv$.

Thus, the compensating variation is a function of the random terms: $cv = cv(\varepsilon', \varepsilon'')$.

The expectation of the compensating variation $E_{k\to m}[cv]$ conditional on the transition from path k to path m is given (by definition of expectation) by:

$$E_{k\to m}[cv] = \int \cdots \int_{\mathrm{R}^{|K'|+K''}} cv(\varepsilon', \varepsilon'') \cdot I\left(\left[\varepsilon', \varepsilon''\right]^T \in S_{k\to m} \right) \cdot h\left(\varepsilon'_1, \ldots \varepsilon'_{|K''|}, \varepsilon''_1, \ldots \varepsilon''_{|K'|} \right)$$
$$\varepsilon''_{|K''|} \ldots d\varepsilon''_1 d\varepsilon'_{|K'|} \varepsilon'_1 \tag{19}$$

To compute the conditional expectation of the compensating variation by simulation we use the sample mean estimator:

$$\hat{E}_{k\to m}[cv] = \frac{1}{T}\cdot\sum_{t=1}^{T} cv\left(\boldsymbol{\varepsilon}_t', \boldsymbol{\varepsilon}_t''\right)\cdot I\left(\left[\boldsymbol{\varepsilon}_t', \boldsymbol{\varepsilon}_t''\right]^T \in S_{k\to m}\right) \tag{20}$$

The sample mean estimator is unbiased and strongly consistent.

The unconditional expectation of the compensating variation $E[cv]$ is given (again, by definition of expectation) by:

$$E[cv] = \int\cdots\int_{\mathrm{R}^{|K'|+|K''|}} cv(\varepsilon', \varepsilon'')\cdot h\left(\varepsilon_1', \ldots \varepsilon_{|K''|}', \varepsilon_1'', \ldots \varepsilon_{|K'|}''\right) \\ d\varepsilon_{|K''|}'' \ldots d\varepsilon_1'' d\varepsilon_{|K'|}' \ldots d\varepsilon_1' \tag{21}$$

and can be computed by simulation using the sample mean estimator:

$$\hat{E}[cv] = \frac{1}{T}\cdot\sum_{t=1}^{T} cv\left(\boldsymbol{\varepsilon}_t', \boldsymbol{\varepsilon}_t''\right) \tag{22}$$

In the case without income effects of Eq. (5) the unconditional expectation of the compensating variation $E[cv]$ is independent of the before-after correlation, i.e. depends only on the marginal distributions of the random terms (i.e. the distribution over the single choice). Thus, in the case of multinomial logit it is given by the logsum formula. This property has been proved by Delle Site and Salucci [5] and is confirmed numerically (see also Zhao et al. [16]). Notice that this property applies only to the unconditional expectation, not to the conditional expectation.

4 Numerical Illustration

4.1 Route Choice Model

We use a multinomial logit route choice model estimated on the basis of data from a stated preference survey which took place in Rome in 2007. The survey (1068 observations) investigated the preferences of the potential users of a new bypass subject to a toll planned in the southern-eastern part of the urban area.

The systematic utilities take the form in Eq. (5), i.e. we consider a case without income effects. The estimate of the marginal utility of income α is 1.52248 (measurement unit of the attribute is EUR/trip; t-statistic is 14.447). The estimate of the time coefficient β is –0.10796 (measurement unit of the attribute is minutes/trip; t-statistic is –10.868).

Table 1 Two-link network, SUE

Link	Before			After		
	Share	Time (minutes/trip)	Cost (EUR/trip)	Share (%)	Time (minutes/trip)	Cost (EUR/trip)
1	100 %	31.6	0	76.1	6.06	0
2	–	–	–	23.9	2.70	1

4.2 Two-Link Network

We consider the classical textbook case of a town bypass with toll. In the before state we have only the town-centre route. The intervention consists in building the bypass. Thus, in the after state we have two routes, the town-centre route and the bypass route, with the bypass subject to a toll.

We assume a total demand of 1000 veh/h. For supply, Bureau of Public Roads (BPR) time-flow functions derived empirically for similar routes are used. The functions (in hours) are $t = 0.057 \cdot [1 + (z/800)^{5.2}]$ for the town-centre route, and $t = 0.045 \cdot [1 + 0.68 \cdot (z/1230)^{4.6}]$ for the bypass route.

The results of the SUE computation for the before and after states are in Table 1. The town-centre route is highly congested in the before state. Congestion is alleviated in the after state. The share of users on the town-centre route is higher than the share on the bypass route due to the combined effect of travel time and toll.

For the disaggregate users' benefits analysis we consider the case where the random terms are independent across the two states, and the case where they are unchanged, i.e. perfect correlation. These are cases where simulation can be carried out based only on the marginal distributions of the random terms (i.i.d. Gumbel across alternatives in multinomial logit). Draws are easily obtained using the inverse cumulative distribution function method applied to the Gumbel distribution.

To compute the expectations of the compensating variation, income, which is not considered in the econometric specification of the route choice model, is needed. Income is assumed to be 1000 EUR/month and is referred to a single trip on the network (the timeframe of the monetary cost in the model estimation) on the basis of a frequency of 40 trips/month (i.e. one outward and one return trip per weekday).

The results of the analysis are in Table 2. In the case of perfect correlation the shifters from the town-centre route to the bypass route gain more than those remaining on the town-centre route (2.56 EUR/trip vs 1.81 EUR/trip). In the case of independent random terms the benefit is the same for shifters and non-shifters and equals the benefit for the entire population (1.99 EUR/trip). The benefit for the entire population, which is provided by the unconditional expectation of the compensating variation, is independent of the before-after correlation assumption.

Table 2 Two-link network, disaggregate users' benefits analysis

		Random terms independent	Unchanged
Transition	Share (%)	$E(cv)$ (EUR/trip)	$E(cv)$ (EUR/trip)
1–1	76.1	1.99	1.81
1–2	23.9	1.99	2.56
All	100	1.99	1.99

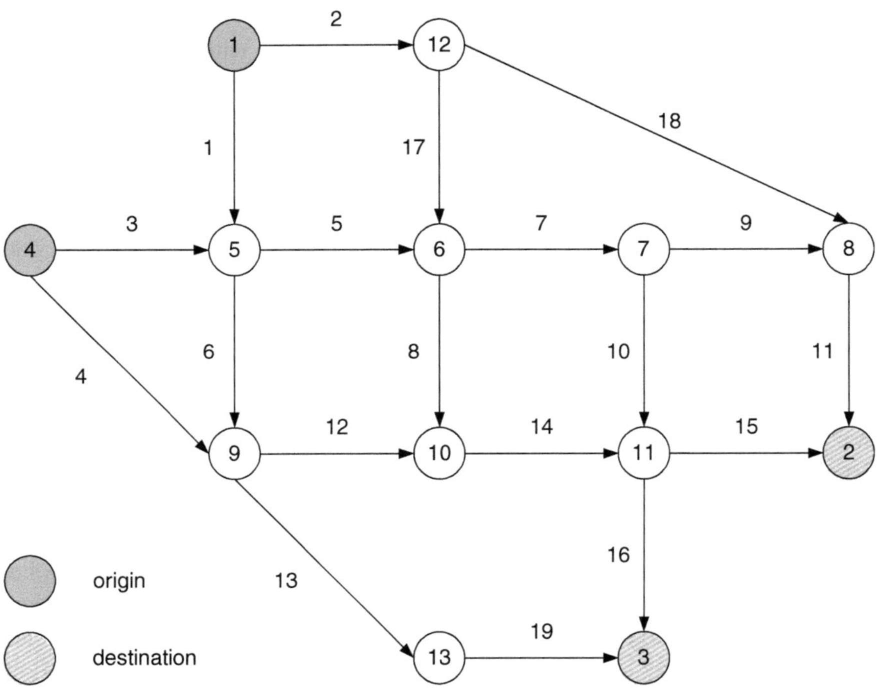

Fig. 1 Nguyen-dupuit network

This result is in line with theoretical and numerical results in the literature (Delle Site and Salucci [5]; Zhao et al. [16]).

4.3 *Nguyen-Dupuit Network*

In the second example, the Nguyen-Dupuis network (Nguyen and Dupuis [12]) is used. The network, which includes 13 nodes, 19 directed links and 4 OD pairs, is shown in Fig. 1. The link-path incidence relationship is shown in Table 3. There is a total of 25 paths.

Table 3 Nguyen-dupuit network, link-path incidence relationship

OD pair	Path	Link sequence	OD pair	Path	Link sequence
(1, 2)	1	2-18-11	(1, 3)	9	2-17-8-14-16
	2	2-17-8-14-15		10	2-17-7-10-16
	3	2-17-7-10-15		11	1-6-13-19
	4	2-17-7-9-11		12	1-6-12-14-16
	5	1-6-12-14-15		13	1-5-8-14-16
	6	1-5-8-14-15		14	1-5-7-10-16
	7	1-5-7-10-15			
	8	1-5-7-9-11			
(4, 2)	15	4-12-14-15	(4, 3)	20	4-13-19
	16	3-6-12-14-15		21	4-12-14-16
	17	3-5-8-14-15		22	3-6-13-19
	18	3-5-7-10-15		23	3-6-12-14-16
	19	3-5-7-9-11		24	3-5-8-14-16
				25	3-5-7-10-16

Table 4 Nguyen-Dupuit network, link characteristics

Link	Free-flow travel time (minutes/trip)	Capacity (veh/h)	Link	Free-flow travel time (minutes/trip)	Capacity (veh/h)
1	7	300	11	9	500
2	9	200	12	10	550
3	9	200	13	9	200
4	12	200	14	6	400
5	3	350	15	9	300
6	9	400	16	8	300
7	5	500	17	7	200
8	13	250	18	14	300
9	5	250	19	11	200
10	9	300			

The OD demand flows (in veh/h) are $q_{1,2} = 660, q_{1,3} = 495, q_{4,2} = 412.5, q_{4,3} = 495$. The following BPR time-flow functions are used (time in minutes, flow and capacity in veh/h): $t_a = t_a^0 \cdot [1 + 0.15(z_a/c_a)^4]$, where the free-flow travel time t_a^0 and the capacity c_a are given, for each link, in Table 4 (the values of the OD flows and of the parameters of the time-flow functions are from Xu et al. [15]).

In the before state we consider the network without link number 10. In the after state the network with the full set of links. There is no toll in the before state and in the after state.

Table 5 shows the results of the SUE computations (obtained with a MSA algorithm) for the OD pair (4,2). In the after state there is a new path, namely path number 18.

Table 5 Nguyen-Dupuit network, SUE results for the OD pair (4,2)

Path	Before		After	
	Share (%)	Time (minutes/trip)	Share (%)	Time (minutes/trip)
15 (4, 12, 14, 15)	28.24	102.15	32.26	93.45
16 (3, 6, 12, 14, 15)	10.63	111.09	11.25	103.09
17 (3, 5, 8, 14, 15)	8.98	112.74	7.29	107.19
18 (3, 5, 7, 10, 15)	–	–	14.30	100.96
19 (3, 5, 7, 9, 11)	52.15	96.44	34.90	92.68

Table 6 Nguyen-dupuit network, disaggregate users' benefits analysis for the OD pair (4,2) and transitions to the new path

Transition	Random terms independent		Unchanged	
	Share (%)	$E(cv)$ (EUR/trip)	Share (%)	$E(cv)$ (EUR/trip)
15–18	4.02	0.53	3.13	1.31
16–18	1.54	0.53	1.27	1.27
17–18	1.28	0.54	1.31	1.12
19–18	7.45	0.53	8.58	1.03

For the disaggregate users' benefits analysis, the same assumptions on the before-after correlation of the random terms (i.e. independence and perfect correlation) and on income as in the two-link network example are made.

Table 6 shows the results of the analysis for the OD pair (4,2) and the sub-populations of shifters to the new path. The benefit tends to converge to a common value across transitions in the case of independence, a result also found in the two-link network example.

5 Conclusions

The chapter has shown how it is possible to carry out a rigorous welfare analysis in a road network and attribute benefits to shifters and non-shifters. Cases where the intervention modifies the set of route alternatives are accounted for. The shares of shifters and non-shifters and the benefits attributed to them can be affected significantly by the assumption on the before-after correlation of the random terms.

The numerical examples have considered the extreme cases of independence and perfect correlation. Intermediate cases can be considered when the joint before-after distribution of the random terms is available. Generally, there is more than one distribution with given marginals and covariance matrix. When the marginals are multinomial logit, it is possible to use the bivariate Gumbel type C distribution (introduced by Tiago de Oliveira [13]; a review is in Garrow et al. [7]; for its use in welfare analysis see Delle Site and Salucci [5]), for which closed-form expressions are available to generate draws.

The identification of the before-after correlation is an area of future research. In principle, it is possible to estimate the correlation as an additional parameter of a RUM that considers the joint probabilities over repeated choices. Two are the research problems in this respect: the data that need to be collected and the estimation procedures.

References

1. Cascetta, E.: Transportation Systems Analysis. Models and Applications. Springer, New York (2009)
2. Daganzo, C., Sheffi, Y.: On stochastic models of traffic assignment. Trans. Sci. **11**(3), 253–274 (1977)
3. Dagsvik, J.K., Karlström, A.: Compensating variation and hicksian choice probabilities in random utility models that are nonlinear in income. Rev. Econ. Stud. **72**(250), 57–76 (2005)
4. De Jong, G., Daly, A., Pieters, M., van der Hoorn, T.: The logsum as an evaluation measure: review of the literature and new results. Transp. Res. Part A **41**(9), 874–889 (2007)
5. Delle Site, P., Salucci, M.V.: Transition choice probabilities and welfare analysis in random utility models with imperfect before-after correlation. Transportation Research Part B-Methodological **58**, 215–242 (2013)
6. De Palma, A., Kilani, K.: Transition choice probabilities and welfare analysis in additive random utility models. Econ. Theor. **46**(3), 427–454 (2011)
7. Garrow, L.A., Bodea, T.D., Lee, M.: Generation of synthetic datasets for discrete choice analysis. Transportation **37**(2), 183–202 (2010)
8. Jara-Díaz, S.R.: Transport Economic Theory. Elsevier, Amsterdam (2007)
9. Kolmogorov A.N.: Foundations of the Theory of Probability. 2nd English Edition. Chelsea Publishing Company, New York (1956)
10. Lerman, S., Manski, C.: On the use of simulated frequencies to approximate choice probabilities. In: Manski, C., McFadden, D. (eds.) Structural Analysis of Discrete Data with Econometric Applications, pp. 305–319. MIT Press, Cambridge MA (1981)
11. McFadden, D.: Computing willingness-to-pay in random utility models. In: Melvin, R., Moore, J.C., Riezman, R. (eds.) Trade Theory and Econometrics—Essays in Honor of John Chipman, pp. 253–274. Routledge, London (1999)
12. Nguyen, S., Dupuis, C.: An efficient method for computing traffic equilibria in networks with asymmetric transportation costs. Trans. Sci. **18**(2), 185–202 (1984)
13. Tiago de Oliveira, J.: Bivariate extremes: foundations and statistics. In: Krishnaiah P.R. (ed) Multivariate Analysis V, pp. 349–366. North Holland, Amsterdam (1980)
14. Train, K.: Discrete Choice Methods with Simulation. Cambridge University Press, Cambridge (2009)
15. Xu, H., Lou, Y., Yin, Y., Zhou, J.: A prospect-based user equilibrium model with endogenous reference points and its application in congestion pricing. Transp. Res. Part B **45**(2), 311–328 (2011)
16. Zhao, Y., Kockelman, K., Karlström, A.: Welfare calculations in discrete choice settings: an exploratory analysis of error term correlation with finite populations. Transp. Policy **19**(1), 76–84 (2012)

Part VII
Mobility, Accessibility and Travel Behavior

Mobility is about moving people and goods from place to place. *Accessibility* involves something which is easily approached, entered, obtainable, or attained. Mobility provides access, but it is not access. Also, accessibility does not provide mobility. Improving both mobility and accessibility is good for society and cities. *Travel behavior* is the study of what people do in space, and how they use transport, and these two concepts appear to be connected in some way. *Eiró and Martinez* develop a new modelling approach using structural equation models, within a path analysis approach which not only clarifies which attributes influence satisfaction but also allows its values to be predicted according to measurable exogenous variables. *Nuzzolo et al.* present a research project to define an advanced trip planner for transit networks, describing both the logical architecture of the trip planner and the theoretical aspects of the path choice model. The module supporting users with personalized pre-trip information based on their preferences are also presented. *Filgueiras et al.* provide a proof-of-concept deployment of sensors using Bluetooth technology to detect traffic flow conditions: besides the traditional method, consisting of a network of stationary sensors, these authors develop a novel approach which applies sensors deployed in moving vehicles. *Li Jie et al.* present a Driving Behavior Questionnaire (DBQ), to collect information about drivers' attitudes and behavior and their possible impact on safety and traffic performance. The authors report that there are significant differences in road traffic performance between drivers in China and The Netherlands, and identify the human factor, which plays an important role in traffic performance, as the primary cause of this difference.

Modelling Daily Mobility Satisfaction Using a Structural Equation Model

Tomás Eiró and Luis M. Martínez

Abstract Recent research has started to focus on understanding the elements that influence the perception of users with the existent mobility options and their impact over the stated performance or satisfaction evaluation. The main methodology that has been applied to extract this information has been confirmatory factor analysis under a Structural Equation Modelling (SEM) framework. With this work, we intend to develop a new modelling approach using structural equation models, under a path analysis approach that not only allows understanding which attributes influence satisfaction but also allows predicting its values based on the exogenous measurable variables. The model obtained, presents an acceptable fit and was able to provide some insightful conclusions on the relation between satisfaction and accessibility, mobility, land use and socio-demographic characteristics.

Keywords Structural equation mode · Path analysis · Mobility satisfaction

1 Introduction

Urban transportation has been identified as one of the critical concerns of modern societies, especially in mega cities. Mobility, while a fundamental element of economic competitiveness and instrument to exert some of the main social rights, may produce large externalities, especially in societies with an abusive use of private cars. The increase in the mobility chain complexity, associated with an

T. Eiró (✉) · L. M. Martínez
CESUR, Department of Civil Engineering, Instituto Superior Técnico, Lisbon Technical University, Avenida Rovisco Pais, Lisboa 1049-001, Portugal
e-mail: tomas.eiro@ist.utl.pt

L. M. Martínez
e-mail: luis.martinez@ist.utl.pt

J. F. de Sousa and R. Rossi (eds.), *Computer-based Modelling and Optimization in Transportation*, Advances in Intelligent Systems and Computing 262,
DOI: 10.1007/978-3-319-04630-3_29,

intense urban sprawl and the inability of many public transport systems to respond to all this requirements has led to this unbalanced mode share that favours private transport options.

In order to try to improve the transportation system, many studies, around the world, have been trying to establish casual relationships between trip characteristics and the levels of satisfaction with the quality of services provided or supplied, that may help explaining the observed modal share.

The majority of these studies are focusing on an individual assessment of each trip instead of analysing the entire trip chain. This trip based assessment might be misleading since the good performance of mode for a specific trip might be penalised by subsequent or precedent trips with a low performance for a given mode, jeopardising the overall mobility level of satisfaction, which can induce a new mode choice selection with a more balanced performance for the whole chain.

A survey conducted in 2011, in the Lisbon Metropolitan Area, under the MIT-Portugal SCUSSE project, which intended not only to examine the current mobility patterns of the LMA but also to assess the potential of implementing new alternative transport alternatives was used. The levels of satisfaction of the respondents' with their mobility chain as the depiction of the main attributes that drive their mode choice were also collected due to their relevance on assessing mode diversion.

With this study we intend to develop a Structural Equation Model (SEM) that tries to elucidate the interrelationship between the observed variables (accessibility, mobility, socio-demographic and land use attributes) and unobserved variables (attitudes towards different mode share attributes and satisfaction with the characteristics of the main mobility mode used) and their impact to the overall individual's trip chain satisfaction.

This assessment will allow a better understanding of the conditions that imply an evaluation of mode choice by users, and identify the situations when users may consider searching for alternative transport options.

This chapter is divided in six different sections: after this brief introduction we will provide a brief literature review about the main research that has been developed in understanding what influences satisfaction in transportation and the main methodologies that have been applied; afterwards, we will present the data collection process and the model development; we end this chapter with a brief analysis of the obtained results and some conclusions.

2 Literature Review

The current trend in developed countries of public transport modal share decline has driven several policy institutions and researchers of this field to investigate how users' satisfaction is formed when travelling. Therefore, transportation systems are evolving to more market oriented services, since users are starting to regard transportation as a service product [20]. This requires the system to start providing services that perform accordingly to its travellers' expectations guarantying acceptable levels of satisfaction. Not only in terms of intrinsic service characteristics (e.g. travel time, service frequency), but also in terms of less easily measured characteristics that depend on costumer tastes (e.g. comfort). Satisfaction has been identified as the main driver of consumer behaviour and loyalty [22].

The assessment of the quality of a service is supported by four main pillars that interact at different levels: expected, perceived, targeted and delivered quality. The first two aspects are related to the customer while the last ones are influenced by the ability of services to match the users' expectations.

In the public transport sector, operators establish their levels of service based on users' expectations. Nevertheless, the service delivered might not be in accordance with the targeted levels, which indicates the measurement of the operator's performance, and might not be perceived by users as they aimed. The relationship between the expected and perceived value of a service is a measure of customer satisfaction.

The understanding of these relationships are really important and might dictate the success on the introduction of a new service in the market.

In the last years, much research has been trying to define measurement indexes of satisfaction levels in services and products.

One of the most well-known indexes is the Service Quality Method developed in Parasuraman et al. [23]. This measurement indicator is defined through a function of the consumers' expectations and their perceived value. It comprises five service quality dimensions and 22 items for measuring service quality that are obtained through a likert scale on seven levels of agreement/disagreement. The index is calculated by the difference between perception and expectation rates expressed for the items, weighted as a function of the five service quality dimensions embedding the items. Nevertheless, this index has the problem of needing to assign a value to each level of judgment, which might not reproduce accurately the reality, since the same level of judgments amongst respondents might imply different values. Other approaches have tried to improve this index like the ServPerf [2] and the Normed Quality Model [26].

Other satisfaction indexes have also tried to find casual relations between land use, accessibility and mobility patterns with travel satisfaction, some containing latent or unobserved variables.

A more direct measure for service quality evaluation is the Customer Satisfaction Index developed by Hill et al. [16], which represents a measure of service quality on the basis of the user/consumer perceptions on service aspects expressed in terms of importance rates, compared with user/consumer expectations expressed in terms of satisfaction rates. This index is simply calculated as a weighted function of the importance weight of each service attribute and the mean of the satisfaction rates expressed by users. Yet, this indicator does not into account the heterogeneity amongst user judgments. So, Eboli and Mazzulla [9] proposed a variation of this index, the Heterogeneous Customer Satisfaction Index (HCSI), that corrected the obtained users' evaluations according to the dispersion of the rates from the average value.

Some researchers have also been developing service quality indexes based on discrete choice models and random utility theory. The Service Quality Index (SQI) was firstly introduced by Hensher and Prioni [14] where they developed a Multinomial Logit (MNL) model based on stated preferences (SP) experiments. The design of these experiments might be very complex where an example can be found in [7]. Since then, many other model specifications have been tested always supported by SP data (e.g. Hierarchical Logit models ([11, 15]), Mixed Logit models ([8, 13])).

Del Castillo and Benitez [4] propose a more complex approach that mixes different analysis tools in an holistic methodology. In their work they use SP data to calibrate three different models (a model based on means, a model based on a statistical distribution and a generalized linear model). By using these three different models they are able to obtain a more robust evaluation of the service's quality.

A more recent approach has been the use of Structural Equation Models (SEMs) that are having a wide application in psychological and social sciences, natural sciences, economics and statistics. SEM has the advantage of being parsimonious models, since they use a linear approach, but, at the same time, they are able to represent complex systems and evaluate non-observable variables while taking measurement errors into account.

The application of this methodology in the evaluation of transport modes is still not widespread.

Eboli and Mazzulla [6] used SEM to measure the impact of several public transportation attributes mainly concerned with connectivity, service reliability and some other factors that include accessibility aspects and pricing. Githui, Okamura and Nakamura [12] also used SEM to investigate the relationship between several public transport service attributes and the satisfaction of travellers. Their results show that increasing Service Quality (related mostly with service reliability and connectivity) and Safety, and decreasing Travel Cost, make the service more attractive and may lead to an increase in ridership.

The current research work on customer satisfaction in the mobility system has been focusing on the assessment of single trips and not focusing on the entire trip chain. The performance of a single trip might not be representative of the entire mobility agenda and it might condition the assessment of satisfaction. Moreover, the confirmatory characteristic of these models only allows evaluating the relative impact of attributes on the satisfaction, and do not estimate absolute values of satisfaction. The estimation of such values would be helpful in predicting the propensity of travellers to search for new mobility solutions, or for more permanent solutions as relocating to places where their mobility needs are better fulfilled.

In this work, we not only intend to develop a SEM that moves from a single trip evaluator model to a model that encompasses the entire trip chain, but also to develop a model that is able to predict the satisfaction given a set of attributes.

3 Data Collection and Processing

As above mentioned, the data used to calibrate our model was based on a survey conducted in the Lisbon Metropolitan Area in 2011 through the MIT-Portugal SCUSSE project [24].

The data retrieving process included an online survey and a domiciliary computed assisted personal interviewing with 1,000 additional responses.

Besides having a socio-demographic section and all the traditional stated and revealed preferences questions of a mobility study, this survey included two specific sections that tried to classify the quality of the existing mobility choices.

One of the sections tried to assess how each respondent was satisfied with his/her current mobility chain using a likert scale from very unsatisfied to very satisfied using seven levels. This variable was used as the main dependent variable of the model.

The other component of the survey included attributes that drive mode choice selection based on the work of [25], where she tried to assess the factors that drive mode choice selection and a natural bias towards private car.

A similar structure was followed in the current study classifying the importance of variables on a likert scale from one to seven. Moreover, the used methodology inspected the differences between systematic or sporadic trips, and the satisfaction of the respondent with these factors in his current mobility options, filtering just the factors considered as relevant in the previous analysis of systematic attributes (classifications greater or equal to five).

The survey contained 14 different characteristics, ranging from attitudinal statements (i.e. freedom, flexibility), mode specific attributes (time and cost) and safety and security issues.

This data was processed through a factorial analysis process which will be explained in the following section.

It was also necessary to generate all the transportation alternatives for all the trips retrieved from the survey. It was used a discretization of the study area in 281 zones, adopted for the survey design to characterise transportation infrastructure and services of the LMA.

4 Development of New Mobility Satisfaction Model

The model developed to assess the satisfaction of citizens with their current mobility was formulated using a Structural Equation Modelling framework.

SEM is a multi-equation method that enables the joint estimation of direct effects and several endogenous characteristics, normally represented through latent variables, which cannot be clearly captured through measured variables. The two more common applications of SEMs are: confirmatory analysis, where the objective is to test whether a set of data fits an a priori hypothesized measurement model; and path analysis that is used to measure the direct dependencies among a set of variables. The majority of applications of SEMs are using confirmatory analysis with latent variables.

When analysing SEMs' results, there are three types of effects that should be taken into consideration: direct effects, indirect effects and total effects. Direct effects are the coefficients of the model obtained from the relation between a dependent and independent variables in each equation of the model, while indirect effects represent the influence of a variable through the mediation of at least a third variable; and, total effects are the sum of the direct and indirect effects and represent the actual effect off each variable on the dependent variable [19].

The estimation of SEMs is performed through the method of moments for continuous variables, where the objective is to minimize the differences between the sample variance–covariance matrix and the model-replicated matrix. The methods more commonly used are maximum likelihood, generalized least squares and weighted least squares.

The developed methodology makes use of a path diagram and has the particularity of using a forward procedure that enables the estimation of the satisfaction levels given a set of observed measurable variables. In our estimation process, we used maximum likelihood including the estimation of means and intercepts. Given the characteristics of our estimation procedure, the main dependent variable, overall satisfaction level of trip chain, was considered to be continuous instead of the discrete Likert scale with seven levels. Nevertheless, more than five levels of ordinal of data have been found to be correctly estimated by continuous formulations ([1, 5, 17, 18]).

Given this objective, the use of latent endogenous variables was avoided in order to ensure the use of the resulting model for prediction.

Instead, using the ratings of the retrieved mode's characteristics as inputs, two factor analysis were performed to extract attitudinal constructs towards mode choice selection, using the principal components as extraction method, and as number of factors threshold an eigenvalue equal to one. Afterwards, the obtained results were rotated through a Varimax procedure.

The first factor analysis considered both the systematic and sporadic evaluation in order to obtain what attribute people consider relevant when choosing a transportation mode. The factor analysis results are summarised in Table 1, showing the variables with higher loads at each factor of the eight computed factors. The statistical tests obtained reveal an acceptable data reduction (Kayser-Meyer-Olkin (KMO) measure of 0.636).

The second factor analysis was based on the responses from the satisfaction levels of the actual transport mode and enabled to identify the quality of the current transport mode of each respondent. The factor analysis results are summarised in Table 2, once again just presenting the variables with higher load in each factor presenting a KMO of 0.868, which indicates significant and efficient data reduction.

Other variables were also computed to be inputted in our structural equation model. These variables were based on the responses of the RP section of the

Table 1 Factor analysis over systematic and sporadic evaluation

Factor	Variables	Trip type	
		Systematic	Sporadic
Factor 1 (SECURITY)	Safety (e.g. Assaults)	0.853	0.845
	Risk of accident	0.820	0.815
	Price	0.529	0.361
Factor 2 (FREEDOM)	Freedom	0.681	0.690
	Control	0.774	0.796
	Flexibility	0.500	0.536
Factor 3 (STATUS)	Status	0.597	0.611
	Pleasure or satisfaction	0.781	0.797
Factor 4 (EXTRA)	Possibility of doing extra tasks on-board	0.817	0.790
Factor 5 (ENVIRON)	Environmental concern	0.857	0.881
Factor 6 (FIT)	Adequate to mobility needs	0.848	0.799
Factor 7 (COMFORT)	Comfort	0.822	0.798
Factor 8 (TIMECOST)	Cost	0.524	0.693
	Flexibility	0.442	0.453
	Travel time	0.369	0.436

Table 2 Factor analysis over satisfaction of actual mode

Factor	Variables	Trip type Actual
Factor 1 (Performance)	Comfort	0.603
	Adequate to mobility needs	0.830
	Freedom	0.815
	Control	0.670
	Flexibility	0.802
	Travel time	0.674
Factor 2 (CurrentStatus)	Stress	0.581
	Status	0.792
	Pleasure or satisfaction	0.672
Factor 3 (CostSafe)	Risk of accident	0.748
	Environmental concern	0.732
	Price	0.693

survey and include socio-demographic variables, land-use variables and general characteristics of the mobility chain reported by the respondents. Although several other variables were computed and tested in our model we will only present the ones that ended up being used:

- 18–25—binary variable that takes the value of one for people with age between 18 and 25 years old;
- More65—binary variable that takes the value of one for people with more than 65 years old;

- ActTime—total activity time in hours;
- ActualStraight—ratio between the total reported travel time and the total travel time calculated through the quotient of the total Euclidean travel distance and a constant speed of 25 km/h;
- AvDist—Average Euclidean travelled distance;
- Basic—binary variables that takes the value of one for people with primary school level or below;
- DistBus—minimum Euclidean distance from a bus stop to the respondent's house location;
- Entropy—continuous variable that ranges from zero (representing a homogeneous area that only has a single type of activity) to one (representing a heterogeneous area where all land-uses are equally distributed). The considered index presents 10 different land-uses that includes residential, retail, education, etc.
- Gender—binary variable that takes the value of one for females and zero for males;
- High—binary variable that takes the value of one for people with a Bachelor or a higher degree;
- Ntrips—number of trips performed;
- OfferActual—ratio between the minimum possible total travel time, including all the transportation options, and the experienced by the respondent according to his/her mode choices;
- OnlyEco—binary variable that takes the value of one if the entire mobility chain only has eco-friendly options (heavy modes or walking);
- OnlyPriv—binary variable that takes the value of one if the entire mobility chain only has private transport modes;
- OnlyTasks—binary variable that takes the value of one if the entire mobility chain only has modes that allow performing extra tasks on board (bus, heavy mode or a mixture of both);
- OwnCar—binary variable that takes the value of one if the respondent has a private car;
- Pass—binary variable that takes the value of one if the respondent has a public transportation pass;
- Poor—binary variable that takes the value of one if the respondent's household has a monthly income level below 2000€;
- RatioTCTI—ratio between the gravity based accessibility in public transportation and private transportation at the respondent's residential location;
- TransfCM—average number of transfers of the respondent's mobility chain;
- TransfPT—minimum average number of transfers that the respondent would experience if only public transport modes were used.

As mentioned above, several model specifications were tested based on the correlation matrix between variables considered for the model. The final model specification is presented in Fig. 1 along with the standardized regression weights and the intercepts, and the correspondent t-statistics.

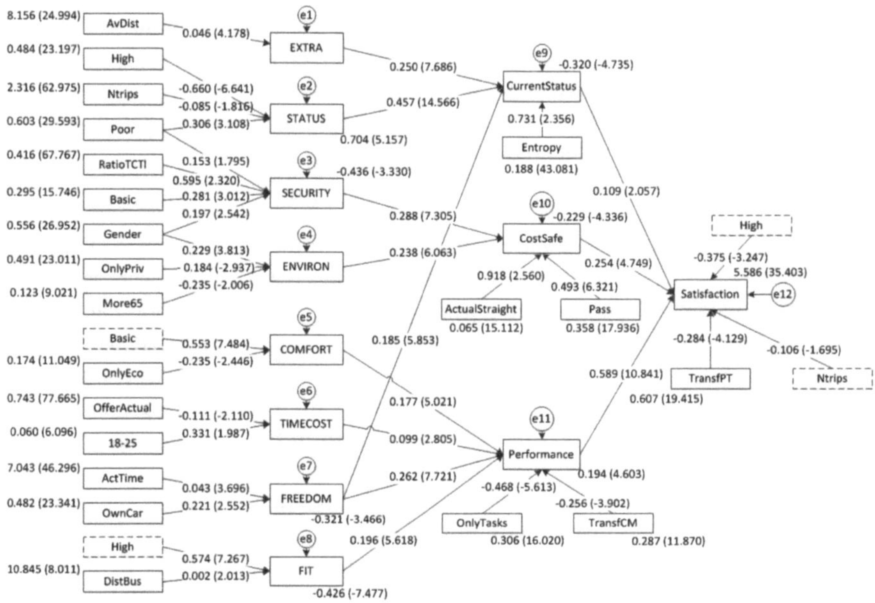

Fig. 1 Model configuration and the standardized regression weights

5 Analysis of Results

In this section, we will present the model's results and focus our analysis on the standardized direct and total effects.

The estimated results show an acceptable fit under the main standard statistical tests. The value of the Chi squared statistic was 860.045 with 383 degrees of freedom presenting a ratio between these two values of 2.246 which is an indication of an acceptable fit. The value of the Root Mean Square Error of Approximation (RMSEA) was 0.046, the value of the Normed Fit Index (NFI) was 0.841 and the Comparative Fit Index (CFI) 0.902 all being indicators of an acceptable fit [10].

All the values presented in Fig. 1 are significant for a 95 % confidence level, except the "Ntrips" and "Poor → Security" variables that are significant at the 90 % level but, due to their meaningful contribution were kept in the model. Also, all the intercepts are significant for a 95 % confidence level.

5.1 Direct Effects

The results presented in Fig. 1 seem to be in accordance to previous research on the field regarding the impacts and the scale of the socio-demographic, land use, accessibility and mobility profile on the main satisfaction components and the overall evaluation [3].

People with long distanced trips tend to value the possibility of performing extra tasks on board ("EXTRA"). Lower income people give more value to status and low educated people demand more "COMFORT", while highly educated people do not seem to give much relevance to "STATUS". Travellers that reside in areas with good public transport accessibility show greater concerns about "SECURITY". Women, as already acknowledged in the literature, are also more thoughtful about environmental and security issues [21]. Private car users and elder people value less about the environment ("ENVIRON"), while car owners demand higher levels of flexibility in their mobility choices ("FREEDOM"). Users between 18 and 25 years old, who might not present high income levels, and that choose slower transport modes, prefer less costly transportation options. Moreover, high educated people and people far away from bus stops prefer transportation modes more adjusted to their mobility needs ("FIT").

The preferences of the respondents, represented in the systematic/sporadic factors, are all positively related with the satisfaction factors of the current modes. This shows a coherent mode choice process, since people are likely to select the alternatives that best suit their preferences.

As expected, a high number of transfers and the choice of heavier modes and/or buses reduce the performance satisfaction levels. On the contrary, people with a public transport pass and using fast transportation mode tend to be satisfied with their current choice in terms of cost and safety. The current mode performance factors are also positively related with the final satisfaction level showing, once again, the decisive contribution of the used mode characteristics.

The overall satisfaction levels are lower in persons with a public transport supply that demands more transfers for their mobility patterns. More educated people seem to be more demanding regarding their mobility performance, since they may have more options available, as well as people with a large number of trips in their mobility agenda.

5.2 Total Effects

The total effects present the aggregation of direct and indirect effects among exogenous and endogenous variables. The evaluation of only the direct effects might be misleading as the indirect effects may change the overall impact mediated through other variables. The total standardized effects can be observed in Table 3.

All the obtained results present the signs and the relative values according to what was expected.

There are some socio-demographic and mobility characteristics, like the number of trips, number of transfers and the level of education, that present a significant impact on the overall satisfaction. Characteristics such as travel time,

Table 3 Total standardized effects on the mobility performance evaluation

	Variables	Satisfaction
Socio-demographic	Gender	0.01
	18–25	0.003
	More65	−0.003
	Poor	0.009
	Basic	0.025
	High	−0.119
Attitudes relevancy	SECURITY	0.048
	FREEDOM	0.125
	STATUS	0.036
	EXTRA	0.019
	ENVIRONMENT	0.04
	FIT	0.08
	COMFORT	0.071
	TIMECOST	0.039
Attitudes performance	Performance	0.39
	CostSafe	0.176
	CurrentStatus	0.074
Accessibility	DistBus	0.006
	Entropy	0.006
	ActualStraight	0.017
	OfferActual	−0.001
	RatioTCTI	0.004
	TransfCM	−0.061
	TransfPT	−0.148
Mobility	Ntrips	−0.068
	AvDist	0.002
	ActTime	0.019
	OnlyEco	−0.006
	OnlyPriv	−0.004
	OnlyTasks	−0.088
	OwnCar	0.013
	Pass	0.042

fitness, flexibility and control, which belong to the factor “Performance”, seem to be more important than status, pleasure and environmental concerns (factors “CurrentStatus” and “CostSafe”). This proves the importance of the design of the transportation system that should provide flexible and fast connections.

6 Conclusions

With this chapter, we were able to present a structural equation model that not only enables the understanding of which factors impact the satisfaction of a user’s mobility chain but also proved to be possible, with an acceptable accuracy, to evaluate the satisfaction of users with their current mobility chain. This new approach intends to go beyond traditional confirmatory analyses that have been widely used in the literature.

The model was built over recently collected data through a survey that was partially designed with the purpose of providing data for such an analysis. The results obtained seem to suggest that the design of the network, at the tactical level (accessibility and connectivity), is rather relevant in users' evaluation. Also, there are some socio-demographic attitudes that affect significantly the levels of satisfaction, providing evidence of a self-selection process that has already been identified in the literature [21]. This bias might affect heavily the correct evaluation of the system.

References

1. Babakus, E., et al.: The sensitivity of confirmatory maximum-likelihood factor-analysis to violations of measurement scale and distributional assumptions. J. Mark. Res. **24**(2), 222–228 (1987)
2. Cronin Jr, J.J., Taylor, S.A.: SERVPERF Versus SERVQUAL: reconciling performance-based and perceptions-minus-expectations measurement of service quality. J. Mark. **58**(1), 125–131 (1994)
3. de Abreu e Silva, J., Goulias, K.G.: Structural equations model of land use patterns, location choice, and travel behavior, transportation research record. J. Transp. Res. Board, no. 2135, 106–113 (2009)
4. Del Castillo, J.M., Benitez, F.G.: A methodology for modeling and identifying users satisfaction issues in public transport systems based on users surveys. Procedia—Soc. Behav. Sci. **54**, 1104–1114 (2012)
5. Dolan, C.V.: Factor analysis of variables with 2, 3, 5 and 7 response categories: a comparison of categorical variable estimators using simulated data. Br. J. Math. Stat. Psychol. **47**(2), 309–326 (1994)
6. Eboli, L., Mazzulla, G.: Service quality attributes affecting customer satisfaction for bus transit. Public Transp. **10**(3), 14 (2007)
7. Eboli, L., Mazzulla, G.: A stated preference experiment for measuring service quality in public transport. Transp. Plann. Tech. **31**(5), 509–523 (2008)
8. Eboli, L., Mazzulla, G.: Willingness-to-pay of public transport users for improvement in service quality. Eur. Transp.\Trasporti Europei **38**, 107–118 (2008)
9. Eboli, L., Mazzulla, G.: A new customer satisfaction index for evaluating transit service quality. J. Public Transp. **12**(3), 21–37 (2009)
10. Fadelmula, F.K.: Assessing power of structural equation modeling studies: a meta-analysis. Educ. Res. J. **1**(3), 37–42 (2011)
11. Gatta, V., Marcucci, E.: Quality and public transport service contracts. Eur. Transp.\Trasporti Europei **36**, 92–106 (2007)
12. Githui, J.N., et al.: The structure of users' satisfaction on urban public transport service in developing country: the case of Nairobi. J. East. Asia Soc. Transp. Stud. **8**, 1288–1300 (2010)
13. Hensher, D.A.: Service quality as a package: what does it mean to heterogeneous consumers. Paper presented to 9th world conference on transport research, Seoul, Korea, 22–27 July 2001
14. Hensher, D.A., Prioni, P.: A service quality index for area-wide contract performance assessment. J. Transp. Econ. Policy **36**(1), 93–113 (2002)

15. Hensher, D.A., et al.: Service quality—developing a service quality index in the provision of commercial bus contracts. Transp. Res. Part A: Policy Pract. **37**(6), 499–517 (2003)
16. Hill, N., et al.: How to Measure Customer Satisfaction. Gower Publishing Limited, Hampshire (2003)
17. Hutchinson, S.R., Olmos, A.: Behavior of descriptive fit indexes in confirmatory factor analysis using ordered categorical data. Struct. Equ. Model.: Multidisc. J. **5**(4), 344–364 (1998)
18. Johnson, D.R., Creech, J.C.: Ordinal measures in multiple indicator models: a simulation study of categorization error. Am. Sociol. Rev. **48**, 398–407 (1983)
19. Kaplan, D.: Structural Equation Modeling: Foundations and Extensions. Sage Publications, Thousand Oaks (2000)
20. Lai, W.-T., Chen, C.-F.: Behavioral intentions of public transit passengers—the roles of service quality, perceived value, satisfaction and involvement. Transp. Policy **18**(2), 318–325 (2011)
21. Mokhtarian, P.L., Cao, X.Y.: Examining the impacts of residential self-selection on travel behavior: a focus on methodologies. Transp. Res. Part B: Methodol. **42**(3), 204–228 (2008)
22. Olsen, S.O.: Repurchase loyalty: the role of involvement and satisfaction. Psychol. Mark. **24**(4), 315–341 (2007)
23. Parasuraman, A., et al.: A conceptual model of service quality and its implications for future research. J. Mark. **49**(4), 41–50 (1985)
24. Santos, G.D., Martínez, L.M., Viegas, J.M., Alves, D.: Design and deployment of an innovative mobility survey for the Lisbon metropolitan area oriented to assess the market potential and obtain the best configuration of new alternative transport modes. In: European Transport Conference, Glasgow, (2011)
25. Steg, L.: Car use: lust and must. Instrumental, symbolic and affective motives for car use. Transp. Res. Part a-Policy Pract. **39**, 147–162 (2005)
26. Teas, R.K.: Expectations, performance evaluation, and consumers' perceptions of quality. J. Mark. **57**(4), 18 (1993)

Advanced Trip Planners for Transit Networks: Some Theoretical and Experimental Aspects of Pre-Trip Path Choice Modeling

Agostino Nuzzolo, Umberto Crisalli, Antonio Comi and Luca Rosati

Abstract The chapter reports the first results of a research project for the definition of an advanced trip planner for transit networks. The project at the current stage has developed the module to support the user with personalized pre-trip information based on his/her preferences. The first part of the chapter describes the user needs and the logical architecture of the trip planner. The second part deals with the theoretical aspects of the path choice model used to support the path choice set individuation, the path utility calculation and the user preference learning procedure. In order to apply the theoretical framework and to show the benefits of the proposed approach, some experimental results of a test case on the transit system of the metropolitan area of Rome are presented.

Keywords Transit · Path choice models · Schedule-based · Personalized information · Single-user · Parameters estimation

A. Nuzzolo (✉) · U. Crisalli · A. Comi
Department of Enterprise Engineering, Tor Vergata University of Rome, via del politecnico 1, Rome, Italy
e-mail: nuzzolo@ing.uniroma2.it

U. Crisalli
e-mail: crisalli@ing.uniroma2.it

A. Comi
e-mail: comi@ing.uniroma2.it

L. Rosati
Department of Civil Engineering and Computer Science Engineering, Tor Vergata University of Rome, via del politecnico 1, Rome, Italy
e-mail: rosati@ing.uniroma2.it

J. F. de Sousa and R. Rossi (eds.), *Computer-based Modelling and Optimization in Transportation*, Advances in Intelligent Systems and Computing 262,
DOI: 10.1007/978-3-319-04630-3_30,

1 Introduction

Trip planners are traveller tools able to suggest the best travel alternatives (path) on transport networks including their relevant information, such as travel time, monetary cost, estimated departure and arrival time, service characteristics, alerts, disruptions. They allow the user to easily access to organized information in order to compare the different alternatives for a rational choice of the transport mode [1].

Trip planners for transit networks have to define travel alternatives (paths) both in space (among stops) and in time (in relation to the user desired arrival/departure time and/or the arrival/departure time of transit vehicles at stops) according to user preferences. In fact, transit networks are usually characterized by different boarding stops and available runs for the same Origin-Destination (O/D) pair *od* and target time τ_{TT_i} (e.g. desired arrival time). For this reason, advanced transit trip planners have to use a path modeling approach that explicitly takes into account the above space–time features of path alternatives, such as the schedule-based approach recalled in Sect. 3.

As the main reason that leads to reject transit as travel mode is the uncertainty about routes and timetable, transit trip planners have to provide accurate pre-trip information to reduce this uncertainty. The state-of-the-art presents many papers [2–6] that demonstrate the added value of ITS (Intelligent Transport Systems) to improve transit ridership.

In order to improve the accuracy of the pre-trip planner information (e.g. in case of interchanges or when the user is constrained to the arrival time at destination), trip planners can benefit from the use of real-time data relative to the transit network. Moreover, advanced trip planners have to provide path alternatives according to single user preferences, for which different path alternatives are suggested to different users according to their different weights on path attributes (e.g. walking time, waiting time, on-board time, transfer time and so on). This aspect highlights the frontier of advanced transit pre-trip planners, which have to provide personalized real-time information by using learning process mechanisms able to track profiled travellers and to give them information according to their personal travel habits [7, 8].

This chapter presents the main theoretical and experimental results of an advanced trip planner aiming at providing real-time personalized pre-trip information to support the user in his/her choice of the best path from the (dis)utility point of view on a multiservice transit network.

Section 2 illustrates the user needs and the logical architecture of the trip planner. Section 3 presents the modeling framework to provide personalized pre-trip information, while Sect. 4 describes the features of learning process on the individual preferences. These theoretical aspects are investigated through some experimental results carried out by a test case on the transit network of the metropolitan area of Rome (Italy). Finally, Sect. 5 reports some conclusions and the future developments of this research.

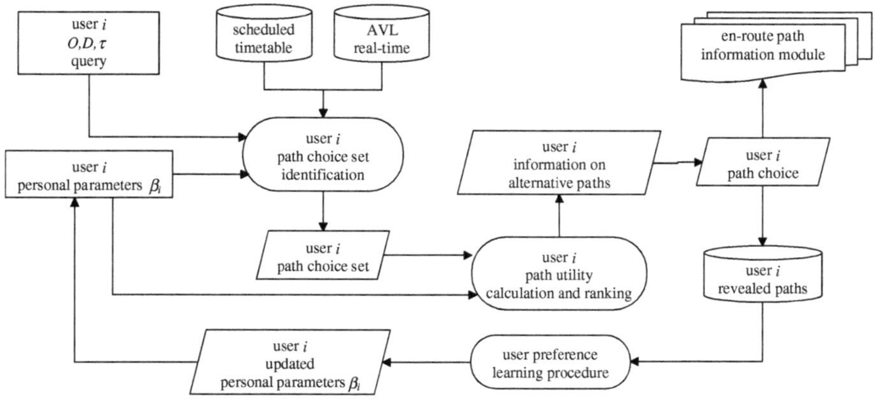

Fig. 1 Logical architecture of the pre-trip personalized information module

2 User Needs and Logical Architecture

In order to provide pre-trip information, this section describes the user needs and the logical architecture of a trip planner [1, 9], which has to provide pre-trip personalized information to transit users. In particular, the trip planner should provide:

- transit path alternatives according to some user preferences (e.g. maximum number of transfers, maximum walking distance and so on);
- path attributes (departure time, walking times and distances, waiting time at stops, on-board travel time, crowding, transfer number and times, and so on) based on the real transit system operations.

The logical architecture of the trip planner implemented to support transit users with personalized pre-trip information, is reported in Fig. 1.

The pre-trip personalized advice module is enabled by a query of the registered user i, who is logged into the system. At time τ in which user i asks for a support to travel from origin O to destination D with a desired arrival time τ_{Ai}, the system identifies and ranks the path choice set of user i based on his/her preferences and the current information on the multiservice transit network (i.e. scheduled time-table and real-time data), by using the path choice set identification and ranking procedure described in Sect. 3.

In order to provide to the user i a ranking of alternative paths, in the framework of the Random Utility Theory [10], personal utility parameters β_i of user i are used to calculate path utility for all paths belonging to the path choice set of user i (see Sect. 3).

The path chosen by user i is added to the personal database of revealed preferences of user i, which updates the personal parameters β_i by using the user preference learning procedure described in Sect. 4.

The information on the actual path choice of user i represents the main input of the en-route path information module, aiming at supporting user i during the trip (Fig. 1). The en-route trip planner development is one of the future perspectives of this research.

3 Transit Path Choice Modeling

In order to be used within a trip planner, path choice models allow us to estimate the utility associated with each path k belonging to a set of possible alternative paths defined according to traveller characteristics, transport system performance (e.g., travel times and costs) and traveller behavioral assumptions.

As the arrival/departure time coordination of runs at interchanges enables path alternatives, a path modeling approach able to explicitly consider the space-time features of both demand and supply has to be used. In this case, the state of the art indicates the schedule-based approach [11] as the unique way to model transit path choice on multimodal networks in presence of ATIS.

The schedule-based approach [12] refers to services in terms of runs (vehicles) using the real vehicle arrival/departure times, and hence all the values of level of service attributes, evaluated at time in which users make their choices, can be explicitly taken into account. This approach allows us to consider the evolution in time of both supply and demand, as well as run loads and level of service attributes, by an explicit treatment of:

- the temporal segmentation of the demand to consider user desired departure or arrival times;
- the supply modeling, in which single run of transit services with its departure/arrival times at stops;
- path choice models, which have to consider attribute time-dependencies.

On the demand side, the temporal segmentation of the demand in relation to a Desired Arrival Time at destination (DAT), represented by τ_{Ai}, and/or a Desired Departure Time from origin (DDT), represented by τ_{Di} is obtained by dividing the generic day t in n elementary time intervals of $\delta\tau$ width (e.g. $\delta\tau_i = 1$ min) to which the user target times τ_{TT_i} are associated.

On the supply side, path attributes and on-board loads for each run of transit services can be calculated using a diachronic network [12]. The *diachronic* graph Ω consists of three different sub-graphs in which each node has an explicit time coordinate:

- a service sub-graph, in which each run of each line is defined both in space, through stops, and in time, according to arrival/departure times at stops;
- a temporal centroids sub-graph, in which each node represents both temporal centroids, in order to simulate the space-time characteristics of trips, and user arrival/departure times;

- an access-egress sub-graph, which allows the connection between centroids and stops, and stops between them.

Deepening on the definition of path reported in Sect. 1, a path k on a transit network is the space-time sequence of transport infrastructures and services used by an user travelling from an origin o at a given origin departure time τ_{Di} to a destination d with the relative arrival time at destination τ_d, which also includes access stop s with relative arrival time at stop τ_{Dis}, line and run with run departure time τ_r from access stop (or sequence of lines and runs including the relative stop interchanges) and egress stop s' allowing the user to reach his/her destination. It can be defined through a sequence of links of the Ω graph. As the diachronic network is represented through a planar graph, we can apply the entire traditional network algorithms (e.g. least cost) to efficiently obtain path search and space–time attributes. Furthermore, the explicit representation of single runs with their characteristics (e.g. capacity) allows us to explicitly treat congestion through vehicle capacity constraints [13].

As the path choice model plays a key role in the entire architecture of an advanced trip planner, the following sections will describe the main features of the presented modeling framework in terms of path choice set generation, path utility calculation and model calibration on the basis of single-user personal preferences.

The presented theoretical modeling framework is accompanied by some experimental results carried out testing the pre-trip planner on the transport system of the metropolitan area of Rome (Italy), which is served by a multiservice transit network (urban bus, tram and metro, regional railways and buses) operated by different companies with an integrated fare structure.

3.1 The Pre-Trip Path Choice Set Generation

Given the user i travelling on the od pair at a given time τ_{TT_i}, a selective approach can be used to individuate the path choice set. This approach is based on a set of rules that allows us to generate feasible paths according to traveller characteristics, transport system performance (e.g., travel times and costs) and traveller behavioral assumptions. The set of rules to reduce the potentially high number of path alternatives can be heuristically calibrated on the basis of a sample of observed choices (maximum coverage factor method).

For example, considering the trip from Frascati (a town near Rome) to the centre of Rome (i.e. Piazza Sempione) for an user travelling with the desired arrival time at 9.30 a.m., the four different paths schematically pictured in Fig. 2 are available to travel on this od pair. Given a distance on the road network of about 25 km, the average travel time on this od pair is about 2 h. Path alternatives differ in terms of travel time, waiting and transfer times, modes to be used (train, metro or bus) and the early/late arrival time.

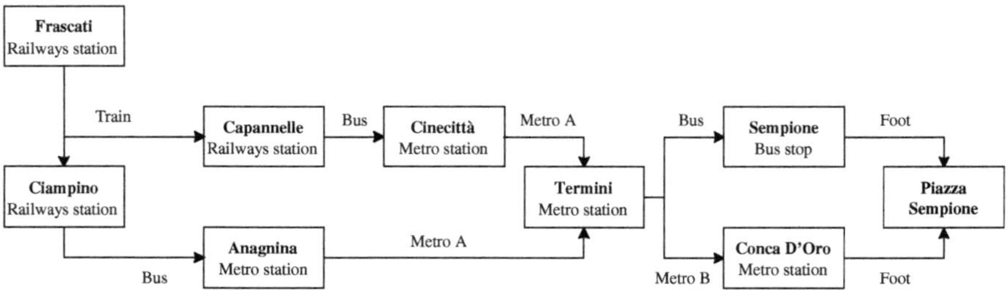

Fig. 2 Test case: transit network

3.2 The Single-User Path Utility Model

Given the path choice set generated as described in Sect. 3.1, the utility $V^{\tau}_{OD,\tau_{TT_i}}[k]$ that user i at time τ associates to the path k identified by departure time τ_{Di}, access stop s and run r, can be expressed as a linear combination of attributes. For example $V^{\tau}_{OD,\tau_{TT_i}}[k]$ can be written as:

$$V^{\tau}_{OD,\tau_{TT_i}}[k] = \beta_{ED} \cdot ED_k + \beta_{AE} \cdot AE_k + \sum_m [\beta_{TW,m} \cdot TW_{m,k} + \beta_{OB,m} \cdot OB_{m,k} + \beta_{CFW,m} \cdot CFW_{m,k} + \beta_{MP,m} \cdot MP_{m,k} + \beta_{NT,m} \cdot NT_{m,k}] \quad (1)$$

where ED_k is the Early or Late arrival time (i.e. the difference between the desired and the actual arrival time at destination) using path k; AE_k is the sum of access and egress times on path k; $TW_{m,k}$ is the waiting time spent for boarding runs of the transit service m (train, metro, tram, bus) belonging to path k; $OB_{m,k}$ is the on-board time spent on the transit service m belonging to path k; $CFW_{m,k}$ is the average on-board crowding degree on runs of transit service m belonging to path k; $MP_{m,k}$ is a preference attribute for transit service m on path k (e.g. expressed as a function of the travel distance on transit service m w.r.t. the total distance on path k); $NT_{m,k}$ is the number of transfers on transit service m belonging to path k; β_i are the model parameters.

Equation (1) is applied on time-dependent choice sets and attributes comprising each alternative at time τ. Most of the above attributes are provided by ATIS on the basis of the real-time information on the current state of the system and the short-term prediction of transit operations.

Given path utilities, it is possible to rank path alternatives based on such utilities and to select some of them (e.g. one or more) that the trip planner will suggest to the user. In order to consider the single-user preferences/attitudes, a personal (individual) set of model parameters β_i is used in (1). These parameters can be estimated as follows.

3.3 The Estimation of Individual Pre-Trip Path Utility Parameters

Calibrating individual path choice parameters implies that the model functional form could be the same for different users, as the same could be the values of attributes considered in path choice, but different set of parameters (different for each registered user) have to be considered and calibrated according to user travel preferences.

The estimation of individual coefficients β_i can be performed using the information collected from a sample of observations. Given a sample of N observations of a single user i, the problem is to find the estimates of coefficients β_i through which the travel planner is able to suggest the user best perceived paths, including their ranking.

Except for laboratory choice experiments in psychology, it is rare to see discrete choice models estimated for single people. After [14], there was little work on ways to measure and model individual choices in survey applications up to now [15]. In fact, demand models are traditionally used to simulate the average number of trips of given characteristics undertaken by homogeneous user groups and it is not easy to obtain large choice samples of single decision-maker (it is easier to have choice samples from many decision-makers). Therefore, instead of disaggregate or individual behavioural models, user groups (homogeneous with respect to their attributes, parameters and the functional form of the models) have been used and aggregate behavioural models have been developed. Different types of aggregate models have been proposed but their performances appear limited to support personal travel advices, because of the dispersion among users and/or the variations in taste/preferences among users. For this reason, some aspects of the individual model development should be better considered, as here presented.

The estimation of discrete choice model parameters with repeated observations for each respondent gives rise to an obvious correlation of disturbances, or heterogeneity (e.g. the parameters can also vary for the same user according to travel purpose), which refers to variations in unobserved contributing factors across behavioural units. If behavioural differences are largely due to unobserved factors, and if unobserved factors are correlated with the measured explanatory variables, then estimates of model coefficients will be biased if correlation heterogeneity and correlation are not properly treated. The problem may be more pronounced in repeated measurement data since unobserved factors may be invariant across these repeated measurements. As the panel data contains multiple observations of the same user, the assumption of independence between choices for the same user may not be appropriate. For this reason, there has been growing interest in the representation of unexplained heterogeneity in choice data, using random coefficients models, such as Mixed Multinomial Logit [16]. Moreover, different model specifications could be used (e.g. nested logit models, error components logit model) to consider that the paths could be not completely independent due to possible overlaps.

Table 1 Model estimation using 4-alternative sets

Attributes	Multinomial logit (MNL)		Mixed logit (MXL)	
	Parameter	*t*-st	Parameter	*t*-st
Waiting time (total)	−0.439	−4.06	−0.469	−3.93
On-board time (train)	−0.078	−1.30	−0.088	−1.38
On-board time (metro)	−0.336	−1.64	−0.327	−1.45
On-board time (bus)	−0.426	−3.24	−0.447	−3.12
Early and late arrival time	−0.291	−3.41	−0.306	−3.30
Standard deviation of early and late arrival time			0.179	1.77
ρ^2	0.71		0.72	
%-of-right	84 %		84 %	
%-of-right including the best and the 2nd best alternatives	99 %			

In order to investigate the best functional form of the individual path choice model, different models forms where tested starting from the simplest multinomial logit model (MNL) to the mixed-logit (MXL) and nested-logit (NSL) ones. They were performed considering the working-day journey from Frascati (a town near Rome) to the center of Rome (i.e. Piazza Sempione) with the desired arrival time at 9.30 am, whose path alternative features are detailed in Sect. 3.1. Path choice models were estimated carrying out a SP survey with 150 sets of four alternatives on a test user (a university student) that was asked to choose the preferred one. The set of 150 scenarios with 4-alternatives was defined by path alternatives randomly extracted from the previous experimented status of the transportation system. We estimated multinomial, nested and mixed logit models (with normal distribution of parameters). Even if further analyses are in progress, the development of nested and error components logit models did not provided satisfactory results. Therefore, in the following only the results obtained by MNL (Multinomial Logit) and MXL (Mixed Logit) estimations are discussed. According to [17], the random parameters were selected assuming that all parameters included in logit models are random and examining their estimated standard deviations, with a zero-based t test for individual parameters and a likelihood-ratio test for establishing the overall contribution of the additional information. Then, only the early/late arrival time was selected as random parameter, while the other parameter estimates remain quite constant between the two specifications.

Although all the attributes displayed in Eq. (1) were tested, only those reported in Table 1 were statistically significant. The estimated parameters corresponding to different components of travel time (e.g. waiting, on-board and delay time) have increasing absolute values for less appreciated components. For example, the time spent on the bus is lower than on the metro and is about twice lower than the early/late arrival time. The lowest value of travel time refers to time spent on the regional train. The *%-of-right* was calculated and we can see that the 4-alternative model predicts at least the 84 % of the chosen paths. If the first and the second paths with the highest estimated utility are considered, the *%-of-right* grows up to the 99 %.

Deepening on values of Table 1, it possible to say that the performances of the MXL and MNL models are quite similar. For this reason, in the following we only use the MNL model as it is simpler to apply.

4 The Learning Process of Single-User Preferences

In order to obtain the individual pre-trip path utility parameters that take into account the single-user preferences, a two-step procedure is implemented.

The first step allows us to initialize the path utility function parameters of a new user on the basis of Stated Preference (SP) Interviews. The second step aims at updating the initial model parameters based on the individual revealed choices acquired by the Traveller Tool during its use.

4.1 The Initialization of Individual Model Parameters

The challenge of the initialization of individual model parameters is to reduce the initialization phase for new users by minimizing the time needed for the estimation of user-tailored model parameters able to provide suggestions that reasonably fit the user preferences.

In order to initialize the path utility function of new users, [18] proposed a method in which the updating starts from initial average parameters obtained through multi-user Stated Preference surveys. In this chapter a possible alternative approach is explored. It requires some data about origin and destination of a typical and well-known transit trip, and desired arrival time of the new user. Then, the system generates some alternatives according to the system past-recorded data. Sequentially, some alternative path scenarios are provided to the user, which has to choose the preferred option among those suggested. The results of this SP survey are then used to estimate the initial parameter values of the path utility function.

Greater attention should be given to the initialization phase, which should not be too long to avoid the user becomes bored during this operation. The investigation of this problem concerns the minimum number of scenarios to be proposed to the new user and the minimum number of attributes that can be used in the initial path utility functions, by considering that the number of quite correct estimated parameters could also depend on the number of available observations.

For this reason, we investigated the possibility to reduce time the user has to spend for the initialization phase by limiting the number of alternatives to be suggested during this operation. This analysis was performed carrying out a further SP survey consisting of 150 scenarios with only 2-alternative paths, which were randomly selected among the same 4 path alternatives described in Sect. 3.1.

Also in this case, different behavioural models were estimated (Table 2) and only MNL and MXL (with on-board time on the bus as random parameter) models gave

Table 2 Model estimation using 2-alternative sets

Attributes	Multinomial logit (MNL)		Mixed logit (MXL)	
	Parameter	*t*-st	Parameter	*t*-st
Waiting time (total)	−0.386	−3.66	−0.594	−2.66
On-board time (train)	−0.285	−3.16	−0.439	−2.43
On-board time (metro)	−0.485	−2.03	−0.741	−1.78
On-board time (bus)	−0.538	−3.40	−0.902	−2.45
Standard deviation of on-board time (bus)			0.185	2.15
Early and late arrival time	−0.223	−2.47	−0.414	−2.08
ρ^2	0.80		0.82	
%-of-right	93 %		96 %	
%-of-right in reproducing 4-alternatives	81 %			

good results. As expected the *%-of-right* increased with respect to model performances of Table 1 (from 84 to 93 % for the MNL model), but the real performances of this model have to be considered by applying to the 4-alternative scenarios. For this reason, the *%-of-right* in reproducing 4-alternatives was calculated (see Table 2). In this case, the MNL models predicted the alternatives chosen by user in 81 % of scenarios. Therefore, with a reasonable reduction of the *%-of-right* (about 3 %), the model which uses the SP scenarios with 2-alternatives instead of that with 4-alternatives, can be used and satisfactory applied to save time in the initialization phase.

Starting from the analysis reported in the Table 2 (2-alternative scenario), different random utility models were estimated in order to verify the minimum number of observations and parameters required to reach a satisfactory suggestion to the user. As satisfactory *%-of-right* it is assumed the 79 %, which is only 2 % less than the previous 81 % reported in Table 2. Moreover, we obtained that the above satisfactory 79 *%-of-right* in the simulation of choices revealed in the 4-alternative dataset can be reached using 10 observations and one parameter (i.e. on-board time). Figure 3 reports the number of observations required to obtain the 79 *%-of-right* related to the number of attributes of the utility path function. From figure, it emerges that about 80 observations are needed if only two attributes are used, then, as detailed in the following section, we simulated the updating process of parameters starting from 10 2-alternative observations.

4.2 The Updating of Individual Model Parameters

After the initialization phase, at the second step, the parameters are updated using a user preference learning procedure based on the choice revealed when the traveller uses the system. The parameter updating procedure can use two different approaches based on: *Bayesian and batch methods*. Within the former approach, the probabilities of parameter estimators are usually represented by a normal distribution and their mean values represent the current best estimate of the user's

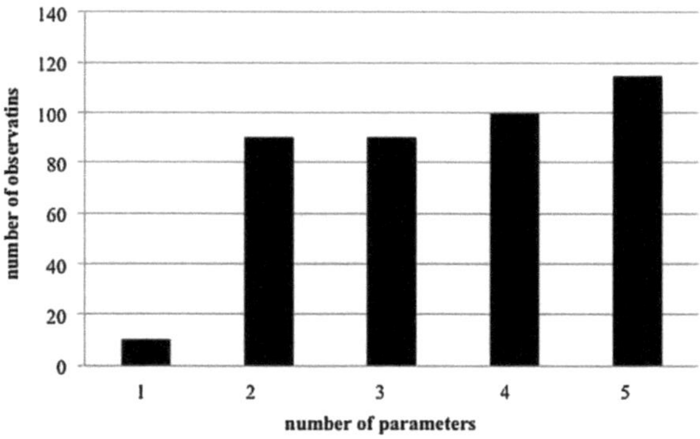

Fig. 3 Number of observations versus number of parameters for reaching the 79 *%-of-right*

value of the parameters. Each time a choice (of the user) is observed, the system updates the model parameters by updating the distribution of each parameter, i.e. by updating means and standard deviations. This updating is based on the well-known theorem of Bayes. Besides, although Bayesian updating is a suitable technique to learn the hidden parameters incrementally, existing Bayesian methods of learning continuous parameters are not feasible for incremental learning due to the long computation times involved in a learning step. Then, [7] proposed a new Bayesian method that reduces computation time by assuming a sequential processing of parameters and a systematic sampling of the parameter space. The *batch method* means that, after a given number of choices (of the user) is observed, the system provides a new estimate of model parameters by using all the available observations within a Maximum Likelihood estimation procedure [10, 19].

Starting from the above results, the learning process was also simulated using the same 150 SP survey data with 4-alternative sets. The parameters of the utility function were estimated varying the number of attributes and the number of observations. The parameter updating was performed including 10 new observations at a time (i.e. batch updating). The first results showed that the improving of learning process is quite slow because too many (more than 50) observations are necessary to obtain a statistically good model. For example, considering a model with only three parameters and an 80 *%-of-right*, more that 50 observations are needed. Further analyses have to be carried out in order to verify if other estimation procedures are more performing in terms of swiftness of convergence (e.g. the Bayesian method).

5 Conclusions

This chapter presents the first results of the theoretical and experimental aspects carried out in order to implement a trip planner able to give pre-trip personalized information to the user about travel alternatives on a transit network by real-time data. It is based on a path choice modelling framework able to provide transit path alternatives on the basis of user personal travel preferences defined according to a learning procedure. The focus was mainly on the investigation of models to be used for suggesting the best paths perceived by each user. Both the initialization and the updating of the model parameters were analysed by pointing out the choices made by some students travelling for leisure.

The obtained results show that the model parameter initialization phase (performed by SP interviews) can use 2-alternative scenarios with a minimum number of 10 observations, through which the estimated parameters of the path utility function allow us to suggest the individual preferred paths with a good reliability. On the other hand, in relation to the updating of individual model parameters, we tested that the process can be quite slow and too many observations are required before reaching a statistically significant stability of utility function. Therefore, different approaches have to be investigated, including the approaches based on Bayesian methods.

Further developments of this research mainly regard the additional investigation of the path choice modelling and the extension of the trip planner to provide en-route personalized information. In particular, advances in path choice modelling concern the exploration of other od pairs, user preferences and model forms. In order to focus on behavioural aspects related to specific user preferences, the design of ad-hoc SP surveys will be studied for the initialization phase.

References

1. Caulfield B., O'Mahony M.: An examination of the Public Transport Information requirements of users. IEEE Trans. Intell. Transp. Syst. **8**(1), 108–120 (2007)
2. Adbel-Aty M.A.: Using ordered probit modelling to study the effect of ATIS on transit ridership. Transp. Res. Part C **9**, 265–277 (2001) Elsevier
3. Grotenhuis J.W., Wiegmans B.W., Rietveld P.: The desired quality of integrated multimodal travel information in public transport: Customer needs for time and effort savings. Transp. Policy **14**(1), 27–38 (2007) Elsevier
4. Kenyon S., Lyons G.: The value of integrated multimodal traveller information and its potential contribution to modal change. Transp. Res. Part F **6**, 1–21 (2003) Elsevier
5. Tang L., Thakuriah P.: Will the psychological effects of real-time transit information systems lead to ridership gain?. In: Proceedings of the Transportation Research Board Annual Meeting, Washington, USA (2011)
6. Zhang L., Li J., Zhou K., Gupta S.D., Li M., Zhang W.B., Miller M.A., Misener J.A.: Design and implementation of a traveller information tool with integrated real-time transit information and multi-modal trip planning. In: Proceedings of the Transportation Research Board Annual Meeting, Washington, USA (2011)
7. Arentze T.A.: Adaptive, personalized travel information systems: A Bayesian method to learn users' personal preferences in multi-modal transport networks. In: Proceedings of the Transportation Research Board Annual Meeting, Washington, USA (2013)
8. Nuzzolo A., Crisalli U., Comi A., Rosati L.: An advanced pre-trip planner with personalized information on transit networks with ATIS. In: Proceedings of 16th International IEEE Conference on Intelligent Transport Systems, The Hague, The Netherlands (2013)
9. ARTIST: ARchitettura Telematica Italiana per il Sistema dei Trasporti Ministero delle Infrastrutture e dei trasporti, http://www.its-artist.rupa.it (2003)

10. Ben-Akiva, M., Lerman, S.: Discrete Choice Analysis. MIT Press, Cambridge (1985)
11. Nuzzolo A., Crisalli, U.: The schedule-based approach in dynamic transit modelling: a general overview. In: Wilson, N.H.M., Nuzzolo, A. (eds.) Schedule-Based Dynamic Transit Modeling. Theory and Applications, pp. 1–24. Kluwer, Dordrecht (2004)
12. Nuzzolo, A., Russo, F., Crisalli, U.: A doubly dynamic schedule-based assignment model for transit networks. Transp. Sci. **35**, 268–285 (2001)
13. Nuzzolo A., Crisalli U., Rosati L.: A schedule-based assignment model with explicit capacity constraints for congested transit networks. Transp. Res. Part C **20**(1), 16–33 (2012) Elsevier. doi:10.1016/j.trc.2011.02.007
14. Chapman R.G.: An approach to estimating logit models of a single decision maker's choice behavior. In: Kinnear T.C. (ed.) Advances in Consumer Research, vol. 11, pp. 656–661. Association for Consumer Research, Provo (1984)
15. Frischknecht B., Eckert C., Louviere J.: Simple ways to estimate choice models for single consumers. Centre for the Study of Choice (CenSoC), Working Paper Series, No. 11-006, University of Technology of Sydney (2011)
16. Hess S., Rose J.M.: Allowing for intra-respondent variations in coefficients estimated on repeated choice data, Transp. Res. Part B **43**, 708–719 (2009) Elsevier
17. Hensher D.A., Greene W.H.: The Mixed Logit Model: The State of Practice, Transportation, vol. 30, pp. 133–176. Kluwer Academic Publishers, Boston (2003)
18. Molin E.J.E., Arentze, T.A.: Travelers' preferences in multimodal networks: design and results of a comprehensive series of choice experiments. In: Proceedings of the Transportation Research Board Annual Meeting, Washington, USA (2013)
19. Lancsar E., Louviere J.: Estimating individual level discrete choice models and welfare measures using best worst choice experiments and sequential best worst MNL. Centre for the Study of Choice (CenSoC). University of Technology of Sydney, Sydney (2008)

Sensing Bluetooth Mobility Data: Potentials and Applications

João Filgueiras, Rosaldo J. F. Rossetti, Zafeiris Kokkinogenis, Michel Ferreira, Cristina Olaverri-Monreal, Marco Paiva, João Manuel R. S. Tavares and Joaquim Gabriel

Abstract Information related to mobility dynamics constitutes an important factor to be considered in traffic management to improve the efficiency of existing systems. We present a proof-of-concept deployment of sensors using the Bluetooth technology to detect traffic flow conditions. Besides traditional method consisting of a network of stationary sensors, we present a novel approach that uses sensors deployed in moving vehicles that allows new type studies and captures new insights of mobility. Both approaches complement the most common methods of traffic sensing while being more cost-effective and easily available. Early experimental results show the variety of information available through both approaches

J. Filgueiras
Instituto de Engenharia de Sistemas e Computadores Investigação e Desenvolvimento (INESC-ID), Lisbon, Portugal
e-mail: jfilgueiras@inesc-id.pt

R. J. F. Rossetti · Z. Kokkinogenis (✉)
LIACC, DEI, Faculty of Engineering, University of Porto, Porto, Portugal
e-mail: zafeiris.kokkinogenis@gmail.com

R. J. F. Rossetti
e-mail: rossetti@fe.up.pt

M. Ferreira · C. Olaverri-Monreal · M. Paiva
Instituto de Telecomunicações, University of Porto, Porto, Portugal
e-mail: michel@dcc.fc.up.pt

C. Olaverri-Monreal
e-mail: cristina.olaverri@dcc.fc.up.pt

M. Paiva
e-mail: marcogouveia.paiva@gmail.com

J. M. R. S. Tavares · J. Gabriel
DEMec, Faculty of Engineering, University of Porto, Porto, Portugal
e-mail: tavares@fe.up.pt

J. Gabriel
e-mail: jgabriel@fe.up.pt

J. F. de Sousa and R. Rossi (eds.), *Computer-based Modelling and Optimization in Transportation*, Advances in Intelligent Systems and Computing 262,
DOI: 10.1007/978-3-319-04630-3_31,

spanning from Origin/Destination matrices and travel times to insights into emerging mobile neighborhoods. These metrics are important to improve traffic management increasing the efficiency of urban mobility networks.

Keywords Urban mobility data · Bluetooth sensing · Traffic monitoring

1 Introduction

The extended use of traditional roadside traffic sensors involves high implementation and maintenance costs and a limited coverage. As a consequence, in the last years, the traffic and transportation research community has shown a great interest in the use of wireless communications and satellite technologies. In combination with traditional traffic sensing approaches, these technologies allow the development of traffic monitoring and advanced traffic management systems at a lower cost and greater coverage [11].

The study of human mobility as a whole is important when trying to comprehend traffic phenomena and design solutions to traffic-related issues. It becomes necessary to account for pedestrian, as well as vehicle, mobility, and to understand the interaction between them.

In the new horizon of the urban system tissue and in the perspective of urban designers, engineers and practitioners, users are becoming the central interest and not mere commodities to be transported. New type of monitoring techniques and new type of sensors must be used to achieve such innovative solutions. Advancements in information and communication technologies made information and data to be ubiquitous and measurable. Vehicles and portable electronic devices, such as smart phones, come out of the factory with more and more sensors. Different kinds of interfaces and services are being devised for the full exploitation of their capabilities. A large-scale urban mobile cloud of data streams is thus formed capable of offering valuable information for city planners, traffic controllers, public transportation authorities, tourists and citizens. In that sense, data can be used to understand how and where people gather, their destinations and the centers of their daily activities. On the other hand, mobility information can be useful for the users themselves when planning their trips and commutes.

Smart phones, in particular, have already been proposed as mobility sensors, and they are currently used by commercial services, such as Google for traffic estimation. But such sensing systems are mostly proprietary and commonly focus on a specific aspect of transportation systems.

The cost-effective solution that is offered by using the Bluetooth wireless communications technology can overcome these difficulties. This clue allows mobility monitoring using autonomous stations that continuously search for nearby Bluetooth devices. These devices that can range from mobile phones to hands-free devices embedded in cars can be tracked over networks of sensors because of their unique hardware address. This solution allows monitoring vehicles as well as pedestrians simultaneously.

This work uses the Bluetooth technology to build and deploy a network of fixed and mobile sensors in order to study and characterize urban mobility conditions in the city of Porto, in Portugal. We present our preliminary results with a fixed network of sensors, as well as results using a novel approach to Bluetooth monitoring by deploying sensors in moving vehicles. Long-term experiments using mobile sensors allow an insight into the concept of *mobile vehicular neighbors*.

Next section revises related work in the areas of traffic monitoring and management using wireless communication technology. Section 3 presents a description of the methodology used to collect and visualize the collected Bluetooth data. Section 4 reports a preliminary evaluation of the monitoring tool. Finally, Sect. 5 concludes the chapter.

2 Related Work

The use of wireless networks to estimate patterns of human mobility and social dynamics has been a subject of great interest in studies over the past decade. Results show that humans usually follow simple reproducible patterns that have a single spatial distribution [4]. Consequently, this behavior model might be reflected in traffic mobility patterns. Information related to Origin/Destination (O/D) matrices, travel times and route inference to characterize an urban mobility system is needed to achieve an effective traffic management, control and flow optimization. Sensors based on traditional methods for data acquisition lead to a limited coverage and high implementation and maintenance costs. Since high quality traffic information in real time is crucial to develop Intelligent Transportation Systems (ITS), a greater interest for the use of wireless communications and satellite technologies has been shown in the last years [11].

Some recent works discuss the last developments about Floating Car Data based collection methods and applications, focusing on the potential and bottlenecks of this technology are shown [11]. In [1] the author used GPS-based FCD from taxis to dynamically estimate O/D matrices and to infer route analysis. The idea behind this study is to use the fleet of taxis as probes to infer mobility patterns. The authors concluded that the data collected through this method was not conclusive enough since it was based on a small sample. Additionally, the original O/D matrix of the studied area was unknown and consequently, a general mobility behavior could not be derived. Additional road traffic conditions by GPS reports via GSM network were studied in [7]. The study concluded that high quality travel time could be produced only with a diffusion of samples above 3 and 5 % in a mid-sized city. Nevertheless, since FCD is real time data, we think that the combination of FCD methods with traditional methods might increase the quality of data resulting on more accurate mobility parameters estimation. Further investigation focusing on the use of cellular network data for inferring real-time mobility information has been conducted through surveys and data classification identifying the advantages and limitations of this method. Particularly, a discussion on the ways to increase the accuracy of known cellular positioning techniques through a traffic congestion estimation service application and studying the trajectory of several thousand of anonymous mobile phone users are found in [20, 21].

A new promising technology has been proposed in the traffic and urban planning community based on the interception of Bluetooth Media Access Control (MAC) addresses for the monitoring of traffic conditions that will open new measurement possibilities [2, 5, 16, 18, 22, 23]. The uniqueness of the detected MAC addresses makes the Bluetooth technology particularly appropriate to monitor the mobility conditions. For example, travel times can be determined through the calculation of the difference in time and space from two measurement sites. Malinovskiy et al. [13] compared MAC address based travel times on a corridor equipped with automatic license plate recognition (ALPR) sensors. Authors conclude that Bluetooth sensor are a satisfactory substitute of the traditional monitoring methods, even though the sample size found is significantly smaller than what can be achieved by ALPR systems, when characterizing the actual traffic conditions.

Moreover, through GPS equipment installed in Bluetooth scanner devices, an accurate measurement of distance between successive Bluetooth data collection sites can be accomplished, that leads to valuable Origin–Destination (OD) information. This information is obtained by tracing a Bluetooth transceivers path through a series of Bluetooth units with known locations. Barcel et al. present a method to forecast travel times and time-dependent matrices, in uncongested freeway networks, using Bluetooth traces of detected devices [2].

Bluetooth technology only captures the travel time of a fraction of vehicles in the traffic stream. As a consequence, the authors in [17] collected an extensive amount of data on several highways from Maryland and Delaware calculating the average detection rate to be 2–7 %. Haseman [6] studied the possibility of using Bluetooth monitoring methods to design work zone traffic control plans. Authors concluded that the measurement on a weekly basis of work zone travel time could provide quantitative data traffic planners can use to evaluate alternative maintenance of traffic techniques, and identify cost-effective practices. They propose also the use of Bluetooth probe tracking as a potential way of traffic contracting methods while strongly advocating for travel mobility when demands warrant.

Tsubota et al. [19] discusses the data filtering aspect of the Bluetooth data collection underlying the importance of the duration that is defined as the time spent by a Bluetooth device to pass through the detection range of a Bluetooth scanner. Young in [24] discusses the insight of the Bluetooth traffic-monitoring project at University of Maryland. Critical issues of design, development, testing, and operations of Bluetooth scanners as real-time traffic sensors are presented in this report.

ONeill in [9, 10, 15] illustrates how Bluetooth data can be used to understand pedestrian mobility behavior and what is the interaction with the urban environment. Bullock et al. [3] verifies the use of Bluetooth sensing technology in tracking pedestrian mobility in public transportation spaces. Specifically, authors show the adaptability of the technique to measure the transit time through a security checkpoint in an airport environment. Kostakos et al. [8] used Bluetooth scanners installed on public bus to estimate passengers' O/D matrices as an inexpensive approach. In their work, authors estimated that 12 % of the passengers carry a Bluetooth equipped device and the travel data thus collected present 80 % accuracy of the daily fluctuation of the actual flows. Moreover, they conclude that their approach allows characterizing passenger mobility through polycentric networks.

In [12] it is demonstrated the use of Bluetooth scanners as mobility sensors is feasible for spatial pattern extraction. The authors have applied the technology at social events to monitor a sample of visitors and extract their route choice. Utilizing the Bluetooth approach Liebig et al. were also able to understand microscopic movement behavior.

This work approaches the Bluetooth sensing technology in a twofold way to characterize urban mobility dynamics. In the next section, we discuss a proof-of-concept deployment of Bluetooth sensors in a static network as well as in moving vehicles in the city of Porto presenting thus a novel way of sensing urban mobility.

3 Methodology

We consider two distinct setups for monitoring urban mobility: a) a network of static Bluetooth sensors deployed on key points of the city, and b) mobile Bluetooth sensors deployed on commuting vehicles. In either case, we implement the workflow depicted in Fig. 1 and with three stages:

- Data collection
- Data filtering
- Data analysis and visualization

The sensors, described in the next section, perform data collection and yields lists of devices found in the proximity of the sensors at a given time. The data is then filtered to remove entries that are not of interest, such as static devices from nearby houses that do not relate to mobility. The final sample is stored on a central database where it can be queried to allow data analysis and visualization.

3.1 Bluetooth Sensors

Both setups, static and mobile, rely on sensors that implement the Bluetooth protocol to perform an "inquiry" (discovery) to find nearby devices. An inquiry consists of sending a message to all devices within range asking for replies. A device that receives an inquiry message will reply with information about itself: unique hardware address, device type and provided services. The sensors perform

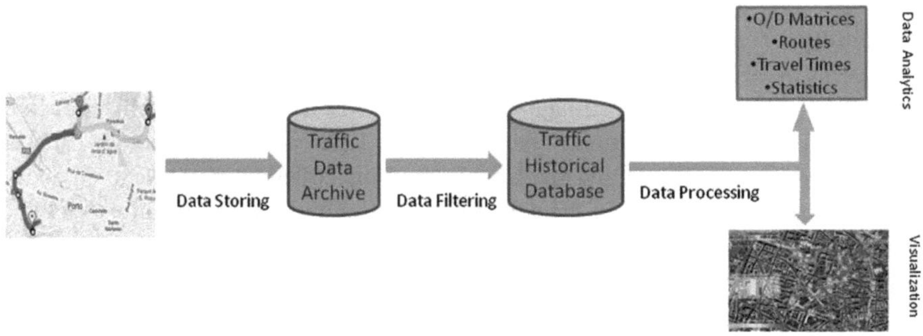

Fig. 1 Data acquisition, filtering and visualization workflow

discoveries continuously with an interval of 10 s as they wait for replies. This behavior can be modified to use just 5 s, which might improve the resolution of our data but may also come at the cost of missing some replies. The data collected by each sensor is stored in a pseudo-XML file as depicted in Fig. 2. Each discovery has a time stamp, a list of captured devices and, optionally, GPS coordinates.

For each device, we store its unique hardware address and device types, for instance, computer/laptop, or phone/Smartphone. We can also discover the manufacturer of the device by looking at the first half of the hardware address that is linked to specific companies. The list of prefixes and correspondent companies can be found at the IEEE website.[1] Having stored discoveries for periods of time, we can then track the movements of any device over space, on each deployed sensor, and over time. Depending on the type of study we want to perform we can filter.

3.2 Mobility Bluetooth Sensing

In this setup, a sensor is deployed on a vehicle, continuously capturing device information while the vehicle moves. This type of monitoring will capture devices inside nearby vehicles, carried by pedestrians or, not very often, stationary devices. The resulting data complements what we obtain in a more traditional fashion and gives us a different point of view over urban mobility. Deploying these sensors on a fleet of vehicles could yield traffic volumes for areas where there are no static sensors. However, there are other uses for this data.

An interesting example would be the study of car-pooling (or car-sharing) system feasibility. Car-pooling stored discoveries for periods a potential stranger during daily commutes with the objective of reducing fuel and other associated to

[1] http://standards.ieee.org/develop/regauth/oui/oui.txt

```
<discovery n="23" time="2011-09-19 20:29:27" Lat="38.721631419" lon="-9.296376806">
        <device>
                <address>00:1F:SD:BF:S7:11</address>
                <devclass>Phone</devclass>
        </device>
        <device>
                <address>00:12:1C:C2:6A:DC</address>
                <devclass>Audio/Video</devclass>
        </device>
        <device>
                <address>70:D4:F2:7D:DB:8D</address>
                <devclass>Phone</devclass>
        </device>
</discovery>
```


Fig. 2 Output sample of the Bluetooth sensor

the journey costs. Daily tracks of commutes collected with this setup may show similar routes taken at similar times by different vehicles. Thus, the drivers of these vehicles share a routine and are likely to benefit from car-pooling. Another example would be the study of reputation-based systems on Vehicular Ad-hoc Networks (VANET). In this case, we want to know the number of devices that we recognize in daily routines, therefore, allowing for an actual reputation to be built. Thus, the reading of the mobile sensor could be used as a first way of measuring the association and bond among neighbors building the reputation-based mechanisms. This idea is inspired by the concept of homophily [14], a tendency for members of a network to have a stronger link if they share something in common. In this case, they would share mobility patterns. A stronger link between two members would mean a higher level of trust in the information (and eventually services) one shares with the other. Leveraging on this concept, drivers would be able to identify the sources fellow-drivers that will provide him with correct information and services in a VANET formed environment.

3.3 Static Network Bluetooth Scanners

Networks of static sensors constitute the traditional approach when using Bluetooth for traffic monitoring [23]. Here, a number of Bluetooth sensors are installed in key points of a traffic network. This setup allows the collection of data that helps characterizing the traffic network macroscopically.

Effective mobility management, control and flow optimization can benefit from valuable travel information such as Origin/Destination (O/D) matrices, travel times and route inference. Since we can track individual devices, we can build O/D matrices and by measuring the time drivers took to reach one sensor after being captured by another, we can estimate travel times. The O/D matrices can be both a) static that is, based simply on device counts, and b) dynamically time-dependent, derived by tracing a Bluetooth device through a series of sensors with known locations.

Considering that a sensor will continuously emit discoveries, we can register the number of consecutive discoveries in which the same device is captured. If a device stays in range for a longer period of time, then it will appear in more consecutive discoveries. Therefore, a higher number of average consecutive discoveries will mean that most devices are spending more time in range, indicating slower local traffic.

4 Preliminary Results

This section presents preliminary results using the two aforementioned setups. First measurements show that approximately one vehicle in five contains some type of Bluetooth device that can be detected. The quantity of data from Bluetooth sensors is typically good enough to estimate O/D matrices and mean travel times. It is also sufficient to estimate the travel time variance of traffic and detect abnormal mobility conditions, such as traffic jams due to accidents.

4.1 Stationary Sensing Proof-of-Concept

A proof-of-concept experiment was performed using a stationary setup with sensors on five locations of the urban area of the city of Porto, Portugal. The experiment took place in key entry points to the city depicted in Fig. 3. The goal was to measure the average travel times between these 5 points. The experiment took place between 8 and 10 AM, during mornings rush hours.

In Table 1, we show information related to the number of devices detected in this experiment. Table 2 presents a partial O/D matrix, while Table 3 shows average travel times and speed between two given points.

A different long-term experiment using a single sensor was placed in S1. Using only one sensor restricts the utility of our data in characterizing the traffic network globally. However, it still provides an insight on local traffic. Figure 4 shows device volumes for a single day. Rush hours are clearly visible.

Since the sensor was placed at a key entry point to the city, we were able to measure the amount of time that people stayed in the city before leaving again, or vice versa. Figure 5 shows time intervals between two captures of devices, for three weekdays. The usual 8-h work schedule is clearly visible. A considerable number of vehicles also stay for short periods of time inside, or outside, the city.

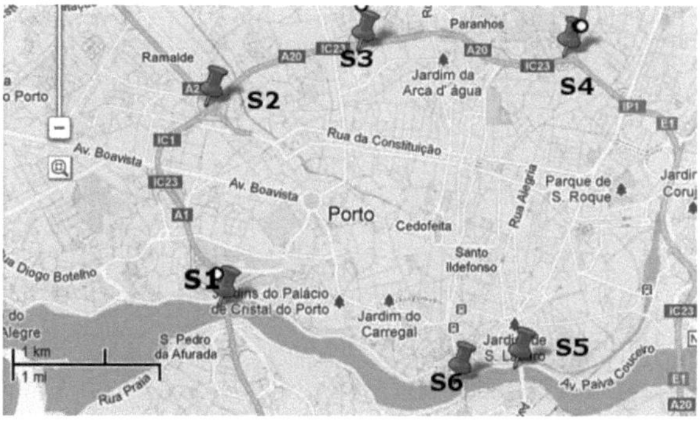

Fig. 3 Monitoring points

Table 1 Information related to discovered devices

Station	Discovered devices	Devices/min
S1	2367	29.4
S2	2367	42.8
S3	929	8.36
S4	4379	40.6
S5	1016	8.40
S6	1016	6.57

Table 2 Example of O/D pairs between stations

From S1 to	Devices	% of Devices
S2	434	18.3
S3	9	0.38
S4	80	3.38
S5	87	3.68
S6	7	0.30

Table 3 Average travel times and speeds between points S1, S2

S1 to S2	
Average time	9.63 min
Average speed	18.07 km/h
Fastest	75.11 km/h–2.32 min
Slowest	1.84 km/h–89.7 min
Outliers (>2*Average)	44 (out of 434)
Without outliers	
Average time	4.93 min
Average speed	35.31 km/h
Fastest	75.11 km/h–2.32 min
Slowest	9.47 km/h–18.38 min

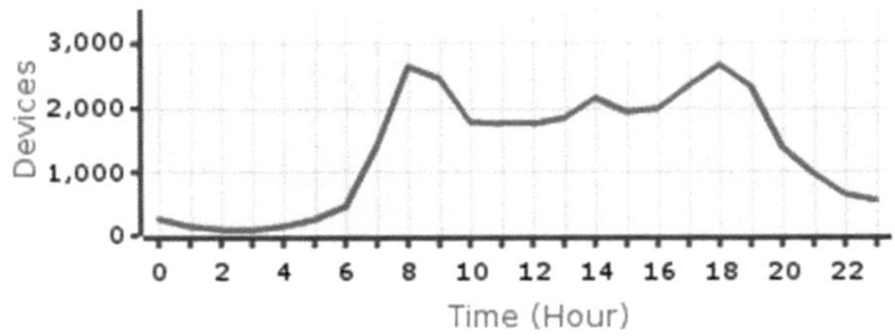

Fig. 4 Device volume for a single day

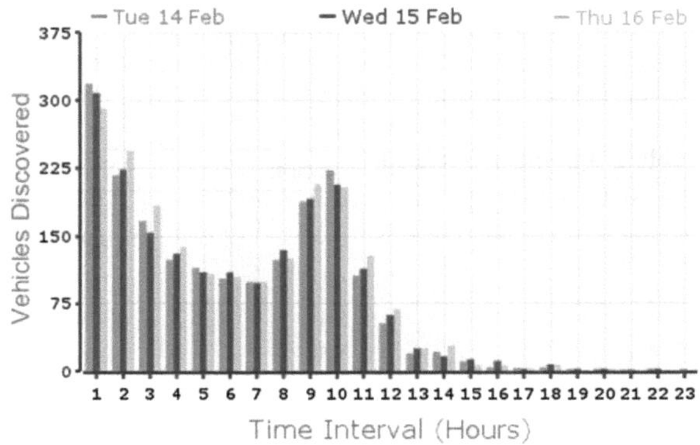

Fig. 5 Time intervals between two detection of a Bluetooth device

4.2 Mobile Sensing Proof-of-Concept

Another proof-of-concept experiment was performed for mobile. We opted to use a vehicle with a fixed 30 km daily commute. The experiment ran for over a month, resulting in 26 h worth of Bluetooth discoveries.

Figure 6 shows the evolution of recognized devices over the course of the experiment. We can see a large considerable increase in the percentage of known devices. By the end of the experiment, the probing vehicle already recognized around 30 % of all devices found on the commuting route. As a curiosity, the 36th commute took an unusual route and the effect is clearly visible.

Depicted in Fig. 7 is the web-based visualization tool developed to analyze data captured with mobile sensors. We use a heat-map to depict the number of devices captured. The movement of the probing vehicle may be played as an animation, showing the amount of devices over time and over space.

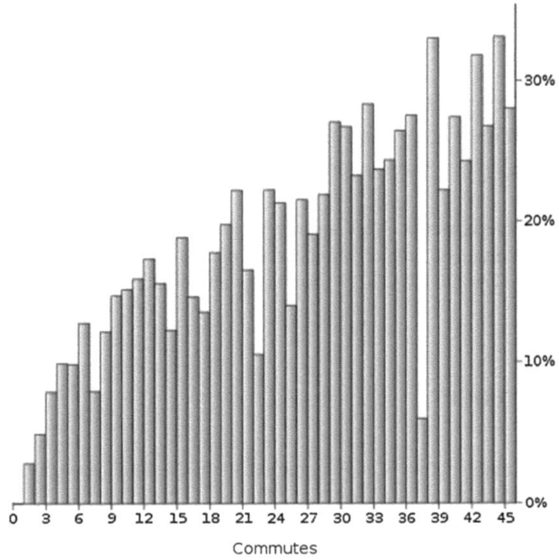

Fig. 6 Percentage of neighbour devices per trip

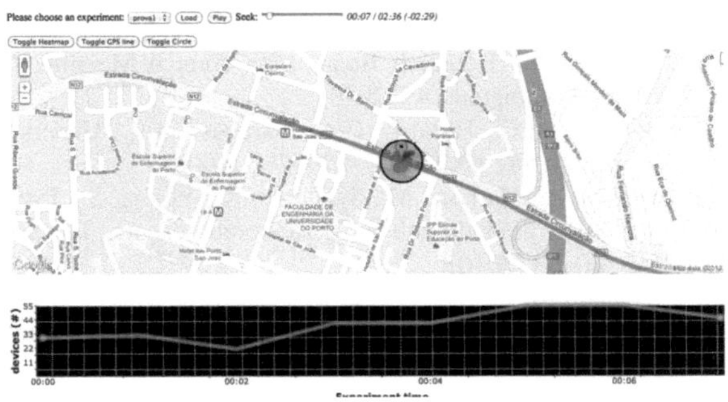

Fig. 7 Mobility activity captured by a mobile Bluetooth scanner

5 Conclusion and Future Work

Bluetooth sensing technology has been proven to be very promising in mobility studies since it complements with great efficiency traditional traffic detection approaches. Bluetooth data is portable and offers great accuracy in estimating travel times and O-D matrices. In turn, these metrics have various applications such as urban planning or calibration of simulated scenarios both in cases of traffic and pedestrian monitoring.

In this chapter, we presented our take on the traditional approach for Bluetooth monitoring as well as an alternative and novel approach. A mobile sensor provides complementary data to static sensors, and also allows for new studies and new insights. How many people share driving routines? How long does it take for a vehicle to know its neighbors? These questions can be answered by further studying this approach. For example, it would be worthwhile studying further the percentage of recognized vehicles over time and how it changes both with minor and major variations in route or in time of day on which they are taken.

Our early experimental results confirmed that Bluetooth sensing technology could be employed in either setup with promising results. We showed that several varied metrics, from travel times to device recognition rates, could be easily derived from the collected data, with a multitude of applications.

Acknowledgments This project has been partially supported by FCT (Fundação para a Ciência e a Tecnologia), the Portuguese Agency for R&D, under the Bluetooth Sensing Technology project IT/LA/01081/2011 and the grant SFRH/BD/67202/2009.

References

1. Asmundsdottir, R.: Dynamic od matrix estimation using foating car data. M.sc thesis, Delft University of Technology (2008)
2. Barcelo, J., Montero, L., Marques, L., Carmona, C.: Travel time forecasting and dynamic od estimation in freeways based on bluetooth traffic monitoring. Transp. Res. Rec. J. Transp. Res. Board. **2175**, 19–27 (2010)
3. Bullock, D., Haseman, R., Wasson, J., Spitler, R.: Anonymous bluetooth probes for measuring airport security screening passage time: the indianapolis pilot deployment. In: Transportation Research Board 89th Annual Meeting. CDROM. Transportation Research Board, Washington DC (2010)
4. González, M.C., Hidalgo, C.A., Barabási, A.L.: Understanding individual human mobility patterns. Nature **453**(7196), 779–782 (2008)
5. Haghani, A., Hamedi, M., Sadabadi, K., Young, S., Tarnoff, P.: Data collection of freeway travel time ground truth with bluetooth sensors. Transp. Res. Rec. J. Transp. Res. Board **2160**, 60–68 (2010)
6. Haseman, R., Wasson, J., Bullock, D.: Real-time measurement of travel time delay in work zones and evaluation metrics using bluetooth probe tracking. Transp. Res. Rec. J. Transp. Res. Board **2169**(1), 40–53 (2010)
7. Karlsson, N.: Floating car data deployment and traffic advisory services. Bridging the European ITS Business Cooperation with China, **40** (2003)
8. Kostakos, V., Camacho, T., Mantero, C.: Towards proximity-based passenger sensing on public transport buses. Pers. Ubiquit. Comput. **17**, 1807–1816 (2013)
9. Kostakos, V., ONeill, E.: Cityware: urban computing to bridge online and real-world social networks. In: Foth, M. (ed.) Handbook of Research on Urban Informatics: The Practice and Promise of the Real-Time City, pp. 195–204. IGI Global (2008)
10. Kostakos, V., O'Neill, E., Penn, A., Roussos, G., Papadongonas, D.: Brief encounters: sensing, modelling and visualizing urban mobility and copresence networks. ACM Trans. Comp. Hum. Interact. **17**(1), 1–38 (2010)
11. Leduc, G.: Road traffic data: collection methods and applications. In: Working Papers on Energy, Transport and Climate Change. pp. 47–67 (2008)
12. Liebig, T., Wagoum, A.U.K.: Modelling microscopic pedestrian mobility using bluetooth. In: 4th International Conference on Agents and Artificial Intelligence. **2**, 270–275 (2012)
13. Malinovskiy Y., Wu Y., Wang Y., Lee U.: Field experiments on bluetooth-based travel time data collection. In: Transportation Research Board 89th Annual Meeting. CD-ROM. Transportation Research Board, Washington DC (2010)
14. McPherson, M., Smith-Lovin, L., Cook, J.M.: Birds of a feather: Homophily in social networks. Ann. Rev. Sociol. **27**, 415–444 (2001)

15. ONeill, E., Kostakos, V., Kindberg, T., Schiek, A., Penn, A., Fraser, D., Jones, T.: Instrumenting the city: Developing methods for observing and understanding the digital cityscape. In: Dourish, P., Friday, A. (eds.) 8th International Conference of Ubiquitous Computing (UbiComp 2006). Lecture Notes in Computer Science, vol. 4206, pp. 315–332. Springer, Heidelberg (2006)
16. Pels, M., Barhorst, J., Michels, M., Hobo, R., Barendse, J.: Tracking people using bluetooth: implications of enabling bluetooth discoverable mode. Final report, University of Amsterdam. http://www.remcohobo.nl/IDS/bluetoothreport.pdf (2005). Accessed 16 Jan 2014
17. Sharifi, E., Hamedi, M., Haghani, A.:Vehicle detection rate for bluetooth travel time sensors: a case study in Maryland and Delaware. Paper presented at the 91st annual transportation research board meeting, Washington DC, US (2010)
18. Tarnoff, P.J., Bullock, D.M., Young, S.E., Wasson, J., Ganig, N., Sturdevant, J.R.: Continuing evolution of travel time data information collection and processing. In: Transportation Research Board 88th Annual Meeting (2009)
19. Tsubota, T., Bhaskar, A., Chung, E., Billot, R.: Arterial traffic congestion analysis using Bluetooth duration data. In: Australasian Transport Research Forum (ATRF), **34** (2011)
20. Valerio, D., D'Alconzo, A., Ricciato, F., Wiedermann, W.: Exploiting cellular networks for road traffic estimation: A survey and a research roadmap. In: IEEE 69th Vehicular Technology Conference. pp. 1–5. IEEE (2009)
21. Waadt, A., Wang, S., Bruck, G., Jung, P.: Traffic congestion estimation service exploiting mobile assisted positioning schemes in gsm networks. Procedia Earth Planet. Sci. **1**(1), 1385–1392 (2009)
22. Wasson, J.S., Sturdevant, J.R., Bullock, D.M.: Real-time travel time estimates using media access control address matching. ITE J. **78**(6), 20–23 (2008)
23. Young, S.: Bluetooth traffic monitoring technology: concepts of operation and deployment guidelines. http://www.catt.umd.edu/sites/default/files/documents/UMD-BT-Brochure_REV3.pdf. Accessed Jan 2014
24. Young, S.E.: Bluetooth traffic detectors for use as permanently installed travel time instruments. Technical Report, Maryland State Highway Administration, University of Maryland, College Park (2012)

The Driver Behaviour Questionnaire: An Investigation Study Applied to Chinese Drivers

Jie Li, Henk van Zuylen and Elisabeth van der Horst

Abstract There is a significant difference in road traffic performance between China and Western countries. One important cause for this difference is the human factor, which has an important influence on the performance of traffic. This study presents a Driving Behaviour Questionnaire (DBQ) to collect information about driving attitude and behaviour and their possible impact on safety and traffic performance. The statistical analysis of 175 returned questionnaires confirms the existence of some special properties of Chinese drivers with respect to driving behaviour, especially their attitude to priority rules, driving behaviour at intersections and interaction with other drivers in congestion.

Keywords DBQ · Driving behaviour · Violations · Priority rules · China · Simulation

1 Introduction

Traffic behaviour of Chinese drivers is remarkably different from that of Western drivers, which is directly visible on urban and interurban roads. The official Highway Code is similar in China as in other countries and Chinese drivers have to study the rules for driving and pass an examination before they can get a driving

J. Li · H. van Zuylen (✉)
Department of Traffic and Road Engineering, Civil Engineering College, Hunan University, Lushan South Road, Changsha 410082 Hunan, People's Republic of China
e-mail: h.j.vanzuylen@tudelft.nl

J. Li
e-mail: ljlj369@msn.com

J. Li · H. van Zuylen · E. van der Horst
Department of Transport and Planning, Faculty of Civil Engineering and Geosciences, Delft University of Technology, Stevinweg 1 2628 CN Delft, The Netherlands

J. F. de Sousa and R. Rossi (eds.), *Computer-based Modelling and Optimization in Transportation*, Advances in Intelligent Systems and Computing 262,
DOI: 10.1007/978-3-319-04630-3_32,

license. However, in China, drivers' knowledge about these traffic rules is very limited in practice and the disregard of rules is more a habit than an exception. Traffic policemen try to enforce the traffic rules, but this task is so huge because of the difficulty in traffic monitoring. Some rules, for example red light discipline and speed limitation can rather easily be enforced, but many other rules, like priority rules, lane discipline etc., are difficult to be monitored in practice.

There is a certain driving culture: drivers influence each other and then reinforce irregular driving habits because everybody does the same. Furthermore, drivers expect that other drivers will not respect the rules of the road, which makes them cautious. They will not choose to drive at higher, possible speeds because they expect that some other drivers may have some irregular driving behaviour, which will threaten their safety. This all makes road traffic inefficient and rather risky [9].

The direct reason to execute a study on the attitudes and behaviour of Chinese drivers was the finding that the road capacity in China was significantly lower than what has been observed in a Western reference country (The Netherlands) [9, 12]. Up to 20–30 % longer headways and a larger standard deviation of headways have been observed at stop lines of the investigated Chinese intersections than at comparable intersections in The Netherlands. The reasons behind these phenomena have to be found before conclusions can be drawn with respect to possible measures to improve traffic conditions in Chinese cities. The difference in driving behaviour can also have implications for the modelling of traffic.

A simple way of measuring driver behaviour is to execute a Driving Behaviour Questionnaire (DBQ) survey on drivers who report how they typically behave, experience, and what their attitudes are in driving. Objectively observed driving behaviour is unfeigned, but is also limited to the usually short time period of data gathering and a limited number of drivers. DBQ has been well-documented in the past decades. The Manchester Driver Behaviour Questionnaire (the MDBQ) is one of the most widely used instruments for measuring driving styles. The MDBQ has been used by many researchers to identify the relation between driver characteristics and the involvement in accidents. This kind of DBQ is based on a theoretical taxonomy of aberrant behaviour categorized into violations, errors and lapses [10]. Lawton et al. distinguished violations in aggressive and ordinary ones as "The aggressive ones involve overtly aggressive acts whereas the ordinary ones consist of deliberately breaking the Highway codes and/or law without aggressive motives". They defined errors as a "failure of planned actions to achieve their intended consequences"; and a 'lapse' represents a mistake resulting from inattention [8].

The results of previous research indicate that the DBQ can be utilised to examine self-reported aberrant driving behaviour [2, 6, 14]. Some researchers also paid attention to the variation in driving behaviour among different driving populations and carried out a cross-culture comparison with a DBQ survey [1, 7]. DBQ surveys have been executed quite a lot for several decades with different research objectives [5, 7, 11].

The main objective of the research reported in this chapter is to understand the mechanisms that make urban traffic in Chinese cities so inefficient. According to this research objective, the item structure in this study had to be modified. It is different from the traditional DBQs that focus on the relation between driving style and traffic accidents. The next section introduces the DBQ survey as held in China. Section 3 presents some results from a further statistical analysis. Section 4 discusses the consequences of the specific behaviour of Chinese drivers for the efficiency of the intersections. It also discusses the applicability of the traffic simulation programs which were developed in Western countries. Section 5 presents some important conclusions, analyses the limitations, and offers recommendations for future study.

2 Methodology

2.1 Design of the Questionnaire

In order to make the questionnaire represent Chinese driving conditions and match with the objective of this study, a DBQ was designed with 50 items based on previous versions [1–3, 6, 7, 14]. This section introduces an overview of the DBQ used in this study.

Most of the previous Manchester Driver Behaviour Questionnaire [7] included 24 items for aberrant driver behaviour. Due to the different research targets, the items related to aberrant driver behaviour in the current study have been reduced to 14 items which include aggressive violations (2 items), ordinary violations (4 items), lapses (3 items), and errors (5 items). Respondents were asked to indicate how often they committed each of the 14 mistakes according to their driving experience on a five-point scale (1 = Never, 2 = Nearly Never, 3 = Seldom, 4 = Sometimes, 5 = Often). Table 1 lays out the details about some of these 14 items.

In addition, the authors of this chapter added some new questions based on their main research objective, i.e. the influence of driving behaviour on traffic performance. These questions are about typical driving behaviour, namely reaction to traffic signals, lane changing, and overtaking. Respondents were also asked to indicate what the meaning is of some priority signs, and their attitudes to cooperative lane changing, etc. The influence of the driving behaviour on traffic performance and furthermore the simulation models are discussed in Sect. 4.

Table 1 An example of the questions about aberrant driving behaviour

No.	Item	Type	Answer distribution (percentage)				
			Never (1)	Nearly never (2)	Seldom (3)	Sometimes (4)	Often (5)
Q1	Using mobile phone by keeping it in the hand while driving	OV	12.0	26.3	14.3	27.4	20.0
Q3	Cross stop line during red phase, knowing there isn't a red-light registration camera	AV	54.3	22.3	12.0	9.7	1.7
Q7	Ignore the maximum speed limit in urban areas when driving on a main road without much traffic	OV	28.0	45.7	19.4	5.7	1.2
Q10	Overtake, even though you see the vehicle already has the left indication on	E	61.7	28.6	6.3	2.8	0.6
Q13	Keep driving on the left most lane even when there is no traffic on the right lane and no need for left-turn	L	8.0	26.9	26.9	21.1	17.1
Q14	Remain in a lane with high speed even knowing that this lane will end soon, making a forced lane changing at the last point	E	24.0	38.9	22.8	12.6	1.7

Note Behaviour types are represented as AV = Aggressive violation, OV = Ordinary Violation, E = Error, L = Lapse

2.2 Procedure and Participants

The questionnaire was published on Internet in February, 2013.[1] Respondents were volunteers and gave their answers online individually. To reduce motivation for socially desirable responses, respondents were explained that researchers wished to investigate driving behaviour in different countries and the international comparison will be expected to improve traffic safety and efficiency. For the sake of privacy, the answers could remain anonymous if a respondent wants that. Respondents were instructed to read all questions carefully and to answer each question veraciously. Until March, 2013, 175 volunteers finished the online questionnaires.

The sample for driving behaviour survey included 175 respondents (79 % males; mean age = 35.2; std.age = 8.2; range 21–60 years). Respondents were located throughout China in different regions. The largest proportion of vehicles driven by respondents were reported to be their own vehicle (77.7 %), small proportions were borrowed (12 %) or leased (2.3 %), etc. On average respondents had driving experience for 5.3 years (std. years = 5.5), with the largest proportion driving 1–3 times per day (29.1 %), and less than 30,000 km per year (93 %).

[1] http://www.sojump.com/jq/2146887.aspx

3 Results from the Analysis

3.1 Socio-Demographic Characteristics and Aberrant Driving Behaviour

The respondents were mainly selected from the personal network of the authors; some professional (taxi) drivers were also invited to fill in the questionnaire. Most respondents in this survey graduated from university: Bachelor 83 (47.4 %), Master or higher than Master 55 (31.4 %), Professional Education 24 (13.7 %), and Others 13 (7.5 %). Most respondents (38.3 %) are Professional Persons; Administrative staff, Full-time students, and Teachers/Researchers constitute 24, 10.3 and 9.1 % respectively. Most respondents (66 %) stated they enjoy driving; only few respondents don't have special feeling (24 %) for driving or dislike (10 %) driving.

Respondents were asked to indicate how often they themselves make each of the violations and errors during driving. Responses were on a five-point scale from 1 to 5 with the meanings: 1 = Never, 2 = Nearly Never, 3 = Seldom, 4 = Sometimes, 5 = Often. Respondents' answers for questions about aberrant driving behaviour are presented in Table 1. Q1 and Q13 were reported to be committed much more often than other violations and errors. In terms of the answers, most respondents stated that they had 'Never' the experience of Q10 and Q3. In order to analyse the relation between driving behaviour and socio-demographic characteristics, for each respondent a mean value of each category of aberrant driving behaviour was calculated: Aggressive violation, Ordinary violation, Error and Lapse.

3.2 Inter-Correlations Between the Variables

In this study, the Chi square test is used to verify whether the answer of every question is independent of other items. The null hypothesis is that the answers for each question are independent, i.e. do not have significant influence on each other. A p value larger than 0.05 indicates that the p value fails to reject the null hypothesis at the 95 % confidence level.

The Chi square tests matrix reveals that most variables are not strongly correlated with each other. It can be confirmed that recorded disobedience, yearly driving experience, driving mileage per year, and self-assessed driver type have significant relation with the number of accidents in which the respondents were involved. Yearly driving experience and driving mileage per year also have a significant relation with accidents or disobedience. Obviously, the exposure of driving, i.e. mileage per year, is a critical factor which has a strong relation with several items: disobedience, accident involvement, age, driving experience, self-assessment of skill and driver type, and ordinary violation. Besides mileage per year, driving experience in years and ordinary violation are two other important factors in driving. Among the DBQ surveyed mistakes, most answers are correlated with each other, except that the aggressive violations and errors are two less correlated items. Gender and age appear to have insignificant influence on other factors, apart from a correlation between gender and errors and between age and experience: female drivers stated less errors and older drivers had more driving experience.

3.3 Disobedience and Accident Involvement

Traffic rules play an important role in managing traffic and in the way communication between drivers is established. Besides the factor of gender, other factors, like age, driving experience, location type, also have been stated to contribute to the driving behaviour with regard to obedience to traffic rules. In this study, the driver types are categorized as: very conservative (VC), somewhat conservative (SC), moderate driver (MD), somewhat aggressive (SA), and very aggressive (VA). Respondents are required to assess which driver type he/she is and how friends and family consider him/her.

Quite large parts of the respondents think they are very conservative drivers, somewhat conservative drivers, or moderate drivers, with the percentages 31, 27, and 30 % respectively. Very few respondents consider themselves as aggressive drivers. Most respondents, i.e. 46 %, think the self-estimated driver type is consistent with the impression given by others. 32 % of the respondents report that they are a more aggressive driver in their friends' impression; on the contrast, 22 % of respondents report they give others the impression to be more conservative in driving.

Based on the questionnaire survey, approximately half of the respondents have had been registered offences against the traffic rules in the past year, and almost the same percentage of respondents have been involved in accidents in the past five years.

3.4 Priority Rule Implementation

In driving practice, the priority rules concern the interaction between drivers. The way in which drivers execute the priority rules can have significant influence on traffic both on macro and micro level. On macro traffic flow level, the implementation of priority rules influences the traffic safety and performance to a rather large extent. On the micro traffic level, the interaction between drivers should be properly predicted and understood by all drivers. In the questionnaire, several questions are related to the implementation of priority rules in reality. It is expected that the analysis of the answers will give insight into drivers' attitudes to priority rules execution.

(a) (b) (c) (d)

Fig. 1 Priority sign: give priority by an obligatory stop (**a**, **c**), and by deceleration (**b**, **d**)

(1) Do you know exactly the meaning of the following traffic sign?

A few questions selected from the driving license examinations concern the meaning of the traffic signs related to priority. These signs are to give priority by an obligatory full stop or by deceleration; and are on the road pavement or on the road side respectively, as shown as in Fig. 1.

The answers from the questionnaire are rather peculiar: only 10 persons from the 175 respondents could give the correct answers to the four priority signs. More details about the answers are shown in the Table 2. Only about 15 % of the respondents could indicate the meaning correctly, and more respondents answered they know the meaning but gave wrong answers or did not give an answer. A quite large part of respondents admitted they do not know the meaning of the sign or have not seen these signs. Other respondents just gave ambiguous answers, like deceleration, carefully driving.

The answers to questions about the signs on the road pavement and beside the road represent some interesting deviations. More than 25 % respondents stated that they have never seen the priority sign on the road pavement, but this percentage is only 3.43 % for the priority sign besides the road. This reflects the fact that most respondents do not have the habit to search for traffic signs on the road pavement, but just along the road or above the road. A quite large part of respondents (42.86 %) gave ambiguous answers for the meaning of the roadside sign 'giving priority by deceleration', which indicates that these respondents know this priority rule, but their notion is still quite vague.

When checking the driving experience of the respondents who could point out the correct meaning of all the priority signs, it is interesting to find that 7 out of 10 respondents had only less than 4 years driving experience. Novice drivers seem to keep the memory of the examination answers better than the more experienced drivers. With the advance of the time, lots of drivers don't reinforce their memory about some of the driving rules and even forget them.

(2) Do you have the habit to search priority signs close to an intersection?

Most respondents stated they have the habit to search priority signs close to an intersection with varying frequencies: 24.57, 29.14, and 12 % to 'Sometimes would, sometimes would not', 'More likely would', and 'Generally would' respectively. There are only 12.57 and 21.71 % respondents admitted 'Generally would not' or 'More likely would not'. Even though this percentage is probably much lower than the value in Western countries, it is still much higher than the

Table 2 Answer distribution (percentage of the total answers 175)

Location	Sign	Correct answer	Ambiguous answer	Just select 'known', no answer	False answer	Unknown	Never seen
Road side sign	Give priority by stop	13.14	0.57	10.86	26.29	24	25.14
	Give priority by deceleration	12	4.57	8	11.43	35.43	28.57
On the pavement	Give priority by stop	15.43	1.71	18.86	46.29	14.29	3.43
	Give priority by deceleration	16	42.86	17.71	6.86	13.14	3.43

percentage of respondents who showed that they know the meaning of priority signs correctly. Therefore, it can be deduced that some respondents even stated that they had the habit to search priority signs, but the inconsistency is that they did not recognize the sign in reality. This inconsistency casts a shadow on the validity of the outcomes of the questionnaire.

(3) The probability to give priority

Three scenarios in the questionnaire are related to priority rules implementations in reality. Respondents were required to estimate the probability (in percentage) to give priority to a conflicting vehicle or pedestrian according to their driving experience:

1. Generally would not (<10 %),
2. Sometimes would but more likely would not (10–40 %),
3. Sometimes would, sometimes would not (40–60 %),
4. More likely would (60–90 %), and
5. Generally would (>90 %).

If the respondent selected 'Generally would not', this means that the respondent stated that the probability to give priority to the conflict flowing is lower than 10 %; if the 'Generally would' was selected, the probability will be higher than 90 %, based on respondents' driving self-assessment. The answers are described in Fig. 2, in which the percentages along the horizontal axis represent the likelihood to give priority to conflict flow.

The majority of respondents denoted that they will give priority to the conflicting flow. The discrepancy in self-reported driving behaviour and what is observed in reality makes it is necessary to have a further discussion with drivers to analyse the reasons.

(4) In which cases, will you not give priority to the conflicting pedestrians?

As an additional question to priority rules, the respondents were required to explain the cases in which they will not give priority to pedestrians when turning right and sharing the green phase with pedestrians. The answers are shown in Table 3.

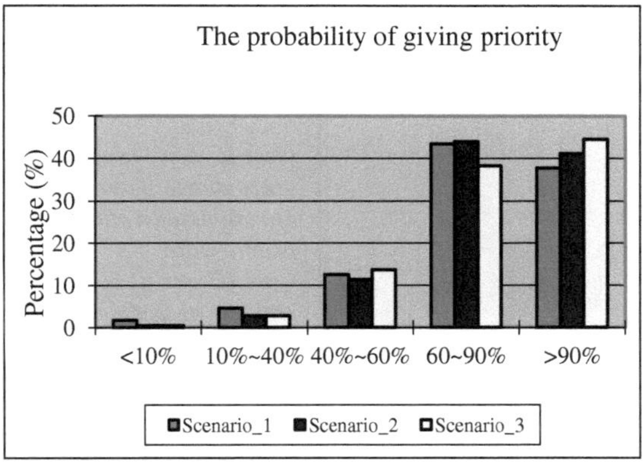

***Scenario 1**: Driving from a minor street conflicting with vehicles on an arterial street at an un-signalized intersection*
***Scenario 2**: Turning left conflicting with through going vehicles at an intersection.*
***Scenario 3**: Turning right conflicting with pedestrians on a zebra at an un-signalized intersection.*

Fig. 2 Three scenarios of priority giving to conflicting vehicles or pedestrians

Table 3 Refusing to give priority to pedestrians when turning right

Options	Data	Percentage (%)
Never happen because I always give priority to pedestrians	79	45.1
Pedestrians never have priority	8	4.6
Congested traffic situation	29	16.6
When I am in a hurry	55	31.4
Too many pedestrians, and too long time waiting make me lose my patience	29	16.6
Others	13	7.4

Table 3 shows that driving behaviour can change according to the traffic conditions, infrastructure and personal status (e.g. hurry, congested traffic).

4 Driving Behaviour Discussion

Driving behaviour in an urban area, especially at signalized intersections, can be simplified into three categories: reaction to signals, longitudinal driving behaviour, and lateral driving behaviour. These behaviours have been described in lots of different models which are essential to microscopic traffic simulation programs. The results of this survey show that such models and also such simulation programs should be adapted to the specific behaviour of Chinese drivers. This section describes some results in this domain.

4.1 Reaction to Signals

Several errors related to driving at signalized intersections are investigated in the questionnaire. Respondents were asked to select the frequency with which they

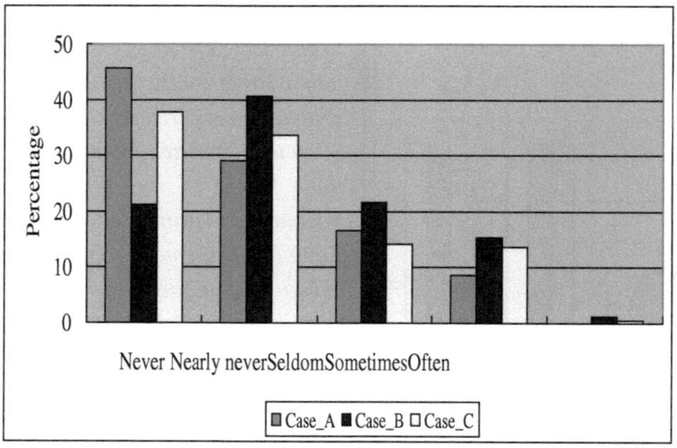

***Case A**: almost be hit by the following car due to your sudden brake in amber time*
***Case B**: don't react to green due to distracted attention, until other people warn*
***Case C**: almost collide with the preceding vehicle due to the sudden braking in amber time*

Fig. 3 Reaction to signals

made these errors, as shown in Fig. 3. These survey results indicate that the reaction to signal not only vary between different drivers, but also comprise some individual behaviour errors, which should also be considered in simulation models. Most simulation models adopt a certain distribution, like normal distribution, uniform distribution, etc., for the values of some driving behaviour related parameters. Inspired by this study, the boundary of these parameters can be extended to a certain 'unsafe' value which can be considered as the consequence of driver's errors. How drivers react to this kind of errors can be simulated as an additional model added to the present models.

4.2 Lane Changing

Lane changing behaviour is another important component of driving behaviour. In most cases, lane changing is distinguished into two classes: mandatory (MLC) and discretionary (DLC). MLC is executed when the driver must leave the current lane. DLC is executed to improve driving conditions. In the questionnaire, two scenarios are developed for respondents to estimate the probability to change lanes. Respondents were asked to choose the level of likelihood that they would change lanes for each scenario according to their driving experience. In Fig. 4, the percentages along the horizontal axis represent the likelihood to change lane.

Figure 4 demonstrates that drivers are more likely to lose patience and try to change lane in the case of following or being followed by a big truck for the sake of safety, compared with the case of following a slowly moving passenger vehicle. Therefore the influence of a big truck on the adjacent vehicles should get more attention in simulation models. The consequences for traffic performance and traffic safety might be important: the disturbance given by trucks to the

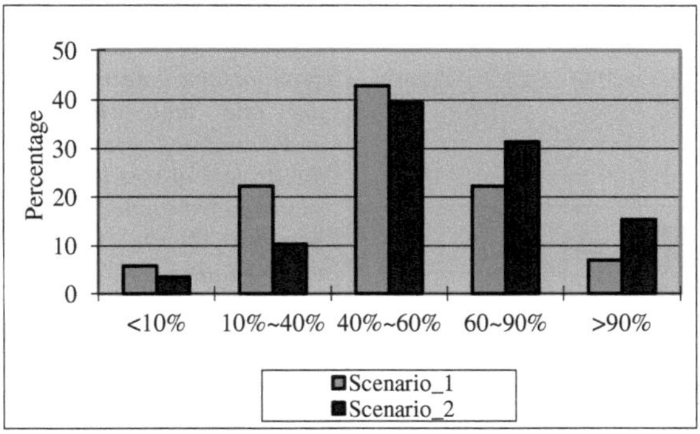

***Scenario 1**: When you are following a slowly moving vehicle, you become impatient and try to change lane as soon as possible.*
***Scenario 2**: When you are followed by or are following a big truck, you feel unsafe and try to change lane as soon as possible.*

Fig. 4 Lane changing

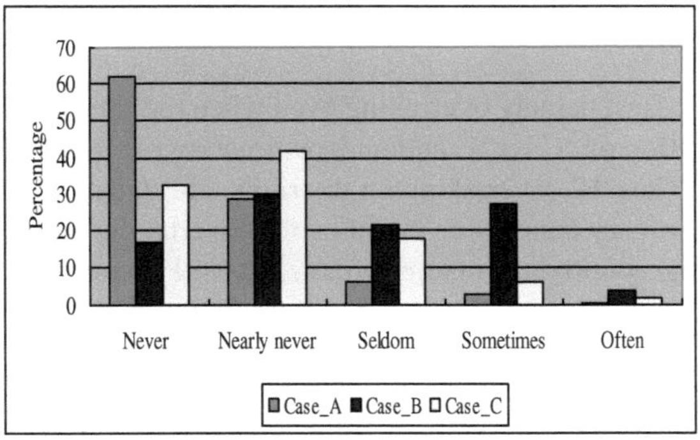

***Case A**: still start overtaking even when the preceding vehicle already has the turning indication on*
***Case B**: overtake a slow driver on the right side in case of no chance in the left side*
***Case C**: underestimate the speed of the oncoming veh cle when overtaking on a two-lane road*

Fig. 5 Overtaking behaviour

traffic flow in China is larger than what present models simulate and lane capacity will become lower by the frequent lane changing stimulated by trucks (Fig. 4).

4.3 Overtaking

Drivers who want to overtake should firstly estimate the 'gap' they need and estimate the available 'gap'. A full overtaking process can be considered as executing lane-changing twice in a given time. Due to the complexity of the overtaking process, inappropriate manoeuvres give conflicts and may cause an accident. In the questionnaire, three typical lapses in overtaking are listed for respondents to estimate the frequency of occurrence, as demonstrated in Fig. 5.

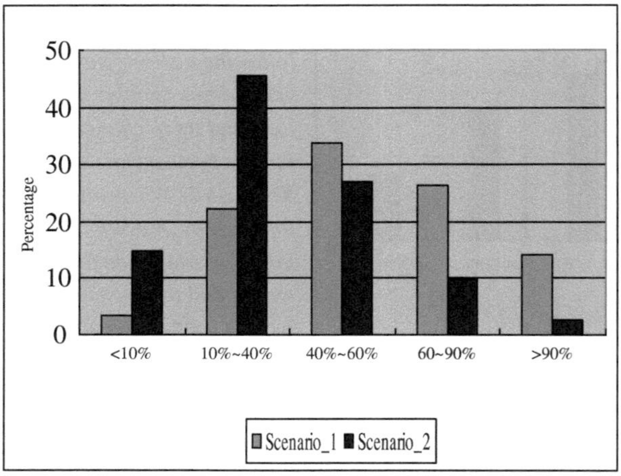

***Scenario 1**: When you drive on the lane next to the one that will be end/closed, will you allow the drivers of that lane to switch to your lane?*
***Scenario 2**: In congestion, another vehicle in the adjacent lane tries to have a forced lane change in front of you. Will you decelerate and provide an available gap?*

Fig. 6 Cooperative in lane changing

Compared with the other two cases, namely overtaking even when the preceding vehicle already has the left indication on 'Case A' and underestimating the speed of the oncoming vehicle 'Case C', 'Case B' (overtaking on the right side) takes place much more frequently. The preliminary conclusion could be that overtaking on the right side (in a right-side driving country) is inappropriate, but still common in China's urban areas. Also this should be considered more in simulation models.

4.4 Interactions During Lane Changing

The interactions between drivers during lane changing are not only important for traffic safety, but also critical to simulation models. Several factors, like driver properties, traffic conditions, etc., are considered to affect the cooperative or forced lane changing decisions. Driving behaviour may change discontinuously between traffic regimes [4, 13]. When the traffic density is high, drivers are inclined to accept a small critical gap and behave more aggressively in lane changing. Figure 6 describes the probability of cooperative lane changing in two different scenarios. The percentages along the horizontal axis represent the likelihood to have a cooperative lane changing according to personal driving experience.

When the adjacent lane will end or be closed and the drivers in this lane try to execute a lane change, most respondents are willing to cooperate. However, in the case of congestion, a small number of respondents would like to decelerate and make a gap available to the driver who tries to execute a forced lane change from the adjacent lane. This demonstrates that drivers are more aggressive in congestion compared with normal traffic situations, and accordingly the relevant simulation models should be modified for dense traffic situations.

5 Conclusions

A first round questionnaire survey has been carried out in China. The stated driving behaviour and socio-demographic variables were analysed to identify correlation among them. By analysing the answers, some special characteristics in driving behaviour have been revealed, and the further influence on simulation models has been discussed. Several important conclusions can be drawn from this study:

- Accident and disobedience have been identified to be mainly effected by driving experience and driving exposure.
- Age and gender appear to be uncorrelated with most of the other items in the DBQ survey in this study.
- Regarding the questions on the 'meaning of priority signs', many respondents gave wrong answers, especially for the signs on the pavement. Only 10 out of 175 respondents indicated exactly the difference between giving priority by an obligatory stop and by deceleration. It can be concluded that most drivers in China are quite ignorant about the way priority rules are implemented, which can have strong negative influence on traffic safety and performance. Most respondents admitted they might violate the priority rules in case of congestion or hurry. This trend should get the attention from the traffic management department, and the driving education and examination system should stimulate drivers to develop better driving habits, rather than just driving skills.
- Discussion about vehicle interactions shows that traffic conditions and personal status affect drivers' driving behaviour. Most respondents indicated they would be more aggressive under dense traffic conditions or when they are in a hurry. In congestion, if another vehicle in the adjacent lane initiated a forced lane change, most respondents would not be willing to decelerate to make a feasible gap. However, if the adjacent lane will end or be closed, most respondents would cooperate with the driver from the adjacent lane in lane changing.
- An important conclusion is that microscopic simulation programs as developed for Western driving conditions need important modifications and calibration before they are valid and applicable in Chinese cities. For instance the following features of driving behaviour should be represented in the models:
  - Influence of congestion on the mechanism of driver decisions,
  - Right overtaking against the rules,
  - Offences of priority rules by a considerable fraction of the drivers can result in grid-locks in a conflict situation until no traffic can move anymore,
  - Reaction of drivers of passenger cars to big trucks,
  - Variations in the reaction time to transitions in the traffic signals.

Some of these features can be implemented in a simulation program by a suitable parameter calibration. Several limitations should be taken into account when considering the results of this study.

- Similar to all research in this field using self-reporting questionnaires, there is criticism with respect to the reliability of this survey methodology. Even if respondents were asked to respond accurately and veraciously, a self-report may still be different from their actual actions due to the recollection discrepancy or the propensity of respondents to give social desirable responses.
- The representativeness of the sample is also doubted. Quite a large part of volunteers are authors' classmates, colleagues and their driving styles may not fully represent the whole driving population.
- In order to generalize the findings of this study—if they can be validated—to the simulation models modification, a more quantitative analysis should be made.

According to the conclusions and limitations of the present study, we will extend the study by focus group discussions—to get better understanding of the decision processes in driving—and in-car driving tests to verify the self-reported driving behaviour.

Acknowledgments This research has been made possible by grants for Changsha Science and Technology Commission under number K1001010-11 and K1106004-11. Also the financial support of vZC is acknowledged.

References

1. Bener, A., Özkan, T., Lajunen, T.: The Driver Behaviour Questionnaire in Arab Gulf countries: Qatar and United Arab Emirates. Accid. Anal. Prev. **40**(4), 1411–1417 (2008)
2. Davey, J., Wishart, D., Freeman, J., Watson, B.: An application of the driver behaviour questionnaire in an Australian organisational fleet setting. Transp. Res. Part F: Psychol. Behav. **10**(1), 11–21 (2007)
3. de Winter, J.C., Dodou, D.: The driver behaviour questionnaire as a predictor of accidents: a meta-analysis. J. Safety. Res. **41**(6), 463–470 (2010)
4. Dijker, T., Bovy, P.H.L., Vermijs, R.G.M.M.: Car-following under congested conditions: empirical findings. Transp. Res. Rec: J. Transp. Res. Board **1644**, 20–28 (1998)
5. Hatakka, M., Keskinen, E., Katila, Lapotti, S.: Self-reported driving habits are valid predictors of violations and accidents. In: Talib, R., Enrique, C.V. (eds) Traffic and Transport Psychology, Pergamon (1997)
6. Kircher, K., Andersson, J.: Truck drivers' opinion on road safety in tanzania—a questionnaire study. Traffic Inj. Prev. **14**(1), 103–111 (2013)
7. Lajunen, T., Parker, D., Summala, H.: The manchester driver behaviour questionnaire: a cross-cultural study. Accid. Anal. Prev. **36**(2), 231–238 (2004)
8. Lawton, R., Parker, D., Manstead, A.S.R., Stradling, S.: The role of affect in predicting social behaviours: the case of road traffic violations. J. Appl. Soc. Psychol. **27**, 1258–1276 (1997)
9. Jie, Li, van Zuylen, H.J., Chen, Y.S., Lu, R.H.: Comparison of driver behaviour and saturation flow in China and the Netherlands. IET Intel. Transport Syst. **6**(3), 10 (2011)
10. Reason, J.T., Manstead, A.S.R., Stradling, S., Baxter, J., Campbell, K.: Errors and violations on the roads. Ergonomics **33**, 1315–1332 (1990)
11. Sun, J., Elefteriadou, L.: Information categorization based on driver behavior for urban lane-changing maneuvers. In: Transportation Research Record, Transportation Research Board of the National Academies, vol. 2249, pp. 86–94. Washington (2011)

12. Van, Z., Henk J., Chen, Y., Li, M.: Optimizing traffic control for intersections with diminishing saturation flow. In: Sadayuki T., Masayoshi A. (eds) CTS2003, 10th IFAC Symposium on Control in Transportation Systems, Tokyo (2003)
13. Zhang, Y., Mahlawat, M.: Driver Risk-Taking Behavior as a Function of Congestion Level: Analysis Using Adopted Headways in Traffic Stream, Paper presented in Transportation Research Board 87th Annual Meeting, Washington DC (2008)
14. Zhao, N., Mehler, B., Reimer, B., D'Ambrosio, L.A., Mehler, A., Coughlin, J.F.: An investigation of the relationship between the driving behavior questionnaire and objective measures of highway driving behavior. Transp. Res. Part F: Psychol. Behav. **15**(6), 676–685 (2012)

Part VIII
Vehicle Routing

The problem of *vehicle routing* is important in the fields of transportation, distribution and logistics, and many methods have been developed to search for good solutions to it. Several variations and specializations of vehicle routing exist. *Gonzalez-Martin et al.* discuss the arc routing problem using a non-smooth cost function, and propose a randomized algorithm for solving it. *Sicilia et al.* present a hybrid algorithm based on meta-heuristic methods and local improvements to solve the problem of goods distribution in large urban areas. *Yıldırım and Çatay* propose a matheuristic approach based on Ant Colony Optimization to solve the set partitioning formulation for vehicle routing, using parallelization.

Solving Non-smooth Arc Routing Problems Throughout Biased-Randomized Heuristics

Sergio Gonzalez-Martin, Albert Ferrer, Angel A. Juan and Daniel Riera

Abstract In non-smooth optimization problems the objective function to minimize or maximize is non-smooth and usually non-convex either, which is a frequent characteristic of real-life optimization problems. In this chapter we discuss the arc routing problem with a non-smooth cost function, and propose a randomized algorithm for solving it. Our approach employs non-uniform probability distributions to add a biased random behavior to the well-known savings heuristic. By doing so, a large set of alternative good solutions can be quickly obtained in a natural way and without complex configuration processes. Since the solution-generation process is based on the criterion of maximizing the savings, it does not need to assume any particular property of the objective function. Therefore, the procedure can be especially useful in problems where properties such as non-smoothness or non-convexity lead to a highly irregulars solution space, for which the traditional optimization methods -both of exact and approximate nature- may fail to reach their full potential. The results obtained so far suggest that using biased probability distributions to randomize classical heuristics can be successfully applied in non-smooth optimization.

Keywords Randomized algorithms · Combinatorial optimization · Heuristics · Arc routing problem

---

S. Gonzalez-Martin · A. A. Juan (✉) · D. Riera
Open University of Catalonia, Barcelona, Spain
e-mail: ajuanp@uoc.edu

S. Gonzalez-Martin
e-mail: sgonzalezmarti@uoc.edu

D. Riera
e-mail: drierat@uoc.edu

A. Ferrer
Technological University of Catalonia, Barcelona, Spain
e-mail: alberto.ferrer@upc.edu

J. F. de Sousa and R. Rossi (eds.), *Computer-based Modelling and Optimization in Transportation*, Advances in Intelligent Systems and Computing 262,
DOI: 10.1007/978-3-319-04630-3_33,

1 Introduction

Many real life problems can be modeled as combinatorial optimization problems, what has brought into the scene new challenges for the scientific community. Usually they have a well-structured definition consisting of an objective function that needs to be minimized or maximized on a set of constraints. A considerable number of methods and techniques that search the solution space and try to find the optimum have been developed. In few cases, the solution space can easily be explored due to certain properties of the functions involved, such as convexity. For those instances, the problem can often be solved efficiently and exactly. However, in other circumstances, the solution space is highly irregular and finding the optimum in a reasonable amount of time is generally impossible. An exhaustive method that checks every single point in the solution space would be of very little help in these scenarios since it would take exponential time. Also, some approaches are fairly complex while others need to take into account the particular features of the problem. Therefore, designing such approaches usually takes a substantial amount of time and the methodology has a limited application time. This chapter is structured as follows: Sect. 2 presents an introduction to non-convex and non-smooth problems with a short literature review on Sect. 3. Section 4 states the Capacitated Arc Routing Problem and its non-smooth variant. Section 5 defines the proposed methodology for solving the problem. Section 6 presents some results obtained in a well-known problem dataset adapted to the non-smooth ARP. Finally, Sect. 7 extracts some conclusions from the current work.

2 Non-convex and Non-smooth Problems

Optimization problems can be classified, from a high-level perspective, as either convex or non-convex. In general, convex optimization problems (COPs) have two parts: a series of constraints that represent convex regions and an objective function to be minimized that is also convex. The dual problem, in which the objective the goal is to maximize the objective function, for the purpose of this chapter will be considered a member of the convex-like class of problems. COPs are worth studying because they have a wide variety of applications and many problems can be reduced to them via change of variables. Linear Programming is one well-known example, since linear functions are trivially convex [5]. The main idea in convex optimization problems is that every constraint restricts the space of solutions to a certain convex region. By taking the intersection of all these regions we obtain the set of feasible solutions, which is also convex. Due to the structure of the solution space, every single local optimum is a global optimum too. This is the key property that permits us to solve COPs exactly and efficiently up to very large instances. However, almost none of the algorithms applied for COPs can be extended to non-convex case. In non-convex optimization (NCOPs) the objective function, or even the feasible region, are not convex, which results in a far more complex solution space than the case of the COPs. In NCOPs we have many disjoint regions, and multiple locally optimal points within each of them. As a result, if a traditional local search is applied, there is a high risk of ending in the vicinity of a local optimum that may still be far from the global optimum. Another drawback is that it can take exponential time in the size of the input to determine that the NCOP is infeasible, that the objective function is unbounded, or that one of the solutions found so far is the actual global optimum. A function is smooth if it is differentiable and it has continuous derivatives of all orders. Therefore, a non-smooth function is one that is missing some of these properties. Non-smooth optimization problems (NSPs) are similar to NCOPs in the sense that they are much more difficult to solve than traditional smooth and convex problems. The function for which a global optimum needs to be computed is now non-smooth and the solution space might contain again multiple disjoint regions and many locally optimal points within each of them. The computational techniques that can be used to solve these types of problem are often fairly complex and depend on the particular structure of the problem. While in convex optimization it is possible, sometimes, to explore the problem structure, and build solution methods that provide the global optimum, non-convex optimization problems are often intractable and have to rely on heuristic algorithms that produce only local optima. As a result, developing such techniques is in general time consuming, and the resulting application range is very limited. However, most real-life objective functions are either non-convex, non-smooth or both. Therefore, combinatorial optimization under these complex but common circumstances is an important field to explore.

3 Literature Review on Non-smooth and Non-convex Problems

In the context of combinatorial optimization, probabilistic or randomized algorithms make use of pseudo-random numbers or variants during the construction or local search phases. In addition to the problem's input data, a probabilistic algorithm use random bits to do random choices during its execution. An important property is that for the same input the algorithm can produce different outputs in different runs. Probabilistic algorithms have been widely used to solve many combinatorial optimization problems. Examples are Vehicle Routing Problems [16], Location and Layout Problems [7], or Covering, Clustering, Packing and Partitions Problems [6]. Despite the great success of application of these methods to the aforementioned combinatorial problems, there exist only a few documented applications of these algorithms to the NCOPs or NSPs. Some of the existing references are reviewed next. Bagirov and Yearwood [2] present a formulation of the Minimum Sum-of-Squares clustering problem, which is a non-smooth, non-convex optimization problem. The goal of clustering problems is to separate a large set of objects into groups or clusters based on certain criteria. The authors point out that a large number of approaches, like branch and bound or K-means algorithms, have been used for the clustering problem, but they are efficient only in certain special settings. The author remarks that, in general, better results are obtained when metaheuristics are used for the clustering problem. Al-Sultan [1] proposed a Tabu Search approach that outperforms the K-means. However, this algorithm requires of three parameters, so an extensive study was necessary to find the best settings. The issue of Optimal Routing in Communication Networks has also received a lot of attention from researchers. The objective is to find the best path for data transmission in short amount of time. The routing strategy can greatly affect the system performance, so there is a high demand for efficient algorithms. Numerous methods that deal with this challenge have been designed. Hamdan and El-Hawary [12] proposed a method which combined Genetic Algorithms with Hopfield networks. Oonsivialai et al. [17] proposed an approach based on Tabu Search. The main drawbacks of most of these methods are either their inability to efficiently explore the solution spacer or very long computational times. Bagirov et al. [3] present a non-smooth formulation for the Location Problem in Wireless Sensor Networks. In general, a wireless sensor network can be defined as a distributed collection of nodes that have limited resources and operate autonomously. The goal is to find o accurately estimate the position of the nodes. Most proposed approaches have assumed accurate range measurements, which is unrealistic for Radio Frequency signal strength measurements. Ramadurai and Sichitiu [18] show that a probabilistic approach can be adopted to deal with range measurements inaccuracy. Finally, in the transportation and logistics arena, Juan et al. [14] presented a non-smooth formulation for the Vehicle Routing Problem. To solve this problem they proposed a hybrid algorithm for solving the problem.

4 The Capacitated Arc Routing Problem

The Capacitated Arc Routing Problem (CARP) is a combinatorial optimization problem originally introduced by Golden and Wong [9]. It is defined over a undirected incomplete graph $G = (V, E, C, Q)$, where:

- V is a set of nodes including the one representing the depot o distribution center.
- E is a set of edges or arcs connecting some of nodes from V.
- C is a costs matrix representing the positive costs of moving from one node to another; these costs arc usually based on distances between pairs of nodes.
- Q is a demands vector representing the non-negative demands associated with each arc.

Consider also a set of K identical vehicles (homogeneous fleet), each of them with a maximum loading capacity $W \gg \max\{q_i \in Q\}$. Under these circumstances, the usual goal is to find a set of routes which minimizes the total delivering costs,

$$\min \sum_{k=1}^{K} \gamma_k, \tag{1}$$

computed as the sum of the costs of all the K routes, which are equals to the sum of the costs c_{ij} associated to each traversed arc in which x_{ij}^k is a binary variable which denotes whether the arc is traversed by the k-route,

$$\gamma_k := \sum_{e_{ij} \in E} e_{ij} x_{ij}^k. \tag{2}$$

This minimization is subjected to the following constraints:

1. Every route starts and ends at the depot node so every route is a round-trip.
2. All the demands are satisfied.
3. Each arc with positive demand is served exactly by one vehicle. However, an arc can be traversed as many times as required by any vehicle.
4. The total demand to be served in any route does not exceed the vehicle loading capacity W.

As mentioned before, one of the main goals of this chapter is to fill the gap in the CARP literature regarding the discussion and solving of non-smooth objective functions, and to show the efficiency of our approach to deal with this kind of functions in the CARP context. In order to test the effectiveness of our procedure and its efficiency in relation to other existing approaches, we relaxed the constraints by violating some conditions, if necessary. We considered soft constraints, which allow conditions to be violated, by incurring in some penalty costs that must be added to the objective function rather than considering hard constraints, which constraint the problem to never exceed the maximum route costs. According to Hashimoto et al. [13], *"in real-world simulations, time windows and capacity constraints can be often violated to some extent"*. Of course, the same analysis can be applied to constraints associated with maximum route costs. In practice, if a given route exceeds a threshold cost or length, then some penalty cost must be added to the total route costs, and these penalty costs are likely to be defined by a piecewise non-smooth function. These costs will depend on the size of the gap between the actual route costs and the threshold. In this chapter we will define the non-smooth arc routing problem by assuming that the cost of a route is given by:

$$\begin{cases} \gamma_k & : \quad \gamma_k \leq C_{max} \\ \lambda(\gamma_k, C_{max}) & : \quad \gamma_k > C_{max} \end{cases} \tag{3}$$

where λ represents a non-smooth function, for example, a piece-wise function representing a variety of penalties.

5 Multi-start Biased Randomization of a Classical Heuristic

The proposed methodology is a probabilistic algorithm which is constructed of a Multi-start procedure with biased Randomization, making use of a classical Heuristic (MIRH). The algorithm starts with the solution generated by a classical heuristic and slightly perturbs it by means of a random biased behavior in order to obtain alternative results. It uses random bits to do random choices during the execution of the algorithm. But, instead of using the uniform distribution as most metaheuristics and probabilistic algorithms do, we consider non-uniform and non-symmetric (biased) distributions (Fig. 1). Examples of these distributions are the geometric distribution, or the decreasing triangular distribution. The use of this biased randomization guides the search process, since the most promising movements are the ones with higher priority to be selected at each step of the MIRH. The algorithm is an improvement and adaptation to the non-smooth case of the algorithm proposed in Gonzalez-Martin et al. [11], which was able to provide good and robust solutions for the classical CARP. MIRH is related to other metaheuristics proposed in the literature. The closest ones are the Hybrid GRASPs or reactive GRASP [19], the Heuristic Biased Stochastic Sampling (HBSS) of Bresina [4] or the Probabilistic Tabu Search by Fleurent and Glover [8]. The common aspects of our MIRH algorithm with GRASP are the construction of an initial solution using randomization and afterwards the application of a local search. But there are relevant differences, as the MIRH does not use a Restrictive Candidate List (RLC), one main characteristic of the GRASP algorithm, and it uses a biased and adaptive non-uniform distribution to select the next element to be included in the solution, while most GRASP implementations only consider uniform distributions (Fig. 1). On the other hand, the HBSS algorithm proposed by Bresina is similar to the MIRH since it uses a biased distribution function combined with a sampling methodology. In fact, the MIRH methodology can be seen as a natural extension of the HBSS. Our approach is similar to the HBSS in the use of non-uniform distribution; however we incorporate a local search step after each solution obtained by the biased sampling. Also, the HBSS was only applied to scheduling problem, but never to Arc Routing Problems. The probabilistic Tabu Search applied to the Quadratic Assignment Problem described by Fleurent and Glover [8], uses a non-uniform biased distribution, but the structure of the algorithm is quite different and there is the need to set various parameters, meanwhile MIRH contains very few parameters to set. The MIRH algorithm is composed mainly of two parts (Fig. 2): (a) the construction of an initial solution using a classical heuristic; and (b) a biased randomization process applied to the construction of a random solution. To construct the random solution we apply a classical greedy heuristic. We have chosen classical heuristics as starting point because of several reasons. First of all, there are efficient heuristics for almost every combinatorial optimization problem. They usually are able to compute competitive solutions in reasonably short amount of time. In addition, classical

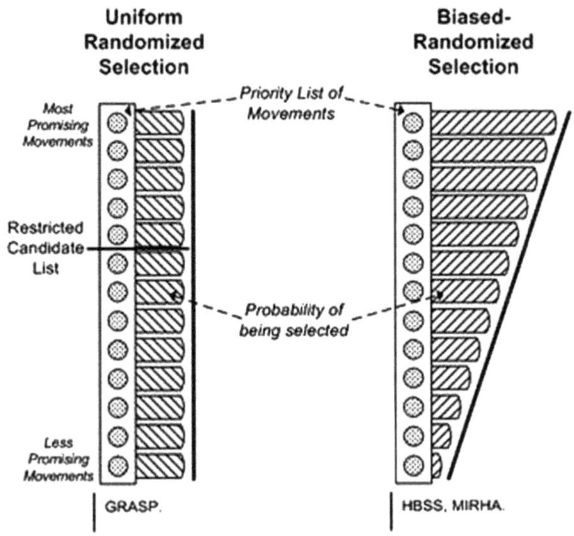

Fig. 1 Uniform randomization versus biased randomization

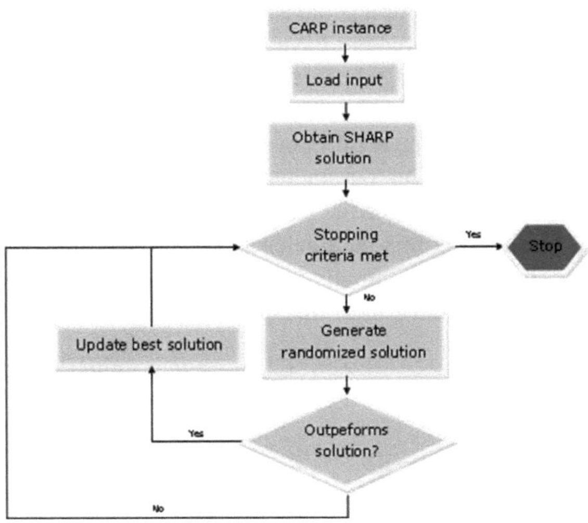

Fig. 2 Flow diagram of the MIRH algorithm for the CARP

heuristics build solutions incrementally using well-tested strategies instead of directly using the objective function itself. With that, issues like non-convexity or non-smoothness of the objective function are not likely to have a significant impact on their efficiency. The main idea of these heuristics is to select the next step from a list of possible options of movements, usually following a selection criteria. For the case of the current chapter, we have chosen the SHARP (Gonzalez-Martin et al. [11]) heuristic for the CARP as a base heuristic. This heuristic mainly is an adaption of the classical Clarke and Wright heuristic for the Vehicle Routing Problem (VRP), to the CARP.

6 Computational Experiments

To evaluate the performance of the proposed algorithm, we have implemented it as a computer program. Java SE6 over Netbeans IDE was used to develop it for several reasons: (a) being an object-oriented programming language with advanced memory management features such as the garbage collection, and with readily-available data structure, it allows a somewhat faster development of algorithmic software; (b) it offers immediate portability to different platforms; and (c) it offers better replicability and duplicability than other languages. However, a downside of using Java instead other languages such as C or C++ is the reduction on code execution performance, mainly due to the fact that Java is executed over a virtual machine and it is not a complied language and to the lack of pointer-based optimization. A standard personal computer was used to perform all tests, an Intel ® Core2® Quad CPU Q9300 @ 2.50 GHz and 8 GB of RAM running the Windows ® 7 Pro operating system. For the generation of random number we have employed the L'Ecuyer [15] SSJ library for Java, concretely the LFSR113 random number generation, which offers a period value approximately equal to 2113. To assess the performance and the quality of the solutions obtained with the proposed algorithms, a complete dataset originally proposed for the standard CARP problem was adapted to make it have a non-smooth objective function. In concrete, the *gdb* (Golden et al. [10]) dataset was used. This dataset consists of 23 instances of small to medium size with a mixture of dense and sparse graph networks. For these instances, we have introduced parameter which determines the maximum route cost allowed. This parameter was defined considering the results obtained by the MIRH algorithm for the CARP instance, rounded to a multiple of 10. But, instead of considering this parameter as a hard constraint, we have considered it as a soft one that could eventually be violated. Following the penalty costs function (3), for our tests we have used the specific non-linear and non-smooth function:

$$\lambda(\gamma_k, C_{max}) := \gamma_k + \min\{\theta(\gamma_k, C_{max}), 8\} \tag{4}$$

with

$$\theta(\gamma_k, C_{max}) := 0.5 + 100\left(\frac{\gamma_k - C_{max}}{\gamma_k}\right)^4. \tag{5}$$

It is worth to mention that these non-smooth functions have been selected for the problem instances which are being tested in our experiments. These values depend on the magnitude of the costs of the instance. In this case the maximum route penalty due to exceeding C_{max} is equal to 8, which approximately is the 10 % of the average cost of a route, when considering the MIRH solution for the *gdb* CARP instances. The results obtained are displayed in Table 1. The table is structured in two halves: the first one showing the characteristics of every problem dataset, and the second one which contain the results. On the first half we display the instance *gdb*, the number of arcs, |*E*|, and nodes, |*V*|, in the problem instance,

Table 1 Evaluated problem instances and obtained results

gdb	|*E*|	|*V*|	*W*	*MRC*	BKS	MIRH-*C*	Gap	MIRH-*S*	Gap	MIRH-*H*	Gap
					(1)	(2)	(1)–(2) (%)	(3)	(1)–(3) (%)	(4)	(1)–(4) (%)
1	12	22	5	60	316	316	0.00	317.87	0.59	343.02	8.55
2	12	26	5	50	339	339	0.00	341.76	0.81	422.56	24.65
3	12	22	5	50	275	275	0.00	276.60	0.58	340.09	23.67
4	11	19	5	50	287	287	0.00	289.08	0.72	483.71	68.54
5	13	26	5	60	377	377	0.00	378.85	0.49	467.00	23.87
6	12	22	5	60	298	298	0.00	299.08	0.36	351.51	17.96
7	12	22	5	60	325	325	0.00	325.74	0.23	356.50	9.69
8	27	46	27	30	348	350	0.57	359.43	3.28	594.91	70.95
9	27	51	27	30	303	313	3.30	318.75	5.20	433.71	43.14
10	12	25	10	60	275	275	0.00	276.53	0.56	283.50	3.09
11	22	45	50	80	395	395	0.00	400.01	1.27	409.00	3.54
12	13	23	35	60	458	468	2.18	464.89	1.50	739.19	61.40
13	10	28	41	80	536	536	0.00	545.00	1.68	580.70	8.34
14	7	21	21	60	100	100	0.00	104.00	4.00	104.00	4.00
15	7	21	37	50	58	58	0.00	58.00	0.00	58.00	0.00
16	8	28	24	30	127	127	0.00	127.76	0.60	129.00	1.57
17	8	28	41	20	91	91	0.00	91.00	0.00	91.00	0.00
18	9	36	37	30	164	164	0.00	167.24	1.98	182.00	10.98
19	11	11	27	20	55	55	0.00	55.50	0.91	63.00	14.55
20	11	22	27	30	121	121	0.00	121.53	0.44	123.00	1.65
21	11	33	27	30	156	156	0.00	158.00	1.28	158.00	1.28
22	11	44	27	30	200	200	0.00	201.00	0.50	202.00	1.00
23	11	55	27	30	233	233	0.00	235.00	0.86	235.00	0.86
Avg.	13	29	23	77			0.26		1.21		17.23

the vehicle capacity W, and the maximum route costs, MRC, parameter which we have defined. On the second half, the columns contain the following information: best-known solution for the original CARP problem instance. The solution obtained with MIRH when applied to the normal CARP problem instance, MIRH-C, and the gap of this result with respect to the best-know solution. The solution obtained by MIRH in the non-smooth ARP when considering soft-constraint during the design phase of the algorithm, MIRH-S, and its gap with respect the best-known solution of the CARP. Finally, the solution of the MIRH algorithm when considering hard-constraints during the design phase, MIRH-H, and its gap with respect the best-known solution of the CARP.

From the results we can notice first of all that MIRH has a good performance with the original CARP problem instance CARP problem (without maximum route costs constraint and with a smooth objective function). In addition, as it considers soft-constraints during the design phase of the routes, it is able to minimize the effect of having a non-smooth objective function, showing a result closest to the BKS and to the solution obtained by the same algorithm when considering only the CARP. Additionally, we can also notice that when considering hard-constraints in the design of the MIRH algorithm, the performance falls down dramatically. This is due to the fact that considering hard-constraints makes the solution to have more routes required, which means that more overload is obtained in the solution due to the round trips to the depot for refilling. Notice that the gap with respect to the best known solution showed in the table is computed as follows:

$$\mathrm{Gap}\left(C_{MIRH}, C_{BKS}\right) = 100\left(\frac{C_{MIRH} - C_{BKS}}{C_{BKS}}\right). \qquad (6)$$

7 Conclusions

In this chapter an overview of non-convex and non-smooth optimization problems has been presented. We have also discussed how different approaches have been used different non-smooth and non-convex problems in the existing literature. Among others we can find the GRASP, HBSS or Tabu Search. As has been pointed out, our methodology has similarities with some methods already reported in the literature but, at the same time, maintains significant different as previously discussed. In addition, we have defined the non-smooth Arc Routing Problem and described the objective function characteristics which make our approach a good candidate for solving it. We have also presented the multi-start biased randomization of classical heuristics (MIRH) for solving the problem. The key idea of our approach is to employ probability distributions such as the geometric one to add a random biased behavior to a classical heuristics like our SHARP heuristic for the Arc Routing Problem. In this way we obtain a large set of alternative good solutions that outperform the initial solution produced by the heuristic.

Acknowledgments The research of the first, third and fourth authors has been partially supported by the Spanish Ministry of Science and Innovation (TRA2010-21644-C03) and by the Ibero-American Programme for Science, Technology and Development (CYTED2010-511RT0419) in the context of the ICSO-HAROSA Programme of the IN3 (http://dpcs.uoc.edu).

The research of the second author has been partially supported by the Spanish Ministry of Science and Technology (MTM2011-29064-C03-01).

References

1. Al-Sultan, K.S.: A tabu search approach to the clustering problem. Pattern Recogn. **28**(9), 1443–1451 (1995)
2. Bagirov, A.M., Yearwood, J.: A new nonsmooh optimization algorithm for minimum sum-of-squares clustering problem. Eur. J. Oper. Res. **170**, 578–596 (2006)
3. Bagirov, A.M, Lai, D.T.H., Palaniswami, M.: A nonsmooth optimization approach to sensor network location. In: Palaniswami, M., Marusic, M., Law, Y.W. (eds) Proceedings of the 2007 International Conference on Intelligent Sensors, Sensor Networks and Information Processing, pp. 727–732 (2007)
4. Bresina, J.L.: Heuristic-biased srochastic sampling. In: Proceeding of the Thirteenth National Conference on Artificial Intelligence and the Eighth Innovative Applications of Artificial Intelligence Conference, pp. 271–278 (1996)
5. Boyd, S., Vandenberghe, L.: Convex Optimization. Cambridge University Press, Cambridge (2004)
6. Chaves, A.A., Lorena, L.A.N.: Clustering search algorithm for the capacitated centered clustering problema. Comput. Oper. Res. **37**(3), 552–558 (2010)
7. Drezner, Z., Hamacher, H. (eds.): Facility Location: Applications and Theory. Springer, New York (2010)
8. Fleurent, C., Glover, F.: Improved constructive multistart strategies for the quadratic assignment problem using adaptive memory. INFORMS J. Comput. **11**(2), 198–204 (1999)
9. Golden, B.T., Wong, R.T.: Capacitated arc routing problems. Networks **11**(3), 215–305 (1981)
10. Golden, B.T., Dearmon, J.S., Baker, E.K.: Computational experiments with algorithms for a class of routing problems. Comput. Oper. Res. **10**(1), 47–59 (1983)
11. Gonzalez-Martin, S., Juan, A.A., Riera, D., Castella, Q., Perez-Bonilla, A, Muñoz, R.: Development and assessment of the SHARP and RandSHARP algorithms for the arc routing problema. AI Commun. **25**(2), 173–189 (2012)
12. Hamdan, M., El-Hawary, M.E.: Hopfield-genetic approach for solving the routing problem. In: Computer Networks. Proceedings of the 2002 IEEE Canadian Conference on Electrical and Computer Engineering, pp. 823–827 (2002)
13. Hashimoto, H., Ibaraki, T., Imahori, S., Yagiura, M.: The vehicle routing problema with flexible time Windows and traveling times. Discrete Appl. Math. **154**(16), 2271–2290 (2006)
14. Juan, A.A., Faulin, J., Ferrer, A., Lourenço, H.R., Barrios, B.: MIRHA: multi-start biased randomization of heuristics with adaptive local search for solving non-smooth routing problems. TOP **21**, 109–132 (2013)
15. L'Ecuyer, P.: Random Number Generation in Simulation. Elsevier, Amsterdam (2006)
16. Laporte, G.: Fifty years of vehicle routing. Transp. Sci. **43**(4), 408–413 (2009)
17. Oonsivilai, A., Srisuruk, W., Marungsri, B., Kulworawanichpong, T.: Tabu search approach to solve routing issues in communication networks. In: World Academy of Science, Engineering and Technology, pp. 1174–1177 (2009)

18. Ramadurai, V., Sichitiu, M.L.: Localization in wireless sensor networks: a probabilistic approach. In: International Conference on Wireless Networks (ICWN03), pp. 275–281 (2003)
19. Resende, M.G.C.: Metaheuristic hybridization with greedy randomized adaptive search procedure. In: Chen, S.L., Raghavan, S. (eds.) Handbook of Metaherustics. International Series in Operations Research Management Science, 146(2), pp. 227–264. Kluwer Academic, Dordrecht (2008)

A Hybrid Algorithm for Solving the General Vehicle Routing Problem in the Case of the Urban Freight Distribution

Juan-Antonio Sicilia, David Escuín, Beatriz Royo, Emilio Larrodé and Jesús Medrano

Abstract This chapter presents a hybrid algorithm based on metaheuristic methods (Variable Neighbourhood Search and Tabu Search) and local improvements to solve the problem of the distribution of goods in large urban areas taking into account the characteristics encountered in real life. The logistics of the short distance transport of goods has an essentially urban dimension so that distribution requires efficient algorithms and the process between warehouses and customers must be effective and clean. Therefore, it is necessary to efficiently optimize urban logistics and improve connections between urban and interurban freight transport in order to ensure effective distribution. Due to the great variety of constraints and complexities of the problem, known as the General Vehicle Routing Problem, the algorithm proposes feasible solutions in order to achieve the main objective of reducing costs based on minimizing distances and reducing the number of vehicles used as long as the service quality to customers is optimum and a load balance between vehicles is maintained. This article arises from a research project carried out for a large Spanish distribution company aiming to optimally manage its resources in urban areas by reducing as much as possible costs caused by inefficiency and ineffectiveness.

J.-A. Sicilia (✉) · B. Royo · E. Larrodé · J. Medrano
Department of Mechanical Engineering, University of Zaragoza, María de Luna 3, 50018 Zaragoza, Spain
e-mail: jsicilia@unizar.es

B. Royo
e-mail: broyoa@unizar.es

E. Larrodé
e-mail: elarrode@unizar.es

J. Medrano
e-mail: jmedrano@unizar.es

D. Escuín
Instituto Tecnológico de Aragón, María de Luna 1, 50018 Zaragoza, Spain
e-mail: descuin@ita.es

J. F. de Sousa and R. Rossi (eds.), *Computer-based Modelling and Optimization in Transportation*, Advances in Intelligent Systems and Computing 262,
DOI: 10.1007/978-3-319-04630-3_34,

Keywords General vehicle routing problem · Hybrid algorithm · Metaheuristic methods · Local improvements · Load balance

1 Introduction

In the European Union, over 60 % of the population resides in urban areas, generating just under 85 % of the EU's gross domestic product [1]. For this reason, there is great potential for freight transport in these areas. Freight transport by means of vehicles over 3.5 tonnes represents about 10 % of the total traffic within urban areas. If lighter vehicles such as vans are included, the percentage increases considerably. This is a significant volume of daily operations (in Barcelona, according to the city council, the transport of goods accounts for 16 % of daily trips). It is therefore necessary to find logistics solutions for urban freight distribution consistent with the restrictions imposed by the authorities to protect the interests of the public [2]. Furthermore, pickup and delivery operations in urban areas represent about 40 % of the total cost of transport activities carried out at home. These costs are further increased by the reduction of stocks, the smaller size of goods and the increase in the number of requests.

Most problems encountered in industry, particularly in city logistics, are multi-objective in nature. This makes it difficult to apply classical routing models to real-life problems because these models are implemented with the single objective of minimizing the cost of the solution. However, in real life there may be several requirements associated with a single tour such as balancing workloads, time, distance, cost, etc. that need to be taken into account simply by adding new constraints.

This chapter arises from a research project carried out for a large Spanish distribution company aiming to optimally manage its resources in urban areas by reducing as much as possible costs caused by inefficiency and ineffectiveness. A hybrid algorithm that solves a General Vehicle Routing Problem (GVRP) is proposed. It is based on the constraints and complexities currently encountered in the urban distribution of goods, using the Variable Neighbourhood Descent (VND), General Variable Neighbourhood Search (GVNS) and Tabu Search (TS) metaheuristic methods, and local improvements to improve the solution.

The structure of this chapter is as follows. Section 2 presents the state of the art, Sect. 3 describes the problem with its characteristics, Sect. 4 provides a formal mathematical definition of the problem, Sect. 5 proposes a hybrid algorithm to resolve the problem using metaheuristics, Sect. 6 presents the results of the computational experiments and the conclusions are set out in Sect. 7.

2 Literature Review

Vehicle Routing Problem (VRP) is one of the most extensively studied problems in operations research and is one of the most significant in the field of distribution, transport and logistics. It was first introduced by Dantzig and Ramser [3]. The problem is based on the distribution of goods to a number of customers by means of a fleet of vehicles finding the optimal routes, with the objective of minimizing the total cost. As the VRP is a very complex NP-hard problem, solving and optimizing real-life VRPs is often not possible within the limited computing time available in practical situations. Therefore, most research has focused on optimization algorithms, heuristic and metaheuristic methods designed to produce high quality solutions in a limited time.

The problem described is the GVRP with the multiple requirements and complexities encountered in urban freight distribution in real life. Although the literature related to the VRP is extensive and varied, there is little that includes all real-life variations. Most studies focus on only one or two variants such as the time window or capacity. Goel and Gruhn [4] describes a generalization of the classical problem which introduces several real-life complexities (time windows, heterogeneous fleet, order/vehicle compatibility and different locations for vehicles). It is solved by iterative improvement approaches based on the idea of changing the neighbourhood structure during the search. Pisinger and Ropke [5] presents an adaptive large neighbourhood search heuristic to solve up to five variants of the VRP. Jozefowiez et al. [6] provides an overview of the research into routing problems with several objectives.

The most important variants of the VRP can be found in [7]. In the Capacitated VRP (CVRP), the routes, which start and terminate at the depot, have to deliver goods, with minimum cost, to a set of customers with known demands. The vehicles are assumed to be homogeneous and have a certain capacity [8]. On the other hand, in VRP with Time Windows (VRPTW) each customer must be serviced in a specified time interval [9].

The Site Dependent VRP (SDVRP) is another extension of the VRP. In this case, due to the characteristics of the goods, certain orders can only be served by specific vehicles, so that the fleet must be heterogeneous [10]. Besides, the case where vehicles are not required to return to the warehouse after visiting the last customer of the route, and where the route may finish near the driver's home, is known as the Open VRP (OVRP) [11].

Nowadays, the competitiveness of a transport company depends not only on minimizing costs and distances to obtain higher profit margins, but also depends on its ability to treat employees fairly. Thus, an important objective is to balance the workload among vehicles as much as possible [12].

In the GVRP, customers may simultaneously receive and send goods. The VRP with Simultaneous Pickups and Deliveries (VRPSPD) is another variant in which each load has to be transported by one vehicle from its origin to its destination without any transshipment at other locations [13]. Also, the variant VRP with

Backhauls (VRPB) occurs when the goods that leave the warehouse must be delivered to customers and the goods that are picked up from customers must be transported to the warehouse [14]. Furthermore, in this problem there are dependent orders where vehicles have to pick up goods at pickup nodes and deliver them to delivery nodes without having to return to the warehouse.

The existent literature about the metaheuristic methods used in this chapter is extensive. Variable Neighbourhood Search (VNS) is a metaheuristic based on the idea of systematic change of neighbourhood within a local search [15]. A variant is VND where the best neighbour of the current solution is considered instead of a random one [16]. GVNS is an extended variant which uses more than one neighbourhood in a local search [17]. TS is a metaheuristic that uses the memory of the search process to improve the solutions obtained. TS is a local search strategy where the best solution in the neighbourhood of the current solution is selected as the new current solution, even if it leads to an increase in solution cost [14].

3 Problem Description

The problem is based on the optimization of urban freight transport. There is one depot at the beginning of the routes to which the vehicles do not have to return after serving the last customer. An order is defined as the customer request for the transport of goods. There are three types of orders. An order may consist of only one pickup, or only one delivery or one pair pickup-delivery dependent nodes. The nodes have to be visited in a specific sequence by the same vehicle with the time window imposed by the customer. Each customer can be visited more than once by different vehicles, since a customer can request more than one order. Additionally, there are dependent orders where vehicles have to pickup goods at a pickup node and deliver them to a delivery node without having to return to the warehouse.

In addition, constraints of vehicle capacity and compatibilities between orders and vehicles mean that the possibility of allocating orders to certain vehicles may be excluded or that specific orders can only be transported by specific vehicles. This situation requires a fleet heterogeneous in capacity and in technical properties. The vehicles may have different capacities and belong to one or more load type (normal, refrigerated, isotherm). It is supposed that the available vehicles are enough to cover all demand.

All orders have a service time (load/unload time of goods) and a time window in which the operation must be performed. It is important to note that the time interval does not prevent any vehicle from arriving before the lower time limit, but this causes losses due to the inactivity of the vehicle during a period of time. The intention is for the arrival of the vehicle to be within the time interval.

One route is defined as an ordered sequence of orders that the same vehicle has to dispatch. Each vehicle is assigned a point location both at the beginning and at the end of the route. The end node may be the warehouse or the driver's home, while the initial node corresponds to the warehouse regardless of where the vehicle had finished the previous day. In this problem, everyday are different since they are not related with the previous day.

An important premise to be met is the best load balance between drivers. A maximum limit of orders per vehicle is imposed so a route is carried out independently of its costs. A similar balance between the distance and the operations to be performed by each driver must be achieved in all routes.

A Geographic Information System (GIS) is used to compute the routes. It is necessary for calculating the distance from the GPS position of each node to be visited, since the number of orders might vary each day, the number of customers (geographic nodes) is not fixed and the distances have to be exact in urban areas as the routes are short. The distance is only calculated between orders that may go on the same route. Previously, a check of constraints (capacity, compatibility and time) is carried out to obtain compatible sets of orders.

The objective is to obtain the set of optimal routes that minimizes the total cost of the transport operations by means of minimizing the total distance traveled and reducing the number of vehicles. At the same time, a load balance between vehicles must be maintained, the available resources must be maximized and there must be a commitment to quality customer service.

4 Mathematical Formulation

The problem can be described as follows. Let G = (N, A) be a complete graph with N $= \{1\} \cup O \cup \{o+1\}$ the set of nodes and $O = \{2, \ldots, o\}$ the set of orders to be served. The nodes $\{1\}$ and $\{o + 1\}$ represent the depot and are added to the network to allow empty tours. The aim is to design a set of routes for a heterogeneous fleet of V vehicles (each has a capacity, a load type and a limited number of orders to be served), servicing a set of O orders. Each arc $(i,j) \in A$ has a cost c_{ij}. The interval $[a_i, b_i]$ denotes the time windows for order $i \in O$ and t_i represents the time service. A vehicle can not start servicing node i before a_i and after b_i. A vehicle can arrive before a_i and wait for service. Some nodes represent drivers' homes and vehicles have to serve them just before returning to the depot. There are three kinds of nodes:

- Deliveries: vehicles depart from depot with goods.
- Pickups: vehicles pick up goods at these nodes and transport them to depot.
- Pickup-delivery dependents: vehicles have to pick up goods at pickup node and deliver them at delivery node.

Routes must satisfy that the total quantity of goods picked up and delivered can not exceed vehicle capacity. Each node i represents both a delivery node, d_i, and a pickup node, p_i. r_i denotes the pickup node dependent of a delivery node i. Besides, a node may have a demand to be picked up and to be delivered. Different nodes $i \in N$ may correspond to the same geographical location.

V denotes the set of available vehicles, being Q_k the capacity of vehicle $k \in V$ and T_k the maximum number of nodes that vehicle k can serve. h_k represents the last node to be served (driver's home) before returning to depot. In addition, m_{ik} represents the compatibility between node i and vehicle $k, \forall i \in N, \forall k \in V$. Node i may be served by vehicle k if and only if $m_{ik} = 1, m_{ik} = 0$ otherwise. The GVRP can be formulated as follows:

Objective function:

$$min \sum_{k \in V} \sum_{i \in N} \sum_{j \in N} c_{ij} x_{ij}^k - \sum_{k \in V} x_{1h_k}^k M \tag{1}$$

Subject to:

$$\sum_{j \in N} \sum_{k \in V} x_{ij}^k = 1 \quad \forall i \in O \tag{2}$$

$$\sum_{i \in N} x_{ih}^k - \sum_{j \in N} x_{hj}^k = 0 \quad \forall h \in O, \forall k \in V \tag{3}$$

$$x_{hh}^k = 0 \quad \forall k \in V, \quad \forall h \in N \tag{4}$$

$$x_{o+1i}^k = 0 \quad \forall k \in V, \quad \forall i \in N \tag{5}$$

$$\sum_{i \in N} x_{io+1}^k = 1 \quad \forall k \in V \tag{6}$$

$$\sum_{j \in N} x_{1j}^k = 1 \quad \forall k \in V \tag{7}$$

$$s_i^k + t_i - s_j^k + c_{ij} \leq \left(1 - x_{ij}^k\right) M \quad \forall k \in V, \forall i \in N, \forall j \in N \tag{8}$$

$$s_1^k = 0 \quad \forall k \in V \tag{9}$$

$$a_i \leq s_i^k = 0 \quad \forall k \in V, \forall i \in N \tag{10}$$

$$b_i \geq s_i^k = 0 \quad \forall k \in V, \forall i \in N \tag{11}$$

$$m_{ik} + m_{jk} \geq 2x_{ij}^k \quad \forall k \in V, \quad \forall i \in O, \forall j \in O \tag{12}$$

$$l_i^k \leq Q_k \quad \forall k \in V, \forall i \in N \tag{13}$$

$$\sum_{i \in N} \sum_{j \in N} d_i x_{ij}^k = l_1^k \quad \forall k \in V \tag{14}$$

$$\sum_{i \in N} \sum_{j \in N} p_i x_{ij}^k = l_{o+1}^k \quad \forall k \in V \tag{15}$$

$$l_i^k - d_j + p_j - l_j^k \leq \left(1 - x_{ij}^k\right)M \quad \forall k \in V, \forall i \in N, \forall j \in N \tag{16}$$

$$T_k + 1 \geq \sum_{i \in N} \sum_{j \in N} x_{ij}^k \quad \forall k \in V \tag{17}$$

$$1 - \sum_{h \in N} x_{r_j h}^k \leq \left(1 - x_{ij}^k\right)M \quad \forall k \in V, \forall i \in N, \forall j \in O \tag{18}$$

$$s_{r_i}^k \leq s_i^k \quad \forall k \in V, \forall i \in O \tag{19}$$

$$x_{h_k n+1}^k = 1 \quad \forall k \in V \tag{20}$$

Decision variable:

$$x_{ij}^k = \begin{cases} 1 & if\, arc(i,j) \in Aistraveledbyvehiclek \in V \\ 0 & otherwise \end{cases} \tag{21}$$

The objective function (1) states that the distance traveled and the number of vehicles should be minimized being M a sufficiently large constant. Flow conservation is ensured by constraints (2, 3 and 4). Equation (7) states that all vehicles leave the depot while (6) represents all vehicles returning to depot. Equation (5) ensures no vehicles leave the virtual depot and vehicles end their routes at node $\{o + 1\}$. Demand constraint is ensured by (13, 14, 15 and 16) with l_i^k to denote the load of the vehicle k at node i after serving node i. Compatibility between vehicles and orders is represented by (12). Time window constraints are ensured by constraints (8, 9, 10 and 11) with s_i^k to denote when the vehicle k starts the service at node i. Equations (10 and 11) ensure that node i cannot be served before a_i and after b_i. Equation (17) ensures that the number of orders served by vehicle k is not greater than the maximum permitted. Equations (18 and 19) represent precedence constraints for pickup-and-delivery dependents. Equation (20) ensures vehicles serve their driver's home prior to return to depot. The domain of decision variables is set by (21).

5 Hybrid Algorithm

The difficulty of the problem consists in the fact that if the number of elements to be combined is relatively large, the resulting combinatorial grows exponentially turning it into an NP-hard problem. This requires the application of optimization techniques to obtain a high-quality solution in reasonable computational times. This section presents an efficient algorithm to solve the problem as formulated so that the objective is met and the constraints are satisfied. The problem is solved by means of a powerful route construction method that starts with the determination of a feasible initial solution, the VND metaheuristic and the GVNS + TS hybrid method to obtain an optimum solution.

5.1 Routes Construction Method

The route construction method consists of allocating orders to vehicles based on certain parameters, taking into account the limits of load balance between routes. This procedure has been implemented by a factor of randomness to avoid always obtaining the same solutions and to diversify the total space of the same. Initially there are a heterogeneous fleet of vehicles and a set of orders to be distributed. The algorithm will generate as many routes as vehicles available, taking into account the restrictions and compatibilities.

At the beginning there is no route, and the system proceeds to create routes choosing the first order according to several criteria. It is important to mention that this does not mean that this order is the first order of the final route. It has been chosen only to start the insertion process. In this process, orders are inserted until the route is not feasible or it has reached the maximum number of orders allowed per route. The process finishes when there are no orders without a route, at which point the initial solution has been created. The insertion criteria of the first order are:

- Latest hour of arrival of the vehicle to the depot.
- Close earliest of the start time to serve a customer.
- Smallest width of time window after the arrival to the customer.
- Effective distance from the customer to depot.
- Best average of the above criteria.

The insertion process of the rest of orders is based on the following criteria:

- Greatest time margin: It is the extra time available to distribute other orders taking into account the time taken to distribute orders that have already inserted.
- Shortest distance of the route: Choice of a nearby customer. The order will be inserted in the position of the route with minimum distance.

5.2 Local Improvements and Metaheuristics

Local improvement operators are methods applied to a set of one or more routes in order to improve the evaluation function. The basic process consists of movements and exchanges of segments of orders. It is necessary to apply the operators to improve certain aspects of the problem in the majority of the calculations. These operators are the well known intra-route improvements applied on a single route and that are computationally very fast, OR-OPT and IOPT, and the inter-route improvements that exchange segments of orders between two routes, ICROSS and 2-OPT*.

Initialization: Select the set of neighbourhood structures N_k that will be used in the descent Find an initial solution s Repeat: The following until no improvement is obtained: (1) Choose $k \in \{1, \dots, k_{max}\}$ (2) Exploration of neighbourhood: Find the best solution $s' \in N_k(s)$ (3) Move or not: If the new solution s' is better than s, move there $(s \leftarrow s')$

Fig. 1 VND algorithm

Once an initial solution has been constructed with the smallest number of routes possible, it is attempted to improve the solution based on the application of metaheuristics using VNS. To define a neighbourhood search, a sequence of operators with their respective configurations is specified. The metaheuristic process is based on the use of a neighbourhood that exchanges segments. The final state of each operator of a neighbourhood determines the subsequent execution of the following neighbourhood. Therefore, a good configuration of operators helps to obtain one solution or another.

5.2.1 Variable Neighbourhood Descent Method

The VND consists of iteratively replacing the current solution with the result from the local solution if there is an improvement. Every time a local minimum is reached, there is a deterministic change in the structure of neighbourhoods defined by different operators. The VND algorithm can be outlined as illustrated in Fig. 1.

The algorithm starts with the selection of the set of neighbourhood structures defined by means of the local operators and the determination of an initial solution s which is obtained by the route construction method. The algorithm repeats the following steps until an improvement is obtained. First, the next neighbourhood $N_k(s)$ to be considered is chosen. Then, the best solution s' is generated. This new solution is accepted as the next current solution if the objective is improved. The aim of using multi-operators is to explore the solution space more extensively.

5.2.2 Hybrid Method: General Variable Neighbourhood Search with Tabu Search

Since the systematic change of neighbourhood structure is a simple and very powerful technique, another way to extend the VNS is incorporating it into other metaheuristics, resulting in hybrid methods. The metaheuristic chosen is the TS that uses a neighbourhood structure to execute up and down movements working with different memory types. The algorithm of the new hybrid method can be outlined as illustrated in Fig. 2.

The algorithm starts with the selection of the set of neighbourhood structures both for the TS phase and for the local search. The algorithm repeats the following

Initialization: Select the set of neighbourhood structures N_k that will be used in phase 2
Select the set of neighbourhood structures N_j that will be used in phase 3
Find an initial solution s
Choose a stopping condition
Repeat: The following until the stopping condition is met:
(1) Choose $k \in \{1, \dots, k_{max}\}$
(2) Generate a new solution $s' \in N_k(s)$ by Tabu Search
(3) Local search: Choose $j \in \{1, \dots, j_{max}\}$
Apply VND with s' as initial solution
Denote with $s'' \in N_j(s')$ the new solution obtained
(4) Move or not: If the new solution s'' is better than s, move there ($s \leftarrow s''$)

Fig. 2 GVNS + TS hybrid algorithm

steps until a stopping condition is met. First, the next neighbourhood $N_k(s)$ to be considered is chosen. Then, TS is used to generate a new solution s' replacing the shaking phase of the basic GVNS algorithm. This has been carried out to eliminate randomness and to diversify the exploration neighbourhood, avoiding the decrease in local minimums and allowing the search for better solutions. Second, apply VND to obtain the new solution s''. This new solution s'' is accepted as the next current solution if the objective is improved. If no stopping condition is met, the algorithm continues with the next iteration.

6 Computational Experiments

In order to evaluate our algorithm, test problems have been generated incorporating most of the complexities found in real-life problems. Our computational experiments have been extracted from real cases experienced by a large transport company in Spain. To create an instance, all the necessary characteristics of the elements constituting the problem were extracted from the historical data. The first elements were the orders including the geographical position of customers, the quantity and the load type, the time required to complete the operation and the time windows in which each customer had to be served. The second element was the heterogeneous fleet with its capacity limit and the load type that it was able transport. General constraints imposed for all test problems are based on satisfying time constraints imposed by each customer and compatibilities between goods and vehicles, and maintaining a load balance between vehicles (for each instance, all vehicles take the same number of maximum orders but this number vary to analyze the performance of the algorithm).

The test problem generation is defined by the number of orders (O), the number of available vehicles (V) and the maximum number of orders per vehicle (L). The experiments have been performed on a PC Pentium Dual at 2 GHz with 2 GB of

Table 1 Algorithm results of the test problems (50–100 orders)

P	O	V	L	Constructor		VND		GVNS + TS	
				Δ(d)	Δ(v)	Δ(d)	Δ(v)	Δ(d)	Δ(v)
P1	50	7	10	700.66	7	698.85	7	700.66	7
P2			12	670.72	7	668.92	7	670.72	7
P3			14	670.72	7	668.92	7	670.72	7
P4		8	10	785.15	8	785.08	8	785.08	8
P5			12	755.22	8	755.14	8	794.24	6
P6			14	755.22	8	755.14	8	756.98	6
P7		9	10	871.32	9	871.25	9	876.47	8
P8			12	835.80	9	835.73	9	847.15	7
P9			14	835.80	9	835.73	9	835.73	9
P10	100	14	10	1497.62	14	1414.18	14	1395.64	12
P11			12	1509.37	14	1430.12	14	1452.43	12
P12			14	1533.78	14	1455.45	14	1470.26	13
P13		16	10	1698.94	16	1616.12	16	1649.32	13
P14			12	1672.37	16	1626.70	16	1635.46	15
P15			14	1672.37	16	1626.70	16	1645.23	15
P16		18	10	2063.54	18	1994.90	18	1993.72	15
P17			12	2037.88	18	1993.30	18	1986.72	17
P18			14	2037.88	18	1993.30	18	1992.81	17
P19	200	26	10	3156.34	26	2868.32	26	2865.26	24
P20			12	2995.92	26	2821.34	26	2880.28	24
P21			14	2990.20	26	2818.59	26	2833.39	25
P22		28	10	3256.71	28	3006.51	28	3043.65	27
P23			12	3293.72	28	3025.08	28	3073.58	27
P24			14	3324.40	28	3024.63	28	3047.86	25
P25		30	10	3354.54	30	3163.60	30	3138.64	26
P26			12	3329.70	30	3153.73	30	3157.61	29
P27			14	3424.26	30	3169.83	30	3168.94	30
P28	500	64	10	4618.51	61	3915.05	61	4618.51	61
P29			12	4735.81	61	3932.03	61	3990.26	57
P30			14	4786.16	61	3875.15	61	4786.16	61
P31		66	10	4683.28	63	3964.06	63	4683.28	63
P32			12	4821.26	63	3993.59	63	4100.81	59
P33			14	4986.74	63	4005.54	63	4986.74	63
P34		68	10	4726.20	64	4046.15	64	4042.66	64
P35			12	4891.40	65	4068.07	65	4113.26	63
P36			14	4839.39	65	3981.25	65	3993.13	64

RAM using Windows XP. The computing time did not exceed 4 min in any case and the calculation was immediate with few orders.

The results of the experiments are listed in Table 1, obtained by multiple runs of the algorithm. The first four columns define the settings of the experiment. The rest of the columns measure the efficiency of all the solutions by means of the total distance traveled, Δ(d) and the number of used vehicles, Δ(v). The results are shown for the three methods that compose the algorithm in order to evaluate them (routes construction, VND and hybrid).

For a small volume of orders, it can be seen that the construction method obtains the best solutions mainly due to the high number of constraints imposed on the problem initially. However, for large volumes the metaheuristic methods encounter better solutions especially in terms of a reduction in vehicles used.

The VND metaheuristic is performed to minimize the distance of the initial solution calculated, while the GVND + TS metaheuristic prioritizes the reduction in the number of vehicles used although the distance is greater, provided that the maximum of orders per vehicle is not exceeded.

It can also be appreciated that the algorithm initially uses the greatest number of available vehicles, mainly for few orders, since the construction method only reduces the available vehicles for 500 orders. As the GVRP centralizes most of the variants of the classical problem, the methods presented in this chapter may also be used for instances of these problems.

7 Conclusions

This chapter presents a hybrid algorithm that solves the GVRP problem. This problem is based on constraints and complexities actually encountered in urban freight. The problem has been solved by modeling several methods taking into account their more particular characteristics: time windows, capacity constraints, compatibility between orders and vehicles, maximum number of orders per vehicle, pickup-delivery dependent orders and not returning to the depot. Although attracting increasing interest in recent years, the relatively modest number of publications would seem to indicate that the domain of vehicle routing problems with real-life constraints and complexities is still young.

Much of the power of the proposed methods are due to the definition of neighbourhood structures, which are based on reversing segments of routes and exchanging segments between routes. In addition, the algorithm can solve real-world instances with hundreds of orders to optimality in reasonable time. The methods have been checked by means of several instances extracted from real data provided by a large transport company in Spain.

Future research could address the possibility of using the proposal outlined in this chapter to solve more general forms of the VRP, which would contain more real-world objectives and constraints (multiple depots, periodic and dynamic VRP, inventory management, hub and spoke strategy, etc.). The powerful capacity of the algorithm to find excellent solutions to a difficult combinatorial optimization problem should make it a useful model for solving many other problems in transportation and logistics.

References

1. European Commission: Green Paper: Towards a New culture for Urban Mobility. European Union, Brussels (2007)
2. European Commission: Urban Freight Transport and City Logistics: Research for Sustainable Mobility. European Union, Brussels (2003)
3. Dantzig, G.B., Ramser, J.H.: The truck dispatching problem. Manag. Sci. **6**, 80–91 (1959)
4. Goel, A., Gruhn, V.: A general vehicle routing problem. Eur. J. Oper. Res. **191**, 650–660 (2008)
5. Pisinger, D., Ropke, S.: A general heuristic for vehicle routing problems. Comput. Oper. Res. **34**, 2403–2435 (2007)
6. Jozefowiez, N., Semet, F., Talbi, E.G.: Multi-objective vehicle routing problems. Eur. J. Oper. Res. **189**, 293–309 (2008)
7. Toth, P., Vigo, D.: The Vehicle Routing Problem. SIAM Publishing, Philadelphia (2002)
8. Lin, S.W., Lee, Z.J., Ying, K.C., Lee, C.Y.: Applying hybrid meta-heuristics for capacitated vehicle routing problem. Expert Syst. Appl. **36**, 1505–1512 (2009)
9. Escuín, D., Millán, C., Larrodé, E.: Modelization of time-dependent urban freight problems by using a multiple number of distribution centers. Netw. Spat. Econ. **12**, 321–336 (2012)
10. Imran, A., Salhi, S., Wassan, N.A.: A variable neighborhood-based heuristic for the heterogeneous fleet vehicle routing problem. Eur. J. Oper. Res. **197**, 509–518 (2009)
11. Fleszar, K., Osman, I.H., Hindi, K.S.: A variable neighbourhood search algorithm for the open vehicle routing problem. Eur. J. Oper. Res. **195**, 803–809 (2009)
12. Kritikos, M.N., Ioannou, G.: The balanced cargo vehicle routing problem with time windows. Int. J. Prod. Econ. **123**, 42–51 (2010)
13. Wang, H.F., Chen, Y.Y.: A genetic algorithm for the simultaneous delivery and pickup problems with time window. Comput. Ind. Eng. **62**, 84–95 (2012)
14. Brandão, J.: A new tabu search algorithm for the vehicle routing problem with backhauls. Eur. J. Oper. Res. **173**, 540–555 (2006)
15. Hansen, P., Mladenović, N.: A tutorial on variable neighborhood search. Technical Report G-2003-46, Les Cahiers du GERAD, HEC Montreal and GERAD, Canada (2003)
16. Chen, P., Huang, H.K., Dong, X.Y.: Iterated variable neighborhood descent algorithm for the capacitated vehicle routing problem. Expert Syst. Appl. **37**, 1620–1627 (2010)
17. Mladenović, N., Dražić, M., Kovačevic-Vujčić, V., Čangalović, M.: General variable neighborhood search for the continuous optimization. Eur. J. Oper. Res. **191**, 753–770 (2008)

A Parallel Matheuristic for Solving the Vehicle Routing Problems

Umman Mahir Yıldırım and Bülent Çatay

Abstract In this chapter, we present a matheuristic approach for solving the Vehicle Routing Problems (VRP). Our approach couples the Ant Colony Optimization (ACO) algorithm with solving the Set Partitioning (SP) formulation of the VRP. As the ACO algorithm, we use a rank-based ant system approach where an agent level-based parallelization is implemented. The interim solutions which correspond to single vehicle routes are collected in a solution pool. To prevent duplicate routes, we present an elimination rule based on an identification key that is used to differentiate the routes. After a pre-determined number of iterations, the routes accumulated in the solution pool are used to solve the SP formulation of the problem to find a complete optimal solution. Once the optimal solution is obtained it is fed back to ACO as an elite solution that can be used in the pheromone reinforcement procedure. Our experimental study using the well-known VRP with Time-Windows benchmark instances of Solomon shows that the proposed methodology provides promising results.

Keywords Vehicle routing problem · Matheuristic · Ant colony optimization

1 Introduction

This chapter deals with one of the most widely known combinatorial optimization problem, namely the Vehicle Routing Problem (VRP). The basic VRP aims to serve a set of geographically dispersed customers with known demands, using a

U. M. Yıldırım (✉) · B. Çatay
Faculty of Engineering and Natural Sciences, Sabanci University, 34956 Istanbul, Turkey
e-mail: mahiryldrm@sabanciuniv.edu

B. Çatay
e-mail: catay@sabanciuniv.edu

J. F. de Sousa and R. Rossi (eds.), *Computer-based Modelling and Optimization in Transportation*, Advances in Intelligent Systems and Computing 262,
DOI: 10.1007/978-3-319-04630-3_35,

homogeneous fleet of capacitated vehicles located at a central depot. The objective is to determine the best set of routes that minimizes either the total distance traveled or the number of routes while complying with the following constraints: (i) every route starts and ends at the central depot, (ii) each customer is assigned to a single route, and (iii) the vehicle capacity is not exceeded. In the vast literature on the VRP and its variants, exact methods, heuristics and metaheuristics are widely used. In addition, hybridization of these heuristics/metaheuristics as well as the exact methods has received notable attention. Yet, articles presenting matheuristic approaches for solving the VRPs are recently gaining momentum.

Matheuristics may be considered as a special case of hybrid heuristics. Boschetti et al. [1] claim that the interoperation of metaheuristics and mathematical programming techniques yields the matheuristics and the features derived by the mathematical model of the problem are further exploited by the metaheuristic. On the other hand, [2] define a matheuristic as any heuristic that utilizes mathematical programming in one of its solution steps. In this notion, the mathematical model can be embedded in the solution procedure in several ways such as solving sub-problems, solving parts of an instance, restricting the search space and exploring neighborhoods. Some recent matheuristic approaches and applications can be found in [3].

Goerner and Schmid [4] classify the matheuristics for the VRP under three categories based on local branching, decomposition and set-partitioning/set-covering formulations. Our approach falls within the last category. In this category, first a heuristic/metaheuristic method generates preferably high quality solutions. Also, giving more importance to the solution diversification could be preferred as it may help to escape local optima and also generate a high quality solution. Then, these solutions are fed as columns for the set-partitioning/set-covering formulation of the problem. This approach has been adopted for solving different VRPs such as the capacitated VRP [5, 6], the VRP with time-windows (VRPTW) [7], the periodic VRPTW [8] and the stochastic VRP [9].

For the split delivery VRP [10] implemented a Tabu Search (TS) approach. They identified the promising parts of the solution space with the TS and further explored them using the integer programming (IP). Gulczynski et al. [11] developed an IP-based heuristic for the periodic VRP. In their parallel algorithm, [12] combined a local search heuristic with IP for solving the VRP. Recently, [13] and [14] have coupled iterated local search (ILS) with mixed IP (MIP) in a matheuristic environment.

Matheuristic approaches have been implemented for many other routing problem variants such as the truck and trailer routing problem [15], the dial-a-ride problem [16], the traveling salesman problem [17] and the technician routing and scheduling problem [18].

In this study, we present a parallel matheuristic approach, namely MathAnt, for solving the VRP. Our approach couples the ACO approach with solving the SP formulation of the VRP. To the best of our knowledge, this is the first attempt to integrate these two methods to solve a combinatorial optimization problem. The remainder of this chapter is structured as follows. The Sect. 2 contains a general description of our algorithmic approach. Section 3 proposes an elimination method to handle duplicate routes. The computational results are presented in Sect. 4. In Sect. 5 we finally give the concluding remarks and the future research directions.

2 A Parallel Matheuristic Method: MathAnt

The proposed method is based on the idea that the solutions generated by an algorithm may contain a subset of partial solutions which, when combined, can yield a better solution. Nevertheless, to generate such a promising subset, the solution method itself should be able to produce both distinct and good partial solutions. One such method is the ACO which is a constructive algorithm that builds diverse solutions at each iteration by using the foraging behavior of ants. For the VRP, the algorithm has the potential to find high quality partial solutions, i.e. vehicle routes that can be combined to obtain improved complete solutions. Building upon this potential, MathAnt integrates the ACO with IP in an attempt to efficiently solve the VRPs. It basically solves the SP formulation of the VRP at certain iterations of the ACO using the routes constructed by ant colonies. Gendreau and Potvin [19] claim that running several threads concurrently in a parallel exploration context seem to be very promising compared with the various implementations reported in the literature. In addition, the foraging behavior of the ants in the ACO is suitable for parallelization. So, to further enhance the performance of the algorithm, even if not in terms of the solution quality, we implemented an agent level-based parallelization.

The general scheme of the algorithm is given in Fig. 1. In our implementation we use IBM ILOG CPLEX for solving the SP formulation. In the ACO phase, we use the Time-based Ant System (TbAS) presented in [20]. TbAS may only be applied to VRPTW since it uses the time-window nature of the problem in the visibility mechanism. It has a multi-layer pheromone network structure to distinguish the pheromone levels belonging to different time intervals and utilizes the timing of the visit as implicit heuristic information in the route construction phase. Basically, it takes into account time-wise desirability to travel from one customer to the next within the random selection rule. In TbAS, each foraging individual ant of the colony moves independently until reaching the food source, which stands for a complete solution. In the pheromone update procedure only the best-so-far ant and the elite ants are allowed to deposit pheromone. The amount of the pheromone deposited is inversely proportional to the rank of the ant in the colony in terms of solution quality. In other words, for the $w - 1$ elite ants and the best-so-far ant, the pheromone amounts of the kth elite ant and the best-so-far ant (bs) are $(w - k)/L^k$ and w/L^{bs} respectively. Here, L^k and L^{bs} denote the total length of the complete solution of the kth elite ant and bs. We refer the interested reader to [20] for the details of the TbAS approach.

In the MathAnt implementation, in addition to the best-so-far ant and the elite ants (referred to as ACO-ants) the optimal solution obtained by solving the SP

Fig. 1 Flow chart of MathAnt

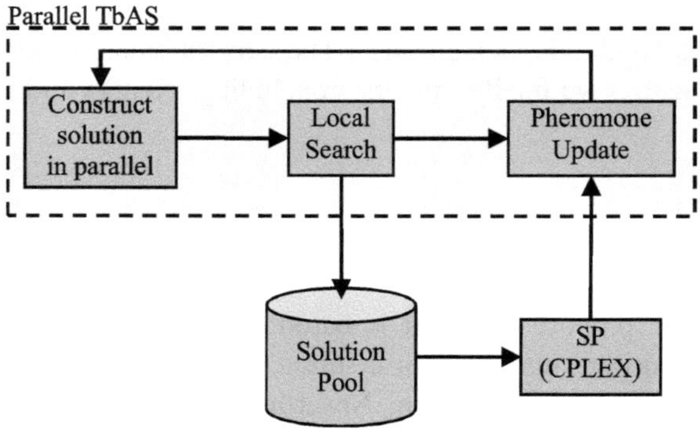

formulation is also used to further enhance the pheromone trails. The optimal solution is achieved by the so-called CPLEX-ant. When the CPLEX-ant is used to update the pheromone network, it is given a weight proportional to the weight of *bs*. Hence, to intensify the search near the CPLEX-ant, one can amplify its relative weight with respect to that of *bs* since it corresponds to the highest quality solution.

3 Handling Duplicate Routes

In the ACO, the ants in the colony do not necessarily construct distinct routes within the same iteration or in different iterations. The same route can be found multiple times by different ants, particularly when the algorithm is converging or stagnating. These duplicate routes can be handled in two ways. One alternative is to eliminate them while the pool of routes is being updated at the end of each iteration by comparing a newly constructed route against the existing ones in the pool. In this case, the whole route information should be recorded to make a full comparison. This will obviously be very time consuming; however, the SP problem will be solved using only unique routes, which may reduce the presolve processing time of CPLEX significantly. The other alternative is to add the constructed routes to the pool without any comparison and let CPLEX perform the elimination using its presolve process. In this case, the CPLEX presolve process may be more efficient in CPU time, nevertheless, keeping the duplicate routes will increase the size of the pool, which may require significant additional memory.

In the former case, each route should be assigned a unique value (referred to as identification key), if possible. The goal is to minimize the number of false eliminations, i.e. counting two different routes as the same. Matching each customer number with a unique prime number and multiplying the corresponding prime numbers of the customers in the route will perfectly and uniquely represent any route and will prevent false eliminations. However, as the number of customers in the route increases, this multiplication becomes intractable. The multiplication of the first 20 prime numbers only yields a value of 5.58E + 26, which is far larger

Table 1 Criteria used in eliminating duplicate routes

Criteria	Description
I1	$31^4D + 31^3FC + 31^2LC + 31NC$
I2	$31^4D + 31^3FC + 31^2LC + 31NC + TT$
S1	D_S + '-' + FC_S + '-' + LC_S + '-' + NC_S
S2	D_S + '-' + FC_S + '-' + LC_S + '-' + NC_S + '-' + TT

than the maximum value that can be stored in any programming language. Using any single characteristic of the route such as the total distance, total time, the number of customers, etc. as the discrimination criterion may yield many false eliminations. So, we considered and analyzed four different criteria as summarized in Table 1.

As illustrated in Table 1, we utilized five integer representative discriminative characteristics, namely the total distance (D), first customer ID number (FC), last customer ID number (LC), total number of customers (NC), and total tour time (TT). Any characteristic with a subscript S denotes its string counterpart. Using these characteristics in the given order, we have mainly two criteria groups based on integer (I) and string (S) values. In the integer subgroup, the criterion values are multiplied with a certain power of a prime number. Using a prime number in hashing is traditional. Here, we used 31 as the prime number. In addition to being an odd prime, 31 has a nice property that the multiplication can be replaced by a shift and subtraction for better performance [21]. So, the coefficients of D, FC, LC, NC and TT are set to 31^4, 31^3, 31^2, 31 and 1 respectively. In the string subgroup, the criterion values are concatenated with a hyphen. The hyphen is used as a separator to differentiate routes such as "0-1-23-0" and "0-12-3-0", where 0 denotes the depot and the remaining numbers are the customer ID numbers which show the sequence of their visits. The performances of these four criteria are tested in Sect. 4.1.

It is noteworthy to remark that the inclusion of each additional criterion most often decreases false eliminations, if not always. In other words, including an additional information does not always help to differentiate two routes. The example given in Table 2 shows how the inclusion of the tour time as additional information prevents distinguishing two different routes. The second and the third rows in this Table provide information related with two distinct vehicle routes: 0-5-12-9-0 and 0-35-42-45-33-40-0. The identification keys for these two routes according to I1 yields different values as shown in the sixth column. On the other hand, the identification keys according to I2 returns the same value for both routes, which will result in a false elimination.

4 Computational Analysis

We have tested the performance of the proposed approach on the well-known VRPTW instances of [22]. These instances have three main sets which differ by the distribution of the customers over a 100×100 grid. The customers are

Table 2 Elimination criteria: false elimination

D	FC	LC	NC	TT	Identification key I1	Identification key I2
183	5	9	3	362	169162040	169162402
182	35	40	5	300	169162102	169162402

clustered (C), randomly distributed (R) or both clustered and randomly distributed (RC). Each set is also divided into two subsets as type 1 and type 2 which have different time window lengths and vehicle capacities. The parameters of TbAS have been set as described in [20]. We have used an Intel Core2 Quad 2.33 GHz computer with 8.0 GB RAM and 64-bit operating system. The IP-solver is IBM ILOG CPLEX version 12.2.

4.1 Comparing the Elimination Methods for Duplicate Routes

To evaluate the four criteria described in Sect. 3 we performed a single run using the first and the last instances of each subset of Solomon data C1, C2, R1, R2, RC1 and RC2. To better observe how the elimination methods reduce the pool size, we set the iteration limit to 300 to accumulate a large number of routes. The results are summarized in Table 3. The first column shows the instance. The second and third columns report the total number of routes in the pool and the total number of unique routes, respectively. The last four columns correspond to the number of routes in the reduced pool after applying the four elimination criteria.

We first analyze the average number of routes by taking all 12 instances into consideration. We observe that on the average 1.091 million routes are obtained, out of which 372 thousand (35 %) are unique. The number of all routes found in type 1 instances (C1, R1 and RC1) is nearly twice the number of routes found in type 2 instances (C2, R2 and RC2): 1.397 million routes compared to 735 thousand routes, respectively. This is basically the result of longer routes involving more customers typically obtained in type 2 problems. On the other hand, the total number of unique routes found in type 1 and type 2 instances are 329 thousand and 424 thousand, respectively. So, we see that wider time windows in type 2 instances extend the size of the solution space, as expected.

Applying any of the elimination criteria decreases the size of the pool but at the expense of eliminating some unique routes as well. Integer criteria I1 and I2 eliminate 12.2 and 4.1 % of the unique routes, respectively. On the other hand, the false eliminations by using string criteria S1 and S2 are 10.0 and 4.2 %, respectively. So, we observe that the number of different discrimination criteria plays a more important role compared to the main group of the criteria (integer or string). Both I2 and S2 involving five different characteristics are able to keep more than 95 % of the unique routes. Nevertheless, as the number of nodes increases, a single string

Table 3 Comparing elimination criteria

Instance	Number of routes	Number of unique routes	Number of routes using I1	Number of routes using I2	Number of routes using S1	Number of routes using S2
C101	1,458,513	154,676	133,329	148,605	135,726	148,876
C109	972,487	492,785	427,913	464,680	436,073	465,016
R101	2,075,507	50,698	45,860	48,329	46,594	48,556
R112	1,005,643	644,038	480,661	563,178	516,548	546,686
RC101	1,735,879	91,973	79,206	89,300	81,502	89,459
RC108	1,136,336	537,137	405,375	473,012	434,536	474,441
Type 1 average	1,397,394	328,551	262,057	297,851	275,163	295,506
C201	936,673	393,816	346,336	381,549	354,753	382,534
C208	602,936	402,030	383,520	400,883	386,667	400,972
R201	961,235	492,038	440,795	486,130	448,121	486,361
R211	428,969	392,145	376,448	390,833	379,762	390,913
RC201	1,002,893	460,026	409,595	452,793	420,498	453,093
RC208	481,032	405,270	386,936	400,252	389,806	400,323
Type 2 average	735,623	424,221	390,605	418,740	396,601	419,033
Total average	1,091,961	372,706	321,387	353,646	331,211	352,518

representation of a node uses more memory compared to that of integer (4 bytes). Thus, taking the memory usage into account we decided to implement I2 criterion.

4.2 Elimination of Routes: Elimination Method Versus CPLEX Presolve Process

In this section, we analyze how to eliminate the duplicate routes, either via the proposed elimination method or CPLEX presolve. All the tests in this section are conducted using the 39 instances in R1, R2, RC1 and RC2 sets of Solomon. Since the optimal solutions can be easily obtained for the clustered instances of C1 and C2 sets, they are omitted as their sensitivity to parametric changes cannot be evaluated.

The detailed computational time analysis is given in Table 4. All time units are in seconds. The number of CPLEX calls directly affect the solution quality (analyzed in detail in Sect. 4.3) and the computational time. Thus, we tested three different CPLEX call frequency settings in a run with 100 iterations: 1, 2, and 5. Note that CPLEX is run only once at the end of the ACO procedure when CPLEX Call Frequency = 1 whereas CPLEX Call Frequency = 5 represents that optimization using CPLEX is performed five times, after every 20 iterations.

Table 4 Elimination of duplicate routes

CPLEX call frequency	Set	Duplicate route elimination		CPLEX presolve	
		Number of routes	Average CPU time (s)	Number of routes	Average CPU time (s)
5	R1	37,991.60	165.45	147,806.18	148.25
	R2	59,170.24	545.20	72,515.71	525.72
	RC1	36,755.78	118.40	146,083.93	122.40
	RC2	55,443.15	361.48	79,038.03	355.21
	Average	47,291.36	303.12	112,110.84	291.87
2	R1	39,730.20	122.92	148,734.05	117.82
	R2	57,823.65	466.97	73,009.76	463.39
	RC1	37,311.23	100.65	146,683.90	94.38
	RC2	55,988.98	336.80	78,833.80	321.22
	Average	47,672.42	259.26	112,616.86	252.2
1	R1	37,266.07	110.56	149,681.03	105.75
	R2	57,569.75	445.76	73,347.18	438.98
	RC1	35,964.50	92.71	146,624.80	84.34
	RC2	54,822.25	321.16	78,938.78	314.19
	Average	46,327.03	244.64	113,012.82	238.10

Table 5 Solution quality for different parameter combinations

δ	CPLEX call frequency		
	5	2	1
1	1097.91	1098.51	1099.47
2	1098.80	1099.01	1099.72
5	1097.06	1098.28	1099.04

We observe that the computational time of the algorithm using duplicate route elimination is 2.96 % longer compared to the elimination through CPLEX presolve. Taking into consideration this small margin one can question the benefit of implementing the duplicate route elimination method. However, when the size of the solution pool increases, the memory requirement and the time spent by CPLEX also increase and leaving the route elimination procedure to CPLEX may not be favorable. On a sample run with instance R101, when the number of the routes in the solution pool reached up to 8.5 million, CPLEX failed to solve the SP problem because of excessive memory requirements. Nonetheless, applying the elimination criteria beforehand kept the size of the solution pool at most 24,049 routes, which in turn allowed finding the optimal solution in seconds.

4.3 *Effect of Parameters on the Solution Quality*

The frequency of the CPLEX calls and the pheromone reinforcement weight of the SP-ant affect the solution quality. Table 5 reports the average solution quality of

Table 6 Comparison of results for type 1 problems

Instance	BKS	Ref[a]	TbAS	MathAnt	Gap (%) (BKS)	Gap (%) (TbAS)
R101	1642.87	AMT	1642.88	1642.88	0.00	0.00
R102	1472.62	AMT	1472.81	1472.82	0.01	0.00
R103	1213.62	JM	1213.62	**1213.62**	0.00	0.00
R104	976.61	JM	977.55	**976.61**	0.00	−0.10
R105	1360.78	JM	1360.78	**1360.78**	0.00	0.00
R106	1240.26	YÇ	1240.26	**1239.37**	−0.07	−0.07
R107	1073.01	YÇ	1073.01	1075.14	0.20	0.20
R108	944.44	YÇ	944.44	**938.20**	−0.66	−0.66
R109	1151.84	JM	1151.84	**1151.84**	0.00	0.00
R110	1072.41	JM	1072.41	1072.42	0.00	0.00
R111	1053.50	JM	1053.50	**1053.50**	0.00	0.00
R112	953.63	RT	959.58	955.68	0.21	−0.41
R1 average	1179.63		1180.22	1179.40	−0.03	−0.09
RC101	1623.58	RT	1638.00	1623.59	0.00	−0.88
RC102	1461.23	JM	1461.44	**1461.23**	0.00	−0.01
RC103	1261.67	S	1262.68	**1261.67**	0.00	−0.08
RC104	1135.48	CLM	1141.66	1135.83	0.03	−0.51
RC105	1518.58	JM	1518.58	**1518.58**	0.00	0.00
RC106	1376.99	YÇ	1376.99	**1376.99**	0.00	0.00
RC107	1212.83	JM	1212.83	**1211.11**	−0.14	−0.14
RC108	1117.53	JM	1117.53	**1117.53**	0.00	0.00
RC1 average	1338.49		1341.21	1338.32	−0.01	−0.20
Total average	1243.17		1244.62	1242.97	−0.02	−0.13

[a] *AMT* Alvarenga et al. [7], *JM* Jung and Moon [25], *YÇ* Yıldırım and Çatay [20], *RT* Rochat and Taillard [28], *CLM* Cordeau et al. [24], *S* Shaw [29]

five runs for different parameter combinations. The average solution quality does not show a significant difference across different parameter settings. Nevertheless, the best solutions are obtained when $\delta = 5$. Intensifying the search near the CPLEX solution in the solution space generates better solutions compared to equally exploring the solution space near the ACO and CPLEX solutions. Among different CPLEX frequency call values five yields the best results. The increasing frequency of the CPLEX calls helps better improve the solution as expected. This comes at the expense of an increase in the computational effort. Calling CPLEX every 20 iterations increases the computational time by 23 % compared to a single call at the end of the algorithm. In light of these results, we set $\delta = 5$ and CPLEX call frequency = 5 in the following experiments.

4.4 Performance Against the Best Heuristic Solutions

In this section we compare the performance of MathAnt against the best performing heuristics and metaheuristics in the literature as well as TbAS of [20] to

Table 7 Comparison of results for type 2 problems

Instance	BKS	Ref[a]	TbAS	MathAnt	Gap (%) (BKS)	Gap (%) (TbAS)
R201	1147.8	OV	1155.8	1149.39	0.14	−0.55
R202	1034.35	JM	1036.6	1034.58	0.02	−0.20
R203	874.87	JM	875.62	877.23	0.27	0.18
R204	735.8	OV	746.98	740.98	0.70	−0.80
R205	954.16	ORH	964.64	957.33	0.33	−0.76
R206	879.89	JM	892.95	883.92	0.46	−1.01
R207	797.99	OV	805.7	810.91	1.62	0.65
R208	705.45	JM	711.37	712.93	1.06	0.22
R209	859.39	JM	876.33	**859.39**	0.00	−1.93
R210	910.7	JM	915.49	915.48	0.52	0.00
R211	755.82	OV	773.51	765.04	1.22	−1.09
R2 average	877.84		886.82	882.47	0.58	−0.48
RC201	1265.56	JM	1272.63	1267.16	0.13	−0.43
RC202	1095.64	JM	1104.92	1096.75	0.10	−0.74
RC203	926.89	OV	938.19	937.76	1.17	−0.05
RC204	786.38	JM	800.48	789.26	0.37	−1.40
RC205	1157.55	JM	1157.55	**1157.55**	0.00	0.00
RC206	1054.61	JM	1072.08	1055.77	0.11	−1.52
RC207	966.08	JM	972.74	967.07	0.10	−0.58
RC208	779.31	JM	792.65	783.93	0.59	−1.10
RC2 average	1004		1013.91	1006.91	0.32	−0.73
Total average	930.96		940.33	934.86	0.47	−0.59

[a] *OV* Oliveira and Vasconcelos [26], *JM* Jung and Moon [25], *ORH* Ombuki et al. [27]

investigate the benefit of hybridizing TbAS with IP. Tables 6 and 7 summarize the results for type 1 and type 2 problems, respectively. Note that [20] reported the results of four different implementations of their algorithm. In these Tables, we consider the best results achieved.

In both Tables, the first column identifies the problem and the fifth column shows the results achieved by MathAnt. The second and the third columns give the best-known solutions (BKS) from the literature and the corresponding articles, respectively. The best results found by TbAS are given in the fourth column. Column six and seven report the percentage gaps between MathAnt and best-known solutions and TbAS results, respectively. A negative number shows an improvement. In general, we observe that the proposed MathAnt method is able to generate good solutions. Combining TbAS with IP improves the solutions of type 1 and type 2 problems by 0.13 and 0.59 %, respectively, compared to using TbAS alone. The average improvement of the MathAnt matheuristic over TbAS is 0.35 %. When we compare MathAnt results against the best-known results from the literature, we see that the performance of MathAnt is better in type 1 problems as this was the case for TbAS as well. The average gap for type 2 problems is 0.47 % whereas it is −0.02 % for type 1 problems. The average results on type 1 problems reveal that MathAnt is the best performing method in the literature. Note that MathAnt improved the best-known solutions of R106, R108 and RC107 instances (as shown in bold-italic in Table 6) and matched the best-known results in 10 instances (as shown in bold in Table 6). In type 2 problems, MathAnt could match the best-known result in only 2 instances (as shown in bold in Table 7).

5 Conclusions

In this chapter we presented MathAnt, a parallel matheuristic that combines the ACO and the IP, for solving the VRP and its variants. We used TbAS [20] as the ACO approach and CPLEX as the IP solver. An agent level parallelization was implemented and the pheromone reinforcement procedure was adapted so as to incorporate the CPLEX solution in TbAS. After determining the frequency of CPLEX calls we conducted experiments on the VRPTW instances to test the performance of MathAnt. The comparison results against the published best distances in the literature show that MathAnt is capable of generating good solutions. MathAnt had superior performance on type 1 instances in particular where the time-window lengths are narrow and it improved three best-known results in the literature. Furthermore, an elimination method was investigated to cope with the duplicate routes generated by TbAS. We observed that our elimination method effectively decreases false eliminations and keeps the size of the pool of routes minimum. Although the computational time of MathAnt equipped with our elimination method is longer compared to leaving the elimination to CPLEX, the elimination method becomes advantageous especially when the solution pool size increases and CPLEX fails to generate a solution.

MathAnt can be easily implemented to solve any VRP variant using any ACO approach. In this paper, we only considered the VRPTW. Further research will address the other VRP variants. Russell and Chiang [23] state that using the set covering formulation instead of the set partitioning model in a VRP context may lead to an improved solution. We will investigate the impact of this relaxation on the performance of our algorithm as well. Moreover, we utilized parallelism for only reducing the computational time. However, a parallel implementation by devising multiple ant colonies evolving on different processors may lead to improved performance with respect to the solution quality as well as processor load balance.

References

1. Boschetti, M.A., Maniezzo, V., Roffilli, M., Röhler, A.B.: Matheuristics: optimization, simulation and control. In: Hybrid Metaheuristics. Lecture Notes Computer Science, vol. 5818, pp. 171–177 (2009)
2. Bertazzi, L., Speranza, M.G.: Matheuristics for inventory routing problems. In: Montoya-Torres, J.R., Juan, A.A., Huatuco, L.H., Faulin, J., Rodriguez-Verjan, G.L. (eds.) Hybrid Algorithms for Service, Computing and Manufacturing Systems: Routing and Scheduling Solutions. IGI Global, Hershey (2011)

3. Maniezzo, V., Stützle, T., Voß, S.: Matheuristics: Hybridizing Metaheuristics and Mathematical Programming. Springer, New-York (2010)
4. Doerner, K.F., Schmid, V.: Survey: matheuristics for rich vehicle routing problems. In: 7th International Workshop on Hybrid Metaheuristics. Lecture Notes in Computer Science, vol. 6373, pp. 206–221 (2010)
5. Groër, C., Golden, B., Wasil, E.: A library of local search heuristics for the vehicle routing problem. Math. Program Comput. **2**, 79–101 (2010)
6. Kelly, J.P., Xu, J.: A set-partitioning-based heuristic for the vehicle routing problem. Informs J. Comput. **11**, 161–172 (1999)
7. Alvarenga, G.B., Mateus, G.R., Tomi, G.: A genetic and set partitioning two-phase approach for the vehicle routing problem with time windows. Comput. Oper. Res. **34**, 1561–1584 (2007)
8. Pirkwieser, S., Raidl, G.R.: Multiple variable neighborhood search enriched with ILP techniques for the periodic vehicle routing problem with time windows. In: Hybrid Metaheuristics. Lecture Notes Computer Science, vol. 5818, pp. 45–59 (2009)
9. Mendoza, J.E., Villegas, J.G.: A multi-space sampling heuristic for the vehicle routing problem with stochastic demands. Optim. Lett. (2012) (Available online)
10. Archetti, C., Speranza, M., Savelsbergh, M.: An optimization-based heuristic for the split delivery vehicle routing problem. Transp. Sci. **42**, 22–31 (2008)
11. Gulczynski, D., Golden, B., Wasil, E.: The period vehicle routing problem: new heuristics and real-world variants. Transp. Res. E-Log. **47**, 648–668 (2011)
12. Groër, C., Golden, B., Wasil, E.: A parallel algorithm for the vehicle routing problem. Informs J. Comput. **23**, 315–330 (2011)
13. Subramanian, A., Penna, P.H.V., Uchoa, E., Ochi, L.S.: A hybrid algorithm for the heterogeneous fleet vehicle routing problem. Eur. J. Oper. Res. **221**, 285–295 (2012)
14. Subramanian, A., Uchoa, E., Ochi, L.S.: A hybrid algorithm for a class of vehicle routing problems. Comput. Oper. Res. **40**, 2519–2531 (2013)
15. Villegas, J.G., Prins, C., Prodhon, C., Medaglia, A.L., Velasco, N.: A matheuristic for the truck and trailer routing problem. Eur. J. Oper. Res. **230**, 231–244 (2013)
16. Calvo, R.W., Touati-Moungla, N.: A matheuristic for the dial-a-ride problem. In: Network Optimization. Lecture Notes Computer Science, vol. 6701, pp. 450–463 (2011)
17. Rodríguez-Martín, I., Salazar-González, J.J.: The multi-commodity one-to-one pickup-and-delivery traveling salesman problem: a matheuristic. In: Network Optimization. Lecture Notes Computer Science, vol. 6701, pp. 401–405 (2011)
18. Pillac, V., Guéret, C., Medaglia, A.: A parallel matheuristic for the technician routing and scheduling problem. Optim. Lett. (2012) (Available online)
19. Gendreau, M., Potvin, J.-Y.: Metaheuristics in combinatorial optimization. Ann. Oper. Res. **140**(1), 189–213 (2005)
20. Yıldırım, U.M., Çatay, B.: A time-based pheromone approach for the ant system. Optim. Lett. **6**, 1081–1099 (2012)
21. Blosch, J.: Effective Java, 2nd edn. Addison-Wesley, Boston (2008)
22. Solomon, M.M.: Algorithms for the vehicle routing and scheduling problems with time window constraints. Oper. Res. **35**, 254–265 (1987)
23. Russell, R.A., Chiang, W-C.: Scatter search for the vehicle routing problem with time windows. Eur. J. Oper. Res. **169**, 606–622 (2006)
24. Cordeau, J.-F., Laporte, G., Mercier, A.: A unified tabu search heuristic for vehicle routing problems with time windows. J. Oper. Res. Soc. **52**, 928–936 (2001)
25. Jung, S., Moon, B-R.: A hybrid genetic algorithm for the vehicle routing problem with time windows. In: Proceedings of the 2002 Genetic and Evolutionary Computation Conference, pp. 1309–1316. GECCO 2002, New York (2002)
26. Oliveira, H.C.B., Vasconcelos, G.C.: A hybrid search method for the vehicle routing problem with time windows. Ann. Oper. Res. **180**, 125–144 (2008)
27. Ombuki, B., Ross, B.J., Hanshar, F.: Multi-objective genetic algorithms for vehicle routing problem with time windows. Appl. Intell. **24**, 17–30 (2006)

28. Rochat, Y., Taillard, E.D.: Probabilistic diversification and intensification in local search for vehicle routing. J. Heuristics **1**, 147–167 (1995)
29. Shaw, P.: Using constraint programming and local search methods to solve vehicle routing problems. In: Principles and Practice of Constraint Programming (CP'98). Lecture Notes Computer Science, vol. 1520, pp. 417–431 (1998)

MIX
Papier aus verantwortungsvollen Quellen
Paper from responsible sources
FSC® C105338

If you have any concerns about our products,
you can contact us on
ProductSafety@springernature.com

In case Publisher is established outside the EU,
the EU authorized representative is:
Springer Nature Customer Service Center GmbH
Europaplatz 3, 69115 Heidelberg, Germany

Printed by Libri Plureos GmbH
in Hamburg, Germany